Connectivity Conservation

One of the biggest threats to the survival of many plant and animal species is
the destruction or fragmentation of their natural habitats. The conservation of
landscape connections, where animal, plants, and ecological processes can move
freely from one habitat to another, is therefore an essential part of any new
conservation or environmental protection plan. In practice, however, maintaining,
creating, and protecting connectivity in our increasingly dissected world is a
daunting challenge. This fascinating volume provides a synthesis on the current
status and literature of connectivity conservation research and implementation.
It shows the challenges involved in applying existing knowledge to real-world
examples and highlights areas in need of further study. Containing contributions
from leading scientists and practitioners, this topical and thought-provoking
volume will be essential reading for graduate students, researchers, and
practitioners working in conservation biology and natural resource management.

KEVIN CROOKS is an assistant professor in the Department of Fish, Wildlife,
and Conservation Biology and the Graduate Degree Program in Ecology at Colorado
State University. His research investigates the effects of habitat fragmentation,
urbanization, and landscape connectivity on the behavior, ecology, and conservation
of wildlife.

M. SANJAYAN is a lead scientist for The Nature Conservancy. His current work
focuses on ensuring connectivity in applied conservation projects worldwide and
on understanding the role of ecosystem services in human well-being and
conservation.

Conservation Biology

Conservation biology is a flourishing field, but there is still enormous potential for making further use of the science that underpins it. This new series aims to present internationally significant contributions from leading researchers in particularly active areas of conservation biology. It will focus on topics where basic theory is strong and where there are pressing problems for practical conservation. The series will include both single-authored and edited volumes and will adopt a direct and accessible style targeted at interested undergraduates, postgraduates, researchers and university teachers. Books and chapters will be rounded, authoritative accounts of particular areas with the emphasis on review rather than original data papers. The series is the result of a collaboration between the Zoological Society of London and Cambridge University Press. The series ethos is that there are unexploited areas of basic science that can help define conservation biology and bring a radical new agenda to the solution of pressing conservation problems.

Connectivity Conservation

Edited by

KEVIN R. CROOKS
Colorado State University

AND

M. SANJAYAN
The Nature Conservancy

CAMBRIDGE UNIVERSITY PRESS
Cambridge, New York, Melbourne, Madrid, Cape Town, Singapore, São Paulo

Cambridge University Press
The Edinburgh Building, Cambridge CB2 2RU, UK

Published in the United States of America by Cambridge University Press, New York

www.cambridge.org
Information on this title: www.cambridge.org/9780521857062

First published 2006

Printed in the United Kingdom at the University Press, Cambridge

A Catalog record for this publication is available from the British Library

ISBN-13 978-0-521-85706-2 hardback
ISBN-10 0-521-85706-6 hardback

ISBN-13 978-0-521-67381-5 paperback
ISBN-10 0-521-67381-X paperback

Contents

List of Contributors

SHELLEY M. ALEXANDER
Department of Geography,
University of Calgary, Calgary,
Alberta, T2N 1N4, Canada

KEITH ALGER
Center for Applied Biodiversity
Science, Conservation International,
1919 M Street NW, Suite 600,
Washington, DC 20036, USA

PAUL BEIER
School of Forestry,
Northern Arizona University,
Flagstaff, AZ 86011, USA

ANDREW F. BENNETT
School of Life and Environmental
Sciences, Deakin University,
221 Burwood Highway, Burwood,
Vic 3125, Australia

KAREN A. BJORNDAL
Archie Carr Center for Sea Turtle
Research and Department of Zoology,
University of Florida, Gainesville,
FL 32611, USA

CLINT CABAÑERO
South Coast Wildlands, PO Box 1102,
Idyllwild, CA 92549, USA

JUSTIN M. CALABRESE
Department of Biology,
University of Maryland,
College Park, MD 20742, USA

CARLOS CARROLL
Klamath Center for Conservation
Research, PO Box 104, Orleans,
CA 95556, USA

ANTHONY P. CLEVENGER
Western Transportation Institute,
Montana State University,
PO Box 174250, Bozeman,
MT 59717, USA

JEFFREY A. CROOKS
Tijuana River National Estuarine
Research Reserve, Imperial Beach,
CA 91932, USA

KEVIN R. CROOKS
Department of Fish, Wildlife, and
Conservation Biology, Colorado
State University,
Fort Collins, CO 80523, USA

KATHLEEN M. DALY
Wildlands Project,
PO Box 455, Richmond,
VT 05477, USA

CHRIS T. DARIMONT
Department of Biology,
University of Victoria, PO Box 3020,
Victoria, BC V8W 3N5, Canada

WILLIAM C. DENNISON
University of Maryland Center for
Environmental Science, PO Box 775,
Cambridge, MD 21613, USA

CLAUDIO DIBACCO
Department of Earth and Ocean
Sciences, University of British
Columbia, 1461-6270
University Boulevard, Vancouver,
British Columbia, V6T 1Z4, Canada

ANDY DOBSON
Department of Ecology and
Evolutionary Biology, Princeton
University, Princeton, NJ 08544, USA

DON DRISCOLL
School of Biological Sciences,
Flinders University, GPO Box 2100,
Adelaide, SA 5001, Australia

JASON DUNHAM
US Forest Service, Rocky Mountain
Research Station, Suite 401, 322
E. Front St, Boise, ID 83702, USA;
USGS Forest and Rangeland
Ecosystem Science Center,
Corvallis Research Group,
3200 SW Jefferson Way,
Corvallis, OR 97331, USA

WILLIAM F. FAGAN
Department of Biology,
University of Maryland,
College Park, MD 20742, USA

LENORE FAHRIG
Department of Biology, Carleton
University, Ottawa, Ontario,
K1S 5B6, Canada

GUSTAVO A. B. DA FONSECA
Center for Applied Biodiversity
Science, Conservation International,
1919 M Street NW, Suite 600,
Washington, DC 20036, USA

RICHARD FRANKHAM
Key Centre for Biodiversity and
Bioresources, Department
of Biological Sciences,

Macquarie University,
NSW 2109, Australia;
Department of Ecology and Genetics,
University of Aarhus,
DK-8000 Aarhus C, Denmark;
Australian Museum, 6 College Street,
Sydney NSW 2010, Australia

CARLOS GALINDO-LEAL
Coordinador Programa Bosques
Mexicanos, Fondo Mundial para la
Naturaleza (World Wildlife
Fund–México), Av. México 51,
Col. Hipódromo, México,
D.F. 06100, México

NICK M. HADDAD
Department of Zoology, PO Box 7617,
North Carolina State University,
Raleigh, NC 27695, USA

SUSAN M. HAIG
US Geological Survey Forest and
Rangeland Ecosystem Science Center,
3200 SW Jefferson Way, Corvallis,
OR 97331, USA

ILKKA HANSKI
Metapopulation Research Group,
Department of Biological and
Environmental Sciences, PO Box 65
(Biocenter 3), FI-00014 University of
Helsinki, Finland

AUTUMN-LYNN HARRISON
Department of Ecology and
Evolutionary Biology, University of
California Santa Cruz, Long Marine
Laboratory, 100 Shaffer Road,
Santa Cruz, CA 95060, USA

MARCEL HOLYOAK
Department of Environmental
Science and Policy, University of
California, 1 Shields Avenue, Davis,
CA 95616, USA

GARY R. HUXEL
Department of Biology,
University of South Florida,
4202 East Fowler Avenue,
SCA110, Tampa,
FL 33620, USA

VICTOR HUGO INCHAUSTY
Director, Areas Protegidas,
Ministerio Desarrollo Agropecuario
Rural y Medio Ambiente,
Loayza Edif. Lara Bisch,
La Paz, Bolivia

MENNA E. JONES
School of Zoology,
University of Tasmania,
Private Bag 5, Hobart,
Tas 7004, Australia

PETER KAREIVA
The Nature Conservancy, Seattle,
WA 98105, USA

LISA A. LEVIN
Integrative Oceanography Division,
Scripps Institution of Oceanography,
9500 Gilman Drive,
La Jolla, CA 92093, USA

THOMAS LOVEJOY
The H. John Heinz III Center for
Science, Economics and the
Environment, Washington, DC
20004, USA

CLAUDIA LUKE
Bodega Marine Laboratory and
Reserve, University of California,
PO Box 247, Bodega Bay,
CA 94923, USA

BRENDAN G. MACKEY
School of Resources,
Environment and Society,
Faculty of Science,

The Australian National University,
Canberra, ACT 0200, Australia

PETER P. MARRA
Smithsonian Migratory Bird Center,
National Zoological Park,
3001 Connecticut Avenue NW,
Washington, DC 20008, USA

MARGRARET M. MAYFIELD
Department of Ecology, Evolution
and Marine Sciences, University
of California Santa Barbara,
Santa Barbara, CA 93103, USA

HAMISH MCCALLUM
School of Integrative Biology,
The University of Queensland,
Brisbane, Qld 4072, Australia

ATTE MOILANEN
Metapopulation Research Group,
Department of Biological and
Environmental Sciences,
PO Box 65 (Biocenter 3), FI-00014
University of Helsinki, Finland

KARL MORRISON
Center for Applied Biodiversity
Science, Conservation International,
1919 M Street NW,
Suite 600, Washington,
DC 20036, USA

SCOTT A. MORRISON
The Nature Conservancy,
201 Mission Street, 4th Floor,
San Francisco, CA 94105, USA

HELEN NEVILLE
Department of Biology, University
of Nevada, Reno, NV 89557, USA;
US Forest Service, Rocky Mountain
Research Station, Suite 401,
322 E. Front St,
Boise, ID 83702, USA

D. RYAN NORRIS
Department of Integrative Biology,
University of Guelph, Guelph,
Ontario, Canada N1G 2W1

REED F. NOSS
Department of Biology, University
of Central Florida, Orlando,
FL 32816, USA; Wildlands Project,
PO Box 455, Richmond,
VT 05477, USA

PAUL C. PAQUET
Faculty of Environmental Design,
University of Calgary, Calgary,
Alberta, T2N 1N4, Canada;
Raincoast Conservation Society,
PO Box 26, Bella Bella,
British Columbia,
V0T 1B0, Canada

MARY PEACOCK
Department of Biology,
University of Nevada,
Reno, NV 89557, USA

KRISTEEN L. PENROD
South Coast Wildlands,
PO Box 1102, Idyllwild,
CA 92549, USA

CATHERINE PRINGLE
Institute of Ecology,
University of Georgia, Athens,
GA 30602, USA

HARRY F. RECHER
School of Natural Sciences,
Edith Cowan University, Joondalup,
WA 6027, Australia

MARK D. REYNOLDS
The Nature Conservancy, 4th Floor,
201 Mission Street, San Francisco,
CA 94105, USA

TAYLOR H. RICKETTS
Conservation Science Program,
World Wildlife Fund,
1250 24th Street NW,
Washington, DC 20037, USA

J. ANDREW ROYLE
US Fish and Wildlife Service,
Division of Migratory Bird
Management, 11510 American Holly
Drive, Laurel, MD 20707, USA

ANTHONY RYLANDS
Center for Applied Biodiversity
Science, Conservation International,
1919 M Street NW, Suite 600,
Washington,
DC 20036, USA

ENRIC SALA
Center for Marine Biodiversity
and Conservation, Scripps
Institution of Oceanography,
9500 Gilman Drive,
La Jolla, CA 92093, USA

JAMES SANDERSON
Center for Applied Biodiversity
Science, Conservation International,
1919 M Street NW,
Suite 600, Washington,
DC 20036, USA

M. SANJAYAN
The Nature Conservancy,
4245 North Fairfax Drive,
Arlington, VA 22203, USA

MICHAEL E. SOULÉ
PO Box 1808, Paonia,
CO 81428, USA

WAYNE D. SPENCER
Conservation Biology Institute,
815 Madison Avenue,
San Diego, CA 92116, USA

ANDREW V. SUAREZ
Department of Entomology and
Department of Animal Biology,
School of Integrative Biology,
University of Illinois at
Urbana–Champaign,
Urbana, IL 61801, USA

PATRICIA L. SWAN
Department of Geography,
University of Calgary, Calgary,
Alberta, T2N 1N4, Canada

DREW M. TALLEY
Department of Environmental
Science and Policy, University
of California, 1 Shields Avenue,
Davis, CA 95616, USA;
San Francisco Bay National
Estuarine Research Reserve,
San Francisco State
University/Romberg Tiburon Center,
3152 Paradise Drive, Tiburon,
CA 94920, USA

PHILIP D. TAYLOR
Department of Biology,
Acadia University,
Wolfville, Nova Scotia,
B4P 2R6, Canada

JOSH J. TEWKSBURY
Department of Biology,
PO Box 351800,
24 Kincaid Hall,
University of Washington,
Seattle, WA 98195, USA

DAVID M. THEOBALD
Natural Resource Ecology
Laboratory and Department of
Recreation and Tourism,
Colorado State University,
Fort Collins, CO 80523, USA

JEFF A. TRACEY
Graduate Degree Program in Ecology
and Program for Interdisciplinary
Mathematics, Ecology, and Statistics,
Colorado State University,
Fort Collins, CO 80523, USA

MIKE WEBSTER
School of Biological Sciences,
Washington State University,
PO Box 644236, Pullman,
WA 99164, USA

JOHN A. WIENS
The Nature Conservancy,
Arlington, VA 22203, USA

JACK WIERZCHOWSKI
Geomar, PO Box 1843,
Grand Forks, British Columbia,
V0H 1H0, Canada

JANN E. WILLIAMS
Centre for Sustainable Regional
Communities, LaTrobe University,
PO Box 199, Bendigo,
Vic 3550, Australia

NEAL M. WILLIAMS
Department of Biology,
Bryn Mawr College, 101 N Merion
Avenue, Bryn Mawr,
PA 19010, USA

KIMBERLY A. WITH
Division of Biology,
Kansas State University,
Manhattan, KS 66506, USA

JOHN C. Z. WOINARSKI
Biodiversity Section,
Natural Systems,
Department of Infrastructure,
Planning and Environment,
PO Box 496, Palmerston,
NT 0831, Australia

Acknowledgements

We first would like to thank all chapter authors for their excellent contributions to this volume. Their collective breadth and depth of knowledge, as well as their enthusiasm, diligence, and patience throughout the publication process, made this volume possible. We are particularly grateful to A. L. Harrison for her tireless work in helping organize, compile, and format the volume prior to submission.

All chapters in this volume received at least two peer reviews and we sincerely thank the reviewers for their thoughtful comments, edits, and suggestions: B. Allan, L. Angeloni, S. Baruch-Mordo, C. Baxter, S. Bayard de Volo, P. Beier, L. Bernatchez, C. Carroll, T. Clevenger, H. Crockett, J. Crooks, K. Daly, C. DiBacco, B. Dickson, M. DiGiorgio, B. Fagan, J. Forester, H. Fox, M. Freeman, B. Grenfell, S. Griffin, A. Gonzalez, N. Haddad, A. L. Harrison, B. Hastings, P. Kareiva, M. Kavermann, J. Kelly, P. Kevan, J. Kintsch, R. Lavier, S. Magle, B. Mila, T. Mildenstein, M. Miller, S. Mills, A. Moilanen, S. Morrison, M. Neel, H. Neville, C. Pague, H. Possingham, J. Powell, T. Ricketts, E. Ruell, J. Sanderson, R. Scherer, R. Shaw, P. Snelgrove, D. Smith, D. Spiering, A. Suarez, D. Talley, S. Tartowski, P. Taylor, T. Tear, M. Tilton, J. Tracey, and M. Wonham. Special thanks also to the EY 592 graduate seminar at Colorado State University (led by K. Crooks), and a graduate seminar at Scripps Institution of Oceanography (led by L. Levin), for their helpful reviews of earlier drafts of this book.

We thank Cambridge University Press for their assistance in publication of the volume, specifically A. Crowden for his continued interest and encouragement, A. Hodson for her meticulous copy-editing skills, and C. Georgy, D. Lewis, and D. Preston for their help with the contract, editing, production, and design stages. Additionally, we acknowledge the funding and support of The Nature Conservancy, The Society for Conservation Biology, University of Wisconsin–Madison, and Colorado State University during the compilation of the volume.

We are also indebted to our advisor and teacher, Michael Soulé, for his guidance, wisdom, and inspiration during our careers. Finally, we extend our deepest appreciation to our families for their continued love and encouragement. K. Crooks would like to thank his wife and daughter, Lisa and Elena, his parents, Bob and Mary, and his brothers, Jeff and Matt. M. Sanjayan would like to thank his parents, Muttu and Ranji, and his sister, Vaithi. We dedicate this volume to you.

Connectivity conservation: maintaining connections for nature

KEVIN R. CROOKS AND M. SANJAYAN

For the first time in Earth's long history, one species – *Homo sapiens* – completely dominates the globe. Over 6 billion people now inhabit the planet, and that number is growing at an alarming rate. In sharp contrast to previous eras, we drive, fly, telephone, and e-communicate from even the most remote regions of Earth. With the aid of technology, no significant segment of our population is truly isolated. Today, it is possible to make a cell phone call from the Serengeti, or live "off the grid" from in-holdings within the midst of national forests or wilderness areas in North America. With the notable exception of weedy species that thrive amidst the disturbances we create, predictably, the more "connected" we become, non-human life with which we share this planet becomes increasingly disconnected.

The vast reach of humans and the resulting parcelization of natural landscapes are of major concern to conservation scientists. Indeed, horror stories about habitat fragmentation appear in every book about conservation biology, make appearances in high-school textbooks, and are featured regularly in our leading newspapers and magazines. And conservation biologists are not alone in their concern about massive habitat destruction and fragmentation. Members of the public also have been inspired to promote special efforts for connecting landscapes in our increasingly dissected world.

While the vision of connected landscapes may be compelling, the practice of preventing fragmentation and conserving connectivity is not a simple matter. Conservationists often fail to articulate clearly what they

Connectivity Conservation eds. Kevin R. Crooks and M. Sanjayan. Published by Cambridge University Press. © Cambridge University Press 2006.

mean by connectivity and what will be gained through their efforts to protect it, and often are insensitive to the logistical and economic costs associated with conserving landscape connections. Many questions remain unanswered. How essential is connectivity for biodiversity conservation? Do corridors — a primary tool employed to enhance connectivity — actually function as intended? Under what situations have human actions created too much connectivity, and how can we prevent it? Might there be better uses of conservation funds than attending to linkages that may or may not work? Such debates have raged within academic circles, not surprisingly spilling over to the rest of the conservation community as well. Nevertheless, the one point that remains clear is that the concern for connecting landscapes is increasingly becoming a part of land management worldwide. Despite its importance, however, the concept of connectivity is currently a loose amalgamation of related topics with little synthesis between them. The need to elucidate this concept therefore is more timely than ever.

In this introductory chapter, we explore the concept of connectivity and review the recent history of connectivity research in the scientific literature. Then, we briefly overview the benefits and challenges inherent in research and implementation of connectivity. Finally, we introduce the organization and content of the volume, and examine how linkages among the theoretical, empirical, and applied efforts discussed in this volume are essential for effective connectivity conservation.

WHAT IS CONNECTIVITY?

Within both the scientific and conservation community, confusion has existed as to what exactly connectivity is, how to define it, and how to measure successes of conservation programs attempting to protect it. At its most fundamental level, connectivity is inherently about the degree of movement of organisms or processes — the more movement, the more connectivity. Perhaps even more critical for those committed to biodiversity conservation is the converse — less movement, less connectivity. Movement in nature can take many forms: soil, fire, wind, and water move; plants and animals move; ecological interactions, ecosystem processes, and natural disturbances move, or elements move through them. All require, to different degrees and at different scales, connectivity in nature.

The idea of connectivity, then, is actually rather straightforward. What is not straightforward is the process of translating and quantifying the concept for scientific analysis and practical application. Connectivity is an entirely scale and target dependent phenomenon — definitions, metrics, functionality, conservation applications, and measures of success depend on the taxa or processes of interest and the spatial and temporal scales at which they occur. This fact is at the heart of differing perceptions and numerous academic debates regarding connectivity. As a result, one single, all-encompassing definition of connectivity has proven elusive, and authors throughout this volume tackle the concept of connectivity from a variety of perspectives, from metapopulations (Moilanen and Hanski Chapter 3) to landscape ecology (Taylor *et al.* Chapter 2), from the flow of energy, material, organisms, or information across dissimilar habitats (Talley *et al.* Chapter 5) to the flow of genetic material within and among populations (Frankham Chapter 4). As emphasized by Fagan and Calabrese (Chapter 12), clear, replicable, and pragmatic metrics of connectivity are vital if conservationists are to invest limited time and resources wisely.

Broadly, we can identify two primary components of connectivity (Bennett 1999; Tischendorf and Fahrig 2000; Taylor *et al.* Chapter 2): (1) the structural (or physical) component: the spatial arrangement of different types of habitat or other elements in the landscape, and (2) the functional (or behavioral) component: the behavioral response of individuals, species, or ecological processes to the physical structure of the landscape. Structural connectivity is often equated with the spatial contagion of habitat, and is measured by analyzing landscape structure without any requisite reference to the movement of organisms or processes across the landscape. Functional connectivity, however, requires not only spatial information about habitats or landscape elements, but also at least some insight on movement of organisms or processes through the landscape. Fagan and Calabrese (Chapter 12) further distinguish functional connectivity into two types based on the extent of available movement data: (1) potential connectivity: metrics that incorporate some basic, indirect knowledge about an organism's dispersal ability, and (2) actual connectivity: metrics that quantify the actual movement of individuals through a landscape and thus provide a direct connectivity estimate. As highlighted by Taylor *et al.* (Chapter 2), although structural connectivity may be easier to measure than functional connectivity, this does not mean that connectivity is a generalized feature of a landscape. That is, structural connectivity does not imply that the same landscape would have the same connectivity for multiple species or processes. Instead, a structurally connected landscape may be functionally connected for some species and not for others.

The discipline of landscape ecology, the study of the effects of landscapes on ecological processes (Turner 1989; Turner *et al.* 2001), has been instrumental in shaping our ideas about connectivity. Drawing from principles of landscape ecology, one of the first ecological papers to explicitly define connectivity was Merriam (1984), who introduced the concept of "landscape connectivity" and defined it as "the degree to which absolute isolation is prevented by landscape elements which allow organisms to move among patches." Taylor *et al.* (1993) later modified this definition to "the degree to which the landscape impedes or facilitates movement among resource patches"; this has since become one of the most frequently used definitions of connectivity in the scientific literature. Likewise, With *et al.* (1997) defined landscape connectivity as "the functional relationship among habitat patches, owing to the spatial contagion of habitat and the movement responses of organisms to landscape structure." Thus, by explicitly describing movement across landscapes, these definitions combine the structural and functional components of connectivity. In this volume, Taylor *et al.* (Chapter 2) return to the basics of landscape connectivity, revisiting these definitions, refining the concept, and providing advice for future applications.

Like landscape ecology, metapopulation ecology also has contributed much to our understanding of connectivity (Moilanen and Hanski 2001; Moilanen and Hanski Chapter 3). Some species naturally exist in meta-populations – a set of local populations within some larger area, linked together by occasional immigration (Hanski and Gilpin 1997; Hanski 1999). Because metapopulation persistence depends on sufficient coloni-zation to compensate for local extinction of subpopulations, the concept of connectivity is a key feature of metapopulations. In this context, connectivity is generally related to migration rate, colonization rate, or gene flow among discrete patches in a metapopulation. Connectivity therefore is focused at the patch level, and is often measured as the distance to the nearest patch or nearest occupied patch. To distinguish this patch scale from the landscape scale discussed above, Tischendorf and Fahrig (2001) have proposed to label connectivity among patches, as used in metapopulation ecology, as "patch connectivity," and con-nectivity of the entire landscape, as used in landscape ecology, as "landscape connectivity." As described by both Taylor *et al.* (Chapter 2) and Moilanen and Hanski (Chapter 3), recent work is attempting to better integrate these various approaches to connectivity.

Notwithstanding the definitional nuances, the crucial point here is that the concept of connectivity, as viewed in the scientific literature, in conservation circles, in the media, and by the general public, is entirely dependent on the scale, species, or process in question. A landscape facilitating movement for an elephant may not do the same for a mouse – and vice versa. Connectivity also is by no means a static concept; rather, it is highly dynamic and often unpredictable; examples of the dynamic nature of connectivity are provided throughout this volume (e.g., Taylor *et al.* Chapter 2; Talley *et al.* Chapter 5; Harrison and Bjorndal Chapter 9; Ricketts *et al.* Chapter 11; Soulé *et al.* Chapter 25). Connectivity that is here today may be gone tomorrow – and vice versa. Lose sight of these facts, and confusion, and perhaps controversy, results.

RECENT HISTORY OF CONNECTIVITY RESEARCH

What are the historic trends of scientific research on connectivity? To explore this, we conducted a literature review (*Biological Abstracts*, February 21, 2005) to search for the keyword "connectivity" in scientific papers published from 1980 to 2004 in 23 major journals in conservation biology, ecology, landscape ecology, wildlife biology, and general science. Because corridors have tended to be a primary, and at times controversial, conservation tool to promote connectivity, we repeated this literature review searching for the term "corridor(s)." To correct for publication volume, we standardized the number of connectivity or corridor papers published each year by the total number of articles published annually in the 23 journals. Although limiting the search to keywords certainly underestimates the number of studies that actually investigated connectivity or corridor topics (see Moilanen and Hanski 2001), the results do provide some insight on patterns in connectivity research over the past two decades.

As evident by Fig. 1.1, research focused explicitly on connectivity is a relatively recent trend, one that mirrors our heightened concern of the impacts of accelerating habitat fragmentation. Out of about 67 500 papers published in these 23 journals during this time period, 328 and 352 papers had "connectivity" and "corridor" keywords, respectively. Corridor studies were relatively rare in the 1980s, but increased in prominence during the 1990s; this trend parallels the controversy within the scientific literature regarding the pros and cons of corridors as conservation tools (see below).

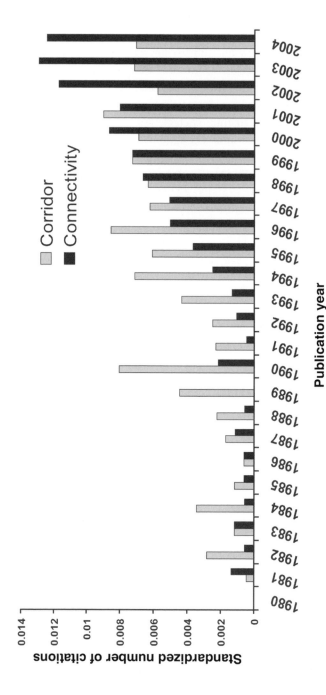

Fig. 1.1. Scientific papers published each year from 1980 to 2004 in 23 major landscape ecology, conservation biology, wildlife biology, and ecology journals with the terms "connectivity" or "corridor(s)" in their titles or keywords. The annual numbers of connectivity or corridor citations are standardized by the total number of citations in the 23 journals each year.

Studies focusing on connectivity have rapidly increased throughout the 1990s, and today outnumber corridor studies nearly two to one. Similarly, in a literature review of major freshwater ecology and management journals, Pringle (Chapter 10) found that freshwater connectivity studies, although a more recent trend than in landscape ecology and conservation biology journals, have proliferated within the last decade.

When analyzing connectivity and corridor studies by specific journal (Fig. 1.2), *Landscape Ecology* devoted more of its content to connectivity issues than any other journal; this result is intuitive given the history of connectivity research in the discipline of landscape ecology. Connectivity and corridor studies were next most frequent in conservation biology journals, again expected considering the conservation implications of connectivity research.

Although connectivity studies are on the rise, there is still much work to be done. For example, of the growing number of studies that have focused on movement through corridors, most have been descriptive and not experimental, thus limiting inference about the efficacy of corridors as conservation tools (Beier and Noss 1998; see below). Further, in an extensive review of the literature, Haddad and Tewksbury (Chapter 16) conclude that studies of corridor effects on population viability, community structure, and biological diversity are but in their infancy.

THE IMPORTANCE OF CONNECTIVITY CONSERVATION

Habitat fragmentation and the need for connectivity

How vital is connectivity for biodiversity conservation? We know that habitat destruction and fragmentation are the primary proximal threats to biodiversity (Wilcove *et al.* 1998). Fragmentation not only reduces the total amount of habitat available, but also simultaneously isolates the habitat that remains, preventing movement of organisms and processes in previously connected landscapes. Without natural levels of connectivity, native biodiversity is in jeopardy. Many studies have documented species loss in isolated habitats; even the largest protected areas existing today in western North America (Newmark 1987), eastern North American (Gurd *et al.* 2001), and even East Africa (Newmark 1995), are too small or too isolated to maintain viable populations for many wide-ranging species. As evident in this volume, while the effects of isolation are often most immediately noticeable for larger animals such as wide-ranging terrestrial carnivores (Paquet *et al.* Chapter 6; Tracey Chapter 14; Carroll Chapter 15; Theobald Chapter 17; Clevenger and Wierzchowski Chapter 20) and

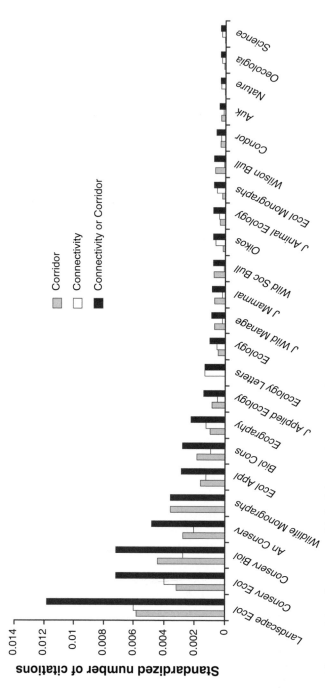

Fig. 1.2. Scientific papers published during the period 1980–2004 with the terms "connectivity" or "corridor(s)" in their titles or keywords, listed by each of 23 major journals in landscape ecology, conservation biology, wildlife biology, and ecology. The total number of connectivity or corridor citations are standardized by the total number of citations in the 23 journals.

migratory oceanic species (Harrison and Bjorndal Chapter 9), smaller animals such as freshwater shrimp (Pringle Chapter 10), marine invertebrates (Dibacco et al. Chapter 8), insect crop pollinators (Ricketts et al. Chapter 11), birds (Marra et al. Chapter 7), small mammals (Frankham Chapter 4), fish (Neville et al. Chapter 13), and butterflies (Moilanen and Hanski Chapter 3; Frankham Chapter 4; Haddad and Tewksbury Chapter 16) can all suffer when natural levels of connectivity are severed.

There are many, often synergistic, mechanisms by which isolation can lead to the extirpation of populations and the extinction of species. Demographic, environmental, and genetic forces, whether random or deterministic, can act independently or in concert to create a "vortex" of extinction in fragmented, isolated populations (Gilpin and Soulé 1986). Extinction vortices may be best repelled by preventing fragmentation and isolation in the first place, ideally by maintaining large populations in large contiguous blocks of quality habitat. Often, however, we must also attempt to maintain connectivity by protecting or restoring linkages in areas where fragmentation has already occurred. Indeed, although connectivity between reserves should certainly not be considered a substitute for the conservation of large core areas (Taylor et al. Chapter 2; Noss and Daly Chapter 23; Bennett et al. Chapter 26), connecting protected areas with linkages might be an effective way, and at times the last remaining option, to increase the effective area of some reserves and the population size of species in crisis.

Benefits and challenges of connectivity conservation

The preservation of natural levels of connectivity undoubtedly lends strength to efforts to protect species and habitats (Noss 1987, 1992; Hudson 1991; Saunders and Hobbs 1991; Noss et al. 1996; Noss and Soulé 1998; Bennett 1999; Soulé and Terborgh 1999; this volume). Possible benefits are many (Table 1.1). For example, connectivity may be essential to allow for the natural ranging behavior of animals among foraging or breeding sites and for the dispersal of organisms from their natal ranges (e.g., Tracey Chapter 14). Such movements may be critical to facilitate the exchange of genetic material among otherwise isolated populations (Frankham Chapter 4; Neville et al. Chapter 13); in the short term, genetic variability may be essential to mitigate the potential deleterious effects of inbreeding depression, and in the long term, to allow species to adapt and evolve to changing environmental conditions. Further, at large spatial and temporal scales, maintaining natural levels of connectivity may be essential to allow for natural range shifts in response to long-term

Table 1.1. *Potential advantages and disadvantages of the use of corridors as conservation tools to facilitate connectivity.*

Potential advantages	Potential disadvantages
(1) Increase immigration rate to a reserve, which could:	(1) Increase immigration rate to a reserve, which could:
(a) Increase or maintain species diversity	(a) Facilitate the spread of infectious diseases
(b) Provide a "rescue effect" to small, isolated populations by augmenting population sizes and decreasing extinction probabilities	(b) Facilitate the spread of exotic predators and competitors
(c) Permit recolonization of extinct local populations, potentially enhancing persistence of metapopulations	(c) Facilitate the spread of weedy or pest species
(d) Prevent inbreeding depression and maintain genetic variation within populations	(d) Decrease the level of genetic variation among subpopulations
	(e) Cause "outbreeding depression" by disrupting local adaptations and coadapted gene complexes
(2) Permit daily or seasonal movements for foraging, breeding, migration, or other behaviors	(2) Facilitate spread of wildfires and other catastrophic abiotic disturbances
(3) Facilitate dispersal of animals from natal ranges to adult breeding ranges	(3) Create a "mortality sink" by increasing exposure of animals in corridors to humans, native and exotic predators and competitors, pollution, and other deleterious "edge effects"
(4) Accommodate natural range shifts due to global climate change	(4) Riparian strips, often recommended as corridors, might not enhance dispersal or survival of upland species
(5) Provide predator-escape cover for movement between patches	(5) High economic cost to purchase, design, construct, restore, maintain, and protect corridors
(6) Provide wildlife habitat for transient or resident animals within corridors	(6) Trade-off costs and conflicts with other conservation acquisitions, including conventional strategies for enlarging core areas and preserving endangered species habitat
(7) Provide alternative refuges from large disturbances (a "fire escape")	(7) Political costs from altering human land-use patterns
(8) Continuance of ecological processes and ecosystem services such as succession, seed dispersal, and flow of water, nutrients, and energy	
(9) Provide "greenbelts" to limit urban sprawl, abate pollution, provide recreational opportunities, and enhance scenery and land values	

Source: Modified from Noss (1987); see also Soulé and Simberloff (1986), Simberloff and Cox (1987), Soulé (1991), Hobbs (1992), Simberloff et al. (1992), McEuen (1993), Rosenberg et al. (1997), Beier and Noss (1998), Bennett (1999), Dobson et al. (1999).

environmental transitions, such as global climate change (Noss and Daly Chapter 23; Soulé *et al.* Chapter 25). Finally, connectivity is also necessary to maintain the continuity of natural disturbance regimes (Soulé *et al.* Chapter 25), ecosystem services such as hydrology (Pringle Chapter 10) and crop pollination (Ricketts *et al.* Chapter 11), and the flow of material, at times across distinct ecosystems (Talley *et al.* Chapter 5). Disruption of such ecological forces in habitat isolates has received relatively limited study, but can have striking effects on biological communities.

Despite some obvious benefits of connectivity, efforts to conserve connectivity have not been without criticism (Table 1.1). Critique has most often focused on the use of corridors as conservation tools to facilitate the movement of organisms among otherwise isolated natural areas (Soulé and Simberloff 1986; Simberloff and Cox 1987; Hobbs 1992; Simberloff *et al.* 1992; McEuen 1993; Rosenberg *et al.* 1997). It is not our intent to rehash this debate in this volume, but we do acknowledge that conservation biologists must be firmly aware of challenges inherent in connectivity conservation. Many questions should be seriously considered. For example, if established, do target species actually use corridors and is this use sufficient to improve the viability of otherwise isolated populations in the core areas that the corridors are intended to connect? If linkages are used, do they provide enough security and/or resources such that transient or resident individuals are not exposed to excessive risk? For example, edge effects can decrease the functionality of corridors, eventually creating a death trap for dispersing animals and forming a "mortality sink" that functions as a net drain on the population (Soulé 1991). Do corridors, or more generally connectivity, facilitate non-intended transmission of disease (McCallum and Dobson Chapter 19), exotic or weedy species (Crooks and Suarez Chapter 18), ecological disturbances (Soulé *et al.* Chapter 25), or genetic material that would disrupt local adaptation of subpopulations (Frankham Chapter 4)? And perhaps most importantly, particularly to those in charge of securing connectivity with limited funds, does connectivity conservation cost too much (Morrison and Reynolds Chapter 21)? Although perhaps frustrating to those seeking global solutions, the answers to these important questions, like many issues in conservation, often turn out to be "it depends" and are largely situation-specific. Nonetheless, given the heightened focus on connectivity worldwide, many have decided that the benefits of connectivity conservation outweigh the costs.

To help evaluate the conservation function of corridors, Beier and Noss (1998) reviewed published experimental and inferential studies that attempted to empirically address whether corridors enhance population viability of species in patches connected by corridors. Beier and Noss (1998) defined a conservation corridor as "linear habitat, embedded in a dissimilar matrix, that connects two or more larger blocks of habitat and that is proposed for conservation on the grounds that it will enhance or maintain the viability of specific wildlife populations in the habitat blocks." Importantly, in their review, Beier and Noss (1998) failed to find any empirical evidence for the hypothetical negative impacts of corridors and concluded that the preponderance of evidence demonstrates that corridors are indeed beneficial in facilitating the movement of animals among otherwise isolated natural areas. Indeed, the precautionary principle would argue that it is usually easier to prevent harm to biodiversity than to repair it later, and that, in scientific assessment of a potentially harmful action (i.e., severing natural levels of connectivity), the burden of proof should lie with proponents of the action, even in the face of our inherent uncertainty about managing ecosystems (Dayton 1998). Evidence for the benefits of connectivity may be sufficient enough to invoke the precautionary principle in many cases (e.g., Soulé *et al* Chapter 25), although it is worth emphasizing that we must also be cautious when *enhancing* connectivity for invasive species (Crooks and Suarez Chapter 18), and that proactive conservation actions need to be balanced with economic, social, and logistical costs (Morrison and Reynolds Chapter 21)

Moving beyond the corridors debate: the key questions

Of course, corridors are but one of many methods to conserve connectivity. As argued by Bennett (1999), the controversy and debate surrounding corridors has been quite narrowly focused, largely ignoring other types of movement and connections, including stepping stones (Schultz 1995; Haddad 2000; Hale *et al.* 2001; Frankham Chapter 4; Soulé *et al.* Chapter 25), migratory stopovers (Winker *et al.* 1992; Marra *et al.* Chapter 7), and habitat mosaics and the permeability of the intervening matrix (Wiens 1989; With *et al.* 1997; Ricketts 2001; Taylor *et al.* Chapter 2; Ricketts *et al.* Chapter 11; Theobald Chapter 17; Sanderson *et al.* Chapter 24). According to Bennett (1999), when considering the necessity of connectivity, key questions to pose are:

1. "Are populations, communities, and natural ecological processes more likely to be maintained in landscapes that comprise an

interconnected system of habitats, than in landscapes where natural habitats occur as dispersed ecologically-isolated fragments?"

2. "What is the most effective pattern of habitats in the landscape to ensure ecological connectivity for species, communities, and ecological processes?"

As Bennett (1999) correctly asserts, few would argue with the first point, although we caution that naturally fragmented landscapes, such as mountain tops or islands, might be exceptions. Instead, much of the focus of efforts to conserve natural levels of connectivity should be on how to best address the second point.

Overall, much of the current discussion regarding the advantages and disadvantages of connectivity seems to be less about the pros and cons of corridors as a specific conservation tool, than about the challenges associated with purchasing, designing, constructing, restoring, maintaining, and protecting natural levels of connectivity. Often, the best argument against connectivity conservation may be the high resource and opportunity costs involved (Morrison and Reynolds Chapter 21). Perhaps the most pertinent question facing conservation biologists, land-use planners, and resource managers today is not why do we need natural levels of connectivity, but rather how should it be achieved, for what target species or ecological process, and at what scale?

VOLUME OVERVIEW

Although habitat loss and fragmentation have been identified as major threats to life on the planet, and although maintaining connections in otherwise fragmented landscapes has become a primary focus of our conservation efforts, relatively few thorough treatments exist on the subject of connectivity conservation. Working as an academic scientist in a university setting (KRC) and as a scientific director for an international non-governmental organization (MS), we both have professional and personal commitments towards studying and implementing connectivity conservation. Through our first-hand experiences with connectivity science and policy, and through many discussions with our colleagues, we recognized that a comprehensive vision was needed to synthesize the growing body of literature on connectivity and to provide directions for future actions. Thus, we felt the time was right to convene a panel of experts to generate a synthetic, and hopefully useful and stimulating, discussion on how to maintain connections for nature. This volume is a product of these efforts.

In the volume, we review the theory of landscape connections and emphasize empirical results to date. Contributing authors explore new material on experimental approaches, modeling, and implementation that will take the concept of connectivity well beyond the old corridors debate. We examine applications and trade-offs inherent in connectivity conservation across multiple spatial and temporal scales, for a variety of targets in a range of systems, and within the constructs of the socio-political environment. To ensure real-world application, authors delve beyond theory to include a variety of case studies of resource management decisions in which landscape connections and corridors were explicitly considered as management options.

This book is organized into three sections, each introduced by a leading expert in the field. Our first section, *Approaches to connectivity research*, highlights the impressive array of perspectives adopted by scientists investigating connectivity, including such diverse topics as metapopulations, landscape ecology, genetics, habitat linkages, migration, aquatic connectivity, and ecosystem services. The second section, *Assessing connectivity*, takes a detailed look at the suite of methods available to actually measure and evaluate connectivity at the genetic, individual, population, community, and landscape levels, a vital goal if we ever hope to evaluate the successes of connectivity conservation actions. Our efforts culminate in a final section, *Challenges and implementation of connectivity conservation*, which addresses some major challenges to connectivity conservation, details how resource managers and conservation practitioners can use a cost–benefit framework to analyze the opportunities for connectivity and to make decisions based on the most current science, and focuses on real-world examples of efforts currently under way to conserve connectivity across local, regional, and continental scales.

CONCLUSIONS

Maintaining connectivity in nature is undoubtedly appealing. It has caught the attention of planners, politicians, and land managers worldwide. Conservation organizations including The Nature Conservancy, Conservation International, The Wildlands Project, and the World Wildlife Fund, as well as numerous local land trusts and government agencies, have embraced the need for connectivity, to maintain it where it exists, and to regenerate it where it has been severed. Focus is shifting away from the debate about the need for connectivity, and is now directed towards putting theory into practice and actually implementing plans for securing connectivity. However, despite its popularity, actually studying, designing, and protecting connected landscapes is hardly an exact science. Why is this so, and what can we do about it?

The issue of connectivity in conservation can be defined by many individual topics derived from three broad and overlapping domains of study — the theoretical, the empirical, and the applied (Fig. 1.3). You are probably reading this book because you are interested in or come from at least one of these domains. Perhaps you are a conservation planner for a major organization trying to prioritize which parcels of land to protect in order to maintain a movement corridor through a fragmented landscape. Perhaps you are a researcher developing a new theoretical model to describe the movement of an organism or ecological process through

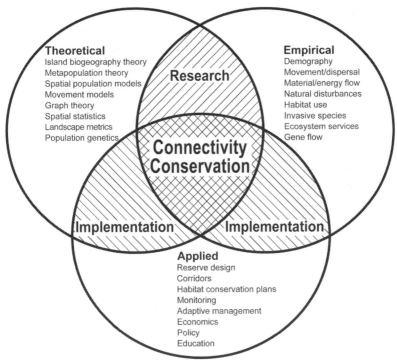

Fig. 1.3. Conceptual model of the theoretical, empirical, and applied domains of connectivity conservation. Synthesis and breakthroughs in connectivity conservation largely occur at the intersection of the three domains (hatched).

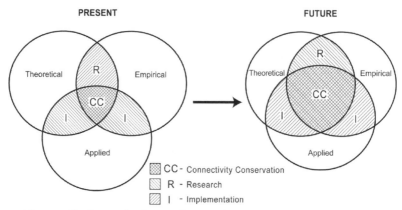

Fig. 1.4. The key to the future of connectivity conservation (CC) is to enlarge the area of overlap, the synergy between domains, of our theoretical, empirical, and applied efforts in connectivity research and implementation.

the landscape. Or perhaps you are a new graduate student planning to collect empirical data on dispersal between habitat remnants of a rare or threatened species.

Collectively viewing each domain of study produces a laundry list of topics, all discussed in the volume, which are vital to our efforts to conserve connectivity (Fig. 1.3). However, we feel that true synthesis and breakthroughs really occur at the intersection of the three domains (the hatched area in Fig. 1.3) — when scientists developing models and metrics make their work accessible to field researchers, when empirical biologists use theory as a foundation to formulate their research questions, and when conservation practitioners work closely with scientists to ensure that the best science guides applied conservation action. We believe that enlarging this area of overlap and synergy between domains is a key to connectivity conservation (Fig. 1.4); the concluding chapter to this volume (Bennett *et al.* Chapter 26) highlights some of the future opportunities for synergy between theoretical, empirical, and applied aspects of conservation science. In reading through this book, it is our goal that the reader will not only learn about individual topics of interest, but also will be able to explore if and how the synthesis of theoretical, empirical, and applied efforts advances connectivity, and hence biodiversity, conservation. In other words, we believe the sections on connectivity approaches and assessments should be valuable not just to research scientists but to land managers as well, and in turn, the more applied section on trade-offs and implementation should be important for academic researchers as well as practicing conservationists. Ultimately, we hope that the volume will challenge both scientists and conservation practitioners as to how best to develop new knowledge and apply existing knowledge to the task of maintaining connections for nature.

REFERENCES

Beier, P., and R. F. Noss. 1998. Do habitat corridors provide connectivity? *Conservation Biology* **12**:1241–1252.

Bennett, A. F. 1999. *Linkages in the Landscape: The Role of Corridors and Connectivity in Wildlife Conservation.* Gland, Switzerland/Cambridge, UK: IUCN.

Dayton, P. K. 1998. Reversal of the burden of proof in fisheries management. *Science* **279**:821.

Dobson, A., K. Ralls, M. Foster, *et al.* 1999. Corridors: reconnecting fragmented landscapes. Pp. 129–170 in M. E. Soulé and J. Terborgh (eds.) *Continental Conservation: Scientific Foundations of Regional Reserve Networks.* Washington, DC: Island Press.

Gilpin, M. E., and M. E. Soulé. 1996. Minimum viable populations: processes of species extinction. Pp. 19–34 in M. E. Soulé, (ed.) *Conservation Biology: The Science of Scarcity and Diversity.* Sunderland, MA: Sinauer Associates.

Gurd, D. B., T. D. Nudds, and D. H. Rivard. 2001. Conservation of mammals in Eastern North American wildlife reserves: how small is too small? *Conservation Biology* **15**:1355–1363.

Haddad, N. 2000. Corridor length and patch colonization by a butterfly, *Junonia coenia. Conservation Biology* **14**:738–745.

Hale, M. S., P. W. W. Lurz, M. D. F. Shirley, *et al.* 2001. Impact of landscape management on the genetic structure of red squirrel populations. *Science* **293**:2246–2248.

Hanski, I. 1999. *Metapopulation Ecology.* Oxford, UK: Oxford University Press.

Hanksi, I., and M. E. Gilpin. 1997. *Metapopulation Biology.* San Diego, CA: Academic Press.

Hobbs, R. J. 1992. The role of corridors in conservation: solution or bandwagon? *Trends in Ecology and Evolution* **7**:389–392.

Hudson, W. E. 1991. *Landscape Linkages and Biodiversity.* Washington, DC: Island Press.

McEuen, A. 1993. The wildlife corridor controversy: a review. *Endangered Species Update* **10**:1–12.

Merriam, G. 1984. Connectivity: a fundamental ecological characteristic of landscape pattern. Pp. 5–15 in J. Brandt and P. Agger (eds.) *Proceedings of the 1ˢᵗ International Seminar on Methodology in Landscape Ecological Research and Planning.* Roskilde, Denmark: Roskilde University.

Moilanen, A., and I. Hanski. 2001. On the use of connectivity measures in spatial ecology. *Oikos* **95**:147–151.

Newmark, W. D. 1987. Mammalian extinctions in western North American parks: a landbridge perspective. *Nature* 325:430−432.

Newmark, W. D. 1995. Extinction of mammal populations in western North American national parks. *Conservation Biology* 9:512−526.

Noss, R. F. 1987. Corridors in real landscapes: a reply to Simberloff and Cox. *Conservation Biology* 1:159−164.

Noss, R. F., and M. Soulé. 1998. Rewilding and biodiversity: complementary goals for continental conservation. *Wild Earth* 8:18−28.

Noss, R. F., H. B. Quigley, M. G. Hornocker, T. Merrill, and P. C. Paquet. 1996. Conservation biology and carnivore conservation in the Rocky Mountains. *Conservation Biology* 10:949−963.

Ricketts, T. H. 2001. The matrix matters: effective isolation in fragmented landscapes. *American Naturalist* 158:87−99.

Rosenberg, D. K., B. R. Noon, and E. C. Meslow. 1997. Biological corridors: form, function, and efficacy. *BioScience* 47:677−687.

Saunders, D. A., and R. J. Hobbs. 1991. *Nature Conservation: The Role of Corridors.* Chipping Norton, NSW, Australia: Surrey Beatty and Sons.

Schultz, C. B. 1995. Corridors, islands and stepping stones: the role of dispersal behavior in designing reserves for a rare Oregon Butterfly. *Bulletin of the Ecological Society of America* 76:240−261.

Simberloff, D., and J. Cox. 1987. Consequences and costs of corridors. *Conservation Biology* 1:63−71.

Simberloff, D., J. A. Farr, J. Cox, and D. W. Mehlman. 1992. Movement corridors: conservation bargains or poor investments? *Conservation Biology* 6:493−504.

Soulé, M. E. 1991. Conservation corridors: countering habitat fragmentation: theory and strategy. Pp. 91−104 in W. E. Hudson (ed.) *Landscape Linkages and Biodiversity* Washington, DC: Island Press.

Soulé, M. E., and D. Simberloff. 1986. What do genetics and ecology tell us about the design of nature reserves? *Biological Conservation* 35:19−40.

Soulé, M. E., and J. Terborgh. 1999. *Continental Conservation: Scientific Foundations of Regional Reserve Networks.* Washington, DC: Island Press.

Taylor, P. D., L. Fahrig, K. Henein, and G. Merriam. 1993. Connectivity is a vital element of landscape structure. *Oikos* 68:571−573.

Tischendorf, L., and L. Fahrig. 2000. On the usage and measurement of landscape connectivity. *Oikos* 90:7−19.

Tischendorf, L., and L. Fahrig. 2001. On the use of connectivity measures in spatial ecology: a reply. *Oikos* 95:152−155.

Turner, M. G. 1989. Landscape ecology: the effect of pattern on process. *Annual Review of Ecology and Systematics.* 20:171−197.

Turner, M. G., R. H. Gardner, and R. V. O'Neill. 2001. *Landscape Ecology in Theory and Practice: Pattern and Process.* New York: Springer-Verlag.

Wiens, J. A. 1989. *The Ecology of Bird Communities*, vol 2, *Processes and Variations.* New York: Cambridge University Press.

Wilcove, D. S., D. Rothstein, J. Dubow, A. Phillips, and E. Losos. 1998. Quantifying threats to imperiled species in the United States. *BioScience* 48:607−615.

Winker, K., D. W. Warner, and A. R. Weisbrod. 1992. The northern waterthrush and Swainson's thrush as transients at a temperate inland stopover site. Pp. 384–402 in J. M. Hagan III, and D. W. Johnston, (eds.) *Ecology and Conservation of Neotropical Migrant Landbirds*. Washington, DC: Smithsonian Institution Press.

With, K. A., R. H. Gardner, and M. G. Turner. 1997. Landscape connectivity and population distributions in heterogeneous environments. *Oikos* 78:151–169.

Part I

Approaches to connectivity research

Introduction: Connectivity research— what are the issues?

JOHN A. WIENS

People began thinking about connectivity in a serious way when they recognized that fragmentation of habitats was a major threat to bio-diversity. This recognition was crystallized in the 1960s by the development of island biogeography theory, which cast the issue in terms of islands immersed in a sea of inhospitable habitat. As the theory was applied to fragments in terrestrial settings, this binary view of habitat vs. not habitat was carried over to a patch-matrix view of the world. Fragments were viewed as clearly bounded elements set in a matrix of unsuitable habitats. Connectivity among such remnant habitat fragments should reduce isolation and provide buffering against local extinction and biodiversity loss. This is the premise of much of the thinking and research about connectivity in an ecological and conservation context.

It is difficult, however, to disentangle the effects of habitat *fragmentation* from those of habitat *loss*. Both empirical and theoretical analyses suggest that there is a *fragmentation threshold* when habitat loss reaches a certain level and a habitat becomes disconnected. Connectivity becomes important once this threshold is passed and fragmentation becomes widespread.

Our thinking about fragmentation and connectivity has been constrained by a strong terrestrial emphasis, and within that by an even stronger bias toward highly fragmented forest and woodland habitat in temperate zones. One of the challenges of connectivity research is to broaden the perspective, to determine how well the ideas developed in highly fragmented areas apply in places that are less fragmented, or in freshwater or marine environments. Another challenge is to move from thinking about connectivity among patches in a landscape to a perspective that encompasses the richness and complexity of entire landscape mosaics. Only in the most heavily modified landscapes can one think of the "matrix" as being homogeneously unsuitable. Instead, the "matrix" often contains a variety of habitats of varying qualities, and the composition and configuration of this landscape mosaic may have profound effects on how we measure, or organisms perceive, connectivity.

Recent thinking and research about connectivity has developed on two rather separate fronts, differing in the questions asked, in the focal study systems, and in how habitats and connectivity are perceived. The focus of metapopulation theory is on populations and the ways in which the isolation of patches of suitable habitat affects the probability of population persistence. Most empirical studies of metapopulations have been on species that are habitat specialists, for which this patch-matrix approach may be suitable. In these systems, connectivity is important only insofar as it affects the import/export dynamics of individuals from and among habitat patches.

The emergence of landscape ecology as a discipline has fostered a different perspective on connectivity. Because the composition and texture of the entire landscape are important, the emphasis is less on how connectivity affects within- and among-patch population dynamics and more on how the characteristics and structure of the entire landscape affect movement among the landscape elements. The focus goes beyond population dynamics to consider such phenomena as the spread of disturbances, the movement of materials, or biogeochemical processes. And because landscape systems occur over the full range of the habitat loss/fragmentation spectrum, one can think of connectivity in the absence of fragmentation. How places in a landscape are connected will influence the probability that an organism, disturbance, or material will move from one place to another.

As thinking about connectivity has progressed, the distinction between *structural* and *functional* connectivity has become more important. Structural connectivity is by far the more easily visualized and measured of the two. It is what we portray on maps or geographic information system (GIS) images and analyze using spatial statistics or programs such as FRAGSTATS. It is the stage on which the dramas of ecology and evolution are played. But like a stage, it may be beautiful and even interesting, but without the players it is sterile and uninformative. The real action is in functional activity—how the structure of a landscape interacts with the properties of organisms, disturbances, or materials to influence how they actually move, and what facilities ("connectivity") or impedes ("barriers") their movements.

Generally speaking, there are two ways to think about (or model) movements through a landscape: diffusion and dispersal. Diffusion is usually considered as a passive, physical process, which may be appropriate as a starting approximation for the movement of inert materials, organismal propagules, or disturbances through a homogeneous environment or as neutral models for animal movements in structured landscapes. The effects of landscape structure and connectivity can be gauged in terms of departures from the predictions of diffusion models. In a diffusion-based view of the world, the configuration of the landscape (and therefore structural connectivity) is paramount.

Because dispersal involves the behavioral and perceptual attributes of the organisms that are moving through a landscape, it relates much more closely to functional connectivity. Indeed, it is these behavioral and perceptual traits that create the differences between structural and functional connectivity, and make maps or GIS images alone inadequate representations of the ecological connectivity that really matters. Moreover, because we increasingly assess the structural configuration of landscapes from what our remote sensors perceive and computer algorithms map, our view of structural connectivity is often limited to a restricted scale of human perception (generally tens of thousands of square meters). Management occurs over an even more confined range of scales. Different organisms, however, perceive and respond to landscape structure at quite different scales—it is trite but true to observe that a beetle and a bunny and a bison "see" the same prairie landscape, and its structural connectivity, at vastly different scales and in different ways. Consequently, assessing connectivity requires that the scales at which we measure landscape structure are commensurate with the scale(s) on which the organisms of interest perceive and respond to the landscape.

Functional connectivity relates to how things move through a landscape in relation to both the structural connectivity of the landscape and the characteristics of what is moving. These movements become important, in context of conservation, through their *consequences*. Although movement always leads somewhere, it may not lead somewhere important unless it affects such things as individual survival and reproduction, population dynamics and persistence, gene flow and genetic diversification of populations, predator—prey dynamics, nutrient redistribution, or disturbance spread. For example, isolation of populations (as by fragmentation) may lead to a loss of genetic diversity, through small effective population size, inbreeding, reduced gene flow, and the like. Connectivity counters these trends by enhancing genetic interchange among otherwise isolated populations. On the other hand, Sewall Wright argued during the 1930s that peripherally isolated populations that are linked to the main body of a species' distribution by only occasional immigration could develop local adaptations that would be the seeds of speciation. A species with a fragmented geographic range would have greater "evolutionary potential" than one more thoroughly bound together. If this model is correct, then too much connectivity might actually counter such evolutionary diversification.

IMPLICATIONS FOR CONSERVATION

Based on the literature of conservation biology over the past decade, one might easily conclude that enhancing connectivity is the panacea for many of the challenges of conservation, and that corridors are the best way to assure connectivity. The contributions to this volume show that the actual situation is much more complex and the conservation implications more nuanced.

The key issue is, when does connectivity really matter? Conservation managers should be concerned with how to invest resources wisely to realize the greatest return, in terms of biodiversity protected or enhanced. Does spending $X on Y ha to enhance connectivity among habitats in a landscape produce an equivalent, or greater, benefit than spending the same amount on an area of suitable habitat? Does it matter if the habitat area is isolated or part of a larger block of the same habitat? Is it sensible to use resources to conserve areas that increase landscape connectivity if the organisms of interest cannot reproduce there? What if the connectivity only serves to link together sink habitats?

Many questions; few simple answers. I'll end, however, with three observations. First, a focus on conserving habitat *area* may be most appropriate where the habitat is either not yet fragmented (and conservation may help to forestall fragmentation) or where it is highly fragmented (and therefore adding even a little more area can make a big difference). Focusing on conserving habitat *connectivity*, on the other hand, may be more critical between these extremes, at and around the fragmentation threshold. Here, increasing connectivity could significantly reduce the deleterious effects of fragmentation.

Second, the value of a landscape perspective is that it acknowledges that the "matrix" can itself have real conservation value. Increasingly, protected areas by themselves will not provide adequate long-term protection to biodiversity unless the surroundings are also considered.

"Managing the matrix" must become central to conservation as the vision expands beyond protected areas alone, and connectivity is a vital part of the matrix. On the other hand, the simpler patch-matrix perspective may be a suitable framework for conservation in highly fragmented land-scapes, especially if the concern is with habitat specialists (as many threatened and endangered species are), so we should not rush to embrace a landscape approach and in the process leave patch-matrix approaches (e.g., metapopulation theory) behind.

Finally, a word about models and theory. Conclusions about the relative importance of habitat amount versus arrangements, for example, have been derived largely from modeling exercises, and much (but by no means all) of our thinking about metapopulations has been derived from theory rather than observations. Models or theories can enlighten, but because they intentionally simplify nature, they can also mislead. Writing in the 1930s, the British evolutionist and geneticist J. B. S. Haldane observed that "no scientific theory is worth anything unless it enables us to predict something which is actually going on. Until that is done, theories are a mere game of words, and not such a good game as poetry." A good thing to remember.

Landscape connectivity: a return to the basics

PHILIP D. TAYLOR, LENORE FAHRIG
AND KIMBERLY A. WITH

INTRODUCTION

In the decade or so since the concept was formalized in landscape ecology
(Taylor *et al.* 1993) the meaning of the term "landscape connectivity" has
become rather diffuse and ambiguous. Many researchers continue to
ignore key elements of the original concept, which greatly diminishes its
potential utility for land management and the conservation of biodiversity.
As originally defined, *landscape connectivity* is "the degree to which the
landscape facilitates or impedes movement among resource patches"
(Taylor *et al.* 1993; see also With *et al.* 1997). This definition emphasizes
that the types, amounts, and arrangement of habitat or land use on the
landscape influence movement and, ultimately, population dynamics and
community structure. Landscape connectivity thus combines a description
of the physical structure of the landscape with an organism's response to
that structure. In contrast, common usage generally emphasizes only the
structural aspect, where landscape connectivity is simply equated with
linear features of the landscape that promote dispersal, such as corridors.
Moreover, most commonly employed measures of connectivity focus
only on how patch area and inter-patch distances affect movement (e.g.,
Moilanen and Hanski Chapter 3); such measures ignore the rich complexity of how organisms interact with spatial heterogeneity that may
ultimately affect dispersal and colonization success (e.g., interactions with
patch boundaries, matrix heterogeneity: Wiens *et al.* 1993; Wiens 1997;
Jonsen and Taylor 2000a). Our aim in this chapter is thus to refine the
concepts inherent in the original definition of landscape connectivity, to
outline why it is important to disentangle landscape connectivity from

Connectivity Conservation eds. Kevin R. Crooks and M. Sanjayan. Published by Cambridge
University Press. © Cambridge University Press 2006.

other (equally important) landscape characteristics, and to advise how a return to the basics may aid land managers charged with managing landscape connectivity as a component of biodiversity.

JUST WHAT IS LANDSCAPE CONNECTIVITY, ANYWAY?!! ISSUES AND CONCEPTS

Landscape connectivity is an "emergent property" of species–landscape interactions, resulting from the interaction between a behavioral process (movement) and the physical structure of the landscape. It is therefore a dynamic property that is assessed at the scale of the landscape (with particular organisms or suites of organisms in mind) and is not simply an aggregate property of a set of patches within the landscape. We can broadly consider two kinds of landscape connectivity: structural and functional connectivity. Structural connectivity ignores the behavioral response of organisms to landscape structure and describes only physical relationships among habitat patches such as habitat corridors or inter-patch distances. It is readily measured with a variety of landscape metrics or spatial analytical approaches (e.g., Gustafson 1998; Moilanen and Nieminen 2002). When physical relationships between habitat patches are tightened, structural connectivity is increased. Functional connectivity, on the other hand, increases when some change in the landscape structure (including but not limited to changes in structural connectivity) increases the degree of movement or flow of organisms through the landscape. The original concept of landscape connectivity thus emphasizes the functional connectivity of landscapes.

This distinction between structural and functional connectivity is not a trivial one. First and foremost, habitat does not necessarily need to be structurally connected to be functionally connected. Some organisms, by virtue of their gap-crossing abilities, are capable of linking resources across an uninhabitable or partially inhabitable matrix (Dale *et al.* 1994; Desrochers *et al.* 1998; Pither and Taylor 1998; Hinsley 2000; Bélisle and Desrochers 2002). Conversely, structural connectivity does not provide functional connectivity if corridors are not used by target species (see also Crooks and Sanjayan Chapter 1; Fagan and Calabrese Chapter 12; Noss and Daly Chapter 23). Nevertheless, structural connectivity is still all-too-frequently equated with functional connectivity in the literature and, in our experience, by land managers.

It is generally not possible to simply extrapolate from measures of structural connectivity (such as distances between patches) to derive a measure of overall landscape connectivity. That is because measures of structural connectivity ignore variability in the behavior of the organism(s) in response to the landscape structure, and ignore broader-scale influences of landscape structure on finer-scale movement decisions. Measures of landscape connectivity could be derived from inter-patch distances in situations where the matrix is invariant and ecologically neutral over the area of study, and where there are no effects of landscape structure on movement decisions at the scale of study, but such situations are likely quite rare.

Some recent work is paving the way to linking these two general frameworks. In particular, incidence function models were originally solely distance-based and so only measured structural connectivity. More recent derivations of these models contain coefficients that represent differential rates of movement of organisms through different habitats (e.g., Roland *et al.* 2000; Moilanen and Hanski Chapter 3). Such coefficients effectively represent the behavioral response of an organism to the physical structure of the landscape, so the models include a better metric of functional connectivity than measures based only on distance. The coefficients capture the "effective isolation" of patches (Ricketts 2001), in which movement may be reduced (or even avoided) in some cover types, and thus patches are effectively more isolated than Euclidean measures of distance would suggest.

ASSESSING LANDSCAPE CONNECTIVITY

Structural connectivity is usually easier to assess than functional connectivity, since the former can be computed using landscape metrics of spatial analyses of maps or within a geographic information system (GIS). This likely explains its prevalence in the literature (including some papers in this volume). However, the relative ease of calculating structural connectivity is not a sufficient reason for defaulting to its use. Using structural connectivity in place of landscape connectivity can (and does) lead to inappropriate land-management strategies, and obfuscates what might be key problems in managing a given landscape. We see this commonly when corridors are proposed as mitigative measures in forested landscapes slated for harvest. In forest landscapes, corridors probably do not improve access to small patches for mobile species, such as songbirds (Hannon and Schmiegelow 2000) and flying squirrels (Selonen and Hanski 2003), because movement among forest patches is actually a function of both gap-crossing abilities (Bélisle and Desrochers 2002) and the successional stage of the intervening habitat (Robichaud *et al.* 2002). Thus, a land manager charged with protecting a given volume of timber within a landscape might be advised to use that allocation to protect areas that serve other functions (e.g., riparian zones) or to increase the sizes of protected areas. Furthermore, attention can then be placed on managing the intervening patches of non-habitat – "managing the matrix" – in such a way as to maximize the speed at which remnant resource patches can be re-accessed by target species. On the other hand, some species may benefit from habitat corridors that physically connect suitable habitat or resource patches (Tewksbury *et al.* 2002; Varkonyi *et al.* 2003). Thus, managing for structural connectivity may, for some taxa, in some situations, also improve functional connectivity.

Another difficulty in measuring functional connectivity is that the effect of a landcover type on movement can also vary depending on type of landscape in which the cover type is embedded. For example, Calopterygid damselflies inhabiting streams in Nova Scotia, Canada, readily cross stream/forest boundaries but move little through forest. When streams are embedded in landscapes that are partially forested, the animals also readily cross stream/pasture boundaries and move through pasture to access the forest resource. However, when forest is almost completely removed damselflies are unlikely to move into pasture (Jonsen and Taylor 2000a, 2000b). Therefore, for these damselflies, the effect of pasture on landscape connectivity depends on the amount of forest in the landscape at a broader spatial scale. Landscape context is thus important for evaluating how land-use change will affect landscape connectivity.

Assessing landscape connectivity requires a species-centered approach (Hansen and Urban 1992). It requires information on species' movement responses to landscape structure (e.g., movement rates through different landscape elements, dispersal range, mortality during dispersal, and boundary interactions) and how those responses differ as a function of broader-scale influences. Such information is typically quite difficult to obtain, but this is changing rapidly owing to advances in satellite-based tracking devices and new methods for analyzing movement data (Jonsen *et al.* 2003; Tracey Chapter 14). Many researchers are now exploiting these technological and analytical advances, which will lead to significant progress in our understanding of how organisms interact with landscape pattern.

Heterogeneity and asymmetrical landscape connectivity

In assessing the functional connectivity of a landscape, one needs to keep in mind that regions that facilitate movement need not be discrete features of the landscape such as habitat corridors, but may occur where the juxtaposition of particular habitats or land uses act to funnel dispersers between habitat patches. Using an individual-based simulation model, Gustafson and Gardner (1996) demonstrated how the structure and heterogeneity of the matrix can affect the transfer of virtual organisms among forest fragments. Although patch size and relative isolation explained most of the variability in dispersal success, with closer and larger patches having the greatest exchange of individuals, the structure of the surrounding matrix also significantly altered transfers among patches. Roads, waterways, or land uses such as plowed fields may act as barriers to dispersal, encouraging dispersing individuals to move away from them. This can create asymmetries in the transfer probabilities among patches or regions of the landscape. There may be a lack of symmetry in the landscape pattern as well; emigrants from an isolated patch may succeed in reaching a neighboring patch, but if the neighboring patch is surrounded by a number of closer patches, then few of its dispersers will likely reach the isolated patch. Such asymmetrical landscape connectivity has been found for the Iberian lynx (*Lynx pardinus*) in southwestern Spain where suitable habitat surrounding one population tends to limit emigration but encourage immigration (Ferreras 2001). Less-suitable habitat can also function as an important conduit for dispersal in regions where landscape structure would otherwise limit structural connectivity among patches or populations (Milne *et al.* 1989). For example, Schultz (1998) suggested that dispersal of the Fender's blue butterfly (*Icaricia icarioides fenderi*) among lupine patches could be facilitated by creating small lupine "stepping stones" because butterflies moved more quickly through the matrix habitat, and were thus more likely to reach isolated patches of lupine in extensively fragmented landscape than if they moved along lupine corridors.

Directional connectivity among patches has also been documented in the case of the cactus bug (*Chelinidea vittiger*), which uses olfaction to locate its *Opuntia* cactus host (Schooley and Wiens 2003). In a release experiment, successful dispersal was determined by the size of the target patch and the structure of the intervening matrix, but there was a strong effect of prevailing wind direction on orientation behavior such that bugs were more likely to orient toward cactus patches located upwind. Similarly, the directionality of water currents may contribute to asymmetrical connectivity of populations in aquatic or marine systems (e.g., Man *et al.* 1995; DiBacco *et al.* Chapter 8; Pringle Chapter 10). Although this clearly complicates assessment of functional connectivity, these examples illustrate why landscape connectivity cannot be captured simply by an index of landscape pattern, but must be determined based on the organisms' perception of, and interaction with, the structure and heterogeneity of the landscape.

Non-linear effects of landscape connectivity are possible

Theory predicts that landscapes will become disconnected abruptly, at a threshold level of habitat availability, which may have some surprising and unexpected consequences for the management of biodiversity. For example, neutral landscape models, derived from percolation theory, are model landscapes in which complex habitat patterns are generated using theoretical spatial distributions (e.g., random or spatially correlated habitat distributions: With 2002). Landscape connectivity is assessed by determining how organisms move and interact with the structural heterogeneity of the resulting landscapes (With and Crist 1995; With *et al.* 1997). Such models may exhibit critical thresholds in landscape connectivity (Gardner *et al.* 1987; With 1997). The specific threshold at which landscapes become disconnected (the *percolation threshold*) depends on both the distribution of habitat and the habitat-specific movement rates of the species (With 2002). *There is thus no single critical threshold value at which a particular landscape becomes disconnected for all species simultaneously,* and a given landscape could be perceived simultaneously as both connected and disconnected by two species that differ in dispersal characteristics (i.e., definitions of landscape connectivity are organism-centered: Pearson *et al.* 1996). Such thresholds have also been identified in other theoretical approaches that quantify connectivity (e.g., Hanski 1999; Urban and Keitt 2001).

It is tempting to assume that thresholds in landscape connectivity precipitate other ecological thresholds, such as in dispersal success, population persistence, species interactions, and community composition, and thus ultimately, system resilience. However, these ecological thresholds may not — and generally do not — coincide with thresholds in landscape connectivity. For example, thresholds in dispersal success were not found to coincide with percolation thresholds in landscape connectivity (With and King 1999a) probably because dispersal success is typically assessed at different spatial and temporal scales than landscape connectivity. It follows then that the critical level of habitat at which landscape connectivity becomes disrupted is generally not the same level as a species' extinction threshold (the minimum habitat area required for population persistence: With and King 1999b). What is important to realize is that a connected landscape does not guarantee species persistence, just as a disruption of landscape connectivity may not result in the immediate extinction of a species. The latter could nevertheless set the stage for delayed extinctions that occur years or decades later (extinction debt: Tilman *et al.* 1994; Hanski and Ovaskainen 2002).

MISAPPLICATION OF CONNECTIVITY CONCEPTS ON THE LANDSCAPE

Land managers are understandably eager to incorporate scientifically based guidelines concerning the maintenance of landscape connectivity into management plans. Unfortunately, this can lead to the well-intentioned but misguided application of principles supposedly derived from theory.

A specific example illustrates this point. The Woodland Trust, the UK's leading conservation organization devoted to the protection of native woodland, had initially established targets for forest cover throughout the UK based on a "30% Rule" derived from random neutral landscape models (Peterken 2002; Woodland Trust 2002). Empirical support for this 30% Rule was also taken from literature that showed persistence thresholds for various species (birds, mammals) in this vicinity (Andrén 1994; Peterken 2002). The implication of the 30% Rule was that connectivity was presumably assured in landscapes with more than 30% forest cover. On the surface, advocating for 30% forest cover would appear to be a good idea, given that only 12% of the country is currently forested (Woodland Trust 2000). But real landscapes are not random; they contain barriers, detrimental matrix habitat, reproductive sinks, and predator pits, and thus simple random neutral landscape models are inappropriate as a basis for developing general guidelines for managing connectivity in these landscapes (although neutral landscapes with greater spatial contagion provided a better model: K. Watts, pers. comm.). Even if 30% woodland cover is above the connectivity threshold for a species of concern, it is not necessarily above its persistence threshold. Suggesting that a goal of 30% woodland cover is "not arbitrary" but is "based on [a] well-established mathematical model" (Woodland Trust 2002) is misleading, in that it lends an aura of scientific rigor and credibility that may not be defensible, even if does help to promote increased protection and restoration of woodlands (i.e., a worthy goal, but for the wrong reason). This is a significant concern when one considers that the concept of forest habitat networks has been incorporated into the current forestry strategies of the Forestry Commission and other conservation organizations in the UK (Peterken 2002; Watts *et al.* 2005). Fortunately, the Woodland Trust is no longer using such simple landcover thresholds as a basis for conservation planning, and instead appears to be exploring alternatives that move beyond such structural-connectivity measures toward more of a functional-based approach that would entail managing the matrix to enhance connectivity among existing woodlands (K. Watts, pers. comm.). Nevertheless, their initial recommendation based on this 30% Rule was apparently so influential (and appealing) that other organizations within the UK, such as the South West Wildlife Trusts, are incorporating the 30% Rule in the development of their own action plans (e.g., the South West Nature Map: South West Wildlife Trusts 2004).

No matter how well-intentioned or advantageous the outcome, such recommendations based on the flawed application of connectivity concepts and theory will erode our credibility as scientists and practitioners. Even worse, we run the risk of mismanaging the lands we have been entrusted to protect. Instead, land managers should focus on the suite of ecological processes that maintain biodiversity in the landscape. Managing landscape connectivity is but a small part of that large task.

LESSONS LEARNED

We summarize this chapter with a set of key lessons that have emerged and should be considered by those studying or implementing concepts of landscape connectivity on the ground.

Landscape connectivity is species-specific

Different organisms interact with landscape structure at different scales and in a variety of ways. Landscape managers must deal with that explicitly. Note that this does not necessarily mean that detailed measures of landscape connectivity are needed for every particular species– landscape combination, but rather that we need to recognize that organisms will exhibit a diversity of responses to any given management intervention, and that we should attempt to manage for a range of responses, across a range of taxa and a range of spatial scales.

Manage the matrix!

Managing the matrix can offer an effective means of managing the landscape to preserve or restore functional connectivity. Recall that there are three components to landscape connectivity (1) species movement patterns and behaviors, (2) the size and arrangement of resource patches, and (3) the matrix. Our ability to manipulate or manage these components varies considerably. We can redirect and manipulate the behavioral responses of species through the use of fences, roadway crossings, and other devices, but we generally cannot directly alter the inherent behavior of a species. It should however be observed that there is surprising plasticity of response to changes in landscape structure in some species. For example, Taylor and Merriam (1995) and Pither and Taylor (2000) showed that wing morphology of a calopterygid damselfly differed between individuals living in landscapes dominated by forest and those in landscapes dominated by pasture. Such responses should be looked for since the need for expensive mitigative measures may be alleviated if species can behaviorally adapt to a disruption in landscape connectivity.

With respect to the second component, there are times when it is possible to alter the size and arrangement of habitat on the landscape, but frequently, economic or social constraints preclude our ability to do so. For example, commercial forest enterprises are often structured such that they require access to fixed amounts of wood fiber on a continuous basis — these economic constraints may preclude some management options. As a further example, it can be costly to reclaim and revegetate landscapes that have already been converted to agricultural use. These examples are not excuses to ignore issues around habitat loss and exploitation, but simply a reflection that at times there is a limited ability for land managers to affect the extent of such human activities.

Managing the matrix is the third means through which we can manage landscape connectivity. The matrix is often more extensive than remaining patches of habitat, so a focus on managing it may make more sense in many circumstances. This is not a new idea but the fundamental relationship between managing the matrix and landscape connectivity is sometimes ignored. This may be partly because managing the matrix can require decisions with high political or economic costs. For example, improvements in landscape connectivity may be best accomplished through removing or moving roadways, restricting urban development, or limiting pesticide use on farmland. Opposition to such decisions by the stakeholders involved may preclude implementation. Conservation plans should nevertheless consider what actions are possible in the matrix along with habitat conservation and restoration.

Real landscapes are not random

Nor are they archipelagos. The use of simple mathematical models to provide guidance for the management of complex, heterogeneous landscape is not always appropriate for practical applications. Their utility lies in generating general expectations about properties of landscapes that may have importance for land management and the conservation of biodiversity (strategic application), rather than providing specific recommendations for the management of a given landscape (tactical application).

Landscape connectivity is a necessary but not sufficient condition for species conservation

Landscape connectivity is important because it influences access to resources and colonization of empty habitat. However, population persistence at the landscape scale ultimately depends on the balance between reproduction and mortality. A change in landscape connectivity can affect reproduction and mortality, for example, through allowing or limiting access to potential breeding sites. However, reproduction and mortality (and therefore population persistence) are mainly determined by the amount and quality of habitats available on the landscape (Fahrig 1997; With and King 1999b; Breininger and Carter 2003; Woodford and Meyer 2003). Note that habitat quality can include the effects of species interactions on reproduction or mortality; for example, forest edge may be low quality habitat if predation rate is higher in edges (Chalfoun *et al.* 2002). Therefore, in conservation, a focus on landscape connectivity to the exclusion of habitat amount and quality (e.g., Stith *et al.* 1996, Keitt *et al.* 1997) will not guarantee species persistence in the landscape.

In fact, landscape connectivity may be either positively or negatively related to population persistence. For example, an increase in habitat amount could lead to a decrease in landscape connectivity if the organism moves less when it is in habitat than when it is in non-habitat (Jonsen and Taylor 2000a; Goodwin and Fahrig 2002). If habitat amount has a large positive effect on population persistence, this behavior would translate into a negative relationship between landscape connectivity and persistence (Tischendorf and Fahrig 2000). Alternatively, human impacts in the matrix, such as roads and pesticide applications, reduce matrix quality and thereby reduce landscape connectivity (Johnson and Collinge 2004; Clevenger and Wierzchowski Chapter 20). Modeling studies suggest that dispersal mortality has a large effect on population persistence (Lindenmayer *et al.* 2000; Fahrig 2001; Cooper *et al.* 2002).

In this situation, the effect of lowering landscape connectivity is thus not only to reduce movement, but also to increase mortality in the population; an increase in landscape connectivity (e.g., through reduced pesticide use or roadway crossing structures) would thus have a positive effect on population persistence.

Our main point is that a focus on landscape connectivity to the exclusion of habitat amount and quality will not guarantee population persistence or maintenance of biodiversity. Landscape connectivity is not a panacea in conservation or land management, but needs to be considered as a component of the suite of interacting factors that influence the demography of a species.

Landscape connectivity is inherently neither good nor bad

Through its effects on ecological processes, connectivity may positively influence population persistence for some organisms in some situations, and negatively influence them in others. Therefore, if one is managing for connectivity in the landscape, one is not trying to "maximize" it in some way — rather, one is trying to understand how altering other elements of landscape structure will affect it, and then assess what the importance of those changes will be to critical ecological outcomes, such as population persistence.

Landscape connectivity is a dynamic concept

Landscape connectivity must be assessed, and therefore managed, in the context of human land-use change. It will change over both short and long timescales. As such, it is prudent to consider assessments of landscape connectivity (and its ultimate effects on population dynamics and persistence) as part of adaptive management and management for resilience (Gunderson and Holling 2000). In many landscapes, connectivity may initially decline with some types of land use (e.g., removal of forest for timber) but then return (for some species) as the matrix undergoes successional change (e.g., Sekercioglu *et al.* 2002). Landscape connectivity will play an increasingly important role in the persistence of many plant and animal populations in the face of global change and resultant shifts and restructuring of species distributions (e.g., Pitelka *et al.* 1997; Warren *et al.* 2001; Soulé *et al.* Chapter 25). Landscape connectivity will be related directly to species' abilities to track shifts in habitat, their abilities to adapt to changing matrix conditions, and their abilities to persist in the modified landscapes that they colonize. Although its role is important, it bears repeating that connectivity does not play an exclusive role in each of these processes. Understanding for which species and under what conditions connectivity is important remains a considerable research challenge.

ACKNOWLEDGEMENTS

PDT acknowledges the support of NSERC and the Atlantic Cooperative Wildlife Ecology Research Network (ACWERN). LF was supported by NSERC; KAW was partially supported by a grant from the EPA-STAR Wildlife Risk Assessment Program (R82-9090) during the writing of this chapter. We thank Kevin Watts (Project Leader for Landscape Ecology, Woodland Ecology Branch of the Forest Research Agency of the Forestry Commission, Surrey, UK) for bringing the "30% Rule" of The Woodland Trust to our attention.

REFERENCES

Andrén, H. 1994. Effects of habitat fragmentation on birds and mammals in landscapes with different proportions of suitable habitat: a review. *Oikos* 71:355–366.
Bélisle, M., and A. Desrochers. 2002. Gap-crossing decisions by forest birds: an empirical basis for parameterizing spatially-explicit, individual-based models. *Landscape Ecology* 17:219–231.
Breininger, D. R., and G. M. Carter. 2003. Territory quality transitions and source–sink dynamics in a Florida scrub-jay population. *Ecological Applications* 13:516–529.
Chalfoun, A. D., F. R. Thompson, and M. J. Ratnaswamy. 2002. Nest predators and fragmentation: a review and meta-analysis. *Conservation Biology* 16:306–318.
Cooper, C. B., J. R. Walters, and J. Priddy. 2002. Landscape patterns and dispersal success: simulated population dynamics in the brown treecreeper. *Ecological Applications* 6:1576–1587.
Dale, V. H., S. M. Pearson, H. L. Offerman, and R. V. O'Neill. 1994. Relating patterns of land-use change to faunal biodiversity in the central Amazon. *Conservation Biology* 8:1027–1036.
Desrochers, A., L. Rochefort, and J. P. L. Savard. 1998. Avian recolonization of eastern Canadian bogs after peat mining. *Canadian Journal of Zoology* 76:989–997.
Fahrig, L. 1997. Relative effects of habitat loss and fragmentation on species extinction. *Journal of Wildlife Management* 61:603–610.
Fahrig, L. 2001. How much habitat is enough? *Biological Conservation* 100:65–74.
Ferreras, P. 2001. Landscape structure and asymmetrical inter-patch connectivity in a metapopulation of the endangered Iberian lynx. *Biological Conservation* 100:125–136.
Gardner, R. H., B. T. Milne, M. G. Turner, and R. V. O'Neill. 1987. Neutral models for the analysis of broad-scale landscape pattern. *Landscape Ecology* 1:19–28.

Goodwin, B. J., and L. Fahrig. 2002. Effect of landscape structure on the movement behaviour of a specialized goldenrod beetle, *Trirhabda borealis*. *Canadian Journal of Zoology* 80:24–35.

Gunderson, L. H., and C. S. Holling. 2002. *Panarchy: Understanding Transformations in Human and Natural Systems*. Washington DC: Island Press.

Gustafson, E. J. 1998. Quantifying landscape spatial pattern: what is state of the art? *Ecosystems* 1:143–156.

Gustafson, E. J., and R. H. Gardner. 1996. The effect of landscape heterogeneity on the probability of patch colonization. *Ecology* 77:94–107.

Hannon, S. J., and F. K. A. Schmiegelow. 2000. Corridors may not improve the conservation value of small reserves for most boreal birds. *Ecological Applications*. 12:1457–1468.

Hansen, A. J., and D. L. Urban. 1992. Avian response to landscape pattern: the role of species' life histories. *Landscape Ecology* 7:163–180.

Hanski, I. 1999. Habitat connectivity, habitat continuity, and metapopulations in dynamic landscapes. *Oikos* 77:209–219.

Hanski, I., and O. Ovaskainen. 2002. Extinction debt at extinction threshold. *Conservation Biology* 16:666–673.

Hinsley, S. A. 2000. The costs of multiple patch use by birds. *Landscape Ecology* 15:765–775.

Johnson, W. C., and S. K. Collinge. 2004. Landscape effects on black-tailed prairie dog colonies. *Biological Conservation* 115:487–497.

Jonsen, I. D., and P. D. Taylor. 2000a. Landscape structure and fine-scale movements of Calopterygid damselflies. *Oikos* 88:553–562.

Jonsen, I. D., and P. D. Taylor. 2000b. *Calopteryx* damselfly dispersions arising from multi-scale responses to landscape structure. *Conservation Ecology* 4:4. Available online at http://www.consecol.org/vol4/iss2/art4

Jonsen, I. D., R. A. Myers, and J. M. Flemming. 2003. Meta-analysis of animal movement using state-space models. *Ecology* 84:3055–3063.

Keitt, T. H., D. L. Urban, and B. T. Milne. 1997. Detecting critical scales in fragmented landscapes. *Conservation Ecology* 1:4. Available online at http://www.consecol.org/vol1/iss1/art4

Lindenmayer, D. B., R. C. Lacy, and M. L. Pope. 2000. Testing a simulation model for population viability analysis. *Ecological Applications* 10:580–597.

Man, A., R. Law, and N. V. C. Polunin. 1995. Role of marine reserves in recruitment to reef fisheries: a metapopulation model. *Biological Conservation* 71:197–204.

Milne, B. T., K. M. Johnston, and R. T. T. Forman. 1989. Scale-dependent proximity of wildlife habitat in a spatially neutral Bayesian model. *Landscape Ecology* 2:101–110.

Moilanen, A., and M. Nieminen. 2002. Simple connectivity measures in spatial ecology. *Ecology* 83:1131–1145.

Pearson, S. M., M. G. Turner, R. H. Gardner, and R. V. O'Neill. 1996. An organism-based perspective of habitat fragmentation. Pp. 77–95 in R. C. Szaro and D. W. Johnston (eds.) *Biodiversity in Managed Landscapes: Theory and Practice*. Oxford, UK: Oxford University Press.

Peterken, G. 2002. *Reversing the Habitat Fragmentation of British Woodlands*. Report, World Wildlife Federation–UK Research Centre, Surrey, UK. Available online at http://www.wwf-uk.org/filelibrary/pdf/reversing_fragmentation.pdf

Pitelka, L. F., and the Plant Migration Workshop Group. 1997. Plant migration and climate change. *American Scientist* **85**:464–473.

Pither, J., and P. D. Taylor. 1998. An experimental assessment of landscape connectivity. *Oikos* **83**:166–174.

Pither, J., and P. D. Taylor. 2000. Directional and fluctuating asymmetry in the black-winged damselfly *Calopteryx maculata* (Beauvois) (Odonata: Calopterygidae). *Canadian Journal of Zoology* **78**:1740–1748.

Ricketts, T. 2001. The matrix matters: effective isolation in fragmented landscapes. *American Naturalist* **158**:87–99.

Robichaud, I., M. A. Villard, and C. S. Machtans. 2002. Effects of forest regeneration on songbird movements in a managed forest landscape of Alberta, Canada. *Landscape Ecology* **17**:247–262.

Roland, J., N. Keyghobadi, and S. Fownes. 2000. Alpine *Parnassius* butterfly dispersal: effects of landscape and population size. *Ecology* **81**:1642–1653.

Schooley, R. L., and J. A. Wiens. 2003. Finding habitat patches and directional connectivity. *Oikos* **102**:559–570.

Schultz, C. B. 1998. Dispersal behavior and its implications for reserve design in a rare Oregon butterfly. *Conservation Biology* **12**:284–292.

Sekercioglu, C. H. 2002. Effects of forestry practices on vegetation structure and bird community of Kibale National Park, Uganda. *Biological Conservation* **107**:229–240.

Selonen, V., and I. K. Hanski. 2003. Movements of the flying squirrel *Pteromys volans* in corridors and in matrix habitat. *Ecography* **26**:641–651.

South West Wildlife Trusts. 2004. *Rebuilding Landscapes for Wildlife and People: An Introductory Primer.* Available online at http://www.swenvo.org.uk/nature-map/solidus

Stith, B. M., J. W. Fitzpatrick, G. E. Woolfenden, and B. Pranty. 1996. Classification and conservation of metapopulations: a case study of the Florida Scrub Jay. Pp. 187–215 in D. R. McCullough (ed.) *Metapopulations and Wildlife Conservation.* Washington DC: Island Press.

Taylor, P. D., and G. Merriam. 1995. Wing morphology of a forest damselfly is related to landscape structure. *Oikos* **73**:43–48.

Taylor, P. D., L. Fahrig, K. Henein, and G. Merriam. 1993. Connectivity is a vital element of landscape structure. *Oikos* **68**:571–572.

Tewksbury, J. J., D. J. Levey, N. M. Haddad, *et al.* 2002. Corridors affect plants, animals, and their interactions in fragmented landscapes. *Proceedings of the National Academy of Sciences of the USA* **99**:12923–12926.

Tilman, D., R. M. May, C. L. Lehman, and M. A. Nowak. 1994. Habitat destruction and the extinction debt. *Nature* **371**:65–66.

Tischendorf, L., and L. Fahrig. 2000. How should we measure landscape connectivity? *Landscape Ecology* **15**:633–641.

Urban, D., and T. Keitt. 2001. Landscape connectivity: a graph-theoretic perspective. *Ecology* **82**:1205–1218.

Varkonyi, G., M. Kuussaari, and H. Lappalainen. 2003. Use of forest corridors by boreal *Xestia* moths. *Oecologia* **137**:466–474.

Warren, M. S., J. K. Hill, J. A. Thomas, *et al.* 2001. Rapid consequences of British butterflies to opposing forces of climate and habitat change. *Nature* **414**:65–69.

Watts, K., M. Griffiths, C. Quine, D. Ray, and J. W Humphrey. 2005. *Towards a Woodland Habitat Network for Wales*, Contract Science Report No. 686. Bangor, UK: Countryside Council for Wales.

Wiens, J. A. 1997. Metapopulation dynamics and landscape ecology. Pp. 43–62 in I. A. Hanski, and M. E. Gilpin (eds.) *Metapopulation Biology: Ecology, Genetics, and Evolution*. San Diego, CA: Academic Press.

Wiens, J. A., N. C. Stenseth, B. Van Horne, and R. A. Ims. 1993. Ecological mechanisms and landscape ecology. *Oikos* **66**:369–380.

With, K. A. 1997. The application of neutral landscape models in conservation biology. *Conservation Biology* **11**:1069–1080.

With, K. A. 2002. Using percolation theory to assess landscape connectivity and effects of habitat fragmentation. Pp. 105–130 in K. J. Gutzwiller, (ed.) *Applying Landscape Ecology in Biological Conservation*. New York: Springer-Verlag.

With, K. A., and T. O. Crist. 1995. Critical thresholds in responses to landscape structure. *Ecology* **76**:2446–2459.

With, K. A., and A. W. King. 1999a. Dispersal success on fractal landscapes: a consequence of lacunarity thresholds. *Landscape Ecology* **14**:73–82.

With, K. A., and A. W. King. 1999b. Extinction thresholds for species in fractal landscapes. *Conservation Biology* **13**:314–326.

With, K. A., R. H. Gardner, and M. G. Turner. 1997. Landscape connectivity and population distributions in heterogeneous environments. *Oikos* **78**:151–169.

Woodford, J. E., and M. W. Meyer. 2003. Impact of lakeshore development on green frog abundance. *Biological Conservation* **110**: 277–284.

Woodland Trust. 2000. *Woodland Biodiversity: Expanding our Horizons*. Available online at http://www.woodland-trust.org.uk/publications/publicationsmore/expandingourhorizons.pdf

Woodland Trust. 2002. *Space for Nature: Landscape-Scale Action for Woodland Biodiversity*. Available online at http://www.woodland-trust.org.uk/publications/publicationsmore/space.pdf

3

Connectivity and metapopulation dynamics in highly fragmented landscapes

ATTE MOILANEN AND ILKKA HANSKI

INTRODUCTION

Connectivity is a key variable in spatial ecology (MacArthur and Wilson 1967; Levin 1974; Hanski 1998; Tishendorf and Fahrig 2000; Moilanen and Hanski 2001; King and With 2002), and the element that turns a conventional population study (or a model) into a spatial one. Broadly speaking, connectivity measures the effect of landscape structure on movements of individuals, but in spite of the pivotal role of this measure in ecology, there is no generally accepted and employed formal definition of connectivity (Crooks and Sanjayan Chapter 1). In particular, the way connectivity is used in metapopulation ecology is different from its use in landscape ecology (Tishendorf and Fahrig 2000; Moilanen and Hanski 2001; With 2004; Taylor et al. Chapter 2). In metapopulation ecology, connectivity is seen as a property of a habitat patch (or a grid cell), and the measure of connectivity is defined via predicted rate of immigration into a patch or via predicted success of migrants leaving a patch.

In landscape ecology, connectivity is viewed as a property of an entire landscape – a measure of how much, or how little, landscape structure hinders movements (Taylor et al. 1993; With 2004; Taylor et al. Chapter 2). This difference in the spatial scale is reflected in the use of connectivity in modeling. Nonetheless, both metapopulation and landscape ecology are concerned with the same phenomena, the effects of landscape structure

Connectivity Conservation eds. Kevin R. Crooks and M. Sanjayan. Published by Cambridge University Press. © Cambridge University Press 2006.

on migration, colonization success, and the persistence of species at the landscape level. Several authors have anticipated better integration of the two fields of metapopulation ecology and landscape ecology (Hanski and Gilpin 1991; Wiens 1997), but so far progress in this direction has been limited (but see Hanski 2001; Hanski and Ovaskainen 2003; Ovaskainen and Hanski 2004).

Metapopulation studies often, though not always, assume that the landscape is characterized by a binary division into suitable habitat versus landscape matrix. Another common assumption, which is made to facilitate data analysis and modeling, is that only a small fraction of the total landscape area represents suitable habitat for the species (Hanski 1999). This chapter is concerned with connectivity measures that have been developed to describe such highly fragmented (patchy) landscapes, but which can also be applied with minor modification to more continuous landscapes with spatial variation in habitat quality. Landscape structure has implications for the construction of spatial models and especially for parameter estimation. For several reasons, it is easier to construct and analyze models for discrete habitat patches than, for example, grid-based models of arbitrary landscapes.

It is worth emphasizing at the outset a key point about the use of connectivity measures in landscape and metapopulation ecology. One reason why much of the connectivity-related research in landscape ecology is based on a landscape-level measure of connectivity is the belief that such a measure would be helpful for assessing the long-term persistence of a species in the landscape. The use of percolation theory for species in fragmented landscapes is a clear example (Gardner *et al.* 1987; With 1997): it is assumed that a sufficient and necessary condition for the long-term persistence of a species is the possibility of moving across the landscape. To us this belief is misguided. If the purpose of research is to make predictions about the persistence of a hypothetical or a real species in a landscape, it is not sufficient to consider just the level of connectivity of the landscape (however defined), one also has to consider the other components of population dynamics, local dynamics and local extinction in particular. And for this purpose a model is required, of the type employed in metapopulation ecology.

Three measures of connectivity have been widely used in studies dealing with highly fragmented landscapes: the nearest neighbor (NN) measure, the buffer (BUF) measure, and the incidence function model (IFM) connectivity measure (Hanski 1994; Moilanen and Nieminen 2002). These measures are widely used because they are simple, intuitive, and practical, but it is not widely appreciated that there are major differences between them. We describe and compare them in the section on "Simple connectivity measures" (pp. 49–51). The IFM connectivity measure has been successfully used in metapopulation modeling (see Hanski 1998; Sjögren-Gulve and Hanski 2001; Moilanen and Nieminen 2002 for references). Next we review the use of this measure at different levels of biological hierarchy. We start by describing how the connectivity measure has been used in the modeling of daily movements of individuals, in the analysis of multi-population mark–release–recapture data, and we demonstrate the use of this connectivity measure and its extensions to predict colonizations in metapopulations. We describe a measure of landscape structure, called the metapopulation capacity (Hanski and Ovaskainen 2000; Ovaskainen and Hanski 2001, 2004), which has been derived from metapopulation theory and which integrates the effects of landscape structure (including connectivity) on metapopulation dynamics. Finally we discuss the use of empirical data to infer parameter values of the IFM connectivity measure. We point out that several types of errors in empirical data can cause large biases in parameter estimates. The chapter by Fagan and Calabrese (Chapter 12) complements our discussion by illustrating how data availability affects the choice of a connectivity measure. We conclude by emphasizing that connectivity is only one element, though a critical one, in assessing the viability of populations in fragmented landscapes.

SIMPLE CONNECTIVITY MEASURES

Connectivity measures have been used for many purposes, for instance for explaining colonization and extinction events, species presence or absence in habitat patches, population density, dispersal success, genetic variation, species richness, and so forth (Moilanen and Nieminen 2002). Below, we define the connectivity measures that are considered in this chapter.

The nearest neighbor (NN) connectivity measure

The simplest and most commonly used connectivity measure is the nearest neighbor measure (NN), which is the smallest distance between the focal site i and any other site j,

$$d_i^{NN} = \min_{j \neq i} d_{ij}, \tag{3.1}$$

where d_{ij} is the distance between focal patch i and any other patch j, calculated either between patch centers or patch edges. The idea of the NN measure is that a site is highly connected if its nearest neighbor is close, and isolated if the nearest neighbor is far away, as locating and reaching a far-away patch would be difficult for a migrating individual.

The buffer connectivity measure

Another commonly used connectivity measure is the buffer measure (BUF), which defines connectivity as the amount (area) of habitat within a buffer (circle) around the focal patch:

$$B_i = \sum_j A_j^b, \quad \text{for all } j \neq i \text{ and } d_{ij} < r. \tag{3.2}$$

Here A_j is the area of patch j, r is the buffer radius, a species-specific distance within which habitat adds to connectivity, and b is an exponent scaling the number of emigrants as a function of patch area. Most often buffer connectivity has been used just for tallying up the amount of suitable habitat within the buffer radius, which implicitly assumes that $b = 1$. As in the NN measure, the buffer measure can be calculated over all suitable habitat or over occupied habitat only, depending on available data and the intended use of the measure. If habitat occupancy is taken into account, one measures population dynamic connectivity (Moilanen and Hanski 2001), which may be used for predicting colonizations or for parameterizing the colonization function of a stochastic metapopulation model. If all habitat patches are taken into account, and hence occupancy is ignored, one measures the connectivity of landscape elements (Moilanen and Hanski 2001), which may be useful for, e.g., predicting the average incidence (probability of occupancy) of habitat patches in a network.

The incidence function model (IFM) connectivity measure

The original IFM measure (Hanski 1994) is defined as

$$S_i = \sum_{j \neq i} p_i \exp(-\alpha d_{ij}) A_j^b, \tag{3.3}$$

where d_{ij} and b are as above, α is a parameter scaling the effect of distance on dispersal, and p_i is 1 and 0 for occupied and empty habitat patches, respectively. If the sum in Eq. (3.3) is extended over all sites by ignoring p_i, one obtains the connectivity of landscape elements. Note that the IFM measure requires more information than the NN or buffer measures, because in addition to the spatial coordinates of the habitat patches one needs an estimate of the value of α, that is, an idea of the typical movement range of the species. While applying any of these measures to grid-based landscape descriptions, exponent b has no relevance and should be set to 1.

The logic and construction of the IFM connectivity measure is worth a closer inspection. Assume that in the context of, say, modeling colonization events we are interested in the expected number of migrants that arrive at the focal patch. Migrants may arrive from any extant population, which suggests that the connectivity measure should sum up the contributions from all the existing populations. More migrants are likely to arrive from nearby populations than from far-away populations, which suggests weighting the contribution of each source population by a decreasing function of distance to the focal patch. Furthermore, large populations (in large patches) should send out more migrants than small populations, which suggests weighting the contribution of each source population by an increasing function of population size (patch area). This is essentially what Eq. (3.3) does. Something that is missing from this equation is the effect of the size of the focal patch on immigration. Often a large patch will be easier to find for a dispersing individual than a small patch, and indeed Moilanen and Nieminen (2002) found that the addition of a term A_i^c to account for the effect of the size of the focal patch consistently improved the performance of the IFM measure when applied to empirical data. Therefore, we use an extended IFM measure of the general form

$$S_i(t) = I(A_i) \sum_{j \neq i} p_i(t)D(d_{ij})E(A_j). \tag{3.4}$$

Here $D()$ is an arbitrary dispersal (migration, redistribution) kernel (see Kot et al. 1996), which most often is a decreasing function of distance d_{ij}, like the exponential function in Eq. (3.3). Individual movements leading to colonization of the currently empty patches and the rescue effect in the currently occupied patches are typically distance-dependent, and this distance effect is taken into account by assuming a distribution of movement distances, which is the dispersal kernel. A definite advantage of the IFM connectivity measure is that it represents one such dispersal kernel — it was originally constructed for that purpose. This is an advantage because the same measure of connectivity can be employed both in data analysis and modeling, and indeed the parameters of the dispersal kernel can be estimated with empirical data.

Function $I()$ in Eq. (3.4), the immigration function, accounts for the relationship between patch area and immigration rate, which most often is a positive relationship. $E()$, the emigration function, is usually an increasing function of area, representing the numbers of migrants leaving the source patch. $E()$ can be further divided into two components, population size and per capita emigration rate. Assuming a linear dependence between population size and patch area, and the decrease in per capita emigration rate by increasing patch area that is predicted by diffusion, we obtain $E(A_j) = A_j^1 A_j^{-0.5} = A_j^{0.5}$. In general, it is reasonable to assume a power law relationship between population size and patch area, and an inverse power law relationship between patch area and emigration, which yields A_j^b, the form used in the original IFM measure (Eq. 3.3). Furthermore, other factors apart from area could affect emigration and immigration and could be incorporated into Eq. (3.4). This equation includes time t to emphasize the fact that population dynamic connectivity changes with the changing pattern of habitat occupancy. As before in Eq. (3.3), connectivity of landscape elements is calculated by omitting the term giving patch occupancy, $p_i(t)$, in Eq. (3.4).

The chief simplification assumed by the basic IFM measure is that landscape matrix is of uniform quality and hence that the effect of landscape structure on migration can be reduced to the effect of distance on migration. However, landscape structure can be partially incorporated into the IFM framework by making the functions $I()$ and $E()$ depend on habitat quality in the surroundings of the patches. There are also ways of accounting for the structure of the matrix in the IFM measure (see Moilanen and Hanski 1998), but the usefulness of such extensions are limited by high data demands in model fitting (Fagan and Calabrese Chapter 12).

Comparing simple connectivity measures

Figure 3.1 uses a patch network for the butterfly *Melitaea diamina* (Wahlberg *et al.* 1996) to compare the three measures of connectivity for a particular patch. Panel A shows the original patch network. Panel B shows the NN connectivity for the focal patch — the distance to the nearest other patch. Panel C shows the patches included in the buffer measure with $r = 1\,\mathrm{km}$ — only patches closer than r from the focal patch affect connectivity. In contrast, the IFM connectivity (panel D) includes the effects of all the potential source populations, but their contributions are scaled both by the distance to the focal patch and the effect of patch area on the number of emigrants. Note that the highest IFM connectivity

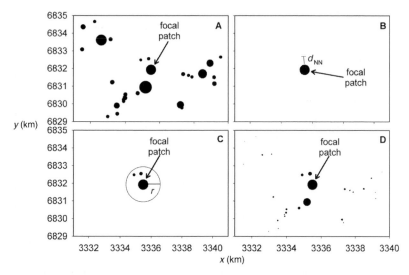

Fig. 3.1. Connectivity measures illustrated with a patch network for the butterfly *Melitaea diamina* (Wahlberg *et al.* 1996). (A) Original network, patch area shown by the size of the symbol. (B) Connectivity of the focal patch as seen by the nearest neighbor (NN) measure. (C) The part of the landscape included in a buffer (BUF) measure with radius *r*. (D) Connectivity of the focal patch as seen by the incidence function model (IFM) connectivity measure. In panel D the sizes of the source populations have been scaled by distance to the focal patch and by emigration rate as a function of patch area (compare to panel A).

is between the focal patch and the large patch 1.5 km to southwest. This patch is outside the radius of the buffer connectivity in panel C.

We shall next review the weaknesses of the simple connectivity measures (Fig. 3.2). The pair of landscapes shown for the NN measure is equal from the viewpoint of connectivity – all patches have the same NN distance. Even so, the landscapes are clearly very different: a cluster of large patches versus a linear arrangement of small patches. All else being equal, it would appear obvious that the landscape in the upper panel should have higher connectivity than the one in the lower panel, but the NN measure is unable to quantify the difference. The landscapes for the buffer measure demonstrate how the buffer measure only takes into account the amount of habitat within the buffer zone, but not the spatial distribution of habitat within the buffer nor any habitat outside the buffer. Incidentally, a graph-theoretical connectivity measure with a fixed critical distance, as in Keitt *et al.* (1997), is identical or very similar to the buffer measure (see also Fagan and Calabrese Chapter 12). Finally, the IFM

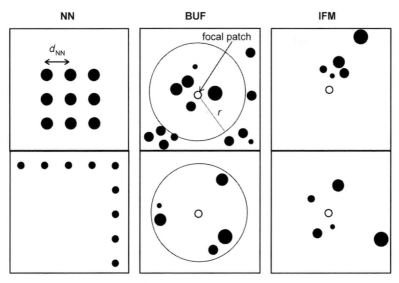

Fig. 3.2. Properties of simple connectivity measures. A pair of landscapes (upper + lower panel) is shown for each measure. The two landscapes shown for the NN measure are identical, the numerical value of connectivity is the same for all patches. For the buffer and IFM measures the landscapes are equal only from the perspective of the focal patch (empty circle).

measure has a more subtle weakness. It does account for all the patches, their sizes, and their distances from the focal patch, but it ignores the spatial relationships of the source populations — you can freely rotate the source patches around the focal patch without changing the numerical value of the IFM measure for the focal patch. This may make a difference because, in reality, patches located close to each other will "compete" for migrants. In the example, the focal patch should in fact be better connected in the lower than in the upper panel, but the IFM measure is unable to quantify this difference.

Moilanen and Nieminen (2002) surveyed the use of connectivity in 74 recent publications in the fields of ecology, landscape ecology, and conservation biology. Table 3.1 summarizes some statistics from their study.

NN measures were used in 44% of the statistical tests involving connectivity. Buffer and IFM measures were also common, while only two out of the 74 studies used a connectivity measure not belonging to one of these three measures. Studies employing NN connectivity frequently failed to find a statistically significant effect of connectivity (on whatever response variable was being studied), whereas studies using the buffer

Table 3.1. *Statistical significance[a] of tests of connectivity as reported in 74 published studies employing different measures of connectivity (Moilanen and Nieminen 2002)*

Measure	n	Significant	Not significant
NN	35	19	9
Buffer	22	16	1
IFM	20	11	0

[a]Studies finding both significant and non-significant results were counted for both columns (the two columns do not add up to n because not all studies included in n tested for significance).

and IFM measures almost invariably found a statistically significant effect. The difference in the outcome of the tests using NN versus BUF/IFM measures was significant (Table 3.1), suggesting that the NN measure is in general less likely to find a significant effect than the other two measures. Moilanen and Nieminen (2002) also tested the performance of the connectivity measures using empirical data for two butterfly species. They found much stronger effects of connectivity with buffer and IFM measures than with the NN measure. IFM and buffer measures could detect significant effects of connectivity with smaller data sets than the NN measure. Similar results were obtained by Bender *et al.* (2003), who found in a simulation study that "area-based metrics," meaning buffer and IFM-type measures, generally outperformed the NN measure and other distance-based measures of connectivity, such as the Voronoi polygon.

CONNECTIVITY IN METAPOPULATION DYNAMICS

In this section, we consider connectivity at different levels of biological hierarchy. First, we employ connectivity to study the daily movements of individuals. This is related to the mechanistic basis of connectivity, because connectivity between populations is made up of individuals moving among habitat patches. Individual movements are necessary for the establishment of new local populations in fragmented landscapes with unstable local populations, and ecologists often assess the role of connectivity, among other factors, in explaining colonization events. Finally, we move on to metapopulation dynamics and persistence of species in fragmented landscapes, and we clarify the role that connectivity plays in this context.

Movements of individuals

Hanski *et al.* (2000) constructed a model that was dubbed the virtual migration (VM) model to analyze mark—release—recapture data gathered from multiple local populations in a metapopulation. We describe here the essentials of this model to demonstrate how IFM-type connectivity can be useful in the context of modeling individual movements (Tracey Chapter 14 reviews the modeling of individual movements more generally). The VM model assumes that habitat patch areas affect the rate of emigration from and the rate of immigration to a patch, while connectivity is assumed to affect survival during migration—an individual leaving an isolated patch is more likely to perish during migration than an individual leaving a well-connected patch. Making this biologically plausible assumption allows one to tease apart mortality during migration and mortality within habitat patches, because the latter is not affected by connectivity. The idea in the VM model is that instead of trying to estimate the movement rate between all pairs of populations, which would be impractical in the case of large metapopulations, we estimate the parameters of a function that relates the probability of successful migration to a measure of connectivity of the source populations.

Figure 3.3 depicts the sequence of daily events for an individual in the VM model. First, the individual either survives in the patch or dies with a constant probability. If the individual survives, it may emigrate, and if it emigrates it may either survive migration or it may die. Survival during migration from patch j, ϕ_{mj}, depends on connectivity according to a sigmoid function

$$\phi_{mj} = \frac{S_j^2}{S_j^2 + \lambda}, \tag{3.5}$$

where λ is a parameter scaling the effect of connectivity on migration mortality. The probability of successful migration increases with increasing connectivity of the source patch j, S_j.

If the individual survives migration, its probability of ending up in a particular patch is proportional to the contribution of that patch to the connectivity of the source population. Connectivity of patch j can indeed be partitioned as

$$S_j = \sum_{k \neq j} S_{jk} = \sum_{k \neq j} p_k \exp(-\alpha d_{jk}) A_k^z, \tag{3.6}$$

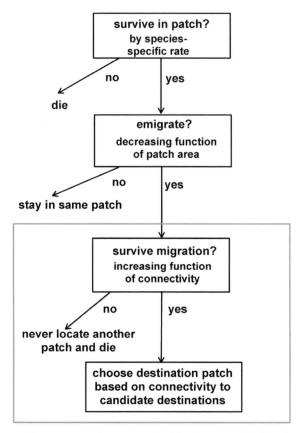

Fig. 3.3. Daily sequence of events in the life of an individual in the virtual migration (VM) model (Hanski *et al.* 2000). The VM model allows the distinction between mortality in a habitat patch and mortality during dispersal, based on the effect of connectivity on the latter. The box marks the connectivity-related part of the model.

where S_{jk} is the contribution of patch k to the connectivity of patch j, d_{jk} is the inter-patch distance, α is the parameter of the dispersal kernel, A_k is the area of patch k, and z is a parameter scaling the immigration rate as a function of patch area. (Parameter z relates to the ability of individuals to detect the target patch.) A reasonable assumption is then that the probability of an individual moving from patch j to patch k is simply S_{jk}/S_j. Note that connectivity is here calculated for the potential target patches rather than for a single focal patch as in metapopulation models. The components described above make it possible to calculate daily probabilities of an individual dying, staying in the present patch, and moving to another patch. By adding a temporally varying recapture (detection) probability into the model, it becomes possible to calculate the likelihood of any observed sequence of recapture histories, which is the basis for parameter estimation (see Hanski *et al.* 2000 for details).

Table 3.2 summarizes connectivity-related results that have been obtained in empirical studies of butterfly metapopulations. The average daily movements range from 100 to 500 m, with a median of 200 m. Patch areas in these studies ranged from 0.1 to 8 ha, and typical daily emigration probabilities from a patch of unit size (ha) were around 0.1. In general, more than 80% of all migration events were estimated to be successful in the study landscapes. However, survival rates for migrants leaving the most isolated patches were < 0.1 for *Parnassius smintheus* and 0.33 for *Melitaea diamina*, much lower than survival rates in the best-connected patches and hence indicating a strong effect of connectivity on migration survival.

Ovaskainen (2004) has recently constructed and analyzed a diffusion approximation of a correlated random walk model that can be parameterized with similar data as the VM model. His model also includes biased behavior of individuals at habitat boundaries, essentially modeling habitat selection. The advantage of Ovaskainen's model is that it is based more mechanistically on individual movement behavior than the VM model, assuming that correlated random walk is an appropriate model in the first place (see Tracey Chapter 14).

Connectivity and colonizations

We illustrate the use of connectivity measures in explaining colonizations with data for the Glanville fritillary butterfly *(Melitaea cinxia)* in the Åland Islands, southwestern Finland (Hanski *et al.* 1996; Hanski 1999). Here the butterfly comprises a large metapopulation living in a network of about 4000 habitat patches, dry meadows with at least one of the obligate larval host plants, *Plantago lanceolata* and *Veronica spicata*. This metapopulation has been surveyed yearly since 1993, and each year around 500 habitat patches have been occupied by the butterfly. The average patch size is only 1600 m², and the annual population turnover rate is very high, with around 100 extinction events and roughly an equal number of colonization events per year. For further details of the biology of the butterfly and the structure of its metapopulation see Hanski (1999) and Nieminen *et al.* (2004). In the following we compare the performance of NN, BUF, and IFM connectivity measures in explaining colonization events in the data for the years 2000–01 (main Åland Island only: see

Table 3.2. *Summary of connectivity-related results of studies using the virtual migration model (Hanski et al. 2000) for multi-population mark–release–recapture data*

Study[a]	α[b]	Daily emigration[c]			Migration survival[d]			Natal %[e]	Reference
		Minimum	1 ha	Maximum	Mean	Minimum	Maximum		
Melitaea diamina, M	4.9	0.15	0.17	0.35	0.81	0.33	0.98	59	Hanski et al. (2000)
Proclossiana eunomia, M	6.0	0.04	0.10	0.5	0.88	NA	NA	57	Petit et al. (2001)
F	4.9	0.07	0.13	0.8	0.95	NA	NA	47	
Euphydryas aurinia, M	2.2	NA	0.03	NA	0.8–0.9	NA	NA	>80	Wahlberg et al. (2002)
Euphydryas maturna, M	8.5	NA	0.09	NA	0.8–0.9	NA	NA	>80	Wahlberg et al. (2002)
Melitaea cinxia, M	6.3	NA	0.13	NA	NA	NA	NA	~50	Wahlberg et al. (2002)
Melitaea athalia, M	4.0	NA	0.10	NA	0.8–0.9	NA	NA	~60	Wahlberg et al. (2002)
Parnassius smintheus, F+M	~2 /4[f]	0.02[g]	0.04	0.07	~0.65	<0.1	>0.9	NA	Matter et al. (2004)

[a]F/M indicates whether the data is for females or males.

[b]α is the parameter of the negative exponential dispersal kernel for daily movements.

[c]The table gives the daily emigration probability from a patch of 1 ha, as well as the minimum and maximum values corresponding to the largest and the smallest patch in the network, respectively.

[d]The survival columns give the mean survival of migrants and the minimum and maximum values corresponding to migrants leaving the least- and the best-connected patches. NA indicates that the information was not available in the publication due to e.g. missing information about the ranges of patch areas or connectivities.

[e]Natal % gives the percentage of individual–days spent in the natal patch.

[f]For meadow habitat and forest, respectively.

[g]Average value for 2 years using a model with host plant abundance replacing patch area.

Hanski 1999). In these data, 472 and 501 patches were occupied in 2000 and 2001, respectively.

The basic IFM connectivity measure (Eq. 3.3) uses patch area as an easily measurable surrogate for population size (Hanski 1994). However, not all patches are generally of equal quality, and if adequate information about habitat quality is available it is possible to replace patch area by a more accurate surrogate of population size. Essentially, A_i will be replaced by Q_j, which is a function of patch area and other environmental variables. Various attributes have been measured for the habitat patches of *M. cinxia*, including information about vegetation (amount and quality of host plants, amount of nectar plants, vegetation height), grazing status, proneness to dry out, edge quality, and so forth (Moilanen and Hanski 1998; Nieminen *et al.* 2004). Considering that the butterfly has obligate larval host plants, the most obvious candidate for a patch quality variable beyond area is some measure of host plant abundance. Here, we use $Q_j = \log(A_{pi}+A_{vi})$, where A_{pi} and A_{vi} are the coverages (m^2) of the two host plants *P. lanceolata* and *V. spicata*, respectively, in the patch. Host plant information for the year 2000, and occupancy information for the years 2000 and 2001, were available for 2400 patches, which are the data used in the following analysis.

Table 3.3 summarizes the results of logistic regression analyses of colonization events. Comparisons of the models were based on the AIC measure (Burnham and Anderson 2002), which is a likelihood-based measure of model fit that includes a penalty for the number of parameters in the model. The AIC is calculated as residual deviance $+2\,k$, where k is the number of parameters in the logistic regression ($=3$) plus one for each additional parameter in the connectivity measure. Models can be evaluated by comparing their relative AIC-values (ΔAIC= AIC of model − AIC of best model). The model with the lowest AIC value is considered the best fit model, and a value of ΔAIC greater than 7 is taken to mean that the model with lower AIC has considerably more support in the data than the alternative model.

In our analyses, all connectivity measures had a highly significant effect on colonizations (Table 3.3). The best model in Table 3.3 is IFM-Q with $b=1$, the extended IFM measure with habitat quality included. The NN measure, with a ΔAIC $= 23.5$, performed poorly compared to the IFM-Q model. The simple buffer measure that just sums up the amount of habitat within a radius ($b=1$) does very badly with all ΔAIC > 115, while the buffer measure with patch area scaled by $b=0.25$ is much improved with ΔAIC $=30$ when $r=1.5$. The large effect of parameter b,

Table 3.3. *Colonization events explained by connectivity measures and patch areas in logistic regression models. Model comparisons are based on AIC values of the fitted models. The analysis has null deviance = 1222.2 and df = 2396*

Measure	K	AIC	ΔAIC	P[correct][a]
NN	3	1061.6	23.5	7.6e−6
BUF, $b=1$				
$r=2/3$	4	1171.0	132.9	1.4e−29
$r=1$	4	1152.9	114.8	1.2e−25
$r=1.5$	4	1155.3	117.2	3.6e−26
BUF, $b=0.5$				
$r=2/3$	5	1120.7	82.6	1.2e−18
$r=1$	5	1091.5	43.4	3.8e−10
$r=1.5$	5	1086.7	48.6	2.8e−11
BUF, $b=0.25$				
$r=2/3$	5	1103.6	65.5	6.0e−15
$r=1$	5	1076.1	38.0	5.6e−9
$r=1.5$	5	1068.0	29.9	3.2e−7
IFM, $\alpha=1.2$				
$b=0.25$	5	1053.9	15.8	3.7e−4
$b=0.5$	5	1073.6	35.4	2.1e−8
IFM-Q, $\alpha=1.2$				
$b=0.25$	5	1040.5	2.4	0.23
$b=0.5$	5	1038.4	0.3	0.46
$b=0.75$	5	1038.6	0.5	0.44
$b=1.0$	**4**	**1038.1**	**0**	

[a] P[correct] is the probability that the model is better than the reference model (shown in bold) in a pair-wise comparison.

the exponent scaling emigration rate with patch area, is evident also with the basic IFM measure, which has a ΔAIC about half as large when $b=0.25$ (ΔAIC$=15.8$) compared to when $b=0.5$ (ΔAIC$=35.4$). All three simple measures (NN, BUF, and IFM) fare badly in comparison with the IFM-Q. Note, however, that the use of IFM-Q requires measurements of the quality variables in all habitat patches, which makes it arduous to use in comparison with the patch area-based measures. These results are in line with those of Moilanen and Nieminen (2002), who found that the NN measure and the unscaled buffer measure performed poorly, while the performances of the IFM and the scaled buffer measures were roughly equal when the parameters were chosen optimally. Finding the best-fitting values of the parameters α, b, and r more accurately than in Table 3.3 would further improve the relative performance of the buffer and IFM measures in comparison with the NN measure. The effect of patch area on immigration was not included in the connectivity measure, because focal patch area was entered into the logistic regression as an independent variable.

Connectivity has a highly significant effect also in explaining extinction events in these data (details not shown; see also Hanski 1999). The negative effect of connectivity on extinctions is often called the rescue effect (Brown and Kodric-Brown 1977), meaning reduced extinction risk of a population in the presence of immigration from other extant populations. In general, extinction events are influenced by a greater number of factors than colonizations, and often extinctions are spatially correlated, due to spatially correlated environmental stochasticity (regional stochasticity: Hanski 1999; Nieminen *et al.* 2004).

We used host plant cover as a measure of patch quality in the above analyses, on the assumption that host plant cover correlates better with population size than patch area. This assumption can be verified using data about population sizes, which are also available for the Glanville fritillary. Host plant cover is indeed the environmental factor that best explains variation in population sizes (results not shown). If independent information on population sizes is available, it is possible to use this information in the construction of the patch quality measure: first a statistical model is built which explains population size as a function of habitat quality variables, then this model is used to calculate the predicted size of populations conditional on the patch being occupied. The predicted population size is used as a measure of habitat quality in subsequent analyses.

Connectivity and modeling metapopulation dynamics

For a population ecologist, the ultimate use of connectivity measures is in the context of metapopulation models, which can be used for analyzing and predicting spatial population dynamics. Classic metapopulation models include two basic components, extinction of local populations and recolonization of empty habitat. Connectivity is the key variable that is used to explain colonizations of currently empty but suitable habitat patches. In metapopulation models with a description of local population dynamics, connectivity will influence the numbers of migrants among local populations. Without going into details of metapopulation modeling, we describe below very briefly one class of classic metapopulation models, stochastic patch occupancy models (SPOMs), from which a useful landscape-level measure can be derived, the metapopulation capacity of a fragmented landscape.

Stochastic patch occupancy models

Stochastic patch occupancy models (SPOMs) describe metapopulation dynamics as an extinction—colonization process, with each habitat patch being either occupied by the focal species or empty at a given time. For a general account of the theory see Ovaskainen and Hanski (2001). SPOMs are often applied in discrete time, most naturally to species with an annual life cycle. SPOMs assume that occupancy at time $t+1$ only depends on occupancy at time t, and hence the models can be treated as first-order linear Markov chains (Moilanen 1999), with a state space of 2^n, where n is the number of patches in the patch network. Combining SPOMs with structural assumptions about how the physical attributes of a patch network are related to the extinction probability of existing populations and to the colonization probability of empty patches leads to what has been called the spatially realistic metapopulation theory (Hanski 2001; Hanski and Ovaskainen 2003). It is typically assumed that the probability of extinction decreases with increasing patch area (a surrogate for population size) and that the probability of colonization increases with IFM-type connectivity, but other landscape features including measures of habitat quality could also be included in the model. SPOMs have been widely applied to metapopulations (Hanski 1999). Ovaskainen and Hanski (2004) present an up-to-date review of SPOMs and their deterministic approximations, while Etienne *et al.* (2004) provide a comprehensive overview of the methods of parameter estimation. Moilanen (2004) describes a software, SPOMSIM, that can be used for parameter estimation and SPOM simulation.

The metapopulation capacity of a fragmented landscape

Metapopulation models employ connectivity measures that are patch-specific and which can be directly related to the movement behavior of individuals in space (see p. 53—54) and to the colonization events in a metapopulation (pp. 55—59). The theory also allows one to derive a landscape-level measure that is a certain kind of average of the connectivities of individual habitat patches, the metapopulation capacity of a fragmented landscape (Hanski and Ovaskainen 2000).

The metapopulation capacity describes the structure of a fragmented landscape in terms of how favorable the landscape is for the persistence of a focal species. As a first approximation, the metapopulation capacity may be viewed as giving just the total amount of suitable habitat in the landscape, in the manner that the familiar environmental carrying capacity gives the amount of resources/habitat available for a single local population. But the metapopulation capacity also takes into account the actual spatial configuration of the habitat in the landscape through the model assumptions about the influence of patch areas and connectivities on rates (or probabilities) of local extinction and recolonization. Thus it makes a difference, in the model, whether a given amount of habitat is distributed in a few large patches or in a large number of small patches, and whether the patches are clustered (giving high average connectivity) or more dispersed (giving low average connectivity). A very convenient property of the metapopulation capacity is that it integrates all the effects of the amount and spatial configuration of habitat into a single real number, which can be used to, e.g., rank multiple landscapes in terms of their capacity to support a viable metapopulation (naturally, "all the effects" refers to those effects that are incorporated in the specific model with which the metapopulation capacity is calculated). The metapopulation capacity, a description of the landscape structure, along with the properties of the species, determines the extinction threshold of the species in a landscape (Ovaskainen and Hanski 2003). Mathematically, the metapopulation capacity is defined as the leading eigenvalue of an appropriate "landscape" matrix, and in this perspective it may be viewed as playing the same role as the population growth rate inferred from age-structured population models (Hanski and Ovaskainen 2003; Ovaskainen and Hanski 2004).

Figure 3.4 compares three landscape-level measures for a large number of semi-independent habitat patch networks inhabited by the Glanville fritillary butterfly in the Åland Islands in southwestern Finland (Hanski 1999; Nieminen et al. 2004): the metapopulation capacity (Meca) of the network, average connectivity of individual patches in the network as measured by the average value of the IFM measure (Eq. 3.3), and the pooled area of habitat in the network. The entire system consists of about 4000 habitat patches, which have been here classified into 59 patch networks as explained in Moilanen (2004), using a connectivity-based clustering method. The metapopulation capacity is rather closely correlated with the average connectivity of the patches in the network. The metapopulation capacity is also correlated with the total area of habitat in the network, but this correlation is not nearly as strong as the correlation between the metapopulation capacity and average connectivity.

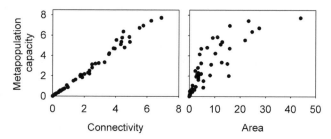

Fig. 3.4. Relationships between the metapopulation capacity and average connectivity, and between the metapopulation capacity and total area of suitable habitat, in 59 habitat patch networks for the Glanville fritillary butterfly (*Melitaea cinxia*) in the Åland Islands in Finland.

Though the metapopulation capacity can be closely approximated by average connectivity in the above example consisting of multiple real patch networks, it is easy to construct examples in which this correlation fails and which therefore demonstrate that the metapopulation capacity is not just the average connectivity of the patches. Consider the following example (Moilanen and Hanski 2001). Assume a landscape A in which patches have certain average connectivity. Now add a few isolated patches to this landscape, resulting in a modified landscape A', which has lower average connectivity than A because the isolated patches that were added to A decrease average connectivity (Moilanen and Hanski 2001). On the other hand, the metapopulation capacity of landscape A' will be slightly higher than that of A, indicating that the persistence of the species in a landscape is not reduced by the addition of a few peripheral patches—a highly reasonable conclusion. To reiterate, when the purpose is to assess the persistence of a species in a landscape, connectivity is a necessary but not a sufficient component to consider, one has to consider also the other components of metapopulation dynamics, which are taken into account in metapopulation capacity. Yet connectivity is an absolutely essential component whenever the spatial scale is such that individual movements are constrained by distance.

ESTIMATION OF CONNECTIVITY AND EMPIRICAL DATA QUALITY

For the purpose of predicting metapopulation dynamics, it is essential to be able to infer the parameters of the connectivity measure from empirical data. In this section we review some of the potential pitfalls in the use of presence–absence data for parameterizing SPOMs of metapopulation dynamics. In principle, parameter estimation is straightforward: a SPOM may be modeled as a Markov chain with 2^n states, where n is the number of habitat patches. It is possible to write down equations for the likelihood of the transition between two patch occupancy patterns as a product of the probabilities of patch-wise transitions, which means that maximum likelihood and Bayesian estimation can be used for parameter estimation (ter Braak *et al.* 1998; Moilanen 1999, 2000, 2002, 2004; O'Hara *et al.* 2002; ter Braak and Etienne 2003; Etienne *et al.* 2004). But problems arise if there are errors in the data. We believe that the errors discussed below are quite common in empirical data, and that they can lead to large errors in the estimation of connectivity in general, not only in the context of the present connectivity measures.

Figure 3.5 illustrates some of the problems that one may encounter with real data. (A) Assume a patch network with two clusters, one occupied by the species at time t and the other one all empty. (B) At time $t+1$ you observe a colonization event, which happens at an isolated patch and gives the impression that the species is capable of frequent long-distance dispersal. But your interpretation of the event may change completely if you recognize the possibility of errors in the data. In panel C the "colonized" patch was actually occupied already at time t as were some neighboring patches. In panel D there are unknown occupied patches in the landscape that might have been the source of colonization. Especially when working with large spatial scales and large patch systems it may be difficult to rule out errors of the type in panels C and D.

Table 3.4 summarizes the consequences of various types of data errors for parameter estimation (Moilanen 2002). Concerning dispersal and connectivity, largest errors are caused by unknown habitat patches within or outside the study area and by errors in the occupancy information. Unknown populations cause observed colonization events to be explained by less than the true connectivity. This leads to overprediction of meta-population occupancy if the model with the estimated parameter values is applied to another network in which all the patches are known. By far the greatest biases in parameter estimates and predictions are caused by false occupancy information. A population missed during a survey translates into a false extinction event followed by a false colonization event. Such statistical outliers bias parameter estimates in many ways including predicting too high colonization rates and too long dispersal distances.

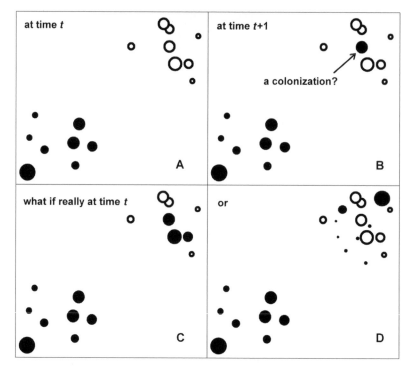

Fig. 3.5. Errors in empirical data. (A) The patch network at time t, closed and open symbols corresponding to occupied and empty patches, respectively. (B) Observation of the network at $t+1$, with an apparent long-distance colonization event. (C) The "colonized" patch was actually occupied at time t — there was false occupancy information. (D) The patch was colonized by a migrant from an unknown nearby source population — the distribution of suitable habitat was incompletely known.

Another question that may arise in the context of connectivity measures concerns the functional form of dispersal kernels and emigration and immigration functions. If the ranges of patch areas and connectivities in empirical data are narrow, practically any functional forms can be fitted to the data. But if one attempts to fit wide ranges, for instance while trying to distinguish between thin-tailed and fat-tailed dispersal kernels, data quality becomes of primary importance.

DISCUSSION

We have reviewed the connectivity measures most commonly used in metapopulation biology, with a special emphasis on the incidence function

Table 3.4. *Consequences of errors in empirical data and biases in parameter estimates for model predictions*

Type of error	Effect on model parameters	Effect on predictions about another metapopulation
Wrongly estimated patch areas, unknown variation in patch quality	Reduced area-dependence of emigration, immigration, and local extinction	Overestimated extinction rates for large populations: overestimated effect of small patches for metapopulation persistence
Missing patches	Overestimated colonization rates and dispersal distances	Overestimated patch occupancy and metapopulation persistence
False occupancy information	Overestimated population turnover rates and dispersal distances; weaker scaling of processes by area; poor model fit and predictive ability	General loss of predictive power; overestimated extinction rates for large populations; underestimated metapopulation persistence

Source: Based on Moilanen (2002).

model (IFM) connectivity measure and its extensions. All these measures are simple in the sense that they ignore much information about the landscape. In particular, the characteristics of the landscape matrix are ignored except for the effect of a single parameter, which scales the effect of distance on migration in the dispersal kernel. The basic IFM measure uses information about patch locations and areas, which can easily be obtained using a geographical information system (GIS). However, it is well known that the quality of the habitat patch and its surroundings can affect emigration and immigration rates (e.g., Wiens *et al.* 1993; Kuussaari *et al.* 1996; Moilanen and Hanski 1998; Lin and Batzli 2001; Matter and Roland 2002; Taylor *et al.* Chapter 2). Extended IFM measures may be constructed that include information about habitat quality (see Eq. 3.4 and p. 57), information about the patch perimeter (length, permeability: Moilanen and Hanski 1998), and information about the surroundings of the patch to obtain more accurate estimates of connectivity. Such extensions of the IFM and buffer measure remain practical as they are based on information that can be obtained by empirical investigation of the habitat patches and their immediate surroundings. Conceivably also some of the landscape indices reviewed by Carroll (Chapter 15) could be used to extend the IFM measure.

Connectivity measures involving information about the landscape matrix can be built with assumptions about habitat type-specific movement and mortality rates, and habitat pair-specific edge behavior. For instance, Matter *et al.* (2004) found that the influence of distance on dispersal in a butterfly species was different in two different habitat types. The cost of such extra features is the rapidly increasing number of parameters, which makes parameter estimation difficult. Recently, Ovaskainen (2004) has developed a mathematically and statistically rigorous method for analyzing dispersal and connectivity in arbitrarily fragmented landscapes. His method is based on spatially structured diffusion in a heterogeneous landscape. The model can be used to estimate movement and mortality rates in multiple habitats.

Prediction using connectivity measures

The most important reason for developing models that are parameterized with empirical data is to be able to make predictions. In the context of spatially realistic models with expressions for connectivity, one would like to be able to predict the dynamics of a particular metapopulation in time and in landscapes that have a different structure from the one used for estimating model parameters. To what extent is this possible? There is no clear-cut answer, but we may outline some of the problems involved.

To begin with, connectivity may be estimated with mark–release–recapture data on individuals, in which case the movement rate may be characterized on a daily basis (as in the VM model on pp. 53–54). To translate this rate to per-generation movement rate and to colonization events involves potential sources of error. Mark–release–recapture studies are typically executed within limited areas, which means that long-distance events cannot be observed and estimates of dispersal distances will be biased towards short distances. Furthermore, it is not straightforward to infer successful colonization rate from the rate of migration. For modeling purposes, more robust estimates of colonization rate are obtained by directly parameterizing colonizations with empirical data, as illustrated by the example on p. 58. But even then projecting model predictions to the future is likely to be problematic because the model most likely has been parameterized with data for a few years only. Data for a few years are unlikely to include observations of rare events, very good and very bad years, and consequently model predictions will underestimate true variability in population dynamics (Sutcliffe *et al.* 1997; Thomas *et al.* 2002). Short-term data are likely to yield parameter estimates that underestimate dispersal distances and connectivity as well as the strength of regional stochasticity.

The possibility of making predictions for another landscape than used for estimating parameter values is the attraction of spatially realistic models. For a good empirical example of successful predictions see Thomas *et al.* (2001). Such predictions are safest when the different landscapes are not too different: i.e., they have similar environmental conditions and distributions of patch sizes and connectivities. If landscapes have very different structures in terms of patch density the question arises as to what extent these differences may influence individual movement behavior and ultimately dispersal distances. Habitat patches "compete" for migrants, and consequently dispersal distances observed in landscapes with a high density of habitat patches may be comparatively short even though the species is capable of longer movements. To address this question properly requires a model that explicitly takes such interactions among habitat patches into account (as is done in the model of Ovaskainen 2004). While using simple connectivity measures to model dispersal, one should be cautious about extrapolating dispersal to much larger spatial scales than was used in parameter estimation. If data do not exist for parameterizing a connectivity measure, it may be possible to use existing information about related and/or ecologically similar species to get reasonable approximations of the parameter values (Wahlberg *et al.* 1996; Vos *et al.* 2001).

SUMMARY AND CONCLUSION

Connectivity is a critical component of any spatially extended population study, and connectivity is the factor that turns a model into a spatial model. There are substantial differences in the properties and performance of commonly used measures of connectivity—it makes a difference which measure is used. It is unfortunate that the worst of the simple measures, the nearest neighbor (NN) distance, is the one that is most commonly used in empirical studies. Much of the empirical literature on connectivity is therefore of questionable quality. We strongly recommend researchers to reread the Section on "Simple connectivity measures" (pp. 46–52) and to consider again which measure of connectivity to use. The type of data partly determines which kind of connectivity measure can be used (Fagan and Calabrese Chapter 12)—more informative connectivity measures require more data. However, while using patch-based connectivity measures, substantial improvement in the quantitative utility of a data set can often be gained with a relatively small increase in the amount of data. It is also important to keep an eye on potential problems in the quality of the data, which may seriously bias estimates of connectivity. Finally, it is critical to realize that if the main interest of research or management is in the persistence of species in fragmented landscapes, calculating any measure of connectivity is not sufficient. Connectivity is meant to reflect the capacity of individuals to move around in the landscape, but there is more to population persistence than just individual movements. Questions about persistence can only be answered with models that address population dynamics.

REFERENCES

Bender, D. J., L. Tischendorf, and L. Fahrig. 2003. Using patch isolation metrics to predict animal movement in binary landscapes. *Landscape Ecology* 18:17–39.

Brown, J. H., and A. Kodric-Brown. 1977. Turnover rates in insular biogeography: effect of immigration on extinction. *Ecology* 58:445–449.

Burnham, K. P., and D. R. Anderson. 2002. *Model Selection and Multi-Model Inference: A Practical Information-Theoretic Approach*, New York: Springer-Verlag.

Etienne, R., C. J. F. ter Braak, and C. C. Vos. 2004. Application of stochastic patch occupancy models to real metapopulations. Pp 105–133 in I. Hanski, and O. Gaggiotti (eds.) *Ecology, Genetics, and Evolution in Metapopulations*, London, UK: Academic Press.

Gardner, R. H., B. T. Milne, M. G. Turner, and R. V. O'Neill. 1987. Neutral models for the analysis of broad-scale landscape pattern. *Landscape Ecology* 1:19–28.

Hanski, I. 1994. A practical model of metapopulation dynamics. *Journal of Animal Ecology* 63:151–162.

Hanski, I. 1998. Metapopulation dynamics. *Nature* 396:41–49.

Hanski, I. 1999. *Metapopulation Ecology*. Oxford, UK: Oxford University Press.

Hanski, I. 2001. Spatially realistic theory of metapopulation ecology. *Naturwissenschaften* 88:372–381.

Hanski, I., and M. E. Gilpin. 1991. Metapopulation dynamics: brief history and conceptual domain. Pp. 3–16 in M. Gilpin, and I. Hanski (eds.) *Metapopulation Dynamics*. London, UK: Academic Press.

Hanski, I., and O. Ovaskainen. 2000. The metapopulation capacity of a fragmented landscape. *Nature* 404:756–758.

Hanski, I., and O. Ovaskainen. 2003. Metapopulation theory for fragmented landscapes. *Theoretical Population Biology* 64:119–127.

Hanski, I., A. Moilanen, T. Pakkala, and M. Kuussaari. 1996. Metapopulation persistence of an endangered butterfly: a test of the quantitative incidence function model. *Conservation Biology* 10:578–590.

Hanski, I., J. Alho, and A. Moilanen. 2000. Estimating the parameters of migration and survival for individuals in metapopulations. *Ecology* 81:239–251.

Keitt, T. H., D. L. Urban, and B. T. Milne. 1997. Detecting critical scales in fragmented landscapes. *Conservation Ecology* 1:4. Available online at http://www.consecol.org/vol1/iss1/art4/

King, A. W., and K. A. With. 2002. Dispersal success on spatially structured landscapes: when do spatial pattern and dispersal behavior really matter? *Ecological Modelling* 147:23–39.

Kot, M., M. A. Lewis, and P. van den Driessche. 1996. Dispersal data and the spread of invading organisms. *Ecology* 77:2027–2042.

Kuussaari, M., M. Nieminen, and I. Hanski. 1996. An experimental study of migration in the Glanville fritillary butterfly *Melitaea cinxia*. *Journal of Animal Ecology* 65:791–801.

Levin, S. A. 1974. Dispersion and population interactions. *American Naturalist* 108:207–228.

Lin, Y. T. K., and G. O. Batzli. 2001. The influence of habitat quality on dispersal demography, and population dynamics of voles. *Ecological Monographs* 71:245–275.

MacArthur, R. H., and E. O. Wilson. 1967. *The Theory of Island Biogeography.* Princeton, NJ: Princeton University Press.

Matter, S. F., and J. Roland. 2002. An experimental examination of the effects of habitat quality on the dispersal and local abundance of the butterfly *Parnassius smintheus*. *Ecological Entomology* 27:308–316.

Matter, S., J. Roland, A. Moilanen, and I. Hanski. 2004. The migration and survival of *Parnassius smintheus*: detecting the effects of habitat for individual butterflies. *Ecological Applications* 14:1526–1534.

Moilanen, A. 1999. Patch occupancy models of metapopulation dynamics: efficient parameter estimation using implicit statistical inference. *Ecology* 80:1031–1043.

Moilanen, A. 2000. The equilibrium assumption in estimating the parameters of metapopulation models. *Journal of Animal Ecology* 69:143–153.

Moilanen, A. 2002. Implications of empirical data quality to metapopulation model parameterization and application. *Oikos* 96:516–530.

Moilanen, A. 2004. SPOMSIM: Software for analyzing stochastic patch occupancy models of metapopulation dynamics. *Ecological Modelling* 179:533–550.

Moilanen, A., and I. Hanski. 1998. Metapopulation dynamics: effects of habitat patch area and isolation, habitat quality and landscape structure. *Ecology* 79:2503–2515.

Moilanen, A., and I. Hanski. 2001. On the use of connectivity in spatial models. *Oikos* 95:147–152.

Moilanen, A., and M. Nieminen. 2002. Simple connectivity measures for metapopulation studies. *Ecology* 83:1131–1145.

Nieminen, M., M. Siljander, and I. Hanski. 2004. Structure and dynamics of *Melitaea cinxia* metapopulations. Pp. 63–91 in P. R. Ehrlich, and I. Hanski (eds.) *On the Wings of Checkerspots: A Model System for Population Biology.* Oxford, UK: Oxford University Press.

Petit, S., A. Moilanen, I. Hanski, and M. Baguette. 2001. Metapopulation dynamics of the bog fritillary butterfly: movements between habitat patches. *Oikos* 92:491–500.

O'Hara, R. B., E. Arjas, H. Toivonen, and I. Hanski. 2002. Bayesian analysis of metapopulation data. *Ecology* **83**:2408–2415.

Ovaskainen, O. 2004. Estimating habitat-specific movement parameters for heterogenous landscapes using spatial mark–recapture data and a diffusion model. *Ecology* **85**:242–257.

Ovaskainen, O., and I. Hanski. 2001. Spatially structured metapopulation models: global and local assessment of metapopulation capacity. *Theoretical Population Biology* **60**:281–302.

Ovaskainen, O., and I. Hanski. 2003. Extinction threshold in metapopulation models. *Annales Zoologici Fennici* **40**:81–97.

Ovaskainen, O., and I. Hanski, 2004. Metapopulation dynamics in highly fragmented landscapes. Pp. 73–104 in I. Hanski, and O. Gaggiotti (eds.) *Ecology, Genetics, and Evolution in Metapopulations*. London, UK: Academic Press.

Sjögren-Gulve, P., and I. Hanski. 2000. Metapopulation viability analysis using occupancy models. *Ecological Bulletins* **48**:53–71.

Sutcliffe, O. L., C. D. Thomas, T. J. Yates, and J. N. Greatorex-Davies. 1997. Correlated extinctions, colonizations and population fluctuations in a highly connected ringlet butterfly metapopulation. *Oecologia* **109**:235–241.

Taylor, P. D., L. Fahrig, K. Henein, and G. Merriam. 1993. Connectivity is a vital element of landscape structure. *Oikos* **68**:571–573.

ter Braak, C. J. F., I. Hanski, and J. Verboom. 1998. The incidence function approach to the modelling of metapopulation dynamics. Pp. 167–188 in J. Bascompte, and R. V. Solé (eds.) *Modelling Spatiotemporal Dynamics in Ecology*. Berlin, Germany: Springer-Verlag.

ter Braak, C. J. F., and R. S. Etienne. 2003. Improved Bayesian analysis of metapopulation data with an application to a tree frog metapopulation. *Ecology* **84**:231–241.

Thomas, C. D., E. J. Bodsworth, R. J. Wilson, *et al.* 2001. Ecological and evolutionary processes at expanding range margins. *Nature* **411**:577–581.

Thomas, C. D., R. J. Wilson, and O. T. Lewis. 2002. Short-term studies under-- estimate 30-generation changes in a butterfly metapopulation. *Proceedings of the Royal Society of London, B* **268**:1791–1796.

Tischendorf, L., and L. Fahrig. 2000. On the usage of landscape connectivity. *Oikos* **90**:7–19.

Vos, C. C., J. Verboom, P. F. M. Opdam, C. J. F. ter Braak, and P. J. M. Bergers. 2001. Towards ecologically scaled landscape indices. *American Naturalist* **157**:24–41.

Wahlberg, N., A. Moilanen, and I. Hanski. 1996. Predicting the occurrence of species in fragmented landscapes. *Science* **273**:1536–1538.

Wahlberg, N., T. Klemetti, and I. Hanski. 2002. Dynamic populations in a dynamic landscape: the metapopulation structure of the marsh fritillary butterfly. *Ecography* **25**:224–232.

Wiens, J. A. 1997. Metapopulation dynamics and landscape ecology. Pp. 43–62 in I. Hanski, and M. E. Gilpin (eds.) *Metapopulation Biology*, San Diego, CA: Academic Press.

Wiens, J. A., N. C. Stenseth, B. Vanhorne, and R. A. Ims. 1993. Ecological mechanisms and landscape ecology. *Oikos* **66**:369–380.

With, K. A. 1997. The application of neutral landscape models in conservation biology. *Conservation Biology* 11:1069–1080.

With, K. A. 2004. Metapopulation dynamics: perspectives from landscape ecology. Pp. 23–44 in I. Hanski, and O. Gaggiotti (eds.) *Ecology, Genetics, and Evolution in Metapopulations*. London, UK: Academic Press.

(4)

Genetics and landscape connectivity

RICHARD FRANKHAM

INTRODUCTION

Most threatened species on the planet are being affected by habitat fragmentation (WCMC 1992). Habitat fragmentation involves two processes, a reduction in total habitat area and creation of separate isolated patches from a large continuous distribution, i.e., reduced connectivity among populations. This chapter deals with the genetic impacts of the latter of these effects.

Fragmentation and reduced connectivity are of major importance in conservation biology, as they affect extinction risk. Genetic factors affecting extinction risk in small fragmented populations are inbreeding depression, loss of genetic diversity, reduced ability to adapt to environmental change, loss of self-incompatibility alleles, mutation accumulation, and outbreeding depression (Frankham et al. 2002).

This chapter discusses the genetic effects of small population size and fragmentation, the impacts of different population structures, assessments of connectivity and measurement of gene flow, and concludes with consideration of the effects of fragmentation in species with diverse breeding systems and the evolutionary consequences of long-term fragmentation. Readers are referred to Frankham et al. (2002 and 2004) for more details on these topics and additional references, the former textbook having detailed referenced reviews and the latter a shorter simpler treatment.

GENETIC IMPACTS OF SMALL POPULATION SIZES AND FRAGMENTATION

Inbreeding and loss of genetic diversity are unavoidable in small isolated populations. For example, random mating populations lose neutral

Connectivity Conservation eds. Kevin R. Crooks and M. Sanjayan. Published by Cambridge University Press. © Cambridge University Press 2006.

genetic diversity and become inbred over generations at greater rates in small than large populations, as described by the following equation (Frankham *et al.* 2002):

$$H_t/H_o = [1 - 1/(2N_e)]^t = 1 - F \tag{4.1}$$

where H_t is heterozygosity at generation t, H_o initial heterozygosity, N_e the effective population size, and F the inbreeding coefficient. The effective population size of a population is the number of individuals that would result in the same inbreeding or loss of genetic diversity or genetic drift (see below) if they behaved in the manner of an idealized population with random mating, Poisson variation in family sizes, and equal number per generation. Most natural populations violate these assumptions and have effective sizes much less than their census sizes; N_e is often only about 10% of the census population size (Frankham 1995). Inbreeding increases at a greater rate than above in species that habitually self-fertilize.

Inbreeding increases homozygosity, exposes deleterious alleles, and thus reduces survival and reproduction (inbreeding depression) in essentially all well-studied naturally outbreeding species (Frankham *et al.* 2002). In spite of early scepticism about its effects, inbreeding depression has been documented in many species, both in captivity (Ralls *et al.* 1988) and in the wild (Crnokrak and Roff 1999). It is a well-established cause of extinctions in captive populations and has been implicated in extinction in two species in the wild (Newman and Pilson 1997; Saccheri *et al.* 1998). Computer projections indicate that it is likely to increase extinction risk in most naturally outbreeding threatened species (Brook *et al.* 2002; O'Grady *et al.* 2006). Inbreeding depression also occurs in species that naturally inbreed, but usually at a lower level than for natural outbreeding species, as natural selection is usually more effective at removing (purging) deleterious recessive alleles in inbred populations (Husband and Schemske 1996).

Species face recurring environmental changes from climate change, changes in disease organisms, predators and competitors, etc. To cope with this they may either physiologically adapt, move, or evolve. When environmental changes are large, usually the only options are to evolve or become extinct. As genetic diversity is required for species to evolve, loss of genetic diversity reduces the ability of species and populations to evolve to cope with environmental change and thus increases extinction risk (Frankham *et al.* 2002).

Self-incompatible species suffer from an additional deleterious impact of loss of genetic diversity (Frankham *et al.* 2002). Self-incompatibility occurs in perhaps one-half of all flowering plant species where it prevents or greatly reduces self-fertilization. It is controlled by one or more genetic loci that have many alleles in large populations. In its simplest form, a plant of genotype S_1S_2 at a self-incompatibility locus cannot be fertilized by pollen bearing S_1 or S_2 alleles, but can be fertilized by S_3, S_4, ... pollen. In small isolated populations, self-incompatibility alleles are lost, reducing mate availability and population fitness and potentially resulting in extinction. For example, the Lakeside daisy (*Hymenoxys acaulis* var. *glabra*) population in Illinois became so small and lost so many self-incompatibility alleles that it did not reproduce for 15 years, in spite of known pollen flow (Demauro 1993). It was only recovered following gene flow from populations elsewhere that had not suffered similarly. Young *et al.* (2000) provided a detailed analysis of the effects of loss of self-incompatibility alleles in an endangered grassland daisy (*Rutidosis leptorrhynchoides*) in southeastern Australia. While this effect has only been described in a handful of species, it will occur in most threatened self-incompatible species.

In small populations, selection is less effective than in large populations, so mildly deleterious alleles become effectively neutral and their fate is determined by random sampling (genetic drift). Some drift to fixation and thus reduce population fitness (Lande 1995; Lynch *et al.* 1995). Mutational accumulation clearly has a role in the long-term decline of asexual species with low population sizes (Zeyl *et al.* 2001). However, its role in extinctions of sexual species is controversial and likely to be less than that of inbreeding (Gilligan *et al.* 1997; Frankham *et al.* 2002; Garcia-Dorado 2003). The risk of mutation accumulation increases with population fragmentation and reduced connectivity (Higgins and Lynch 2001).

Crossing between genetically differentiated populations may reduce reproduction and/or survival (outbreeding depression), as occurs between northern and southern populations of the corroboree frog (*Pseudophryne corroboree*) in the mountains of southeastern Australia (Osborne and Norman 1991). While gene flow is usually genetically beneficial as described above, in this case it is deleterious. The importance of outbreeding depression is controversial and its impacts are likely to be species and population specific (Frankham *et al.* 2002). It is more likely when populations adapted to different environments come into contact. In this case, individuals from hybrid populations may be less adapted to local conditions in either environment and suffer reduced fitness. Outbreeding depression is especially likely when different subspecies or species hybridize. Connecting isolated populations that are strongly genetically differentiated increases the risks of outbreeding depression. As outbreeding depression is known in only a minority of species, it should be much less of a problem than inbreeding depression and loss of genetic diversity, but the possibility should be evaluated when habitat corridors are being contemplated.

EFFECTS OF POPULATION STRUCTURE

The genetic impacts of population fragmentation and loss of connectivity may range from insignificant to severe, depending upon the details of the resulting population structures and gene flow patterns among fragments. Several different population structures can be distinguished (Fig. 4.1):

- Source—sink (or mainland—island) structure
- Island structure where gene flow is equal among all equally sized population fragments

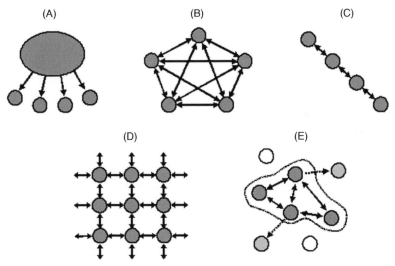

Fig. 4.1. Different fragmented population structures: (A) a source—sink (or mainland—island) structure where the source provides all the input to the sink populations; (B) an island structure where gene flow is equal among equal sized islands; (C) a linear stepping-stone structure where only neighboring populations engage in direct gene flow; (D) a two-dimensional stepping-stone structure where neighboring populations exchange migrants; and (E) a metapopulation (all from Frankham *et al.* 2002). In addition, there are totally isolated island structures and single large population structures.

- Linear stepping-stone structure where only neighboring populations engage in direct gene flow, as in riparian habitat along rivers
- Two-dimensional stepping stone structure where only surrounding populations engage in direct gene flow
- Metapopulations
- Islands: totally isolated fragments with no connectivity and gene flow, and
- Single large population where connectivity is sufficient for adequate gene flow.

Metapopulations differ from the other structures in having regular extinction and recolonization events, while no extinction is assumed in the other structures. These structures have different genetic consequences, with the totally isolated islands and the metapopulations typically having the worst genetic impacts (inbreeding and loss of genetic diversity) for the same total population sizes, especially in the long term (see below). The endangered Glanville fritillary butterfly (*Melitaea cinxia*) metapopulation in Finland (Case Study 4.1) provides a classic example where inbreeding has been shown to contribute to extinction risk.

In evaluating the impacts of isolation and lack of connectivity, I first consider completely isolated fragments with no gene flow and then review partially connected fragments. I then discuss population recovery

---

CASE STUDY 4.1 Inbreeding and extinction risk in butterfly metapopulations in Finland (Saccheri *et al.* 1998)

The endangered Glanville fritillary butterfly population in Finland exists in many fragmented populations with regular extinctions and recolonization. There are about 1600 suitable meadows for the butterfly, 320–524 being occupied in 1993–96. The population turnover rate is high, with an average of 200 extinctions and 114 colonizations per year.

As many of the populations are very small, inbreeding may contribute to extinction risks. Forty-two butterfly populations in Finland were typed for genetic markers in 1995, and their extinction or survival recorded in the following year. Of these, 35 survived to fall 1996 and seven went extinct. Extinction rates were higher for populations with lower heterozygosity, an indication of inbreeding (Eq. 4.1), even after accounting for the effects of demographic and environmental variables (population size, time trend in population size and area) known to affect extinction risk. Overall, inbreeding accounted for 26% of the variation in extinction risk.

following outcrossing of previously isolated populations, and the level of gene flow necessary to connect population fragments. I conclude the section with a summary of the impacts of different population structures on reproductive fitness.

Completely isolated fragments

The most severe effects of fragmentation are expected in isolated fragments with no connectivity or gene flow. Since fragmentation creates "islands" from once-continuous habitat, its effects parallel those in oceanic island populations. Island populations often have reduced genetic diversity, are inbred and have elevated extinction risks compared to mainland populations (Frankham 1997, 1998). Case Study 4.2 provides an

---

CASE STUDY 4.2   **Impact of fragmentation in island populations of rock wallabies in Australia**

Black-footed rock wallabies (*Petrogale lateralis*) in Australia illustrate many of the genetic effects of population fragmentation (Eldridge *et al.* 2001). Rock wallabies are small macropod marsupials (about 1 m tall) that live on rocky outcrops on the Australian mainland and on islands off Western Australia. The Barrow Island population of black-footed rock wallabies has been isolated from the mainland for 8000 years (about 1600 generations) and has a relatively small population size.

The Barrow Island population has very low genetic diversity, as assessed by microsatellites, compared to that for two mainland populations, as shown below. Other island populations also have low genetic diversity, as indicated below.

Population (location)	Proportion of loci polymorphic	Mean number of alleles/locus	Average heterozygosity
Barrow Island	0.1	1.2	0.05
Mainland			
Exmouth	1.0	3.4	0.62
Wheatbelt	1.0	4.4	0.56

The Barrow Island population has an inbreeding coefficient of 0.91. Further, it displays inbreeding depression compared to the mainland

Continued

CASE STUDY 4.2 (cont.)

population. The frequency of lactating females is 92% in mainland rock wallabies, but only 52% on Barrow Island. These rock wallaby populations demonstrate that genetics can have an impact before demographic and environmental stochasticity or catastrophes cause extinctions. They have clearly survived stochastic fluctuations and catastrophes, but they are suffering genetic problems that increase their risk of extinction.

The different island populations have all lost genetic diversity, but have become genetically differentiated since they were isolated by post-glacial sea level rises 8000−15 000 years ago. Alleles present (+) and absent (−) at four microsatellite loci in populations of black-footed rock wallabies on the mainland of Australia and on six offshore islands are shown below. Island populations contain many fewer alleles than mainland populations, but they are usually a subset of alleles found on the mainland. Different island populations often contain different alleles, as expected due to genetic drift.

Locus	Allele	Mainland	BI	SI	PI	MI	WiI	WeI
Pa297	102	+	−	−	−	−	−	−
	106	+	−	−	−	−	−	−
	118	+	−	−	−	−	−	−
	120	−	−	−	+	−	−	−
	124	+	−	−	−	+	−	−
	128	+	−	+	−	−	+	+
	130	+	−	−	−	−	−	−
	136	+	+	−	−	−	−	−
Pa385	157	+	−	−	−	−	−	−
	159	+	−	−	+	−	+	+
	161	+	−	+	−	−	−	−
	163	+	−	−	−	+	−	−
	165	+	−	−	−	−	−	−
	173	−	+	−	−	−	−	−
Pa593	105	+	+	−	−	−	+	+
	113	−	+	−	−	−	−	−
	123	+	−	−	−	−	−	−
	125	+	−	−	+	−	−	−
	127	+	−	−	−	−	−	−
	129	+	−	−	−	−	−	−
	131	+	−	+	−	−	−	−
	133	+	−	−	−	−	−	−
	135	+	−	−	−	−	−	−
	137	−	−	−	−	+	−	−

Locus	Allele	Mainland	Islands					
			BI	SI	PI	MI	WiI	WeI
Me2	216	+	−	−	−	−	−	−
	218	+	−	−	−	+	+	−
	220	+	+	−	−	−	−	+
	222	+	−	+	−	−	−	−
	224	+	−	−	−	−	−	−
	230	−	−	−	+	−	−	−

Island populations have been viewed as ideal sources for restocking depleted or extinct mainland populations, especially in Australasian species. However, as they often have low genetic diversity and are inbred, they may not be good candidates for translocations, if alternative mainland populations still exist. However, the totality of all island populations contains most of the genetic diversity found in the mainland population, so it could provide individuals for reintroduction, following crossing of island populations.

example of the impacts of complete isolation in island populations of rock wallabies in Western Australia.

Replicate population fragments become genetically differentiated with generations at a rate that depends on the inbreeding coefficient (Wright 1969). Observed heterozygosity, measured across the totality of isolated populations, is reduced compared to Hardy–Weinberg expectations, the reduction being proportional to the inbreeding coefficient (the Wahlund effect). These effects depend critically on the effective population size of each isolated fragment, and the duration of isolation in generations. Isolated populations of outbreeding species not only lose genetic diversity, but also suffer inbreeding depression, have reduced ability to evolve to cope with environmental change, and consequently have elevated extinction risk (Frankham et al. 2002; Reed and Frankham 2003; Reed et al. 2003). As inbreeding depression is typically more extreme under stressful environments (Frankham et al. 2002), the genetic effects in isolated fragments are likely to become worse in the future with global climate change, reductions in the ozone layer, accelerated movements of disease organisms across the planet due to human activities, and the increasing degradation of natural areas in general.

The impacts of reduced connectivity in fragmented populations are illustrated by comparing a single large (SL) population with several small (SS) completely isolated populations of the same total population size: the SLOSS comparison, as shown in Fig. 4.2. In the short term when there are no extinctions of SS populations, the SS populations retain greater overall allelic diversity than the SL population (Frankham *et al.* 2002). This expectation has been verified in experiments with fruit flies (Margan *et al.* 1998). In the long term, however, extinction rates will be greater in smaller than in larger population fragments due to environmental and demographic stochasticity, catastrophes, and genetic factors. With extinction of some SS populations (4), the SL population retains more genetic diversity and has higher reproductive fitness than all the SS populations combined (now only two populations).

The genetic impacts of fragmentation are expected to become increasingly deleterious as the degree of fragmentation increases. For a population of total size N_e separated into f totally isolated, equal-sized fragments, the size of each fragment is N_e/f. Each of the fragments will become inbred and lose genetic diversity at a rate dependent upon N_e/f, compared to N_e in a single population of the same total size. The proportion of initial heterozygosity retained after t generations in each of the f small SS fragments, compared to that in the single large population, is

$$\ln(H_t/H_o)_{SS/SL} = (1 - f)t/2N_e. \tag{4.2}$$

Thus, the proportional retention of heterozygosity in several small population fragments, compared to a single large population, declines with the number of fragments and increases with generations. The rate of decline is greater with smaller than larger total effective population size. The effects on inbreeding are similar and can be illustrated with examples all with a total constant effective size of 500 individuals per generation. Following Eq. (4.1), a single population of size 500 loses 5% of its initial heterozygosity over 50 generations, while two populations of effective size 250 each lose 10%, five populations of size 100 each lose 22%, ten populations of size 50 each lose 39% and twenty populations of size 25 each lose 64% of their initial genetic diversity. As expected, reproductive fitness drops at a greater rate in smaller populations than in larger ones (Bryant *et al.* 1999; Woodworth *et al.* 2002).

Loss of genetic diversity has been documented in many small isolated population fragments, including black-footed rock wallabies, greater

Initial populations

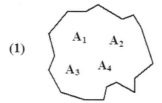

Short-term, no extinctions

Several small   **Single large**

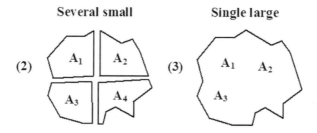

**Long-term, extinction of some
small populations**

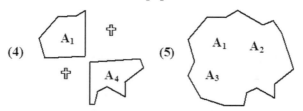

Fig. 4.2. The genetic consequences of a single large population (SL) versus several small (SS) completely isolated population fragments of initially the same total size (SLOSS) over different time frames (from Frankham *et al.* 2002). (1) A_1–A_4 represent four alleles initially present in the population. In the short-term, without extinctions, the several small populations (2) are expected to go to fixation more rapidly, but to retain greater overall genetic diversity than the single large population (3). The chances are greater that an allele will be totally lost from the large population than from all small populations combined. However, the SS populations will each be more inbred than the SL population. In the longer term, when extinctions of small, but not large populations occur, the sum of the small surviving populations (4) will retain less genetic diversity than the single large population (5). A metapopulation with extinctions and recolonizations is similar to (4).

prairie chickens (*Tymphanuchus cupido pinnatus*), adders (*Vipera berus*), Glanville fritillary butterflies, and grassland daisies (Frankham *et al.* 2002). Inbreeding depression has also been documented in these cases and in isolated populations of royal catchfly (*Sabatia angularis*) and scarlet gilia (*Ipomopsis aggregata*) plants.

Partially connected fragments

The impacts of population fragmentation on genetic diversity, inbreeding, differentiation, and extinction risk depend on the connectivity and level of gene flow among fragments. These in turn depend on:

- Number of population fragments
- Distribution of population sizes in the fragments
- Geographic distribution of populations
- Distance between fragments
- Dispersal ability of the species
- Migration rates between fragments
- Survival and reproductive success of migrants in new locations
- Environment of the matrix among the fragments and its impact on dispersal
- Time since fragmentation in generations
- Susceptibility of the species to inbreeding depression.

The genetic impacts in partially connected fragments depends upon rate of gene flow and ranges from that of totally isolated fragments to that of a single large population with complete connectivity. The endangered red-cockaded woodpecker (*Picoides borealis*) in the eastern USA illustrates many of the features and genetic problems associated with habitat fragmentation for a species with reduced gene flow between fragments (Case Study 4.3).

The large number of variables affecting fragmented populations and the stochastic nature of many effects makes them difficult to study in the field or with mathematical models. Consequently, computer simulations seem likely to be the tool that will provide the major insights on the genetic and ecological impacts of different connectivity in fragmented populations. Realistic stochastic computer simulations for real species have been done for several species. Such models predict population size trajectories without significant bias for well-studied species (Brook *et al.* 2000; McCarthy *et al.* 2003). Dobson *et al.* (1992) predicted that there would be substantial benefits of migration on the viability of black rhinoceros (*Diceros bicornis*) populations, especially in very small populations. McCullough *et al.* (1996) predicted benefits from gene flow in tule elk populations (*Cervus elaphus nannodes*). Young *et al.* (2000) showed that small isolated populations of the endangered Australian grassland daisy *Rutidosis leptorrhynchoides* were expected to lose self-incompatibility alleles and suffer reductions in population fitness, in line with observed effects.

Reed (2004) investigated the impact of population fragmentation with different dispersal rates in 30 real species using models that included all relevant variables (including demographic, environmental, and genetic stochasticity and catastrophes). The models included modest inbreeding depression (including purging effects), different values of environmental correlations among fragments, and two different total population sizes (250 and 1000). Dispersal rate was the most important explanatory

---

CASE STUDY 4.3   **Impact of habitat fragmentation on the endangered red-cockaded woodpecker population in southeastern USA**

The red-cockaded woodpecker was once common in the mature pine forests of the southeast USA (Kulhavy *et al.* 1995). It declined in numbers, primarily due to habitat loss, and was placed on the US endangered species list in 1970. It now survives in scattered and isolated sites within the US southeast. There is little connectivity and gene flow among isolated sites. As expected, populations have diverged genetically from each other, and lost genetic diversity, with smaller populations showing the greatest loss of genetic diversity and the most divergence (Stangel *et al.* 1992). Moderate divergences in allele frequencies exist among woodpecker populations. Differentiation, measured as F_{ST} (see pp. 86–87), is 0.14 based on allozyme data, and 0.19 based on randomly amplified polymorphic DNA(RAPD) data, and both show a general tendency for closer genetic similarity among geographically proximate populations (Haig and Avise 1996).

Computer simulations indicate that the smallest woodpecker populations are likely to suffer from inbreeding depression in the near future (Daniels *et al.* 2000). In response to the threats posed by fragmentation, management of the woodpeckers involves habitat protection, improvement of habitat suitability by constructing artificial nest holes, reintroductions into suitable habitat where populations become extinct, and augmentation of small populations to minimize inbreeding and loss of genetic diversity. This is one of the most extensive management programs for a fragmented population anywhere in the world.

variable affecting extinction differences between the fragmented and single large populations, with higher dispersal reducing extinction risk in the metapopulations. The correlation between subpopulations in environmental variation was the second most important variable, with higher correlations increasing the risk of extinction of metapopulations. Population growth rate was the third most important variable, species with low population growth rates being more sensitive to fragmentation than were populations with higher initial fitness. There was also a significant interaction between population growth rate and dispersal. Extinction risk was highly dependent upon the total carrying capacity and the time-frame in generations, decreasing with the former and increasing with the latter. The effects of dispersal, carrying capacity, and time-frame are all predicted from genetic considerations, but are also expected from ecological effects.

Population recovery following outcrossing

The above section presumes that when previously isolated populations are connected and gene flow re-established, inbred populations of naturally outbreeding species recover in fitness. This has been shown in many species, including wolves (*Canis lupus*), greater prairie chickens, snakes, water fleas (*Daphnia*), fruit flies (*Drosophila*), flour beetles (*Tribolium*), houseflies (*Musca domestica*) and plants (Richards 2000; Newman and Tallmon 2001; Ebert *et al.* 2002; Frankham *et al.* 2002; Vilà *et al.* 2003). Even a single immigrant can produce substantial effects and the immigrants can be themselves inbred, provided they are from a different population fragment (Spielman and Frankham 1992; Vilà *et al.* 2003). Migrant alleles are likely to be at a selective advantage, so that they contribute more genetically than expected based on their initial proportion (Ebert *et al.* 2002; Saccheri and Brakefield 2002).

How much gene flow is required to connect population fragments?

Migration reduces the impact of fragmentation by an extent dependent on the rate of gene flow. With sufficient migration, a fragmented population will have the same genetic consequences as a single large population of the same total size.

Sewall Wright obtained the surprising result that a single migrant per generation among idealized populations with an island model (Fig. 4.1B) was sufficient to prevent complete differentiation (and fixation of alleles) (Wright 1969). These results are independent of population size. The deleterious effects of inbreeding on reproductive fitness and extinction risk are also expected to be largely alleviated in populations with one or more migrants per generation. Experimental studies support this prediction (Bryant *et al.* 1999; Newman and Tallmon 2001).

The conclusions above assume that migrants and residents are equally likely to survive and produce offspring, and that all population fragments have idealized structures, apart from the occurrence of migration. In real wild populations where these assumptions are unrealistic, 5–10 migrants per generation may therefore be required to achieve these effects (Frankham *et al.* 2002).

Impacts of different population structures on reproductive fitness

In summary, the overall consequences of different population structures on reproductive fitness will depend primarily on the inbreeding coefficient in each fragment. The single large unfragmented population becomes the standard for comparison. Consequences of different population structures are (Frankham *et al.* 2002):

- In source–sink (or mainland–island) structures, the effective population size will depend on N_e in the source population, rather than that for the total population. Thus, inbreeding and loss of fitness are likely to be much higher with this structure than for SL.
- In the island and stepping-stone models, inbreeding and fitness will depend critically upon the gene flow rates and upon the variation in population sizes on different islands. When there is no gene flow, inbreeding will depend upon the effective population sizes of the individual populations, and will be greater than for SL. Conversely, when there is ample gene flow among populations, inbreeding will depend upon the effective size of the total population, and be similar to SL.
- Metapopulations typically have effective sizes that are markedly less than the number of breeding adults, due to cycles of extinction and recolonization that eliminate ancestors and their alleles (Fig. 4.3). Typically, inbreeding will be greater and fitness lower than for other fragmented and non-fragmented structures. The effective size will approximate the sum of the effective sizes of fragments following extinctions, rather than the sum of all effective sizes. Bottlenecks during recolonization will subsequently reduce N_e still further.

If there are frequent extinctions and recolonization mainly from a few large fragments, the metapopulation structure approaches that of a source–sink, and less genetic diversity is retained than in a single large

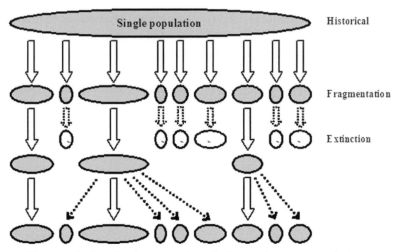

Fig. 4.3. Cycles of extinction and recolonization in a metapopulation, leading to reductions in the effective size of a species (from Frankham *et al.* 2002). The dotted lines indicate bottlenecks during recolonization.

population of the same total size. Conversely, a metapopulation with sufficient migration and low rates of local extinctions approaches the characteristics of a single large population. In general, the higher the rate of extinction and recolonization, the more jeopardized is the metapopulation.

ASSESSING CONNECTIVITY AND MEASURING GENE FLOW

Genetic methods are widely used to assess population connectivity. Migration rates are notoriously difficult to measure by direct tracking of individuals or gametes, and immigrants may not breed in their new habitat. Consequently, gene flow, rather than migration rate, is the critical variable to estimate.

Sewall Wright (1969) partitioned inbreeding of individuals (I) in the total (T) population (F_{IT}) into that due to inbreeding of individuals relative to their subpopulation (S) or fragment, F_{IS}, and inbreeding due to differentiation among subpopulations, relative to the total population, F_{ST} (*F* statistics). With high rates of gene flow among fragments, F_{ST} is low. With low rates of gene flow among fragments, populations diverge and become inbred, and F_{ST} increases.

The F statistics can be calculated from the relationship between heterozygosity for genetic markers such as allozymes or microsatellites and inbreeding (Eq. 4.1) using the following equations (Nei 1987):

$$F_{IS} = 1 - (H_I/H_S) \tag{4.3}$$

$$F_{ST} = 1 - (H_S/H_T) \tag{4.4}$$

$$F_{IT} = 1 - (H_I/H_T) \tag{4.5}$$

where H_I is the observed heterozygosity averaged across all population fragments, H_S is the Hardy–Weinberg expected heterozygosity averaged across all population fragments, and H_T is the expected heterozygosity for the total population. F_{ST} ranges from 0 (no differentiation between fragments) to 1 (fixation of different alleles in fragments). Reservations have been expressed about using F_{ST} with microsatellite data, as they have a different mutational process, so the related measure R_{ST} is often considered more suitable (Slatkin 1995). However, simulation studies indicate that neither statistic has consistently superior performance for microsatellites (Balloux and Goudet 2002). Case Study 4.4 illustrates the calculation of the F statistics based on heterozygosities for the endangered Pacific yew (*Taxus brevifolia*) in western North America. This species exhibits inbreeding within populations ($F_{IS} > 0$) and differentiation among populations ($F_{ST} > 0$).

Equilibrium between migration and inbreeding

With constant population sizes and migration rates, inbreeding and gene flow reach an equilibrium where the reduction in divergence due to migration balances the increase due to drift. The equilibrium inbreeding (F_{ST}) is related to the effective population size and the migration rate (m) in such a population (Wright 1969) as follows:

$$F_{ST} = 1/(4N_e m + 1). \tag{4.6}$$

This equation applies when the migration rate is small. Gene flow can be inferred from patterns of differentiation among populations using this equation. For example, for the Pacific yew, the effective number of migrants per generation $N_e\ m = [(1/F_{ST}) - 1]/4 = [(1/0.078) - 1]/4 = 2.96$. On average about three migrants per generation are entering Pacific yew populations. This value reflects historical evolutionary rates of gene flow in equilibrium circumstances, so it may not reflect current gene flow. This is an approximation based on the island model (Fig. 4.1B), but related expressions have been derived for other models of migration

---

CASE STUDY 4.4   Computation of F statistics for the
rare Pacific yew

The endangered Pacific yew, a conifer found on the Pacific northwest of North America, is the source of the anticancer substance Taxol.

El-Kassaby and Yanchuk (1994) reported genotype frequencies and heterozygosities for 21 allozyme loci in nine Canadian populations. Average observed heterozygosity (H_I) across the nine populations was 0.085, while the average expected heterozygosity for these populations (H_S) was 0.166. Consequently, inbreeding within populations F_{IS} is:

$$F_{IS} = 1 - (H_I/H_S) = 1 - (0.085/0.166) = 0.49$$

The high level of inbreeding is not due to selfing, as the species is dioecious. It is probably due to offspring establishing close to parents and clumping of individuals founded from bird and rodent seed caches. The expected heterozygosity across the nine populations (H_T) was 0.18, so inbreeding due to population differentiation (F_{ST}) is

$$F_{ST} = 1 - (H_S/H_T) = 1 - (0.166/0.180) = 0.078$$

This indicates only a modest degree of population differentiation.

The total inbreeding due to both inbreeding within populations and differentiation among them (F_{IT}) is

$$F_{IT} = 1 - (H_I/H_T) = 1 - (0.085/0.18) = 0.53$$

(Neigel 1996). The exact estimates of migration rates obtained from this equation are not necessarily reliable or current, but they do indicate the relative rates of gene flow that populations would have if they adhered to the island population structure. In this way, they have a role that is analogous to that of effective population sizes.

Assignment tests using multiple genetic loci (e.g., microsatellites) provide a direct means for detecting recent immigrants (Paetkau *et al.* 1995; Neville *et al.* Chapter 13). In brief, analyses are used to ask if each individual originated in the geographic location where it was found or in another population at a different location. Eldridge *et al.* (2001) used an assignment test based on data from 11 polymorphic microsatellite loci to infer that a population of rock wallabies at Gardener's Outcrop in the Wheatbelt on the mainland of Western Australia was re-established by immigrants from the nearest extant population (see Case Study 4.2). Assignment tests can be highly powerful for detecting current movement even when little population differentiation exists. For example, the threatened geometric tortoise (*Psammobates geometricus*) in South Africa shows little genetic differentiation ($F_{ST} = 0.018-0.048$) and retains appreciable genetic variation despite suffering extreme habitat fragmentation, having lost 97% of its habitat and declining from over 15 000 to 3500–5000 individuals. Interestingly, it appears to display some current migration, as indicated by assignment tests using data from eight microsatellite loci (Cunningham *et al.* 2002).

Like F_{ST}, coalescence approaches, based on the genealogies of alleles, can be used to infer historical gene flow and population structure, as illustrated by the work of Tero *et al.* (2003) on the endangered plant *Silene tatarica* in riparian zones in northern Finland. Analyses of amplified fragment-length polymorphism (AFLP) data based on coalescence, plus assignment tests, F_{ST}, and other approaches, revealed a classical metapopulation structure with a high subpopulation turnover rate and low gene flow among subpopulations, consistent with information from ecological surveys. Compared to traditional F_{ST} measures, coalescent methods make fewer assumptions about population dynamics and can provide additional information, such as estimates of asymmetrical dispersal rates among populations with different effective population sizes (Neville *et al.* Chapter 13).

Templeton's (1998) nested clade analysis, based on mapping haplotype networks onto geographic locations, provides a hypothesis-testing framework for delineating the historical processes that led to given patterns of genetic differentiation. He applied this analysis to buffalo (*Syncerus caffer*) and impala (*Aepyceros melampus*) and found different patterns and causes for population differentiation in the two species, in spite of similar F_{ST} values. Genetic distances, such as that of Nei, provide another means for describing genetic differentiation and are used to build phenograms or "trees" to visualize genetic similarity among populations (Nei 1987). Neville *et al.* (Chapter 13) provide further details on several of these methods and several examples from salmonid fish.

Dispersal and gene flow

Since differentiation among populations is dependent on levels of gene flow, we would expect this to be related to the dispersal abilities of species and the degree of isolation among populations. Thus, the degree of genetic differentiation among populations (F_{ST}) is expected to be greater for populations:

- in species with lower versus higher dispersal rates
- in subdivided vs. continuous habitat

- in distant vs. closer fragments
- in smaller vs. larger population fragments
- in species with longer vs. shorter divergence times (in generations)
- with adaptive differences vs. those without.

Observations generally confirm these predictions (Frankham *et al.* 2002).

There is a strong negative correlation between F_{ST} and the dispersal ability of species, as predicted; the average rank correlation was −0.73 in a meta-analysis involving 333 species across 20 animal groups (Bohonak 1999). Examples of mean F_{ST} values for major groups of organisms are given in Table 4.1. Taxa that can fly, such as birds and insects, have lower F_{ST} values than those that do not. Further, F_{ST} is higher in plants that self-fertilize (low pollen dispersal) than in outcrossing plants.

Dispersal rates typically reduce with distance. Consequently, distant fragments are expected to receive fewer migrants than nearby ones. Allozyme variability for island populations generally declines with distance from the mainland in lizards and several species of mammals (Frankham 1997). For mainland population fragments, distant habitat patches are expected to receive fewer migrants than nearby habitat patches, but this effect depends upon the nature of the surrounding matrix and its influence on dispersal rates (e.g., see Taylor *et al.* Chapter 2).

Table 4.1. *Fixation index (F_{ST}) in a range of taxa*

Species	F_{ST}
Mammals (57 species)	0.24
Birds (23 species)	0.05
Reptiles (22 species)	0.26
Amphibians (33 species)	0.32
Fish (79 species)	0.14
Insects (46 species)	0.10
Plants	
Selfing	0.51
Mixed selfing and outcrossing	
animal pollination	0.22
wind pollination	0.10
Outbreeding	
animal pollination	0.20
wind pollination	0.10

Source: After Frankham *et al.* (2002).

Genetic differentiation and gene flow are associated with geographic distance ("isolation by distance") in red-cockaded woodpeckers, bighorn sheep (*Ovis canadensis*), gray wolves, and brown bears (*Ursus arctos*) in North America (Haig and Avise 1996; Forbes and Hogg 1999), and in many other species. However, species with high dispersal rates may not show isolation by distance (e.g., Canada lynx: Schwartz *et al.* 2002), while completely isolated populations are likely to show random patterns of genetic differentiation.

EFFECTS OF FRAGMENTATION AND REDUCED CONNECTIVITY IN SPECIES WITH DIVERSE BREEDING SYSTEMS

The above discussion has concentrated on species that are naturally outbreeding diploids. The effects of fragmentation and reduced gene flow are different for species that are asexual, highly self-fertilizing, haploid, or polyploid (Frankham *et al.* 2002).

Asexual species, such as the critically endangered Meelup mallee tree (*Eucalyptus phylacis*) in Western Australia (Rossetto *et al.* 1999), and haploid species do not suffer from inbreeding depression, but they do suffer from loss of genetic diversity in small populations. Thus, highly isolated fragments with small population sizes are likely to have reduced ability to evolve to cope with environmental change. Selfing species, such as the endangered Malheur wirelettuce (*Stephanomeria malheurensis*: Falk *et al.* 1999), suffer from inbreeding depression, but typically at a lesser level than outbreeding species (Husband and Schemske 1996). They typically have genetic diversity distributed more among than within population, as compared to outbreeders. Consequently, they suffer serious losses of genetic diversity and evolutionary potential in metapopulations with high turnover rates. The taxonomic distribution of asexual and selfing species suggest that they are evolutionary dead-ends due to limited ability to evolve to cope with environmental change (White 1973).

Polyploids, such as the California redwoods (*Sequoia sempervirens*), are less sensitive to loss of genetic diversity than diploids with similar population sizes, seem to be less sensitive to inbreeding depression, and are generally less affected by genetic factors contributing to extinction risk in small populations (Frankham *et al.* 2002).

EVOLUTIONARY CONSEQUENCES OF FRAGMENTATION AND LACK OF CONNECTIVITY

Habitat fragmentation may have long-term evolutionary consequences. If a previously continuous population is fragmented and completely isolated it may change its evolutionary strategy from that of a habitat generalist to become habitat specialists adapted differentially to the conditions of the isolated fragments, provided it persists in the long term. Over even longer time-spans, this could lead to speciation, as habitat isolation is associated with most speciation (Frankham *et al.* 2002).

Fragmented populations without connectivity and gene flow may undergo evolutionary changes in breeding systems or dispersal rates. Small populations of naturally outbreeding plants may become selfers, as in the plant *Isotoma petraea* (James 1970), and some even evolve chromosomal systems that promote permanent heterozygosity. Self-incompatible species may evolve self-compatibility in small populations. Some isolated founder populations of insects, such as Orthoptera, have become parthenogenetic (White 1978). Populations of plants migrating to islands have been initiated by genotypes with high dispersal rates, but then evolved towards low dispersal rates within a few generations (Cody and Overton 1996). In birds, migration rates and patterns are heritable (Berthold and Pulido 1994) and evolutionary changes in migratory patterns have been documented (Berthold *et al.* 1992). Possibly, species with limited connectivity could evolve increased migration rates, but I know of no example of this.

CONCLUSIONS

Habitat fragmentation/connectivity is a fundamental concern in conservation biology as it affects extinction risk. The genetic impacts of population fragmentation depend critically upon connectivity and gene flow among fragments. Lack of gene flow among fragmented populations results in inbreeding and loss of genetic diversity and elevated extinction risk. Population fragmentation usually has deleterious genetic consequences in the long term, compared to a similar-sized unfragmented population. Further, lack of connectivity between populations precludes immigrants re-establishing extinct populations. The genetic impacts of fragmentation and connectivity depend on knowledge of many parameters (gene flows, inbreeding depression, and extinction rates) and so are difficult to investigate in the field, or with mathematical models. Consequently, realistic stochastic computer projections have a major role to play in understanding these effects. Metapopulation structures, with extinctions and recolonization of population fragments, are likely to be particularly deleterious. Genetic methods, such as F_{ST}, assignment tests, and coalescence can define the extent of connectivity and infer rates of gene flow between populations. This allows rational scientific management of threatened species in the wild.

ACKNOWLEDGEMENTS

I thank Jonathan Ballou and David Briscoe for their contributions, as this chapter has relied heavily upon material from our textbook *Introduction to Conservation Genetics* and Kevin Crooks, Scott Mills, Helen Neville, and several anonymous referees for their comments on the manuscript. This chapter was drafted whilst I was at the University of Aarhus in Denmark. Volker Loeschcke kindly hosted my visit and provided funds for travel and accommodation. This is publication number 407 from the Key Centre for Biodiversity and Bioresources, Macquarie University.

REFERENCES

Balloux, F., and J. Goudet. 2002. Statistical properties of population differentiation estimators under stepwise mutation in a finite island model. *Molecular Ecology* 11:771–783.

Berthold, P., and F. Pulido. 1994. Heritability of migratory activity in a natural bird population. *Proceedings of the Royal Society of London B* 257:311–315.

Berthold, P., A. J. Helbig, G. Mohr, and U. Querner. 1992. Rapid microevolution of migratory behaviour in a wild bird species. *Nature* 360:668–670.

Bohonak, A. J. 1999. Dispersal, gene flow and population structure. *Quarterly Review of Biology* 74:21–45.

Brook, B. W., J. J. O'Grady, A. P. Chapman, *et al.* 2000. Predictive accuracy of population viability analysis in conservation biology. *Nature* 404:385–387.

Brook, B. W., D. W. Tonkyn, J. J. O'Grady, and R. Frankham. 2002. Contribution of inbreeding to extinction risk in threatened species. *Conservation Ecology* 6:16. Available online at http://www.consecol.org/vol6/iss11/art16

Bryant, E. H., V. L. Backus, M. E. Clark, and D. H. Reed. 1999. Experimental tests of captive breeding for endangered species. *Conservation Biology* 13:1487–1496.

Cody, M. L., and J. M. Overton. 1996. Short-term evolution of reduced dispersal in island plant populations. *Journal of Ecology* 84:53–61.

Crnokrak, P., and D. A. Roff. 1999. Inbreeding depression in the wild. *Heredity* 83:260–270.

Cunningham, J., E. H. W. Baard, E. H. Harley, and C. O'Ryan. 2002. Investigation of genetic diversity in fragmented geometric tortoise (*Psammobates geometricus*) populations. *Conservation Genetics* 3:215–223.

Daniels, S. J., J. A. Priddy, and J. R. Walters. 2000. Inbreeding in small populations of red-cockaded woodpeckers: insights from a spatially-explicit individual-based model. Pp. 129–147 in A. G. Young, and G. M. Clarke (eds.) *Genetics, Demography and Viability in Fragmented Populations*. Cambridge, UK: Cambridge University Press.

Demauro, M. M. 1993. Relationship of breeding system to rarity in the Lakeside daisy (*Hymenoxys acaulis* var. *glabra*). *Conservation Biology* 7:542–550.

Dobson, A. P., G. M. Mace, J. Poole, and R. A. Brett. 1992. Conservation biology: the ecology and genetics of endangered species. Pp. 405–430 in R. J. Berry, T. J. Crawford, and G. M. Hewitt (eds.) *Genes in Ecology*. Oxford, UK: Blackwell.

Ebert, D., C. Haag, M. Kirkpatrick, *et al.* 2002. A selective advantage to immigrant genes in a *Daphnia* metapopulation. *Science* 295:485–488.

Eldridge, M. D. B., J. E. Kinnear, and M. L. Onus. 2001. Source population of dispersing rock-wallabies (*Petrogale lateralis*) identified by assignment tests on multilocus genotypic data. *Molecular Ecology* 10:2867–2876.

El-Kassaby, Y. A., and A. D. Yanchuk. 1994. Genetic diversity, differentiation, and inbreeding in Pacific yew from British Columbia. *Journal of Heredity* 85:112–117.

Falk, D. A., C. I. Millar, and M. Olwell. 1996. *Restoring Diversity: Strategies for Reintroduction of Endangered Plants*. Washington, DC: Island Press.

Forbes, S. H., and J. T. Hogg. 1999. Assessing population structure at high levels of differentiation: microsatellite comparisons of bighorn sheep and large carnivores. *Animal Conservation* 2:223–233.

Frankham, R. 1995. Effective population size/adult population size ratios in wildlife: a review. *Genetical Research* 66:95–107.

Frankham, R. 1997. Do island populations have lower genetic variation than mainland populations? *Heredity* 78:311–327.

Frankham, R. 1998. Inbreeding and extinction: island populations. *Conservation Biology* 12:665–675.

Frankham, R., J. D. Ballou, and D. A. Briscoe. 2002. *Introduction to Conservation Genetics*. Cambridge, UK: Cambridge University Press.

Frankham, R., J. D. Ballou, and D. A. Briscoe. 2004. *A Primer of Conservation Genetics*. Cambridge, UK: Cambridge University Press.

Garcia-Dorado, A. 2003. Tolerant versus sensitive genomes: the impact of deleterious mutation on fitness and conservation. *Conservation Genetics* 4:311–324.

Gilligan, D. M., L. M. Woodworth, M. E. Montgomery, D. A. Briscoe, and R. Frankham. 1997. Is mutation accumulation a threat to the survival of endangered populations? *Conservation Biology* 11:1235–1241.

Haig, S. M., and J. C. Avise. 1996. Avian conservation genetics. Pp. 160–189 in J. C. Avise, and J. L. Hamrick (eds.) *Conservation Genetics: Case Histories from Nature*. New York: Chapman and Hall.

Higgins, K., and M. Lynch. 2001. Metapopulation extinction caused by mutation accumulation. *Proceedings of the National Academy of Sciences of the USA* 98:2928–2933.

Husband, B. C., and D. W. Schemske. 1996. Evolution of the magnitude and timing of inbreeding depression in plants. *Evolution* 50:54–70.

James, S. H. 1970. Complex hybridity in *Isotoma petrea*. *Heredity* 25:53–77.

Kulhavy, D. L., R. G. Hooper, and R. Costa. 1995. *Red-Cockaded Woodpecker: Recovery, Ecology and Management*. Nacogdoches, TX: Center for Applied Studies, Stephen F. Austin State University.

Lande, R. 1995. Mutation and conservation. *Conservation Biology* 9:782–791.

Lynch, M., J. Conery, and R. Bürger. 1995. Mutational meltdowns in sexual populations. *Evolution* 49:1067–1080.

Margan, S. H., R. K. Nurthen, M. E. Montgomery, *et al.* 1998. Single large or several small? Population fragmentation in the captive management of endangered species. *Zoo Biology* 17:467–480.

McCarthy, M. A., S. J. Andelman, and H. P. Possingham. 2003. Reliability of relative predictions in population viability analysis. *Conservation Biology* 17:982–989.

McCullough, D. R., J. K. Fischer, and J. D. Ballou. 1996. From bottleneck to metapopulation: recovery of the tule elk in California. Pp. 375–403 in D. R. McCullough (ed.) *Metapopulations and Wildlife Conservation*. Washington, DC: Island Press.

Nei, M. 1987. *Molecular Evolutionary Genetics*. New York: Columbia University Press.

Neigel, J. E. 1996. Estimation of effective population size and migration parameters from genetic data. Pp. 329–346 in T. B. Smith, and R. K. Wayne (eds.) *Molecular Genetic Approaches in Conservation*. New York: Oxford University Press.

Newman, D., and D. Pilson. 1997. Increased probability of extinction due to decreased genetic effective population size: experimental populations of *Clarkia pulchella*. *Evolution* 51:354–362.

Newman, D., and D. A. Tallmon. 2001. Experimental evidence for beneficial fitness effects of gene flow in recently isolated populations. *Conservation Biology* 15:1054–1063.

O'Grady, J. J., B. W. Brook, D. H. Reed, *et al.* 2006. Realistic levels of inbreeding depression strongly affect extinction risk in wild populations. *Biological Conservation* in press.

Osborne, W. A., and J. A. Norman. 1991. Conservation genetics of corroboree frogs, *Pseudophryne corroboree* Moore ED: Anura: Myobatrachidae: Population subdivision and genetic divergence. *Australian Journal of Zoology* 39:285–297.

Paetkau, D., W. Calvert, I. Stirling, and C. Strobeck. 1995. Microsatellite analysis of population structure in Canadian polar bears. *Molecular Ecology* 4:347–354.

Ralls, K., J. D. Ballou, and A. Templeton. 1988. Estimates of lethal equivalents and the cost of inbreeding in mammals. *Conservation Biology* 2:185–193.

Reed, D. H. 2004. Extinction risk in fragmented habitats. *Animal Conservation* 7:181–191.

Reed, D. H., and R. Frankham. 2003. Population fitness is correlated with genetic diversity. *Conservation Biology* 17:230–237.

Reed, D. H., E. Lowe, D. A. Briscoe, and R. Frankham. 2003. Inbreeding and extinction: effects of rate of inbreeding. *Conservation Genetics* 4:405–410.

Richards, C. M. 2000. Inbreeding depression and genetic rescue in a plant metapopulation. *American Naturalist* **155**:383–394.

Rossetto, M., G. Jezierski, S. J. Hopper, and K. W. Dixon. 1999. Conservation genetics and clonality in two critically endangered eucalypts from the highly endemic south-western Australian flora. *Biological Conservation* **88**:321–331.

Saccheri, I. J., and P. M. Brakefield. 2002. Rapid spread of immigrant genomes into inbred populations. *Proceedings of the Royal Society of London B* **269**:1073–1078.

Saccheri, I., M. Kuussaari, M. Kankare, *et al.* 1998. Inbreeding and extinction in a butterfly metapopulation. *Nature* **392**:491–494.

Schwartz, M. K., L. S. Mills, K. S. McKelvey, L. F. Ruggiero, and F. W. Allendorf. 2002. DNA reveals high dispersal synchronizing the population dynamics of Canada lynx. *Nature* **415**:520–522.

Slatkin, M. 1995. A measure of population subdivision based on microsatellite allele frequencies. *Genetics* **139**:457–462.

Spielman, D., and R. Frankham. 1992. Modeling problems in conservation genetics using captive *Drosophila* populations: improvement in reproductive fitness due to immigration of one individual into small partially inbred populations. *Zoo Biology* **11**:343–351.

Stangel, P. W., M. R. Lennartz, and M. H. Smith. 1992. Genetic variation and population structure of red-cockaded woodpeckers. *Conservation Biology* **6**:283–292.

Templeton, A. R. 1998. Nested clade analyses of phylogeographic data: testing hypotheses about gene flow and population history. *Molecular Ecology* **7**:381–397.

Tero, B., J. Aspi, P. Siikamaki, A. Jakalaniemi, and J. Tuomi. 2003. Genetic structure and gene flow in a metapopulation of an endangered plant species, *Silene tatarica*. *Molecular Ecology* **12**:2073–2085.

Vilà, C., A.-K. Sundqvist, Ö. Flagstad, *et al.* 2003. Rescue of a severely bottlenecked wolf (*Canis lupus*) population by a single immigrant. *Proceedings of the Royal Society of London B* **270**:91–97.

WCMC. 1992. *Global Biodiversity: Status of the Earth's Living Resources*. London: Chapman and Hall.

White, M. J. D. 1973. *Animal Cytology and Evolution*. Cambridge, UK: Cambridge University Press.

White, M. J. D. 1978. *Modes of Speciation*. San Francisco, CA: W.H. Freeman.

Woodworth, L. M., M. E. Montgomery, D. A. Briscoe, and R. Frankham. 2002. Rapid genetic deterioration in captivity: causes and conservation implications. *Conservation Genetics* **3**:277–288.

Wright, S. 1969. *Evolution and the Genetics of Populations*, vol. 2, *The Theory of Gene Frequencies*. Chicago, IL: University of Chicago Press.

Young, A. G., A. H. D. Brown, B. G. Murray, P. H. Thrall, and C. H. Miller. 2000. Genetic erosion, restricted mating and reduced viability in fragmented populations of the endangered grassland herb *Rutidosis leptorrhynchoides*. Pp. 335–359 in A. G. Young, and G. M. Clarke (eds.) *Genetics, Demography and Viability of Fragmented Populations*. Cambridge, UK: Cambridge University Press.

Zeyl, C., M. Mizesko, J. Arjan, and G. M. de Visser. 2001. Mutational meltdown in laboratory yeast populations. *Evolution* **55**:909–917.

5

Connectivity at the land–water interface

DREW M. TALLEY, GARY R. HUXEL, AND
MARCEL HOLYOAK

INTRODUCTION

There is a growing appreciation in ecology and conservation that even those habitats and ecosystems (we will use these terms interchangeably) traditionally considered "insular" are in fact reticulately connected (Polis and Strong 1996). These connections are mediated by both physical and biological processes spanning a wide range of spatial and temporal scales (Polis and Strong 1996). This is clearly true of linkages between similar habitat types, such as fragmented systems, or patchily distributed communities (e.g., see Moilanen and Hanski Chapter 3; DiBacco *et al.* Chapter 8). However, linkages are also critically important between very distinct habitats, such as forests and grasslands, freshwater and marine habitats, or aquatic and terrestrial ecosystems. In fact, as we will discuss in this chapter, if connected habitats are dissimilar we believe there is at least as much potential for movement of materials to influence the connected systems as when habitats are similar.

Inter-habitat connectivity has been formally recognized at least since 1923 (Summerhayes and Elton 1923), but recognition of the importance of linking systems for altering their productivity goes back much further (e.g., to ancient irrigation schemes, such as those in the Nile Delta of Egypt). Recently, an integration of the landscape and ecosystem ecology concepts of connectivity, and their potential to structure ecosystems, has begun to be examined in detail (e.g., Polis and Hurd 1996). We provide an overview of these linkages, but like Ricketts *et al.* (Chapter 11) and Paquet *et al.* (Chapter 6) we do not address connectivity between similar habitats across

Connectivity Conservation eds. Kevin R. Crooks and M. Sanjayan. Published by Cambridge University Press. © Cambridge University Press 2006.

an intervening matrix (which are reviewed in the literatures on e.g., corridors, metapopulations, island biogeography, etc.). Instead, we concentrate on the mechanisms and importance of connectivity at the boundaries between two very different habitats: aquatic and terrestrial ecosystems. A huge variety of critical functional connections exist between aquatic and terrestrial habitats. These include diseases spreading to marine systems from terrestrial habitats thousands of kilometers away, large marine vertebrates altering terrestrial plant communities, and changes in watersheds altering nearshore benthic communities (Polis *et al.* 1997a). Linkages can also be demographic (e.g., source–sink dynamics, metapopulations), physical (e.g., sedimentation), trophic ("spatial subsidies" – food resources arriving from other habitats), or informational (behavioral). This large diversity creates the need for an organizing framework for land–water connections. Our main aim is to describe such a framework.

In this chapter, we demonstrate that habitat connectivity is a process of overarching importance at the land–water interface. We present several possible frameworks in which to view habitat linkages generally (and those at the land–water interface specifically), and use one of them to provide examples of how these linkages have ramifying effects that can fundamentally structure both aquatic and terrestrial ecosystems. We will then demonstrate the importance of recognizing these connections for both theoretical and conservation biology, including examples of "lost connections" due to anthropogenic changes.

DEFINING HABITAT LINKAGES

For the purposes of this chapter, we will consider two systems to be functionally linked (or "connected") when one system, including biotic and abiotic factors, significantly affects another distinct system. This definition leaves both the question of what constitutes "significant effects" and what constitutes "distinct systems" intentionally vague. We feel this flexibility is appropriate because these terms depend on the system and the spatial and temporal scale of interest. Note also that we will use the terms "linkage" and "connectivity" synonymously.

Habitat linkages in general occur across a wide range of temporal and spatial scales. For instance, for marine fauna connectivity between juvenile and adult habitats has been shown to range from a few meters to thousands of kilometers (Gillanders *et al.* 2003), and to have a broad range of potential effects on population dynamics for the species considered. Examples of connectivity at the aquatic–terrestrial boundary range from short time periods (e.g., lizards foraging in intertidal zones over tidal timescales) to decadal periodicity, such as wind-blown deposition of terrigenous (land-derived) sediments into the ocean. The effects of these linkages range from minor, likely subpopulation level effects (e.g., iguanas occasionally foraging on ghost crabs: Arndt 1999), to being dominant factors structuring ecosystems (e.g., seabird effects on some terrestrial ecosystems: Anderson and Polis 1999).

The heterogeneity among the examples that we have discussed illustrates why we need a way to classify habitat linkages. This need is augmented by the realization that our perception of linkages may also change depending on the system and species studied (e.g., see Crooks and Sanjayan Chapter 1). For instance, if the system under consideration is a salt marsh, then a land—water linkage might be the effects of marine isopods eroding creek-banks, thus changing the shape of the terrestrial landscape (Talley *et al.* 2001). If that same marsh were being viewed relative to large marine mammals that do not use this marsh habitat, there would arguably be no connectivity at all. Therefore our classification scheme needs to be sensitive to this context-dependence of linkages.

A CONCEPTUAL FRAMEWORK FOR HABITAT CONNECTIVITY

We propose a framework for evaluating habitat connectivity that comprises three different factors: units of flux, primary effects, and dynamical features.

Units of flux

Habitat linkages result from the movement of four basic units: organisms, energy, materials (nutrients, chemicals, inorganic materials, etc.), and information. Three of these potential forms of linkage involve physical entities crossing boundaries — movement of energy, material, or organisms between habitats — while the fourth involves only information crossing boundaries, such as when potential prey assess nearby habitats for predation risk.

These units of flux represent the basic currency through which connectivity occurs. These movements are not necessarily independent, and in many cases multiple fluxes will be important simultaneously (and sometimes with opposite signs). For example, inorganic material such as sediment may be moving from the land to the water at the same time emergent insects are moving in the opposite direction. Also note that the structure of the boundary can alter the fluxes between the habitats. Boundaries may be reflecting, absorbing, transforming, or neutral for the different units, potentially affecting each unit type differently (Cadenasso *et al.* 2003; Strayer *et al.* 2003).

Primary effects

Mere movement of material or information across boundaries does not de facto create connectivity between two ecosystems, as defined here. A functional connection exists only when that movement in some (significant) way alters one of the systems. These alterations can take the form of almost any physical or biological interaction, but they can be broadly characterized as belonging to one or more of five classes of primary effect, with strongly overlapping and ramifying indirect effects on communities and ecosystems:

- *Trophic.* Here the linkage between two habitats involves the transfer of food resources, and the primary effect is a change in feeding status. This connection can be formed through the movement of either consumers or resources between habitats – for example, insects emerging from streams can become prey for terrestrial web-building spiders (Collier *et al.* 2002), or terrestrial consumers can move into aquatic habitats to forage (Carlton and Hodder 2003).
- *Demographic.* Demographic or "population" linkages are those that involve a change in population structure or population dynamics resulting from connecting two habitats. These comprise some of the more commonly studied classes of connectivity, and include issues such as metapopulations, source–sink dynamics, island biogeography, nursery or breeding habitats, and wildlife corridors. At the terrestrial–aquatic interface, demographic connections frequently involve ontogenetic (developmental) habitat changes, such as movement of amphibians between water and land.
- *Environmental.* Linkages for which the primary alteration involves a change of abiotic factors in the recipient and donor habitats. For example, woody debris from terrestrial plants can alter flow and sediment properties in rivers and streams (Naiman and Decamps 1997), or terrestrial land-use changes increasing sedimentation to coastal waters (Thrush *et al.* 2004).
- *Behavioral.* Here the primary effect on the habitat involves a change in behavior of an organism based on the state of another habitat. Most examples in the literature involve changes in habitat use related to predation risk or prey availability. Examples at the land–water interface often include changes in fish behavior in response to avian

predator presence (e.g., fishes moving to suboptimal habitat to avoid terrestrial predators: Power 1984; Crowder *et al.* 1997). However, theoretically we would expect that behaviors such as optimal foraging and habitat selection (review: Rosenzweig 1991) could create many other kinds of land—water behavioral linkages.

- *Genetic.* In some cases, the primary effect of connectivity of individuals (or gametes) will not be a change in population structure, density, or presence/absence, but rather a change in gene frequency or genetic structure of the connected populations. While this is an important aspect of habitat linkages generally (Rodriguez-Lanetty and Hoegh-Guldberg 2002; Frankham Chapter. 4), we can find no examples of this form of connectivity at the land—water interface. However, on evolutionary timescales the land—water interface could have profound influences through either acting as a barrier (impeding gene flow) or by creating a corridor for increased genetic exchange between otherwise isolated populations. It is also feasible that there will be cases where correlated life-history traits cause an adaptation in one habitat (e.g., land) to represent a maladaptation in another (e.g., aquatic habitat). There are currently no examples that we are aware of that apply to land—water interfaces. In terrestrial systems there are examples where bird breeding in one "source" habitat is timed to match peak resource availability; birds then disperse to another "sink" habitat where they breed at the same time as in the source habitat, but this is not the time of peak resource availability in the sink habitat, and reduced reproductive success results (Dias 1996).

There are numerous secondary and indirect effects that may cascade from the initial "primary effect." Secondary effects may include individual, population, community, or environment-level impacts (see also Huxel *et al.* 2004). We define secondary effects as being of lower magnitude than primary effects, whereas indirect effects can be of smaller or larger magnitude than direct effects. Essentially, almost any biological or physical interaction that can occur in nature can be modified, primarily or secondarily, by connectivity. As we present examples of connectivity at the aquatic—terrestrial interface in this chapter, we will make reference to the mechanisms and effect(s) involved in each example illustrated.

Dynamical features

Habitat linkages can also be classified on the basis of their dynamical features, and it is on this framework that we will focus our examination of connectivity in this chapter. The specific features of habitat linkages can be evaluated across four main axes — three dichotomous and one continuous.

- *Directionality.* The actual transport of material, organisms, or information can be either unidirectional or bidirectional. As will be seen in the examples that follow, this is in part a scale-dependent feature — often, processes that appear unidirectional at one temporal or spatial scale may be bidirectional across other scales. There is also an observer bias that affects the perception of directionality, which will be addressed later. Known examples mainly involve fluxes of energy, organisms, or materials, but there are also fluxes of information that create connectivity. For example, as described above, prey in one habitat may alter their behavior due to perceived predation risk in an adjoining habitat (Crowder *et al.* 1997).
- *Feedback.* Linkages can also be assessed for feedback: are the connections wholly controlled by one habitat, or are there feedbacks in place between the two interacting systems, whereby the recipient system(s) can affect the linkage from the donor system(s)? The original work on ecosystem subsidy assumed no feedback ("donor control": DeAngelis 1992), whereas recent empirical work (e.g., Nakano and Murakami 2001) and theory about "meta-ecosystems" include feedback (Loreau *et al.* 2003). A meta-ecosystem is defined as a set of ecosystems connected by spatial flows of energy, materials, and organisms across ecosystem boundaries (Loreau *et al.* 2003).
- *Temporal variation.* While all linkages are dynamic when viewed over long enough time periods, it is useful to ask to what extent does the link vary through time and over what time periods does this variation occur (e.g., diel, seasonal, interannual, interdecadal cycles). For example, Nakano and Murakami (2001) discuss a seasonal flux of insects emerging from freshwater streams passing to terrestrial bird communities, whereas some scavengers show diel foraging patterns in visiting beaches to feed on washed-up prey in Baja California (Rose and Polis 1998).
- *Biotic/abiotic mediation.* Connectivity between systems can be mediated by either biotic or abiotic factors. While there are numerous examples of each form of mediation, there are systematic differences between the mediation operating when terrestrial systems are donors

BOX 5.1   Differences in mediation of linkages between freshwater, marine, and terrestrial habitats

The movement of material or organisms mediated by biotic vectors is equally frequent in either direction — organisms, whether prey or predators, will often actively move between terrestrial and aquatic systems (Polis *et al.* 2004). Abiotic mediation, however, more often moves materials, organisms, or nutrients from terrestrial to aquatic systems than in the other direction. One reason for this is that aquatic systems are generally lower in elevation that the terrestrial habitats that border them, and therefore gravity will tend to move material downslope and into the water. Yet this asymmetry of abiotic mediation differs between marine—terrestrial and freshwater—terrestrial habitats. This is due to two important factors. First, in most freshwater systems, land area is greater than inundated area; and terrestrial production outweighs aquatic production (Power 2001), which further increases the probability of the passive (abiotically forced) movement of material from terrestrial to aquatic ecosystems. In most marine—terrestrial ecotones, the area inundated is clearly much larger than the terrestrial area, with a commensurate increase in production. The second factor relates to the physical forcing involved. At the freshwater—terrestrial boundary, physical forcing of material from the aquatic to terrestrial habitats is less frequent, as the dominant force driving the movement of water is gravity. This means that at shorter timescales (e.g., tidal), the passive transport of material from the aquatic to the terrestrial system is less common than at the marine—terrestrial boundary, where physical forcing is also strongly driven by winds, tides, and waves. At longer timescales (e.g., annual or seasonal), freshwater flooding events are often of overarching importance to terrestrial ecosystems, providing nutrients and sediments to terrestrial habitats (e.g., see Likens and Bormann 1974 and references therein). In combination, these factors suggest that abiotic forcing will be less important when examining terrestrial—aquatic connectivity when the donor ecosystem is aquatic, and that at short timescales this is more true in freshwater than marine systems.

versus recipients of linkages with aquatic systems (Box 5.1). Biotic and abiotic mediation are not always mutually exclusive or independent of each other.

The interaction among these dynamical features allows for the development of a conceptual framework in which to study linkages among habitats. We propose to use the first two axes (directionality and feedback) as a framework to illustrate various connectivities at the land—water interface (see Fig. 5.1 below). In addition we will present examples of linkages across mediation types (physical and biological), and will address where the linkage fits into this framework, which unit of flux is producing the linkage, and the primary effects, mediating forces, and timescales associated with the connection.

EXAMPLES OF HABITAT LINKAGES AT THE LAND–WATER INTERFACE

Here we present examples of habitat connectivity at the aquatic–terrestrial interface, paying particular attention to how they fit into the framework described in Fig. 5.1. It should be emphasized that the dichotomization of the processes in Fig. 5.1 is a convenient theoretical construct that simplifies description of real systems. The real world is likely to lie on a continuum between unidirectionality and bidirectionality, and between full and no feedback between connected habitats. Nonetheless, setting aside issues of scale and accepting a dichotomous framework is useful for suggesting a simplified categorization of linkages that serve to illustrate some common types of land–water connection. We describe examples in this simplified framework and then in the next section draw on some common features that aid thinking about linkages in a more flexible framework.

Cell A: unidirectional movement of organisms, material, or energy, without feedback

Many of the habitat linkages at the land–water interface presented in the literature are in this category, with a vast number of them being either donor-controlled subsidies (Polis and Strong 1996), or connectivity

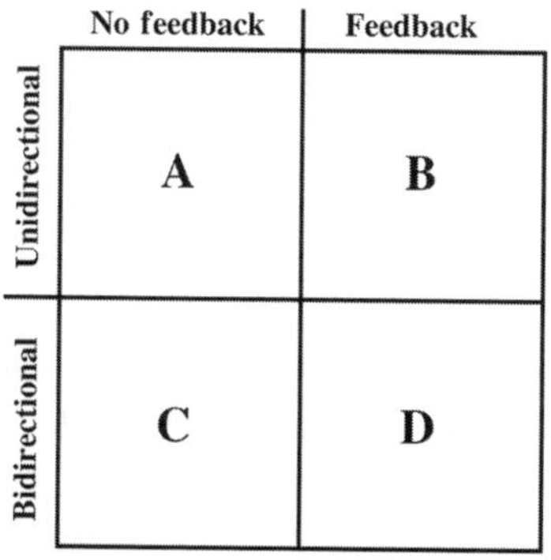

Fig. 5.1. Two axes of dynamical features: feedback and directionality. See text for description and examples.

involving changes to the physical structure of the recipient system. Note, however, that in some cases it is possible that there are actually feedbacks that have gone unrecognized. For example, decades of study were required before the feedbacks between streams and terrestrial systems were fully appreciated.

A clear example of donor-controlled subsidy comes from Gary Polis work on islands in the Gulf of California, which receive a trophic subsidy from the marine ecosystem. In this region, the terrestrial habitats possess very low productivity, while production in the adjacent marine system can be anywhere from 4 to 40 times as high (Polis and Hurd 1996). There are two main conduits through which this marine productivity subsidizes the terrestrial community – through shoredrift (material deposited onshore by tides and waves) of algae and carrion, and through seabird colonies. In this section, we will focus on shoredrift. Trophic subsidy through shoredrift has been shown to lead to densities of terrestrial arthropods nearshore that are as much as 560 times higher than for inland populations (Polis and Hurd 1996). These effects, though, are felt throughout the ecosystem, leading to increased densities of herpetofauna (Barrett et al. 2003) and mammals (Stapp et al. 1999). The primary effect of this subsidy is a trophic one, with food resources that originated from marine production moving unidirectionally to the terrestrial ecosystem. This primary effect leads to a host of ecosystem-wide secondary and tertiary effects, including changes in population persistence (Polis et al. 1997a), source—sink relationships with inland populations (Polis et al. 2004), diversity (Anderson and Wait 2001), or competitive interactions (Polis et al. 2004). This particular subsidy is donor-controlled because in this ecosystem the terrestrial habitat will generally not affect the timing or flow of material, which prevents feedback to the marine system. The unidirectionality of the effect comes about because there is little material produced on the islands to flow to the sea, and the sea is vast in size by comparison to the islands. The dominant dynamics here are at timescales that are tidal, seasonal, and interannual: the tidal timescale arises because more marine material will be washed ashore on high tides, in particular spring high tides; timescales are seasonal because storms and die-offs of marine plants and animals show seasonal periodicity (Bodkin and Jameson 1991); and timescales are interannually related to changes in marine productivity associated with El Niño events. Further, this example is generally an abiotically mediated form of connectivity, in that it is the physical forcing of tides and waves that is moving dead and dying organisms onto the terrestrial landscape.

There are myriad other examples of trophic connectivity at the land–water interface involving unidirectional flow with no feedback. One of the more commonly examined linkages is the input of terrestrial organic matter fueling lake (Gasith and Hasler 1976), stream or riverine (Grimm 1987; Naiman *et al.* 1987; Wallace *et al.* 1997), and marine (Odum 1980; Mann and Lazier 1991) production. In these examples, terrestrial organic matter influences productivity in entire aquatic eco-systems. Here again, the trophic connection is unidirectional, without clearly demonstrated feedback, but in these cases the dominant dynamics occur over timescales more closely tied to production of terrestrial bio-mass and flooding events (e.g., seasonal and interannual timescales). The mediating factors in these examples are also abiotic, being forced by the gravity-driven flow of water and materials.

A very common non-trophic example of unidirectional flow without feedback at the terrestrial–aquatic interface involves the movement of sediment from land to water. In many cases, changes in the terrestrial habitat lead to increased sedimentation, with effects ranging from altering or destroying coral communities (Cortes and Risk 1985) to filling lakes or wetlands (Luo *et al.* 1999) and changing the supply of sand for intertidal and subtidal systems hundreds of kilometers away (Inman 2005). These examples cover a broad range of timescales, from seasonal to geologic, and are generally abiotically mediated, although biotic activity (e.g., plant growth stabilizing sediments) clearly plays a role.

An example that includes marine habitats dramatically altering terrestrial ecosystems comes from studies of the effects of seagrass litter on Saharan desert geomorphology (Hemminga and Nieuwenhuize 1990). Seagrass cast ashore during high tides alters the transport of sand, ultimately leading to the formation of sand dune habitats. Here again there is an interaction between abiotic (tides, wind) and biotic (seagrass growth) mediation of the connection, which occurs over tidal and geologic timescales.

Cell B: unidirectional movement of organisms, material, or energy, with feedback

This is a form of habitat connectivity with few specific examples in the literature applying to the land–water interface. This form of connectivity *does* occur across other habitat boundaries, for example songbirds may disperse to other habitat types only when there is a vacant territory in the alternate habitat type, which causes density-dependent feedback in movement (e.g., Marra and Holmes 1997). More generally this form of movement may involve diffusion with organisms moving from areas of high abundance to areas of lower abundance as a forcing function that creates feedback. Density-dependent dispersal could also lead to this kind of linkage (as in the songbird example above). The commonness of these mechanisms leads us to expect that this will be a frequent form of terrestrial–aquatic linkage if researchers search for it.

One example at the aquatic–terrestrial interface involves large storm events washing woody debris into streambeds (see Naiman and Decamps 1997 and references therein). This in turn can alter the subsequent flow into the stream from the surrounding watershed and may cause damming. There may in fact be an additional feedback from the stream to the terrestrial environment if the damming causes additional flooding, increasing erosion and the likelihood of further deposition of materials causing further blockage. Here the mediation is biotic (vegetation altering flow) and abiotic (gravity-driven movement of water), with dominant temporal signals at seasonal (rainfall or snowmelt) and interannual (El Niño) timescales.

Cell C: bidirectional movement of organisms, material, or energy, without feedback

This interaction requires materials to be moved bidirectionally across boundaries, but in such a way that the movements are decoupled from each other. While there are relatively few examples at the land–water interface, there is a fairly well-studied form of connectivity that potentially fits this category, which we now discuss.

This connection involves the bidirectional movement of nutrients across the land–water interface. Seabirds have been shown to be effective conveyors of marine nutrients into the terrestrial ecosystem, often with dramatic consequences for the recipient communities, such as increases in production or alteration of diversity (Anderson and Polis 1999; Anderson and Wait 2001). In a reciprocal fashion, those nutrients (in the form of guano) have been shown to impact marine intertidal communities when they flow seaward (Bosman and Hockey 1988; Wootton 1991; Wainright et al. 1998). However, there is no evidence for any substantial feedback between these two pathways: guano additions to the terrestrial system have not been shown to increase the rate of bird-derived (ornithogenic) nutrient input, and increased production in the intertidal zone probably has little if any effect on seabird feeding. Note that the water-to-land component of this movement is biotically mediated, while the reciprocal movement from land to sea is generally physically mediated. The primary effects in both cases involve an enhancement of primary production in the recipient habitat.

Despite the relative paucity of examples of bidirectional flux without feedback at the land–water interface, there are numerous good examples of these uncoupled movements across habitats within terrestrial or aquatic systems. For example, larval exchange between embayments and the open ocean may be completely decoupled from each other, with oceanic larval supply being strongly driven by oceanic adult population dynamics, and movement from the embayment seaward being driven by dynamics of reproduction within the bay (DiBacco and Chadwick 2001).

Cell D: bidirectional movement of organisms, material, or energy, with feedback

This quadrant represents another major portion of the habitat connectivity at the land–water interface studied to date. One common form of connection involves organisms undergoing changes in habitat use through development (ontogenetic habitat shifts). Amphibians provide an excellent example of this connectivity.

Anurans (frogs and toads) have complex life cycles, with ontogenetic habitat shifts between aquatic and terrestrial systems. Here, the larval stage is spent in aquatic systems, while many adult anurans dwell largely in terrestrial habitats. This can lead to direct effects on population dynamics between the two systems, where changes to the population in either habitat affects densities in the other (Wilbur 1980). Further, both larval (e.g., tadpole) and adult populations can significantly impact their respective aquatic and terrestrial ecosystems (e.g., Kupferberg 1997; Anderson *et al.* 1999). Consequently, these connections can lead to ecosystem-wide effects of connectivity. In this example, the bidirectional movement of organisms drives the linkage between habitats, at temporal scales focusing around seasonal (breeding) or interannual timescales. In a case like this one, where the bidirectional flux involves different life stages of the same species, there are necessary limits to the extent to which bidirectionality can occur when viewed over long time periods. Specifically, the lack of flux of larvae to terrestrial systems would eventually lead to a loss of terrestrial adults that could return to aquatic habitats to breed.

The anuran example is representative of the broader case of organisms that alternately use the land and the water at different life-history stages, which is a common form of bidirectional connectivity with feedback at the land–water boundary. Other examples at both freshwater and marine interfaces include sea turtles, seabirds, some marine mammals, parasites, and countless insect species (insect ontogenetic shifts are particularly prevalent at the freshwater–terrestrial interface: Cheng 1976). These connections can also be mediated by physical transformation of habitat, such as anurans or insects breeding in pools created by other vertebrates (Gerlanc and Kaufman 2003).

There are also examples of bidirectional movement with feedback that do not require ontogenetic shifts in habitat use. Probably one of the best studied is the connectivity formed by the activities of beavers (*Castor canadensis*). Beavers move back and forth between the terrestrial and aquatic habitats. Through their feeding and dam-building they have been shown to alter hydrology, water biochemistry, and diversity and abundance patterns of plants, birds, fishes, and insects (Wright *et al.* 2004 and references therein). These alterations cause a suite of changes to habitat suitability for beaver populations, often leading to a cyclic pattern of use and disuse of the area by beavers, as food resources are depleted, beavers emigrate or die, and then recolonize as vegetation regrows. This feedback between the terrestrial and aquatic habitats cycles over timescales on the order of several years (Wright *et al.* 2004).

From this framework, we begin to see patterns that indicate that the land–water interface has a number of constraints that are not seen elsewhere, with the exception of saltwater–freshwater connections. The land–water interface represents a strong physiological barrier that limits biotic movement between the two habitats. Over evolutionary timescales, overcoming these barriers from land to water or vice versa has resulted in major evolutionary revolutions in some lineages (such as amphibians, or in the invasion of land by arthropods in the Silurian, 425 million years ago). For most species, however, the land–water interface still represents a strong barrier influencing population size and gene flow. These limitations on the movement of organisms reduce the biotic influence of these linkages and increase the importance of abiotic physical processes (Witman *et al.* 2004). However, many strong biotic linkages do exist and are of great significance in particular systems. In the "Factors promoting and limiting habitat linkages" section below we discuss other factors that enhance or reduce the strength of biotically driven land–water linkages.

SOME COMMON FEATURES OF THE DIFFERENT TYPES OF LINKAGES

Scale dependence of connectivity

It is important to note that many linkages will switch between compartments of the matrix depending upon the temporal or spatial scale over which they are observed. A good example comes from the work of Nakano and Murakami (2001), who examined the food web in a deciduous forest and stream ecotone. They quantitatively and painstakingly measured the biomass of invertebrate prey and vertebrate predators in both terrestrial and stream habitats, cross-habitat prey fluxes, and the percentage contribution of aquatic and terrestrial prey items to terrestrial and stream consumers. Their results showed that reciprocal, cross-habitat prey flux alternately subsidized both forest birds and stream fishes (Nakano and Murakami 2001) (Fig. 5.2). However, short-term within-season observations (e.g., Nakano *et al.* 1999) revealed largely unidirectional flow with no feedback: during the summer, the net flow was from terrestrial to aquatic systems, while during spring the flow was in the

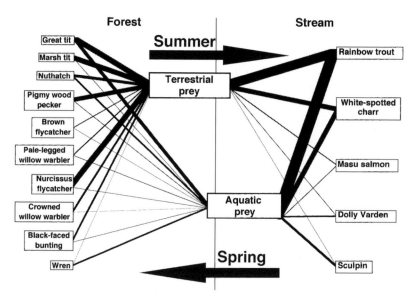

Fig. 5.2. Food web linkage across a forest–stream interface representing predator subsidies by allochthonous, invertebrate prey flux. Relative contributions of terrestrial and aquatic prey to the annual total resource budget of each species are represented by line thickness. Note that the dominant direction of subsidy changes with season. (Adapted from Nakano and Murakami 2001.)

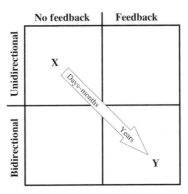

Fig. 5.3. Scale dependence of habitat connectivity. The classification of connectivity between two habitats can depend on the scale over which the dynamics are viewed. Here, if the trophic connection between a riparian and adjacent aquatic habitat are viewed within season (X: Nakano *et al.* 1999), the relationship appears to be one of unidirectional connectivity without feedback. When viewed over longer timescales (Y: Nakano and Murakami 2001) the connection is seen as bidirectional with feedback.

opposite direction. It was only by looking across longer timescales (years) that Nakano and Murakami 2001 observed a bidirectional flow with potential feedback between the two habitats (Fig 5.3).

Scale dependence of linkages is not limited to temporal shifts in perspective. For example, a study of wading birds in wetlands in the delta of the Colorado River found that some nesting birds received no trophic subsidy from adjacent waters on the scale of tens to hundreds of meters, but birds were instead feeding in waters kilometers away (D. Talley, E. Mellink, G. Huxel, S. Herzka, and P. Dayton, unpublished data). Thus, at small spatial scales, connectivity was absent, while at larger scales, it was intense. Such scale dependence is a common feature of habitat connectivity, and requires that investigators pay particular attention to the spatial and temporal scales over which they are assessing connectivity.

Perspective dependence of connectivity

The effect of temporal scale on perceived connectivity patterns brings up a larger issue of the importance of perspective dependence of connectivity studies. The appropriate categorization of effects within our theoretical framework may vary depending on which species, units of flux, and habitats are considered. This applies to all of the aspects of connectivity presented here, including dynamical features, primary effects, or even the existence of connectivity between systems. It is important to ask what are the appropriate time and spatial scales for the process of interest and how do these differ from the scales of observation of that process. For example, one of the best-documented forms of connectivity at the land—water interface involves the movement of vertebrate predators and scavengers into the intertidal zone to feed, with hundreds of cases documented across a broad range of taxa and habitats (Willson *et al.* 1998; Carlton and Hodder 2003). If viewed from the perspective of a littoral prey population, terrestrial vertebrates feeding in the intertidal represent a linkage from terrestrial to aquatic systems, with the primary effect being mortality or changes in abundance for intertidal prey items (e.g., shorebird predation alters intertidal invertebrate communities: Thrush *et al.* 1994). This could reasonably be classified as belonging in Cell A in our scheme for categorizing linkages (Fig. 5.1), with the directionality being terrestrial to aquatic. This perspective results from considering the predator foraging process over short timescales and not considering the longer-term or larger spatial scale dynamics of the invertebrate prey. However, if this same example is viewed from the perspective of a terrestrial predator taking advantage of a periodically available prey resource, the connectivity created by terrestrial vertebrates foraging in the littoral zone represents a trophic subsidy from the aquatic to the terrestrial ecosystem. This creates the potential for feedback, presuming that the predators alter prey availability (Cell D, Fig. 5.1). Again, the literature is rich with examples, including coyotes (*Canis latrans*), bears (*Ursus* spp.), and lizards receiving trophic subsidy from aquatic organisms, leading to population-level responses of the terrestrial consumers (Rose and Polis 1998; Gende *et al.* 2002; Sabo and Power 2002). This second perspective results from focusing on the population dynamics and controls of abundance of the invertebrate prey over longer timescales than those considered in the first perspective. While a holistic perspective would recognize both aspects of these linkages, it is often necessary to restrict one's attention to a particular facet of a linkage. Even in those restricted cases it is nonetheless important to identify the ecological framework in which the connection is embedded, as it profoundly influences the interpretation of the strength, direction, and other parameters of connectivity.

Ramifying and multiple effects

The examples presented in this chapter show that habitat connectivity can yield not only multiple primary effects, but can have impacts that spread reticulately throughout ecosystems. For instance, in the case of trophic subsidies to islands in the Gulf of California, the trophic subsidy provided by the input of guano initially leads to greater terrestrial primary productivity on islands which are heavily used by birds (Anderson and Polis 1999). These initial changes in productivity then have cascading effects that can alter structure, abundance, or even population persistence at all levels of the ecosystem (Stapp *et al.* 1999; Sanchez-Piñero and Polis 2000).

Various studies of the influence of salmon (*Oncorhynchus* spp.) on terrestrial environments in Alaska have found wide-ranging impacts on terrestrial organisms including brown bears (*Ursus arctos*), wolves (*Canis lupus*), riparian plants, bald eagles (*Haliaeetus leucocephalus*), and songbirds (Gende *et al.* 2002; Darimont *et al.* 2003; Willson *et al.* 2004). Many of the impacts are indirect. For example, brown bears feed on salmon carcasses, then defecate nutrients in the upland riparian habitats. This in turn increases riparian vegetation growth providing shelter, nesting sites, and food resources for songbirds (Gende *et al.* 2002; Willson *et al.* 2004). Similar pathways can be followed from bald eagles to riparian trees to riparian insects to songbirds.

Although the focus has been on direct linkages between land and water habitats, one must recognize that these habitats and habitat boundaries often contain large, reticulate, and highly connected food webs (Polis and Strong 1996; Polis *et al.* 2004 and chapters therein). These linkages can have profound influences on the dynamics of the two habitats that extend well beyond the initial linkage pathways (Polis *et al.* 1997a; Huxel and McCann 1998; McCann *et al.* 1998; Huxel *et al.* 2002, 2004). Below we focus on the initial mechanisms and pathways of habitat linkages.

FACTORS PROMOTING AND LIMITING HABITAT LINKAGES

There is a suite of factors that alter the likelihood of connectivity between any two habitats. These include characteristics of the organisms that cross a boundary, of the environment, and of the boundary itself. We consider each of these in turn.

A number of biological parameters may increase the likelihood of connectivity between terrestrial and aquatic habitats. One critical trait involves ontogenetic habitat shifts that allow or require an organism to use aquatic and terrestrial habitats at different life stages. This is a common feature of many insects, birds, mammals, and reptiles, and is a defining characteristic of amphibians (Wilbur 1980). Further, some organisms, such as the many terrestrial vertebrates listed in Carlton and Hodder (2003), have behavioral and/or physiological plasticity and adaptations that allow foraging in either terrestrial or aquatic habitats. In some cases, such as the Galapagos marine iguana (*Amblyrhynchus cristatus*), these cross-boundary foraging excursions are obligate (Wikelski and Thom 2000). Where species exhibiting these habitat shifts are present, additional factors such as home range size, dispersal ability, or trophic position should affect how far the connection penetrates into the recipient habitat. On longer timescales, life-history patterns are expected to be intimately related to linking biological traits and our next category, traits of the environment (e.g., Roff 2002).

Traits of the habitat or environment can also affect the probability of habitat connectivity. It has been proposed that, at least as a null model, strong asymmetries in productivity would be expected through physical processes (such as the Second Law of Thermodynamics) alone to increase cross-boundary connectivity (Laurance *et al.* 2001). Further, large temporal changes in abiotic conditions and physical forcing (waves, currents, winds) should also increase connectivity. This is demonstrated by sea foam and dust transporting nutrients and bacteria between the land and the sea (Blanchard and Syzdek 1970; Griffin *et al.* 2002), or sedimentation under flooding regimes (e.g., Naiman and Decamps 1997 and references therein). Unpredictability of habitats is also expected to lead to selection for dispersal under certain sets of conditions (Southwood 1962; McPeek and Kalisz 1998).

Note that many of these factors, both biological and physical, increase the likelihood of connectivity by creating an inequality between habitats' ability to provide all of the fundamental requirements for an organism — inequality can be a driving force in connectivity. Strong asymmetry between habitats might be expected to decrease biological connectivity (in contrast to the physical mechanisms described above) by decreasing the likelihood of cross-habitat movement of organisms (Laurance *et al.* 2001). However, asymmetries which create trade-offs in habitat function for various life-history or biological needs of an organism may actually increase the likelihood of movement. As we have seen in the examples of allochthonous inputs (resources arriving from outside the system) and ontogenetic shifts in habitat use presented here (Polis *et al.* 1997b; Wikelski and Thom 2000; Nakano and Murakami 2001), inequality between habitats can be a driving force promoting connectivity depending on the interaction between biological and physical traits.

Features of the boundary between habitats will also influence the strength of linkages, therefore necessitating examination of the role of boundaries in understanding the importance of habitat linkages

(Cadenasso *et al.* 2003; Strayer *et al.* 2003). Most studies of habitat linkages treat boundaries as passive entities. This may be appropriate in many situations such as in linkages between nearshore and pelagic habitats, but for land—water linkages the boundary may play an important active role in processing of resources and acting as a zone of absorption, reflection, or transformation of the resources (Strayer *et al.* 2003). Thus, models of boundaries as diffusion zones across which materials are passed unaltered may not be appropriate. For instance, on the islands of Baja California, marine algal wrack (shore-deposited vegetation) may be actively buried in the beach material (the boundary) by heavy wave action and subsequently utilized and converted by the detrital-based communities in the intertidal zones before entering the terrestrial environment. This active incorporation through wave action on beaches contrasts with passive reflection by cliffs, or washing up (often temporarily) on beaches.

Generally, "harder" or less permeable boundaries will tend to lower connectivity between systems, because low permeability will impede the flow of organisms, materials, and information across the boundaries. For example, the transport of marine nutrients to the terrestrial ecosystem by Galapagos sea lions (*Zalophus wollebaecki*) is restricted to shorelines with elevations low enough that sea lions are able to access the land — thus the nature of the boundary between the land and the sea determined the degree of connectivity between the marine and terrestrial system (Farina *et al.* 2003). Boundary permeability (and thus connectivity) is species- and habitat-specific, in that what forms a hard boundary for one organism or connection may not represent a boundary at all for another. In a study of the permeability to shoredrift at the land—sea boundary in the Gulf of California, we examined stable isotopes and terrestrial faunal community data from both hard boundaries (cliffs) and soft boundaries (beaches). Sloping beaches were much more permeable to allochthonous inputs in the form of shoredrift (D. M. Talley *et al.*, unpublished data) (Fig. 5.4). Conversely, marine inputs that are mediated by cliff-roosting seabirds would be expected to show the opposite trend, and seabirds that nest inland would have some level of connectivity that is largely independent of the boundary type at the land—water interface (Fig. 5.4). This shows that there is a complex interaction between the form of connectivity (or mediation) and the boundary, which prevents simple rules from providing more than general guidelines for predicting permeability. It is therefore necessary to study permeability for particular organisms and habitat couplings.

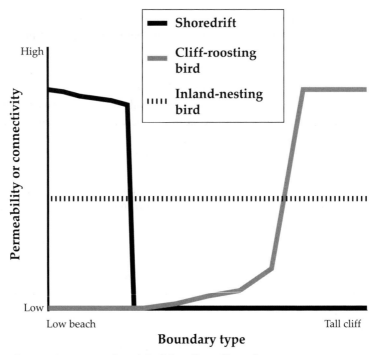

Fig. 5.4. A conceptual model of the effect of boundary structure on connectivity. The *y*-axis represents either permeability or connectivity, assuming a positive correlation between those factors. The *x*-axis is a measure of boundary structure, in this example steepness of the shore from low (beach) to steep and tall (cliffs). Shoredrift input will be high (high connectivity), and decrease slightly with increasing height until attaining some threshold, at which shoredrift will no longer penetrate across the boundary. Cliff-roosting birds, on the other hand, would only rarely use low shorelines, but at a threshold height would preferentially choose those habitats, increasing connectivity. For seabirds that nest inland, there would be some level of connectivity that is independent of the boundary type.

CONSIDERATIONS FOR ECOLOGY

Given the ubiquity of habitat linkages and their potential to alter or structure communities, populations, and environments, it is critical that we incorporate aquatic—terrestrial linkages into ecological thinking. A better understanding of habitat connectivity at the land—water interface will improve both empirical and theoretical ecology, and will inform conservation biology in theory and practice.

Habitat linkages are of importance to both terrestrial and aquatic ecological researchers, since they are often the driving force structuring ecosystems. This is true for linkages that are trophic, demographic, environmental, and behavioral. In many cases, attempting to understand fully the dynamics of a focal habitat will be doomed to failure if external forcing through habitat connectivity is ignored. In the previous example for islands in the Gulf of California, the effects of the extensive marine input lead to extraordinarily high densities of spiders on land, and an apparent trophic cascade, which could not be explained by evaluating in situ production alone. This has been appreciated for some time for terrestrial effects on aquatic systems, with studies outlining phenomena such as watershed connectivity (Likens and Bormann 1974) and the effects of terrestrial food resources in streams (Minshall 1967; Cummins et al. 1973; Wallace et al. 1997). But only recently has the reciprocal case of aquatic effects on terrestrial systems begun to be vigorously pursued, with studies highlighting the role of emerging insects as prey for forest consumers (Jackson and Fisher 1986; Gray 1993; Sanzone et al. 2003) and the role of salmon carcasses in terrestrial ecosystems (Gende et al. 2002).

The importance of linkages between habitats has been explored in great detail theoretically for movement of single species (metapopulation theory: e.g., Levins 1969; reviewed by Hanski and Gilpin 1997) and nutrients (reviewed by DeAngelis 1992). More recently, the influence of habitat linkages on communities has been explored (Polis et al. 1997a; Huxel and McCann 1998; McCann et al. 1998; Huxel et al. 2002). These studies have profoundly affected the way ecologists think about connectivity between habitats. For example, Huxel and McCann (1998) explored the influence of allochthonous inputs across habitat boundaries on food-web stability. They found that low to moderate amounts of allochthonous inputs relative to autochthonous productivity (that origi-nating in the focal habitat) could stabilize various kinds of food-web dynamics. Huxel et al. (2002) also examined whether food-web structure influenced the degree of impact of allochthonous inputs on the stability of food webs. Specialization at the top trophic level (such as with scavengers) on either allochthonous or autochthonous resources tended to limit the indirect effects of allochthonous inputs. In contrast, genera-lists that feed on allochthonous as well as authochthonous prey can exhibit increased densities due to the allochthonous resources, resulting in increased predation pressure on autochthonous prey and possible ramifying effects.

Both of these studies (Huxel and McCann 1998; Huxel *et al.* 2002) also found that differences among trophic levels in linkage among habitats significantly influenced food-web stability. If allochthonous resources are utilized by the second trophic level in a three-trophic-level system, then generalists (herbivores for example) will impact the first trophic level and provide increased resources for the third trophic level. Specialists at the second trophic level (perhaps detritivores) will not directly impact the first trophic level, but will provide increased resources for the third trophic level (Huxel *et al.* 2002). Therefore understanding how allochthonous resources are being used by the recipient community, and in particular what trophic levels are involved in the linkage among habitats, is of considerable importance in discerning the response of the recipient community to these inputs.

CONSIDERATIONS FOR CONSERVATION

Integrating the study of habitat connectivity into our research will improve our understanding of ecological systems and theories, and this will provide various benefits to conservation, in both theory and practice. Two aspects of connectivity and conservation deserve particular attention: situations where habitat linkages at the land–water interface have been altered by humans, and ways in which information about linkages can be used to improve the practice of conservation.

Anthropogenic alterations

Despite a substantial number of studies of terrestrial impacts on aquatic systems, the systematic study of habitat connectivity at the land–water interface is a relatively young field. Thus, questions of alterations to historical connectivity between these two systems remain largely unanswered, at least explicitly. There are nonetheless some clear examples of anthropogenic alteration of connectivity at the land–water interface, such as coastal armoring (seawalls and rip-rap) affecting sand flow to beaches (Runyan and Griggs 2003), or the effects of levee construction on connectivity between riverine and riparian or floodplain habitats (e.g., see Wiens 2002 and references therein).

Habitat linkages that have likely been lost or diminished due to anthropogenic influence are numerous, and include alterations to physical structure as well as human impacts on the biological elements that mediate connectivity. These changes can potentially reduce natural connectivity ("hypoconnectivity"). There are a number of instances where anthropogenic influence has led to hypoconnectivity at the land–water interface. Salmon, for example, represent a significant source of food for a number of terrestrial vertebrates, and significantly affect terrestrial communities and production, both directly and indirectly (see p. 113). However, salmon populations have decreased dramatically in recent decades due both to physical alterations of the landscape, such as damming of rivers and loss of habitat, and biological alterations through harvesting (e.g., Nehlsen *et al.* 1991; Neville *et al.* Chapter 13). This decline in trophic subsidy is affecting terrestrial ecosystems, through diminishing connectivity at the land–water interface (e.g., see Gende *et al.* 2002 and references therein). Similarly, the loss of boundary areas (marshes in particular) between terrestrial and aquatic environments limits the linkages between the land and the sea. Over 90% of Pacific saltwater and Great Lake freshwater marshes have been lost, largely due to direct human habitat alteration (Schoenherr 1992; Ohio Lake Erie Commission 2000), leading to both a loss of water quality and limitation of aquatic resources available to terrestrial consumers.

Channelization of river basins and conversion of lakes to reservoirs has also greatly influenced the connectivity between land and water. This occurs in three major ways (Ward 1998). First, canals and reservoirs tend to be deeper and have steeper banks than their natural counterparts, limiting productivity (Malanson 1993). This results in fewer resources for terrestrial consumers (e.g., raccoons, birds, foxes, bears: Daniels 1960). Second, since the canals tend to be deeper, they also do not warm as much as shallow streams, thus reducing evaporation (Karr and Schlosser 1978). Third, channelization reduces the amount of sediment deposited on the floodplain of the river system. The loss of terrestrial habitat in the Mississippi river basin is a prime example (Gagliano *et al.* 1981). The Kissimmee River in Florida also is a good example in that historically it had a floodplain 1.5–3.0 km wide and covered an area of 180 km^2 (Dahm *et al.* 1995; Toth 1995). Channelization for flood control transformed the natural shallow stream course of 166 km into a straight canal 90 km in length, 9 m in depth, and 100 m in width. The floodplain was deprived of sediments, seasonal patterns of stream flow were reversed, 12 000–14 000 ha of wetlands were lost, and biological communities were disrupted (Koebel 1995).

Further, connectivity between coastal consumers (utilizing aquatic resources) and inland habitats has been dissected in many regions by roads and highways (Forman and Alexander 1998). This can potentially restrict access to these trophic resources for wide-ranging species such as coyotes or small mammals (Carlton and Hodder 2003). This lack of connectivity could have wide-ranging effects on terrestrial communities both through lowering population densities of consumers that are subidized by aquatic resources and by restricting the flow of aquatic nutrients to inland habitats (e.g., Polis *et al.* 1997a; Rose and Polis 1998).

It should be noted here that connectivity is not necessarily a virtue from a conservation standpoint. Anthropogenic changes have often *increased* connectivity between systems ("hyperconnectivity": Crooks and Suarez Chapter 18), with deleterious effects. For example, land-use changes in many regions have led to increases in sedimentation above natural levels, destroying or converting intertidal and subtidal marine habitats (Norkko *et al.* 2002; Hewitt *et al.* 2003). Commercial fishing also creates hyperconnectivity between the land and the sea, moving over 75 million tonnes of marine biomass onto land worldwide, 27 million tonnes of discarded non-target animals (bycatch), as well as 126 500 tonnes (dry weight) of algae (Watson and Pauly 2001; McHugh 2003), with ramifications for the entire oceanic ecosystem (Dayton *et al.* 1995, 2002). Other examples of hyperconnectivity include nutrients moving from land to water creating problems with eutrophication of lakes, rivers, and marine systems (Carpenter *et al.* 1999); the flow of contaminants and diseases between terrestrial and aquatic systems (Harvell *et al.* 1999; Neal *et al.* 2003); and salt water intrusion leading to the loss of trees (Cyrus *et al.* 1997).

Conservation implications

Recognizing and understanding habitat linkages at the land–water interface is necessary for effective conservation. As most examples in this chapter illustrate, there are frequent, strong interactions between the aquatic and terrestrial ecosystems, with one habitat's influence often dominating the dynamics of the other over vast temporal and spatial scales. Besides the conservation benefits inherent in the increased understanding of nature, there are some very specific areas in which connectivity at the aquatic–terrestrial interface will benefit conservation scientists.

Siting decisions for areas of protection for threatened species or habitats will be greatly improved by integrating aquatic–terrestrial connectivity into the decision-making process. For example, streams have traditionally been considered to be recipients, rather than sources, of resources in terrestrial systems (Power 2001). Yet a conservation biologist concerned about protecting temperate forest herpetofauna who ignored riverine inputs might miss a foraging resource that can increase growth rates of their target organism by as much as 700% (Sabo and Power 2002). Clearly habitat location relative to aquatic resources can be critically important. Similarly, our research in Baja California on the effects of boundary type on connectivity suggests that the location of a protected area relative to the coast will profoundly affect the type and degree of trophic subsidy from the marine ecosystem.

A better understanding of functional linkages would also assist in questions about buffer zones and spatial scale of protected areas. Studies of the effects of marine inputs on terrestrial ecosystems (e.g., Polis and Hurd 1995; Stapp et al. 1999) clearly demonstrate that it can be important to account for marine input to understand the dynamics of terrestrial systems. However, there is only inferential information regarding the distance offshore from which these resources originate. Fundamental questions are still unanswered about how far offshore one must protect a habitat to ensure adequate connectivity (in this example, flow of allochthonous input). It is also not clear how far inshore linkages penetrate to impact terrestrial communities. An understanding of the spatial properties of these linkages will provide conservation biologists with better tools to protect threatened habitats and organisms.

CONCLUSION

Even such seemingly distinct systems as land and water are deeply interconnected across a wide range of spatial and temporal scales. As ecologists and conservation biologists, we would benefit from broadening our traditional scope of study to include processes and habitats beyond our focal system, and to explicitly incorporate spatial interconnections at every opportunity.

This is not without its challenges. Terrestrial and aquatic ecosystems are often influenced by processes operating at vastly different spatial and temporal scales, and therefore interactions between the two should be expected to be particularly complex and non-linear (Steele et al. 1993; Carr et al. 2003; Talley et al. 2003). Further, there are differences in evolutionary history, dominant taxa, life-history strategies, and numerous other factors that add complexity to integrating connectivity between these two types of systems into our science. It is clear from the examples presented that the spatial and temporal scales of study can influence our perception, terminology, and appreciation for the dynamical consequences of interhabitat connections in profound ways. There are also more prosaic difficulties — there are very real differences between many aquatic and terrestrial ecologists. They go to different schools, receive funding from different agencies, and attend different meetings, all of which will require greater interdisciplinary effort to integrate connectivity at the land—water interface into our science (Talley *et al.* 2003). Overcoming such difficulties requires collaboration to allow us to include a broad span of study methods and scales. We also need to embrace an integrative conceptual framework that promotes thinking both theoretically and empirically about the full scope of inter-habitat connections.

Despite these challenges, there are some spectacular examples of success studying connectivity at the land—water interface. The work of Gary Polis, Mary Power, Shigeru Nakano, and Mary Willson (e.g., Nakano and Murakami 2001; Power 2001; Polis *et al.* 2004; Willson *et al.* 2004; and references therein) has pushed this field of inquiry forward with elegant and insightful experiments and analyses. It is our opinion that the study of these connections at the terrestrial—aquatic interface complements those better-studied connections relating to source—sink, metapopulation, and corridors being examined in other chapters in this volume. For example, source—sink dynamics emphasize only demographic linkages and assume no feedback between heterogeneous source and sink habitats. By contrast metapopulation studies emphasize connections among similar habitats and allow these to be bidirectional. Corridors may also be bidirectional and can either serve as sources of colonization or may enlarge habitats on much shorter timescales. These issues of directionality, feedback, and temporal variation/timescale are all topics that cover a small part of the framework considered here, but do so in more elaborate detail and often more quantitatively than our descriptive review. The incorporation of abiotic fluxes into concepts like source and sink dynamics will also help to achieve a more integrated and complete ecological framework for considering the importance of connectivity.

The habitats that comprise the land—water interface, such as coastal zones, riparian areas, lakes, and floodplains, are among the most productive and biologically diverse areas on earth (e.g., see Hansen and di Castri 1992; Mitsch and Gosselink 2000; and references therein). The habitats are also a nexus for human activity, vital for transportation, production of energy, water storage, and food resources. Over 50% of all humans live within 60 km of the coast, and that number is expected to grow to 75% by 2020 (DeMaster *et al.* 2001). That these critical biological and human resources overlap so strongly means that human alteration to natural connectivity at the land–water interface will become increasingly important in the years ahead. It behooves us both as ecologists and as conservationists to find further creative ways to understand these phenomena.

ACKNOWLEDGEMENTS

The authors would like to thank M. Sanjayan, K. Crooks, and A.-L. Harrison for their hard work in organizing this volume, as well as the Society for Conservation Biology and The Nature Conservancy for their financial and organizational assistance. We also thank the National Science Foundation for support (D.T. and G.H. were supported by NSF DEB-0079426, G.H. by NSF CHE-0221834 - BE/CBC, and M.H. was supported by NSF DEB-0213026 to M.H. and Alan Hastings). D.T. was also supported in part by the National Sea Grant College Program of the US Department of Commerce's National Oceanic and Atmospheric Administration under NOAA Grant #NA04OAR4170038, project #R/CZ-190A, through the California Sea Grant College Program; and in part by the California State Resources Agency. The views expressed herein do not necessarily reflect the views of any of those organizations. Much of the work that informed and influenced this chapter, as well as inspired us, was that of the late G. Polis, to whom we are deeply indebted. The efforts and support of M. Cortes, the Diaz and Ocaña families, I. Fuentes, C. Godinez, A. Resendiz, and A. Zavala were invaluable to this work. S. Fisler and non-profit Aquatic Adventures provided critical insights and assistance. We would also like to thank the San Francisco Bay National Estuarine Research Reserve, the Romberg Tiburon Center for Environmental Studies, San Francisco State University, the University of South Florida, and the University of California, Davis. This manuscript was greatly improved thanks to the thoughtful comments of C. Baxter, K. Crooks, T. Talley, and an anonymous reviewer.

REFERENCES

Anderson, A. M., D. A. Haukos, and J. T. Anderson. 1999. Diet composition of three anurans from the playa wetlands of northwest Texas. *Copeia* 1999:515–520.

Anderson, W. B., and G. A. Polis. 1999. Nutrient fluxes from water to land: seabirds affect plant nutrient status on Gulf of California islands. *Oecologia* 118:324–332.

Anderson, W. B., and D. A. Wait. 2001. Subsidized island biogeography hypothesis: another new twist on an old theory. *Ecology Letters* 4:289–291.

Arndt, R. G. 1999. Predation by the black iguana (*Ctenosaura similis*) on the painted ghost crab (*Ocypode gaudichaudii*) in Costa Rica. *Florida Scientist* 62:111–114.

Barrett, K., D. A. Wait, and W. B. Anderson. 2003. Small island biogeography in the Gulf of California: lizards, the subsidized island biogeography hypothesis, and the small island effect. *Journal of Biogeography* 30:1575–1581.

Blanchard, D. C., and L. Syzdek. 1970. Mechanism for the water-to-air transfer and concentration of bacteria. *Science* 170:626–628.

Bodkin, J. L., and R. J. Jameson. 1991. Patterns of seabird and mammal carcass deposition along the central California coast, 1980–1986. *Canadian Journal of Zoology* 69:1149–1155.

Bosman, A. L., and P. A. R. Hockey. 1988. The influence of seabird guano on the biological structure of rocky intertidal communities on islands off the west coast of southern Africa. *South African Journal of Marine Science* 7:61–68.

Cadenasso, M. L., S. T. A. Pickett, K. C. Weathers, and C. G. Jones. 2003. A framework for a theory of ecological boundaries. *BioScience* 53:750–758.

Carlton, J. T., and J. Hodder. 2003. Maritime mammals: terrestrial mammals as consumers in marine intertidal communities. *Marine Ecology—Progress Series* 256:271–286.

Carpenter, S. R., D. Ludwig, and W. A. Brock. 1999. Management of eutrophication for lakes subject to potentially irreversible change. *Ecological Applications* 9:751–771.

Carr, M. H., J. E. Neigel, J. A. Estes, *et al.* 2003. Comparing marine and terrestrial ecosystems: implications for the design of coastal marine reserves. *Ecological Applications* 13:S90–S107.

Cheng, L. (ed.) 1976. *Marine Insects*. Amsterdam, Netherlands: North Holland.

Collier, K. J., S. Bury, and M. Gibbs. 2002. A stable isotope study of linkages between stream and terrestrial food webs through spider predation. *Freshwater Biology* 47:1651–1659.

Cortes, J., and M. J. Risk. 1985. A reef under siltation stress: Cahuita, Costa Rica. *Bulletin of Marine Science* 36:339–356.

Crowder, L. B., D. D. Squires, and J. A. Rice. 1997. Nonadditive effects of terrestrial and aquatic predators on juvenile estuarine fish. *Ecology* 78:1796–1804.

Cummins, K. W., R. C. Petersen, F. O. Howard, J. C. Wuycheck, and V. I. Holt. 1973. The utilization of leaf litter by stream detritivores. *Ecology* 54:336–345.

Cyrus, D. P., T. J. Martin, and P. E. Reavell. 1997. Salt-water intrusion from the Mzingazi River and its effects on adjacent swamp forest at Richards Bay, Zululand, South Africa. *Water SA* 23:101–108.

Dahm, C. N., K. W. Cummins, H. M. Valett, and R. L. Coleman. 1995. An ecosystem view of the restoration of the Kissimmee River. *Restoration Ecology* 3:225–238.

Daniels, R. B. 1960. Entrenchment of the Willow drainage ditch. *American Journal of Science* 258:167–176.

Darimont, C. T., T. E. Reimchen, and P. C. Paquet. 2003. Foraging behaviour by gray wolves on salmon streams in coastal British Columbia. *Canadian Journal of Zoology—Revue Canadienne de Zoologie* 81:349–353.

Dayton, P. K., S. F. Thrush, M. T. Agardy, and R. J. Hofman. 1995. Environmental effects of marine fishing. *Aquatic Conservation: Marine and Freshwater Ecosystems* 5:205–232.

Dayton, P. K., S. Thrush, and F. C. Coleman. 2002. *Ecological Effects of Fishing in Marine Ecosystems of the United States*. Arlington, VA: Pew Oceans Commission.

DeAngelis, D. L. 1992. *Dynamics of Nutrient Cycling and Food Webs*. London: Chapman and Hall.

DeMaster, D. P., C. W. Fowler, S. L. Perry, and M. E. Richlen. 2001. Predation and competition: the impact of fisheries on marine-mammal populations over the next one hundred years. *Journal of Mammalogy* 82:641–651.

Dias, P. C. 1996. Sources and sinks in population biology. *Trends in Ecology and Evolution* 11:326–330.

DiBacco, C., and D. B. Chadwick. 2001. Assessing the dispersal and exchange of brachyuran larvae between regions of San Diego Bay, California and nearshore coastal habitats using elemental fingerprinting. *Journal of Marine Research* 59:53–78.

Farina, J. M., S. Salazar, K. P. Wallem, J. D. Witman, and J. C. Ellis. 2003. Nutrient exchanges between marine and terrestrial ecosystems: the case of the Galapagos sea lion *Zalophus wollebaecki*. *Journal of Animal Ecology* 72:873–887.

Forman, R. T. T., and L. E. Alexander. 1998. Roads and their major ecological effects. *Annual Review of Ecology and Systematics* 29:207–231.

Gagliano, S. M., K. J. Meyerarendt, and K. M. Wicker. 1981. Land loss in Mississippi River deltaic plain. *Transaction of the Gulf Coast Association of the Geological Societies* 31:295–300.

Gasith, A., and A. D. Hasler. 1976. Airborne litterfall as a source of organic matter in lakes. *Limnology and Oceanography* 21:253–258.

Gende, S. M., E. D. Edwards, M. F. Willson, and M. S. Wipfli. 2002. Pacific salmon in aquatic and terrestrial ecosystems. *BioScience* 52:917–928.

Gerlanc, N. M., and G. A. Kaufman. 2003. Use of bison wallows by anurans on Konza Prairie. *American Midland Naturalist* 150:158–168.

Gillanders, B. M., K. W. Able, J. A. Brown, D. B. Eggleston, and P. F. Sheridan. 2003. Evidence of connectivity between juvenile and adult habitats for mobile marine fauna: an important component of nurseries. *Marine Ecology—Progress Series* 247:281–295.

Gray, L. J. 1993. Response of insectivorous birds to emerging aquatic insects in riparian habitats of a tallgrass prairie stream. *American Midland Naturalist* 129:288–300.

Griffin, D. W., C. A. Kellogg, V. H. Garrison, and E. A. Shinn. 2002. The global transport of dust: an intercontinental river of dust, microorganisms and toxic chemicals flows through the Earth's atmosphere. *American Scientist* 90:228–235.

Grimm, N. B. 1987. Nitrogen dynamics during succession in a desert stream. *Ecology* 68:1157–1170.

Hansen, A. J., and F. di Castri (eds.) 1992. *Landscape Boundaries: Consequences for Biotic Diversity, and Ecological Flows*. New York: Springer-Verlag.

Hanski, I. A., and M. E. Gilpin (eds.) 1997. *Metapopulation Biology: Ecology, Genetics, and Evolution*. San Diego, CA: Academic Press.

Harvell, C. D., K. Kim, J. M. Burkholder, *et al.* 1999. Review: Marine ecology—emerging marine diseases—climate links and anthropogenic factors. *Science* 285:1505–1510.

Hemminga, M. A., and J. Nieuwenhuize. 1990. Seagrass wrack-induced dune formation on a tropical coast (Banc d'Arguin, Mauritania). *Estuarine, Coastal and Shelf Science* 31:499–502.

Hewitt, J. E., V. J. Cummings, J. I. Ellis, *et al.* 2003. The role of waves in the colonisation of terrestrial sediments deposited in the marine environment. *Journal of Experimental Marine Biology and Ecology* 290:19–47.

Huxel, G. R., and K. McCann. 1998. Food web stability: the influence of trophic flows across habitats. *American Naturalist* 152:460–469.

Huxel, G. R., K. McCann, and G. A. Polis. 2002. Effects of partitioning allochthonous and autochthonous resources on food web stability. *Ecological Research* 17:419–432.

Huxel, G. R., G. A. Polis, and R. D. Holt. 2004. At the frontier of the integration of food web ecology and landscape ecology. Pp. 434–451 in G. A. Polis, M. E. Power, and G. R. Huxel (eds.) *Food Webs at the Landscape Level.* Chicago, IL: University of Chicago Press.

Inman, D. L. 2005. Littoral cells. Pp. 594–598 in M. Schwartz (ed.) *Encyclopedia of Coastal Science.* New York: Springer-Verlag.

Jackson, J. K., and S. G. Fisher. 1986. Secondary production, emergence, and export of aquatic insects of a Sonoran desert stream. *Ecology* 67:629–638.

Karr, J. R., and I. J. Schlosser. 1978. Water resources and the land—water interface. *Science* 201:229–234.

Koebel, J. W. 1995. An historical perspective on the Kissimmee River restoration project. *Restoration Ecology* 3:149–159.

Kupferberg, S. J. 1997. Bullfrog (*Rana catesbeiana*) invasion of a California river: the role of larval competition. *Ecology* 78:1736–1751.

Laurance, W. H., R. K. Didham, and M. E. Power. 2001. Ecological boundaries: a search for synthesis. *Trends in Ecology and Evolution* 16:70–71.

Levins, R. 1969. Some demographic and genetic consequences of environmental heterogeneity for biological control. *Bulletin of the Entomological Society of America* 15:237–240.

Likens, G. E., and F. H. Bormann. 1974. Linkages between terrestrial and aquatic ecosystems. *BioScience* 24:447–456.

Loreau, M., N. Mouquet, and R. D. Holt. 2003. Meta-ecosystems: a theoretical framework for a spatial ecosystem ecology. *Ecology Letters* 6:673–679.

Luo, H. R., L. M. Smith, D. A. Haukos, and B. L. Allen. 1999. Sources of recently deposited sediments in playa wetlands. *Wetlands* 19:176–181.

Malanson, G. P. 1993. *Riparian Landscapes.* Cambridge, U.K: Cambridge University Press.

Mann, K. H., and R. N. Lazier. 1991. *Dynamics of Marine Ecosystems.* Oxford, UK: Blackwell.

Marra, P. P., and R. T. Holmes. 1997. Avian removal experiments: do they test for habitat saturation or female availability? *Ecology* 78:947–952.

McCann, K., A. Hastings, and G. R. Huxel. 1998. Weak trophic interactions and the balance of nature. *Nature* 395:794–798.

McHugh, D. J. 2003. *A Guide to the Seaweed Industry*. FAO Fisheries Technical Paper No. 441. Rome, Italy: Food and Agriculture Organization of the United Nations.

McPeek, M. A., and S. Kalisz. 1998. On the joint evolution of dispersal and dormancy in metapopulations. *Advances in Limnology* 52:33—51.

Minshall, G. W. 1967. Role of allochthonous detritus in the trophic structure of a woodland springbrook community. *Ecology* 48:139—149.

Mitsch, W. J., and J. G. Gosselink. 2000. *Wetlands*, 3rd edn. New York: John Wiley.

Naiman, R. J., and H. Decamps. 1997. The ecology of interfaces: riparian zones. *Annual Review of Ecology and Systematics* 28:621—658.

Naiman, R. J., J. M. Melillo, M. A. Lock, T. E. Ford, and S. R. Reice. 1987. Longitudinal patterns of ecosystem processes and community structure in a sub-arctic river continuum. *Ecology* 68:1139—1156.

Nakano, S., and M. Murakami. 2001. Reciprocal subsidies: dynamic interdependence between terrestrial and aquatic food webs. *Proceedings of the National Academy of Sciences of the USA* 98:166—170.

Nakano, S., H. Miyasaka, and N. Kuhara. 1999. Terrestrial—aquatic linkages: riparian arthropod inputs alter trophic cascades in a stream food web. *Ecology* 80:2435—2441.

Neal, C., G. J. L. Leeks, G. E. Millward, *et al.* 2003. Land—ocean interaction: processes, functioning and environmental management from a UK perspective: an introduction. *Science of the Total Environment* 314:3—11.

Nehlsen, W., J. E. Williams, and J. A. Lichatowich. 1991. Pacific salmon at the crossroads: stocks at risk from California, Oregon, Idaho, and Washington. *Fisheries* 16:4—21.

Norkko, A., S. F. Thrush, J. E. Hewitt, *et al.* 2002. Smothering of estuarine sandflats by terrigenous clay: the role of wind—wave disturbance and bioturbation in site-dependent macrofaunal recovery. *Marine Ecology — Progress Series* 234:23—41.

Odum, W. E. 1980. The status of three ecosystem-level hypotheses regarding salt marsh estuaries: tidal subsidy, outwelling, and detritus-based food chains. Pp. 485—495 in V. S. Kennedy (ed.) *Estuarine Perspectives*. New York: Academic Press.

Ohio Lake Erie Commission. 2000. *Lake Erie Protection and Restoration Plan*. Toledo, OH: Ohio Lake Erie Commission. Available online at http://www.epa.state.oh.US/oleo/index.html/

Polis, G. A., and S. D. Hurd. 1995. Extraordinarily high spider densities on islands: Flow of energy from the marine to terrestrial food webs and the absence of predation. *Proceedings of the National Academy of Sciences of the USA* 92:4382—4386.

Polis, G. A., and S. D. Hurd. 1996. Linking marine and terrestrial food webs: allochthonous input from the ocean supports high secondary productivity on small islands and coastal land communities. *American Naturalist* 147:396—423.

Polis, G. A., and D. R. Strong. 1996. Food web complexity and community dynamics. *American Naturalist* 147:813—846.

Polis, G. A., W. B. Anderson, and R. D. Holt. 1997a. Toward an integration of landscape and food web ecology: the dynamics of spatially subsidized food webs. *Annual Review of Ecology and Systematics* **29**:289–316.

Polis, G. A., S. D. Hurd, C. T. Jackson, and F. S. Piñero. 1997b. El Niño effects on the dynamics and control of an island ecosystem in the Gulf of California. *Ecology* **78**:1884–1897.

Polis, G. A., M. E. Power, and G. R. Huxel (eds.) 2004. *Food Webs at the Landscape Level*. Chicago, IL: University of Chicago Press.

Power, M. E. 1984. Depth distributions of armored catfish: predator-induced resource avoidance? *Ecology* **65**:523–528.

Power, M. E. 2001. Prey exchange between a stream and its forested watershed elevates predator densities in both habitats. *Proceeding of the National Academy of Sciences of the USA* **98**:14–15.

Rodriguez-Lanetty, M., and O. Hoegh-Guldberg. 2002. The phylogeography and connectivity of the latitudinally widespread scleractinian coral *Plesiastrea versipora* in the Western Pacific. *Molecular Ecology* **11**:1177–1189.

Roff, D. A. 2002. *Life History Evolution*. Sunderland, MA: Sinauer Associates.

Rose, M. D., and G. A. Polis. 1998. The distribution and abundance of coyotes: the effects of allochthonous food subsidies from the sea. *Ecology* **79**:998–1007.

Rosenzweig, M. L. 1991. Habitat selection and population interactions: - the search for mechanism. *American Naturalist* **137**:S5–S28.

Runyan, K., and G. B. Griggs. 2003. The effects of armoring seacliffs on the natural sand supply to the beaches of California. *Journal of Coastal Research* **19**:336–347.

Sabo, J. L., and M. E. Power. 2002. Numerical response of lizards to aquatic insects and short-term consequences for terrestrial prey. *Ecology* **83**:3023–3036.

Sánchez-Piñero, F., and G. A. Polis. 2000. Bottom–up dynamics of allochthonous input: direct and indirect effects of seabirds on islands. *Ecology* **81**:3117–3132.

Sanzone, D. M., J. L. Meyer, E. Marti, *et al.* 2003. Carbon and nitrogen transfer from a desert stream to riparian predators. *Oecologia* **134**:238–250.

Schoenherr, A. A. 1992. *A Natural History of California*. Berkeley, CA: University of California Press.

Southwood, T. R. E. 1962. Migration of terrestrial arthropods in relation to habitat. *Biological Reviews* **37**:171–214.

Stapp, P., G. A. Polis, and F. Sanchez-Piñero. 1999. Stable isotopes reveal strong marine and El Niño effects on island food webs. *Nature* **401**:467–469.

Steele, J. H., S. R. Carpenter, J. E. Cohen, P. K. Dayton, and R. E. Ricklefs., 1993. Comparing terrestrial and marine ecological systems. Pp. 1–12 in S. A. Levin, T. M. Powell, and J. H. Steele (eds.) *Patch Dynamics*. New York: Springer-Verlag.

Strayer, D. L., M. E. Power, W. F. Fagan, S. T. A. Pickett, and J. Belnap. 2003. A classification of ecological boundaries. *BioScience* **53**:723–729.

Summerhayes, V. S., and C. S. Elton. 1923. Contributions to the ecology of Spitzbergen and Bear Island. *Journal of Ecology* **11**:214–286.

Talley, D. M., E. W. North, A. R. Juhl, *et al.* 2003. Research challenges at the land–sea interface. *Estuarine Coastal and Shelf Science* **58**:699–702.

Talley, T. S., J. A. Crooks, and L. A. Levin. 2001. Habitat utilization and alteration by the invasive burrowing isopod, *Sphaeroma quoyanum*, in California salt marshes. *Marine Biology* **138**:561–573.

Thrush, S. F., R. D. Pridmore, J. E. Hewitt, and V. J. Cummings. 1994. The importance of predators on a sandflat: interplay between seasonal changes in prey densities and predator effects. *Marine Ecology – Progress Series* **107**:211–222.

Thrush, S. F., J. E. Hewitt, V. Cummings, et al. 2004. Muddy waters: elevating sediment input to coastal and estuarine habitats. *Frontiers in Ecology and the Environment* **2**:299–306.

Toth, L. A. 1995. Principles and guidelines for restoration of river/floodplain ecosystems: Kissimmee River, Florida. Pp. 49–75 in J. Cairns Jr. (ed.) *Rehabilitating Damaged Ecosystems*. Boca Raton, FL: Lewis Publications.

Wainright, S. C., J. C. Haney, C. Kerr, A. N. Golovkin, and M. V. Flint. 1998. Utilization of nitrogen derived from seabird guano by terrestrial and marine plants at St. Paul, Pribilof Islands, Bering Sea, Alaska. *Marine Biology* **131**:63–71.

Wallace, J. B., S. L. Eggert, J. L. Meyer, and J. R. Webster. 1997. Multiple trophic levels of a forest stream linked to terrestrial litter inputs. *Science* **277**:102–104.

Ward, J. V. 1998. Riverine landscapes: biodiversity patterns, disturbance regimes, and aquatic conservation. *Biological Conservation* **83**:269–278.

Watson, R., and D. Pauly. 2001. Systematic distortions in world fisheries catch trends. *Nature* **414**:534–536.

Wiens, J. A. 2002. Riverine landscapes: taking landscape ecology into the water. *Freshwater Biology* **47**:501–515.

Wikelski, M., and C. Thom. 2000. Marine iguanas shrink to survive El Niño: changes in bone metabolism enable these adult lizards to reversibly alter their length. *Nature* **403**:37–38.

Wilbur, H. M. 1980. Complex life cycles. *Annual Review of Ecology and Systematics* **11**:67–93.

Willson, M. F., S. M. Gende, and B. H. Marston. 1998. Fishes and the forest: expanding perspectives on fish–wildlife interactions. *BioScience* **48**:455–462.

Willson, M. F., S. M. Gende, and P. A. Bisson., 2004. Anadromous fishes as ecological links between ocean, fresh water, and land. Pp. 284–300 in G. A. Polis, M. E. Power, and G. R. Huxel (eds.) *Food Webs at the Landscape Level*. Chicago, IL: University of Chicago Press.

Witman, J. D., J. C. Ellis, and W. B. Anderson. 2004. The influence of physical processes, organisms, and permeability on cross-ecosystem fluxes. Pp. 335–358 in G. A. Polis, M. E. Power, and G. R. Huxel (eds.) *Food Webs at the Landscape Level*. Chicago, IL: University of Chicago Press.

Wootton, J. T. 1991. Direct and indirect effects of nutrients on intertidal community structure: variable consequences of seabird guano. *Journal of Experimental Marine Biology and Ecology* **151**:139–154.

Wright, J. P., W. S. C. Gurney, and C. G. Jones. 2004. Patch dynamics in a landscape modified by ecosystem engineers. *Oikos* **105**:336–348.

6

Influence of natural landscape fragmentation and resource availability on distribution and connectivity of gray wolves (*Canis lupus*) in the archipelago of coastal British Columbia, Canada

PAUL C. PAQUET, SHELLEY M. ALEXANDER, PATRICIA L. SWAN, AND CHRIS T. DARIMONT

INTRODUCTION

Connectivity has emerged as an important ecological concept relating to how animals move among habitat patches in fragmented environments. Although connectivity is often considered landscape- and species-specific (Tischendorf and Fahrig 2000; Taylor *et al.* Chapter 2), fundamental ecological and physical processes influence movements of all species. Taylor *et al.* (1993) define connectivity as "the degree to which the landscape facilitates or impedes movement among resource patches." Most research, however, has centered on impediments to movement. For example, Tischendorf and Fahrig (2000) argue that landscape connectivity is "essentially equivalent to the inverse of patch isolation over the landscape." Herein we examine landscape and resource features that not only potentially impede (i.e., isolation) but also might facilitate (i.e., availability of food) movement of wolves (*Canis lupus*) using coastal islands in British Columbia, Canada.

Investigations of oceanic archipelagos have revealed how island communities and species composition are related to area, isolation, and other island characteristics (e.g., MacArthur and Wilson 1967; Abbott 1974; Kadmon and Pulliam 1993; Conroy *et al.* 1999). Biogeographic

Connectivity Conservation eds. Kevin R. Crooks and M. Sanjayan. Published by Cambridge University Press. © Cambridge University Press 2006.

features, however, may also exert influence at the population level, including the mediation of predator–prey dynamics on islands or in other fragmented systems (Kareiva 1990; Kareiva and Wennergren 1995; Dolman and Sutherland 1997; Darimont et al. 2004). The equilibrium theory of island biogeography (MacArthur and Wilson 1967) and metapopulation theory (Levins 1976; Gilpin and Hanski 1991; Hanski and Gilpin 1996; Moilanen and Hanski Chapter 3) postulated several abiotic mechanisms explaining animal distribution and population persistence in patchy landscapes. Now, the original concept of a meta-population as "a population of populations" has expanded to include other spatial population structures (Laurance and Cochrane 2001), including mainland–island (Hanski and Gilpin 1991) and source–sink metapopulations (Pulliam 1998). Again, however, most studies focused on physiographic features thought to impede or limit movement and persistence of species, rather than biotic influences such as availability of food.

According to these theories, suitable habitat often is arranged across the landscape as island-like patches of various sizes (area) and distances from other patches (isolation). The occurrence of animals is thought to reflect direct responses of individuals to this physically varied environment (Kareiva 1990). The availability and productivity of these habitats likely influence animal movements, species composition, population dynamics, and community structure. On large spatial and temporal scales, area and isolation are the most important variables in predicting community structure in island biogeography theory (MacArthur and Wilson 1967) and population dynamics in metapopulation theory (Levins 1976). Collectively, these ideas provide much of the theoretical foundation for conservation biology.

To establish biological priorities for conservation, we need a firm understanding of how geography interacts with species to shape evolution, ecological relationships, and landscape processes. To that end, a long list of studies addressing birds, small mammals, and insects has contributed to the development of contemporary conservation theory (Ehrlich and Hanski 2004). Very few studies, however, have evaluated the response of large terrestrial predators to naturally fragmented landscapes or true island systems. In part, this is due to a lack of pristine sites to carry out such research. Moreover, large mammalian carnivores are rare on isolated or semi-isolated islands (Williamson 1981; Alcover and McMinn 1994). Nevertheless, elucidating the relationship between the geographic and ecological structure of true island systems and distribution of large mammals is a needed link between theory and application. Knowledge about large predators in patchy landscapes is valuable because the planet is becoming increasingly fragmented by human activities (Saunders *et al.* 1991; Fahrig 1997, 2003). Moreover, predators are more likely to decline or become extinct in fragments (Woodroffe and Ginsberg 1998), possibly resulting in mesopredator release and other ecosystem-wide consequences (Crooks and Soulé 1999; Terborgh *et al.* 1999). Consequently, archipelagos may provide model systems in which to predict the effects of size, isolation, resource availability, and other factors on predator persistence.

The archipelago of coastal British Columbia provides an opportunity to examine how a true island environment affects the distribution and abundance of wolves. The region is characterized by forested oceanic islands, mainland mountains, and island mountains that are potential barriers to movement, and diverse marine and terrestrial food resources. This remote and nearly pristine region is naturally fragmented, comprising numerous islands < 100 m to > 13 km apart (Fig. 6.1). Here, the wolf and black-tailed deer (*Odocoileus hemionus*) association forms the dominant mammalian predator−prey system, in which both animals can occupy many islands, at least ephemerally (Darimont and Paquet 2000, 2002). Documenting the response of wolves to true islands (i.e., surrounded by water) provides a reference for comparison with similar studies relating to connectivity in fragmented terrestrial landscapes (Harrision and Chapin 1998; Gaines *et al.* 2000; Carroll *et al.* 2004; Carroll Chapter 15).

This study evaluates and contributes to prevailing theories of meta-population theory and predator−prey dynamics in fragmented landscapes. Herein, we examine the effects of island characteristics on distribution and spatial ecology of gray wolves. Specifically, we predict that geographic and resource characteristics of landmasses influence connectivity for this summit predator. We infer connectivity by documenting wolf presence on islands during a 5-year period. Our overarching research hypothesis is that distances among landmasses, juxtaposition of landmasses, geometry of landmasses, and prey availability influence wolf presence and persistence on islands. We predicted that wolves are more likely to be resident on islands that are largest, roundest, least isolated, and most productive.

GRAY WOLVES

The gray wolf is an elusive, low-density carnivore with an historical distribution comprising most of the Holarctic. The historic range of the species is the largest of any extant terrestrial mammal, with the possible exception of the African lion (*Panthera leo*) (Paquet and Carbyn 2003).

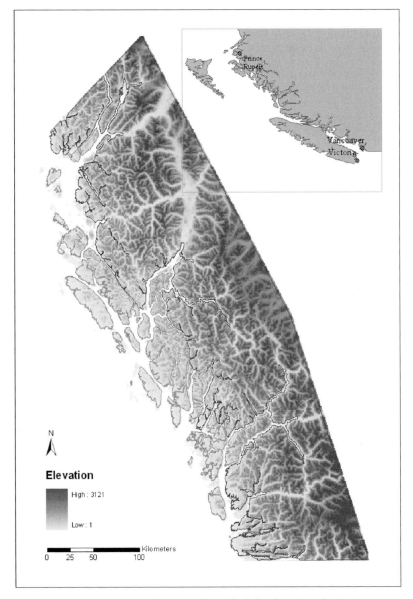

Fig. 6.1. Study area: Pacific coast of British Columbia, Canada (Darimont and Paquet 2002).

Gray wolves occur in a variety of habitats, from dense forest to open grassland and from the Arctic tundra to extreme desert, avoiding only swamps and tropical rainforests. The gray wolf originally occupied all habitats in North America north of about 20° N latitude. On the mainland, wolves were found everywhere except the southeastern United States, California west of the Sierra Nevada, and the tropical and subtropical parts of Mexico. The species also occurred on large continental islands, such as Newfoundland, Vancouver Island, and smaller islands off coastal British Columbia and southeast Alaska, and throughout the Arctic Archipelago and Greenland, but was absent from Prince Edward Island, Anticosti, and the Queen Charlotte Islands (Paquet and Carbyn 2003).

Because wolves are not highly habitat-specific, move long distances, and require large home ranges, the species is regarded as a habitat generalist (Mech 1970; Fuller *et al.* 1992; Mladenoff *et al.* 1995). Populations, however, are adapted to local conditions and specialized concerning den-site use, foraging habitats, physiography, and prey selection (Fritts *et al.* 1995; Mladenoff *et al.* 1995, 1997, 1999; Paquet *et al.* 1996; Haight *et al.* 1998; Mladenoff and Sickley 1998; Callaghan 2002). Thus, wolves are better characterized as ecosystem generalists that are idiosyncratic concerning the surroundings in which they live (Paquet and Carbyn 2003).

The gray wolf is also the most vagile of all large terrestrial predators. Travel and dispersal distances of several hundred kilometers are common and movements of more than 1000 km have been documented (Ballard *et al.* 1983; Fritts 1983; Boyd *et al.* 1995; Mech *et al.* 1995; Walton *et al.* 2001; Paquet and Carbyn 2003). The Canadian west coast wolf typifies this pattern of adaptability. Among regions still populated by wolves, the mainland coast of British Columbia and the associated archipelago of offshore islands are ecologically unique. This remote ocean archipelago hosts what is possibly North America's most pristine wolf population (Paquet *et al.* 2005). There is, however, considerable conservation concern about future abundance and distribution in the area, given large scale and rapid industrial clearcut logging and associated effects (Darimont and Paquet 2000, 2002; Paquet *et al.* 2005).

Wolves are widely distributed in the study area (Darimont and Paquet 2000, 2002). Although territorial behavior is not well documented, coastal wolves in nearby southeast Alaska have average home ranges of about 230 km² (Person *et al.* 1996). Wolves can swim in open ocean between landmasses as distant as 13 km (Darimont and Paquet 2002). Movements, however, are thought to be limited by distance, wind, water temperature, and water currents (Darimont and Paquet 2002). Observations from our studies and southeast Alaska (Person *et al.* 1996; D. Person, pers. comm.) confirm that some packs and individuals include a constellation of islands and mainland areas within their home range.

STUDY AREA

Our study area encompasses the island archipelago north of Vancouver Island (51° 46′ N, 127° 53′ W) to Prince Rupert, British Columbia (55° 37′ N, 129° 48′ W). The region is approximately 29 700 km², of which 19 300 km² is land (Fig. 6.1). This wet, nearly roadless, and biologically productive area is isolated from continental North America by the Coast Mountain Range to the east and Pacific Ocean to the west, resulting in a unique ecological and evolutionary environment largely free from industrial development. The few human settlements consist primarily of First Nation's people.

Coastal temperate rainforest dominates the region (MacDonald and Cook 1996), constituting the largest remaining expanse of ancient temperate rainforest in the world (Shoonmaker *et al.* 1997). Most of the low elevation forest is within the Coastal Western Hemlock biogeoclimatic zone (Pojar and Mackinnon 1994). Climate is temperate and wet. Most areas receive more than 350 cm of annual precipitation (Environment Canada 1991). Habitat heterogeneity in these temperate rainforests corresponds to landscape variability, which includes the following general regions: mountainous mainland, topographically complex inner islands, and flatter outer islands. The region's innumerable islands are separated by fiords, channels, other waterways, and open ocean. Islands we sampled ranged in size from 0.7 km² (Moore) to 2295 km² (Princess Royal), distances (*sensu* Conroy *et al.* 1999) to mainland 250 m to 13.1 km, distances between islands 100 m to 13.0 km, and inter-landmass distances 50 m to 7.3 km. This complex physiography mediates the interaction of marine and terrestrial systems (see Talley *et al.* Chapter 5), creating many different kinds of environments in close proximity. The study region, combined with southeastern Alaska, supports the highest concentration of endemic species and populations for the temperate rainforest region of Pacific North America (Cook and MacDonald 2001).

The region is important to formerly wide-ranging species such as grizzly bears (*Ursus arctos*) and wolves now exterminated from most of their former ranges (Paquet *et al.* 2005). Genetically distinct populations of species such as Kermode black bear (*U. americanus kermodei*) are found on the Central Coast, providing a valuable opportunity for scientists studying evolutionary processes under natural conditions (e.g., Byun *et al.* 1997; Ritland *et al.* 2001; Marshall and Ritland 2002). Basic ecological information, however, such as the distribution, density, and movements of large mammals, including gray wolves and their prey, is largely unknown.

Potential prey base for wolves is diverse, including Sitka black-tailed deer, moose (*Alces alces*), mountain goat (*Oreamnos americanus*), beaver (*Castor canadensis*), black bear (*Ursus americanus*), river otter (*Lontra canadensis*), plus smaller mustelids, rodents, and birds. Five species of spawning salmonids (*Onchorynchus* spp.), crustaceans, mollusks, and marine mammals are also available to wolves (Darimont and Paquet 2000, 2002; Darimont *et al.* 2003, 2004, 2005).

METHODS

Distributional data

We assembled the ecological and geographic information used in our analyses from a variety of sources. During summer and fall 2000 to 2004, we surveyed islands in the study area for the presence of wolves and other mammals (Fig. 6.1). During summers 2000 and 2001, we surveyed 36 islands and 42 mainland watersheds for wolves (Darimont and Paquet 2002). We surveyed 14 additional islands during 2002 to 2004 and resampled many of the islands visited in 2000 and 2001. In total, 50 islands were surveyed, ranging in size from 0.71 km² to 2295 km².

Sampling sites on islands were selected non-randomly but influenced by our knowledge of the area and its resources. Local First Nations colleagues and the traditional ecological knowledge they shared (Huntington 2000; Pierotti and Wildcat 2000; Turner *et al.* 2000) also guided us. At each location, we surveyed beaches, estuaries, and forests of the beach fringe, often on wildlife trails. We also surveyed logging roads when encountered, circumnavigated beaver ponds, and walked forest ridgelines. Surveys rarely extended more than 5 km inland. Presence of wolves was determined by occurrence of wolf feces, tracks, scrapings, vocalizations, or direct observations (Darimont and Paquet 2002). All data were imported into Geographic Information System (GIS) for analyses.

Landscape data

Using a Geographic Information System (ArcView3.2 and ArcMap), we extracted landscape configuration metrics from LANDSAT-7 and a digital elevation model (DEM). Landsat 7 images were taken from 1999 through 2001 between the months of July and October. Mean-subtraction was used to mosaic the images, whereby the pixel value of one image was subtracted from the mean of all pixel values of a second image. This smoothed pixel values and accounted for time discrepancy between images.

A 50-m resolution DEM provided the basis for an island/mainland polygon layer. We calculated area of islands and shape indices from a digital map derived from the DEM using the X-Tools extension in ArcView3.2. Shape index was calculated as (0.25 × Perimeter)/sqrt(Area), where 0.25 accounts for square pixel shape. Smaller shape index values represent rounder islands. An index of 0.099 is perfectly round. Island distance to nearest island and island distance to mainland (cf. Conroy *et al.* 1999) were calculated using the Nearest Feature extension for ArcView3.2.

Prey resource availability

Deer and salmon are the primary prey of wolves in this region (Darimont *et al.* 2004), so we expected that the relative abundance of these resources among islands could predict occurrence of wolves. Accordingly, we conducted deer pellet surveys within a 3000-km² portion of our study area using 1-km paired transects at 60 locations. Paired transects were at least 200 m apart and consisted of 50 continuous 1 × 20 m belt plots. We recorded the number of pellet groups in each plot (Kirchhoff and Pitcher 1988; Kirchhoff 1990). We used hand-held Geographic Position System (GPS) to locate the beginning and end of transects, as well as each 100-m interval. All GPS locations were input into GIS to obtain a spatial map of deer pellet locations.

We obtained information on distribution of chum (*O. keta*), pink (*O. gorbuscha*), coho (*O. kisutch*), chinook (*O. tschawytscha*), and sockeye (*O. nerka*) salmon from Department of Fisheries and Oceans FISS inventory data (Government of Canada 2005). Abundance of salmon fluctuates among years owing to ecological variability and exploitation by humans (Groot and Margulis 1991; National Resources Council 1996). Moreover, the reliability of spawning salmon enumerations by government inventories has not been rigorously assessed, nor have all potential salmon-bearing watersheds been inventoried (Thomson and MacDuffee 2002). Therefore, we used salmon richness, or the number of known salmon species per island, as a proxy for availability and abundance. Because each species spawns at different peak times and in different riparian habitat, richness reflects the temporal and spatial breadth of this resource for island wolves.

Deer sub-model

We created a sub-model to estimate the relative abundance of deer on each island. Using our field data on pellet-groups, we derived an equation that predicts the relative distribution (Loft and Kie 1988; Edge and Marcum 1989; Weckerly and Ricca 2000) and density (Rowland *et al.* 1984; Patterson *et al.* 2002; cf. Fuller 1992; White 1992) of deer based on topographic slope. We assumed a linear positive relationship between numbers of pellet-groups and numbers of deer. Observed pellet-group densities were normalized to account for representation of available slopes and sampling bias. Using a curve-fitting regression method (SPSS 11.0, SPSS Inc., Chicago, USA), we generated a curvilinear function describing the relationship between mean pellet-group density and slope that was highly predictive ($R^2 = 0.92$). This was then converted to a spatial probability layer in ArcView using the following equation:

$$\text{Relative pellet-group density} = 0.0000094[\text{slope}^3]$$
$$- 0.0009[\text{slope}^2] + 0.0231[\text{slope}] - 0.0022 \tag{6.1}$$

Statistical analyses

We used an information-theoretic approach to evaluate which models and parameters best explained observed patterns in our field data (Burnham and Anderson 1998; Anderson *et al.* 2001). We formed exploratory a priori hypotheses to explain how biogeographic features and availability of prey affect the presence of wolves on coastal islands. These hypotheses were based on our knowledge of the area, previous studies, ecological theory, and discussions with First Nations colleagues. Specifically, we postulated that wolves are most likely to occur on the most productive, largest, and roundest islands that are closest to the mainland or other islands. From these hypotheses, we developed a set of candidate logistic regression models (Hosmer and Lemeshow 1989) to predict the probability of wolves being either present or absent on islands during our surveys. These were restricted to combinations of 1 to 6 of the identified (and untransformed) parameters (Table 6.1), and two-way interaction terms we felt were useful based on theory and our previous empirical work (AREA × MAIN, DEER × MAIN, and DEER × AREA: MacArthur and Wilson 1967; Darimont *et al.* 2004). In addition, we assessed the potential bias of our survey effort by including the number of visits to each island as a parameter (SURVEFFORT). A Hosmer–Lemeshow goodness-of-fit statistic based on the global model showed the data did not depart from a logistic-regression model ($P = 0.871$). Multicollinearity diagnostics suggested only weak

Table 6.1. *Metric type, data source, derived variables, and parameter names used in logistic regression analyses to predict the presence of wolves on the islands of the British Columbia's coastal archipelago, 2000 to 2004; see "Methods" for detailed description of variables and data sources*

Metric	Source	Derivative	Parameter
Biogeographic	Digital elevation models	Island area[a]	AREA
	LANDSAT-7	Distance to mainland	MAIN
		Distance to island $> 75\,\text{km}^2$ [a,b]	NEAR
		Island shape	SHAPE
Food resource	Field data	Deer abundance model	DEER
	Government database	Salmon species richness	SALMONRICH
Analytical	Field data	Survey effort	SURVEFFORT

[a]See Conroy *et al.* (1999) for details regarding the measurement of distances.
[b]See Darimont *et al.* (2004) for details regarding the 75 km² threshold.

interdependencies among predictor variables (Variance Inflation Factors range: 1.3 to 2.3).

For each model, we calculated Akaike Information Criteria, adjusted for small sample sizes (AIC$_c$), following the formula:

$$\text{AIC}_c = -2(\log \text{ likelihood}) + 2K + 2K(K + 1)/(n - K - 1), \qquad (6.2)$$

where K is the number of parameters and n the number of sampled islands. We then evaluated ΔAIC_c to select best approximating model(s) and make appropriate inference, using $\Delta\text{AIC}_c < 4$ to describe the top model set (offering substantial level of empirical support: Burnham and Anderson 1998). Finally, we summed Akaike weights (ω_i) across the top model set for each variable to rank them by importance (Burnham and Anderson 1998; Anderson *et al.* 2001). We ran models in SPSS 11.0 (SPSS Inc., Chicago, USA).

RESULTS

Over five field seasons, we observed wolves or sign of wolves on 42 of 50 islands surveyed. The maximum distance between landmasses where wolves were recorded was 13 km (on the most isolated islands in our study area). Based on continuous monitoring and visual re-identification of conspicuous individuals, some packs occupied constellations of islands.

Biogeographic and food resource parameters contributed to highly predictive models to explain the presence of wolves on islands. Moreover, relationships among parameters and island occupancy by wolves were in the direction predicted by theory and our a priori hypotheses (except for island shape) (Tables 6.2, 6.3). Although both were important, biogeographic parameters were more useful than food resources in predicting occupancy by wolves. Specifically, model selection and multi-model inference suggest that AREA was clearly the most important parameter influencing detection of wolves, whereas other biogeographic features (isolation and shape) and food resources were less important.

Table **6.2.** *Top logistic regression model set to predict the probability of wolf occupancy on islands (n = 50) of British Columbia's archipelago*

Model form[a]	Deviance	ΔAIC_c	Akaike weight (ω_i)	Nagelkerke R^2
AREA	20.657	0.00	0.28	0.654
AREA, SURVEFFORT	18.358	2.01	0.10	0.717
AREA, SHAPE	18.952	2.61	0.08	0.706
DEER, SHAPE	19.557	3.21	0.06	0.695
AREA, DEER	19.630	3.29	0.05	0.693
AREA, NEAR	19.970	3.63	0.05	0.687
DEER, SALMONRICH	20.108	3.76	0.04	0.684
AREA, SALMONRICH	20.302	3.96	0.04	0.680

[a]AREA is island area, SURVEFFORT is the number of occasions we surveyed an island, SHAPE is island shape, DEER is estimated deer abundance from a model we generated from field data, NEAR is island distance to another landmass > 75 km^2, and SALMONRICH is the number of salmon species present on the island.

Table **6.3.** *Sum of Akaike weights ($\Sigma\omega_i$) for each parameter included in the top model set (0–4 ΔAIC_c); AIC_c values calculated from logistic regression models to predict the presence of wolves on islands of British Columbia's archipelago, 2000 to 2004*

Parameter	$\Sigma\omega_i$	Association[a]
AREA	0.61	+
DEER	0.16	+
SHAPE	0.13	+
SURVEFFORT	0.10	+
SALMONRICH	0.08	+
NEAR	0.05	−

[a]Association between parameter and presence of wolves on islands (i.e., direction).

Six of eight models in the top model set contained AREA (Table 6.2). The leading model (Akaike weight or $\omega_i = 0.28$) predicted wolf distribution as a function of island size (and intercept) alone. The eight leading models accounted for a collective Akaike weight of 0.71. Remaining models had very little support in the data. Some ambiguity, however, existed among top models, among which Akaike weights were not particularly widely distributed ($\omega_i = 1$ to $8 = 0.28$ to 0.04) (Table 6.2). The top model, containing AREA and the intercept only, was roughly three times more informative than the next leading model ($\omega_i = 0.28$ vs. 0.10). Moreover, these top models explained a similar proportion of the variance (Nagelkerke R^2 range $= 0.654$ to 0.717) (Table 6.2). In cases when the data do not strongly support a single best model, the model with fewest parameters is often worth most consideration, following the rule of parsimony (Burnham and Anderson 1998). Accordingly, we consider model 1, containing only intercept and AREA, as a preferred model (Table 6.2). The probability of wolf occupancy on islands declined with smaller island area.

The top model set can still make robust multi-model inference (Burnham and Anderson 1998); summing the Akaike weights for each parameter across top models (Table 6.3) ranked the variable AREA ($\Sigma\omega_i = 0.61$) roughly 4 and 15 times higher than other biogeographic parameters SHAPE and NEAR ($\Sigma\omega_i = 0.13$ and 0.05 respectively). Collectively, food resources also appear important in attracting wolves to islands. The estimated abundance of deer (DEER) and salmon richness (SALMONRICH) occurred multiple times in top models (Table 6.2). Deer abundance was twice as valuable as salmon richness ($\Sigma\omega_i$ of 0.16 vs. 0.08) (Table 6.3).

Some parameters were not important in predicting island occupancy by wolves. We infer that the interactive properties we examined are not important, as no interaction terms were included in the top model. Likewise, MAIN, or isolation from the mainland, appeared to be of little importance in determining the presence of wolves on islands.

DISCUSSION

Consumers often forage in spatially complex environments where resources are distributed unevenly among patches that differ in productivity, size, isolation, and other physical traits. Therefore, resource acquisition, and ultimately fitness, can depend on patch choice and the ease in which animals are able to move among patches (i.e., connectivity). From an optimal foraging perspective (Stephens and Krebs 1986), "patch choice" is an important aspect of foraging ecology because it is the first decision relating to patch use (Kevan and Greco 2001). Yet, the influence of patch choice and food resources on connectivity has received far less attention than other components of foraging ecology, such as residence and departure times (Kevan and Greco 2001). Moreover, most connectivity and metapopulation models assume that patches of habitat are embedded in a matrix that has little value for the species considered, even though the surrounding matrix may be of varying quality (Taylor *et al.* Chapter 2).

Herein, we address some of these potential limitations. Specifically, we demonstrate that both physical features and patch quality of fragments are important aspects of connectivity for island-dwelling wolves. We inferred patch quality from estimates of prey resource availability. With the proper caveats, these preliminary findings are relevant to understanding the effects of isolation and fragmentation in other ecosystems. We caution that our analytical design may have influenced our results. For example, one reasonable model suggested the probability of detecting wolves increased with survey effort.

Well adapted to the marine environment, many coastal wolves are island-dwellers whose territories can include groups of islands. Consequently, movement within territories requires traveling on land and among landmasses, which means swimming in open ocean (Darimont and Paquet 2002). Dispersing and traveling animals also need to cross expanses of inhospitable terrestrial habitat. Moreover, many of the prey upon which wolves depend for their survival and other carnivores with which they compete (e.g., black bears, grizzly bears), are influenced similarly. We expected, therefore, that isolation, island size, and availability of prey would govern the many potential influences that combine to determine accessibility and use of coastal islands by wolves. Our assessment suggests that island size is the overwhelmingly dominant influence in determining wolf presence, followed by availability of primary prey (deer and salmon), and island shape. In comparison, isolation, expressed as island-to-island distance and island-to-mainland distance, seems unimportant.

Size matters

Island area is clearly the dominant factor in determining wolf residency. In nearby southeast Alaska, wolves also are typically absent on small islands. Wolf scat or wolf-killed deer were observed on only six of 97 islands surveyed repeatedly between 1989 and 1993. These very small islands ranged in size from 0.2 to 134 ha (Kirchhoff 1994; Emlen *et al.* 2003).

We believe island size influences the quantity and quality of resources available to wolves. For example, larger islands are likely to have greater food resources overall, and those prey populations are probably more stable (Burkey 1995; Newmark 1995). Thus, wolves can maintain longer residency times on these islands, which would increase the probability of their detection during our surveys. Moreover, owing to topographic features, a larger island may have disproportionately more habitat than predicted by two-dimensional area calculations alone.

Isolation

The effects of isolation are important for consumers with high resource demands because they relate to the economics of time and energy. Giving up on a site (Kareiva et al. 1989) and moving to another imposes time and energetic costs (Stephens and Krebs 1986; Ritchie 1998). Swimming in marine environments probably carries considerably more energetic costs than travel on land and may impose costs on fitness (e.g., drowning). Moreover, the influences of currents, which vary between landmasses, may be as important as straight-line distances in affecting dispersal (Cameron 1958; Williamson 1981). Sightability may affect orientation (MacArthur and Wilson 1967) and influence which islands wolves select, especially during foggy periods. Finally, in any landscape, animals run a risk that the new patch is less productive than the current one. Therefore, in an environment where patches are unevenly distributed, foragers might prefer to occupy less isolated patches.

Although few in number, other studies carried out in coastal British Columbia and Alaska found isolation on islands to be an important factor affecting species presence and diversity (Craig 1990; MacDonald and Cook 1996; Conroy et al. 1999). We expected wolf presence to be related to island isolation because water poses a considerable barrier to movement for most terrestrial species. Surprisingly, wolves were not limited by island isolation as inferred from distance among landmasses. Likely, however, a threshold exists where isolation does limit or preclude movements. We believe the excellent swimming ability of wolves allows them to move unrestricted among all landmasses within our study area. For example, one pack colonized a 240-km² cluster of islands 13 km from the nearest landmass (Darimont and Paquet 2002), which is the most isolated landmass in the study area. Apparently, these wolves first arrived in 1996 (Tshimsian Nation family, pers. comm.). Moreover, we continue to monitor three packs that move frequently among multiple landmasses. Notably, we rarely observe evidence of these wolves on the smallest of nearby islands. We believe the costs wolves incur traveling among islands are outweighed by the benefits accrued exploiting the most productive sites (Stephens and Krebs 1986).

Although not addressed by our study, connectivity for all coastal mammals that travel through water corridors can be adversely affected by human disturbances such as boat traffic. These disturbances are analogous with the adverse influences associated with roads, railways, and other infrastructure (Paquet *et al.* 1996; Clevenger and Waltho 2000; Duke *et al.* 2001; Clevenger *et al.* 2002; Alexander *et al.* 2004; Clevenger and Wierzchowski Chapter 20). Human presence and boat wash can stress and disturb swimming animals. In addition, humans occasionally harass and kill deer, bear, and wolves as these animals travel between landmasses (C. Darimont, unpublished data; P. Paquet, pers. obs.). Ocean channels, coastlines, and river systems provide humans access to remote areas and opportunities for disrupting connectivity via disturbance. In southeastern Alaska, for example, humans who gained access by boat to areas otherwise secure were responsible for more than 50% of all wolves killed by hunters and trappers (Person *et al.* 1996). Moreover, predation by non-humans may occur, if only rarely. Killer whales (*Orcinus orca*), for example, are known to prey on moose and deer swimming between islands in the study area (Ford and Ellis 1999).

Island shape
In theory, narrower islands should pose greater risk for wildlife than rounder islands of similar size. Rounder islands provide more security because interiors are more difficult to reach, and compared with narrow islands, proportionately less coastline is exposed. Thus, we expected that wolves, being security conscious, would be less likely to occupy islands with more exposed edge (Woodroffe and Ginsberg 1998). We found, however, a positive association between narrow islands and occupancy (Table 6.3). This could reflect survey bias related to our failure to detect wolves using the interior of rounder islands, but we consider this unlikely for the following reasons. First, wolves in this area are only lightly persecuted. We estimated annual human-caused mortality at 2-5% (Darimont and Paquet 2000). Because this danger is concentrated primarily around human settlements, it is unlikely to repel wolves from using island edges.

Second, our foraging data suggest wolves spend considerable time along island edges, which could make narrower islands more attractive to them. A high percentage of the diet of coastal wolves is composed of foods that occur on the beach fringe (Darimont *et al.* 2004). Therefore, we postulate that narrower islands may have food resource benefits greater than rounder islands of similar size. Other canids are known to capitalize on marine food along seashores. The diet of coastal coyotes (*C. latrans*) in Baja California includes a significant proportion of marine food (Rose and Polis 1998; Talley *et al.* Chapter 5). Likewise, Angerbjörn *et al.* (1994) found significantly enriched marine isotope values in coastal compared with interior arctic foxes (*Alopex lagopus*), suggesting substantial use of marine resources.

Finally, the orientation of islands relative to the mainland and other islands might interact with shape to influence the presence of wolves. It seems reasonable that longer islands are more likely to be intercepted by swimming wolves because they are more visible, appear larger, or make a larger target to a wolf if the major axis (simplifying the shape of the island to an ellipse) is perpendicular to the wolf's line of sight. Assuming that movement patterns of wolves are mediated by their awareness of geography (Peters and Mech 1975, 1978), then the ability of wolves to detect and cognitively map landmasses might be an important factor determining presence of wolves on islands. Accordingly, a wolf's perceptual range would vary with the animal's physical position relative to sea level and wind direction. We are uncertain, however, whether the maximum distances between islands classified as occupied in our study exceeded the perceptual range of wolves.

Food resources

Results of our study suggest strongly that physical characteristics of islands, particularly area, are more important than availability of food in influencing occurrence of wolves on islands and travel among islands. This, however, needs to be considered in the proper context. Specifically, we emphasize that in most of our top models, deer and to a lesser extent salmon occur as variables that predict the presence of wolves on islands. For large carnivores such as wolves, energetic needs are substantial, particularly while raising young (Mech and Boitani 2003; Paquet and Carbyn 2003). Although moderated by the physical landscape, demands for food likely motivate the island-hopping behavior of wolves in our study area. In short, food may facilitate connectivity but not as much as biogeographic factors such as water barriers impede it. In any case, we suggest the physical landscape, food productivity, and food availability are inextricably linked.

We believe the relationship between use of food resources and connectivity is important but poorly understood. Island ecosystems may impose unique constraints on activities of predator and prey (Peterson and Page 1988; Peterson *et al.* 1998). Our analysis of wolf foraging behavior on coastal islands could provide helpful insights (Darimont *et al.* 2004). Specifically, isolation (as measured by distance between landmasses) seems to affect wolf–prey dynamics. The probability that wolves forage for deer on islands is determined primarily by distance to the mainland. Wolves are more likely to consume deer on islands close to the mainland rather than islands that are more distant (Fig. 6.2). Thus, we believe that predator–prey dynamics on isolated islands are potentially less stable, characterized by frequent declines in prey populations (Darimont *et al.* 2004)

We suspect deer cannot reproduce or immigrate fast enough to remote islands to replace those killed by wolves (Darimont *et al.* 2004). Because connectivity for deer is restricted, these islands may become temporary mortality sinks, resulting in ephemeral populations of deer and wolves. Without wolves, deer slowly recolonize isolated islands and the cycle of depletion repeats when wolves return. Consequently (and contrary to predictions based on abiotic factors alone, which would suggest reduced movement with isolation), we suspect that coastal wolves are compelled to

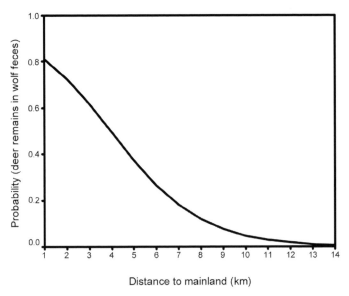

Fig. 6.2. Probability of deer remains occurring in wolf feces on islands as a function of their distance to the mainland. Samples collected in coastal British Columbia, summers 2000 and 2001. (Reproduced from Darimont *et al.* (2004) with permission from Blackwell.)

move frequently among isolated landmasses, assessing the potential of islands to support them while exploiting those that can.

We note that wolves might persist on some isolated islands when deer and other large mammals like goat and moose are scarce or absent by relying on smaller prey such as mink (*Mustela vison*), river otter, and birds (Darimont *et al.* 2004). Many of these taxa are either aquatic or volant and not likely as affected by isolation that limits migration of larger, terrestrial prey. In addition, recent investigations have revealed the coupled nature of marine—terrestrial ecosystems (Polis and Hurd 1995; Rose and Polis 1998; Reimchen 2000; Talley *et al.* Chapter 5). For some consumers, foraging constraints on small or isolated oceanic islands can be moderated by nutrient subsidies from the ocean. Our research suggests the ocean augments food available to wolves inhabiting islands (Darimont and Reimchen 2002; Darimont *et al.* 2003, 2004). In this respect, many coastal islands are not impoverished fragments, as some oceanic islands have been described (Brotons *et al.* 2003; see also Dunning *et al.* 1992; Fahrig 1997). For example, coastal wolves feed on mollusks, crustaceans, and marine carrion such as beached mammals (e.g., *Phoca vitulina*) and whales (e.g., *Megaptera novaeangliae*). In the fall, spawning salmon, having traveled thousands of kilometers in ocean corridors, return to rivers and creeks of the islands of coastal British Columbia, and constitute a considerable part of the diet of coastal and even interior wolves (Szepanski *et al.* 1999; Darimont and Reimchen 2002; Darimont *et al.* 2003, 2004). Notably, these are the same rivers and creeks used by wolves, bears, and other terrestrial species to travel among estuaries and access inland forests. Like bears and river otters (Ben-David *et al.* 1998; Reimchen 2000), wolves act as vectors by transporting marine nutrients from waterways along networks of intersecting trails into the regions' ancient forests. Abandoned salmon carcasses, feces, and urine feed a diversity of users and become important fertilizers in nutrient-limited coastal ecosystems (Reimchen 2000, 2002; Darimont *et al.* 2003).

In addition, data constraints in our analysis might have underestimated the importance of food relative to physical factors. Using GIS and satellite imagery, we were able to measure accurately landscape features of islands. Conversely, the availability of food resources was estimated using gross proxies. Specifically, we employed a habitat model for deer using indices of relative abundance, and used species richness for salmon. In addition, we did not account for how islands may differ in availability of other foods regularly consumed by coastal wolves such as beach carrion, marine mammals, and small terrestrial animals (Darimont *et al.* 2004).

CONCLUSIONS AND FUTURE DIRECTIONS

Our earlier research showed deer are the primary prey of coastal wolves and the probability that wolves eat deer on islands is negatively related to isolation as measured by distance between landmasses (Darimont *et al.* 2004). We postulated that was because deer occupying remote islands cannot immigrate and reproduce fast enough to compensate for mortality. Here we conclude the time wolves spend on islands depends primarily on size of the islands, which we argue is linked inextricably to availability of prey. So, the movement and distribution of island wolves are not significantly affected by what we expected was poor connectivity (Darimont and Paquet 2002), at least within the geographic confines of our study area. The time between population recovery by deer following predation by wolves in isolated fragments, the island's subsequent recolonization by wolves, plus the availability and seasonality of salmon and other marine resources, define a unique predator–prey system with different dynamics than most terrestrial systems.

In addition to spatial influences and food availability, we believe use of islands by wolves reflects, in part, evolved behaviors and life-history traits that confer a rare ability to adapt to environmental diversity at various temporal and spatial scales (Weaver *et al.* 1996). Coastal islands support only a subset of the adjacent mainland's mammalian carnivores, likely biased towards fecund species adept at crossing large expanses of water. For example, grizzly bears, black bears, and cougars (*Felis concolor*) occur on far fewer islands than wolves, and usually close to the mainland (C. Darimont and P. Paquet, unpublished data). In addition, we expect that physiography, vegetation, and human disturbance alter the amount of habitat available to wolves on islands and the mainland. Finally, characteristics of water channels (e.g., tides, depth, water temperature, and currents) separating landmasses may influence species movement through the landscape.

Ongoing studies combining stable isotope and fecal analyses, and occurring over several seasons, might provide better insight into connectivity and predator–prey dynamics in marine archipelagos. Notably, if combined with microsatellite genetic markers, we may learn how food resources influence presence, movements, and fates of individuals and populations over time, and assess how water barriers affect metapopulation dynamics (Hanski 1991; Hanski and Gilpin 1991). Similar frameworks for other large mammals in habitat patches have recently been developed (e.g., Elmhagen and Angerbjörn 2001).

We have presented several working hypotheses and charted a course for future research. As one of the few remaining large blocks of comparatively unmodified landscapes on Earth, we consider the Great Bear Rainforest, to which the area is commonly referred, a precious academic, aesthetic, and spiritual resource (Paquet *et al.* 2005). The Great Bear harbors one of humankind's last opportunities for studying the outcome of long-term evolution on a geographic scale, and observing highly specialized and co-evolved interactions that are being replaced elsewhere with invasive species or managed landscapes (McKinney and Lockwood 1999; Crooks and Suarez Chapter 18). No amount of money or efforts in restoration ecology can recapture the geographic mosaics of these long-term experiments in evolution. Sadly, the rate of human-induced environmental change is so rapid in the Great Bear (Moola *et al.* 2004) that many species must be tracking their changing environment with a noticeable lag. Consequently, connectivity for wolves and other inhabitants of these ancient temperate rainforests will likely emerge as a more pressing concern.

ACKNOWLEDGEMENTS

This study took place in the Traditional Territories of several First Nation groups, from whom we sought permission before research began. We are particularly indebted to the Heiltsuk. We are extremely grateful to the Raincoast Conservation Society for financial and logistical support. We are obliged to Chris Genovali, Ian and Karen McAllister, Misty MacDuffee, Gudrun Pfleuger, Anita Rocamora, Chester Starr, and Erin Urton for fieldwork, and skippers Jean-Marc Leguerrier and Dave Lutz. The McCaw Foundation, National Geographic Society, University of Montana Paquet Wildlife Fund, Wilburforce, Vancouver Foundation, World Wildlife Fund Canada, and private donors kindly provided funding. While preparing the manuscript, CTD was supported by a Natural Sciences and Engineering Research Council (NSERC) Industrial Postgraduate Scholarship.

REFERENCES

Abbott, I. 1974. Numbers of plant, insect and land bird species on nineteen remote islands in the Southern Hemisphere. *Biological Journal of the Linnean Society* 6:143–152.
Alcover, J. A., and M. McMinn. 1994. Predators of vertebrates on islands. *BioScience* 44:12–18.

Alexander, S. M., N. M. Waters, and P. C. Paquet. 2004. A probability-based GIS model for identifying focal species linkage zones across highways in the Canadian Rocky Mountains. Pp. 233–256 in J. Stillwell, and G. Clarke (eds.) *Applied GIS and Spatial Analysis*. Chichester, UK: John Wiley.

Anderson, D. R., W. A. Link, D. H. Johnson, and K. P. Burham. 2001. Suggestions for presenting the results of data analyses. *Journal of Wildlife Management* 65:373–378.

Angerbjörn, A., P. Hersteinsson, K. Liden, and E. Nelson. 1994. Dietary variation in Arctic foxes (*Alopex lagopus*): an analysis of stable carbon isotopes. *Oecologia* 99:226–232.

Ballard, W. B., R. Farnell, and R. O. Stephenson. 1983. Long distance movement by gray wolves, *Canis lupus*. *Canadian Field Naturalist* 97:333.

Ben-David, M., R. T. Bowyer, L. K. Duffy, D. D. Roby, and D. M. Schell. 1998. Social behavior and ecosystem processes: river otter latrines and nutrient dynamics of terrestrial vegetation. *Ecology* 79:2567–2571.

Boyd, D. K., P. C. Paquet, S. Donelon, *et al.* 1995. Transboundary movements of a recolonizing wolf population in the Rocky Mountains. Pp. 135–141 in L. N. Carbyn, S. H. Fritts, and D. R. Seip (eds.) *Ecology and Conservation of Wolves in a Changing World*, Occasional Publication No. 35. Edmonton, Alberta, Canada: Canadian Circumpolar Institute.

Brotons, L., M. Mönkkönen, and J. L. Martin. 2003. Are fragments islands? Landscape context and density–area relationships in Boreal forest birds. *American Naturalist* 162:343–357.

Burkey, T. V. 1995. Extinction rates in archipelagoes: implications for populations in fragmented habitats. *Conservation Biology* 9:527–541.

Burnham, K. P., and D. R. Anderson. 1998. *Model Selection and Inference: A Practical Information-Theoretic Approach*. New York: Springer-Verlag.

Byun, S. A., B. F. Koop, and T. E. Reimchen. 1997. North American black bear mtDNA phylogeography: implications for morphology and the Haida Gwaii glacial refugium controversy. *Evolution* 51:1647–1653

Callaghan, C. J. 2002. The ecology of gray wolf (*Canis lupus*): habitat use, survival, and persistence of gray wolves in the central Rocky Mountains. Ph.D. dissertation, University of Guelph, Guelph, Ontario, Canada.

Cameron, A. W. 1958. Mammals of the islands in the Gulf of St. Lawrence. *National Museum of Canada Bulletin* 154:1–164.

Carroll, C., R. Noss, P. C. Paquet, and N. H. Schumaker. 2004. Extinction debt of protected areas in developing landscapes. *Conservation Biology* 18:1110–1120.

Clevenger, A. P., and N. Waltho. 2000. Factors influencing the effectiveness of wildlife underpasses in Banff National Park, Alberta, Canada. *Conservation Biology* 14:47–56.

Clevenger, A. P., J. Wierzchowski, B. Chruszcz, and K. Gunson. 2002. GIS-generated, expert-based models for identifying wildlife habitat linkages and planning mitigation passages. *Conservation Biology* 16:503–514.

Conroy, C. J., J. R. Demboski., and J. A. Cook. 1999. Mammalian biogeography of the Alexander Archipelago of Alaska: a north temperate nested fauna. *Journal of Biogeography* 26:343–352.

Cook, J. A., and S. O. MacDonald. 2001. Should endemism be a focus of conservation efforts along the North Pacific Coast of North America? *Biological Conservation* **97**:207–213.

Craig, V. 1990. *Area–Distance Relationships for Mammals on Islands of British Columbia*, Independent Study Paper Submitted to Supervisory Committee. Vancouver, British Columbia, Canada: Simon Fraser University.

Crooks, K. R., and M. E. Soulé. 1999. Mesopredator release and avifaunal extinctions in a fragmented system. *Nature* **400**:563–565.

Darimont, C. T., and P. C. Paquet. 2000. *The Gray Wolves (Canis lupus) of British Columbia's Coastal Rainforests: Findings from Year 2000 Pilot Study and Conservation Assessment.* Victoria, British Columbia, Canada: Raincoast Conservation Society. Available online at http://www.raincoast.org

Darimont, C. T., and P. C. Paquet. 2002. The gray wolves, *Canis lupus*, of British Columbia's Central and North Coast: distribution and conservation assessment. *Canadian Field Naturalist* **116**:416–422.

Darimont, C. T., and T. E. Reimchen. 2002. Intra-hair stable isotope analysis implies seasonal shift to salmon in gray wolf diet. *Canadian Journal of Zoology* **80**:1638–1642.

Darimont, C. T., T. E. Reimchen, and P. C. Paquet. 2003. Foraging behaviour by gray wolves on salmon streams in coastal British Columbia. *Canadian Journal of Zoology* **81**:349–353.

Darimont, C. T., M. H. H. Price, N. N. Winchester, J. Gordon-Walker, and P. C. Paquet. 2004. Predators in natural fragments: foraging ecology of wolves in British Columbia's coastal archipelago. *Journal of Biogeography* **31**:1867–1877.

Darimont, C. T., P. C. Paquet, T. E. Reimchen, and V. Crichton. 2005. Range expansion by moose into coastal temperate rainforests of British Columbia, Canada. *Diversity and Distributions* **11**:235–239.

Dolman, P. M., and W. J. Sutherland. 1997. Spatial patterns of depletion imposed by foraging vertebrates: theory, review and meta-analysis. *Journal of Animal Ecology* **66**:481–494.

Duke, D. L., M. Hebblewhite, P. C. Paquet, C. Callaghan, and M. Percy. 2001. Restoration of a large carnivore corridor in Banff National Park, Alberta. Pp. 261–276 in D. S. Maeher, R. F. Noss, and J. L. Larkin (eds.) *Large Mammal Restoration.* Washington, DC: Island Press.

Dunning, J. B., J. B. Danielson, and H. R. Pulliam. 1992. Ecological processes that affect populations in complex landscapes. *Oikos* **65**:169–175.

Edge, W. D., and C. L. Marcum. 1989. Determining elk distribution with pellet-group and telemetry techniques. *Journal of Wildlife Management* **53**:621–624.

Ehrlich, P. R., and I. Hanski. 2004. *On the Wings of Checkerspots: A Model System for Population Biology.* Oxford, UK: Oxford University Press.

Elmhagen, B., and A. Angerbjörn. 2001. The applicability of metapopulation theory to large mammals. *Oikos* **94**:89–100.

Emlen, J. M., D. C. Freeman, M. D. Kirchhoff, *et al.* 2003. Fitting population models from field data. *Ecological Modelling* **62**:119–143.

Environment Canada. 1991. *1961–1990 Weather Normals for British Columbia: Temperature and Precipitation.* Ottawa, Ontario, Canada: Environment Canada Atmospheric Environment Service.

Fahrig, L. 1997. Relative effects of habitat loss and fragmentation on species extinction. *Journal of Wildlife Management* **61**:603–610.

Fahrig, L. 2003. Effects of habitat fragmentation on diversity. *Annual Review of Ecology and Systematics* **34**:487–515.

Ford, J. K. B., and G. M. Ellis. 1999. *Transients: Mammal-Hunting Killer Whales of British Columbia, Washington, and Southeastern Alaska*. Seattle, WA: University of Washington Press.

Fritts, S. H. 1983. Record dispersal by a wolf from Minnesota. *Journal of Mammalogy* **64**:166–167.

Fritts, S. H., E. E. Bangs, J. A. Fontaine, W. G. Brewster, and J. F. Gore. 1995. Restoring wolves to the northern Rocky Mountains of the United States. Pp. 107–126 in L. N. Carbyn, S. H. Fritts, and D. R. Seip (eds.) *Ecology and Conservation of Wolves in a Changing World*, Occasional Publication No. 35. Edmonton, Alberta, Canada: Canadian Circumpolar Institute.

Fuller, T. K. 1992. Do pellet counts index white-tailed deer numbers and population change?: a reply. *Journal of Wildlife Management* **56**:613.

Fuller, T. K., W. E. Berg, G. L. Radde, M. S. Lenarz, and G. B. Joselyn. 1992. A history and current estimate of wolf distribution and numbers in Minnesota. *Wildlife Society Bulletin* **20**:42–55.

Gaines, W. L., P. Singleton, and A. L. Gold. 2000. Conservation of rare carnivores in the North Cascades Ecosystem, western North America. *Natural Areas Journal* **20**:366–375.

Gilpin, M., and I. Hanski (eds.) 1991. *Metapopulation Dynamics: Empirical and Theoretical Investigations*. London: Academic Press.

Government of Canada. 2005. *Fisheries Information Summary System*. Available online at ftp://gis.luco.gov.bc.ca/pub/coastal/aquaculture/; http://www.bcfisheries.gov.bc.ca/fishinv/db/default.asp

Groot, C., and L. Margulis (eds.) 1991. *Pacific Salmon Life Histories*. Vancouver, British Columbia, Canada: University of British Columbia Press.

Haight, R. G., D. J. Mladenoff, and A. P. Wydeven. 1998. Modeling disjunct gray wolf populations in semi-wild landscapes. *Conservation Biology* **12**:879–888.

Hanski, I. 1991. Single-species metapopulation dynamics. Pp. 17–38 in M. Gilpin, and I. Hanski (eds.) *Metapopulation Dynamics: Empirical and Theoretical Investigations*. London: Academic Press.

Hanski, I., and M. Gilpin. 1991. Metapopulation dynamics: brief history and conceptual domain. Pp. 3–16 in M. Gilpin, and I. Hanski (eds.) *Metapopulation Dynamics: Empirical and Theoretical Investigations*. London: Academic Press.

Hanski, I., and M. Gilpin. 1996. *Metapopulation Biology: Ecology, Genetics, and Evolution*. London: Academic Press.

Harestad, A. S., and F. L. Bunnell. 1987. Persistence of black-tailed deer fecal pellets in coastal habitats. *Journal of Wildlife Management* **51**:33–37.

Harrison, D. J., and T. G. Chapin. 1998. Extent and connectivity of habitat for wolves in eastern North America. *Wildlife Society Bulletin* **26**:767–775.

Hosmer, D. W., and S. Lemeshow. 1989. *Applied Logistic Regression*. New York: John Wiley.

Huntington, H. P. 2000. Using traditional ecological knowledge in science: methods and applications. *Ecological Applications* **10**:1270–1274.

Kadmon, R., and H. R. Pulliam. 1993. Island biogeography: effects of geographical isolation on species composition. *Ecology* **74**:977–981.

Kareiva, P. 1990. Population dynamics in spatially complex environments: theory and data. *Philosophical Transactions of the Royal Society of London B* **330**:175–190.

Kareiva, P., and U. Wennergren. 1995. Connecting landscape patterns to ecosystem and population processes. *Nature* **373**:299–302.

Kareiva, P., D. H. Morse, and J. Eccleston. 1989. Stochastic prey arrivals and crab spider residence times: simulations of spider performance using two simple 'rules of thumb'. *Oecologia* **78**:547–549

Kevan, P. G., and C. D. H. F. Greco. 2001. Contrasting patch choice by immature ambush predators, a spider (*Misumena vatia*) and an insect (*Phymata americana*). *Ecological Entomology* **26**:148–153.

Kirchhoff, M. D. 1990. *Evaluation of Methods for Assessing Deer Population Trends in Southeast Alaska*, Federal Aid in Wildlife Restoration, Progress Report, Project W-22-6, W-23-2, W-23-1. Juneau, AK: Alaska Department of Fish and Game.

Kirchhoff, M. D. 1994. *Effects of Forest Fragmentation on Deer in Southeast Alaska*, Federal Aid in Wildlife Restoration, Job 2.10. W-23-3,4,5, W-24-1,2. Juneau, AK: Alaska Department of Fish and Game.

Kirchhoff, M. D., and K. W. Pitcher. 1988. *Deer Pellet-Group Surveys in Southeast Alaska 1981–1987*, Federal Aid in Wildlife Restoration, Final Report, Job 2.9, Objective 1. W-22-6, W-23-1. Juneau, AK: Alaska Department of Fish and Game.

Laurance, W. F., and M. A. Cochrane. 2001. Synergistic effects in fragmented landscapes. *Conservation Biology* **15**:1488–1489.

Levins, S. 1976. Population dynamics models in heterogeneous environments. *Annual Review of Ecology and Systematics* **7**:287–310.

Loft, E. R., and J. G. Kie. 1988. Comparison of pellet-group and radio triangulation methods for assessing deer habitat use. *Journal of Wildlife Management* **52**:524–527.

MacArthur, R. H., and E. O. Wilson. 1967. *The Theory of Island Biogeography*. Princeton, NJ: Princeton University Press.

MacDonald, S. O., and J. A. Cook. 1996. The land mammal fauna of southeast Alaska. *Canadian Field Naturalist* **110**: 571–598.

Marshall, H. D., and K. Ritland. 2002. Genetic diversity and differentiation of Kermode bear populations. *Molecular Ecology* **11**: 685–697.

McKinney, M. L., and J. L. Lockwood. 1999. Biotic homogenization: a few winners replacing many losers in the next mass extinction. *Trends in Ecology and Evolution* **14**:450–453.

Mech, L. D. 1970. *The Wolf: The Ecology and Behavior of an Endangered Species*. Garden City, NY: Natural History Press.

Mech, L. D., and L. Boitani (eds.) 2003. *Wolves: Behavior, Ecology, and Conservation*. Chicago, IL: University of Chicago Press.

Mech, L. D., S. H. Fritts, and D. Wagner. 1995. Minnesota wolf dispersal to Wisconsin and Michigan. *American Midland Naturalist* **133**:368–370.

Mladenoff, D. J., and T. A. Sickley. 1998. Assessing potential gray wolf restoration in the northeastern United States: a spatial prediction of favorable habitat and potential population levels. *Journal of Wildlife Management* **62**:1–10.

Mladenoff, D. J., T. A. Sickley, R. G. Haight, and A. P. Wydeven. 1995. A regional landscape analysis and prediction of favorable gray wolf habitat in the northern Great Lakes region. *Conservation Biology* **9**:279–294.

Mladenoff, D. J., R. G. Haight, T. A. Sickley, and A. P. Wydeven. 1997. Causes and implications of species restoration in altered ecosystems: a spatial landscape projection of wolf population recovery. *BioScience* **47**:21–31.

Mladenoff, D. J., T. A. Sickley, and A. P. Wydeven. 1999. Predicting gray wolf landscape recolonization: logistic regression models vs. new field data. *Ecological Applications* **9**:37–44.

Moola, F. M., D. Martin, B. Wareham, *et al.* 2004. The coastal temperate rainforests of Canada: the need for ecosystem-based management. *Biodiversity* **9**:5–15.

National Resources Council. 1996. *Upstream: Salmon and Society in the Pacific Northwest*. Washington, DC: National Academy Press.

Newmark, W. D. 1995. Extinction of mammal populations in western North American national parks. *Conservation Biology* **9**:512–526.

Paquet, P. C., and L. N. Carbyn. 2003. Wolf, *Canis lupus* and allies. Pp. 482–510 in G. A. Feldhamer, B. C. Thompson, and J. A. Chapman (eds.) *Wild Mammals of North America: Biology, Management, and Conservation*. Baltimore, MD: Johns Hopkins University Press.

Paquet, P. C., J. Wierzchowski, and C. Callaghan. 1996. Summary report on the effects of human activity on gray wolves in the Bow River Valley, Banff National Park, Alberta. Ch. 7 in J. Green, C. Pacas, S. Bayley, and L. Cornwell (eds.) *A Cumulative Effects Assessment and Futures Outlook for the Banff Bow Valley*. Ottawa, Ontario, Canada: Department of Canadian Heritage.

Paquet, P. C., C. T. Darimont, C. Genovali, and F. Moola. 2005. In the Great Bear Rainforest: island-hopping gray wolves give new insights into island biogeography. *Wild Earth* **Fall/Winter**:20–25.

Patterson, B. R., B. A. Macdonald, B. A. Lock, D. G. Anderson, and L. K. Benjamin. 2002. Proximate factors limiting population growth of white-tailed deer in Nova Scotia. *Journal of Wildlife Management* **66**:511–521.

Person, D. K., M. D. Kirchhoff, V. Van Ballenberghe, G. C. Iverson, and E. Grossman. 1996. *The Alexander Archipelago Wolf: A Conservation Assessment*, General Technical Report PNW-GTR-384. Portland, OR: US Department of Agriculture Forest Service.

Peters, R. P., and L. D. Mech. 1975. Scent-marking in wolves. *American Scientist* **63**:628–637.

Peters, R. P., and L. D. Mech. 1978. Scent-marking in wolves. Pp. 133–147 in R. L. Hall and H. S. Sharp (eds.) *Wolf and Man: Evolution in Parallel*. New York: Academic Press.

Peterson, R. O., and R. E. Page. 1988. The rise and fall of Isle Royale wolves, 1975–1986. *Journal of Mammalogy* **69**:89–99.

Peterson, R. O., N. J. Thomas, J. M. Thurber, J. A. Vucetich, and T. A. Waite. 1998. Population limitation and the wolves of Isle Royale. *Journal of Mammalogy* **79**:828–841.

Pierotti, R., and D. Wildcat. 2000. Traditional ecological knowledge: the third alternative. *Ecological Applications* **10**:1333–1340.

Pojar, J., and A. Mackinnon 1994. *Plants of Coastal British Columbia.* Vancouver, British Columbia, Canada: Lone Pine Publishing.

Polis, G. A., and S. D. Hurd. 1995. Extraordinarily high spider densities on islands: flow of energy from the marine to terrestrial food webs and the absence of predation. *Proceedings of the National Academy of Sciences of the USA* 92:4382–4386.

Pulliam, R. 1996. Sources and sinks: empirical evidence and population consequences. Pp. 45–69 in O. E. Rhodes, R. K. Chesser, and M. H. Smith (eds.) *Population Dynamics in Ecological Space and Time.* Chicago, IL: University of Chicago Press.

Reimchen, T. E. 2000. Some ecological and evolutionary aspects of bear–salmon interactions in coastal British Columbia. *Canadian Journal of Zoology* 78:448–457.

Reimchen, T. E. 2002. Some considerations in salmon management. Pp. 93–96 in B. Harvey, and M. MacDuffee (eds.) *Ghost Runs: The Future of Wild Salmon on the North and Central Coasts of British Columbia.* Victoria, British Columbia, Canada: Raincoast Conservation Society. Available online at http://www.raincoast.org

Ritchie, M. E. 1998. Scale-dependent foraging and patch choice in fractal environments. *Evolutionary Ecology* 12:309–330.

Ritland, K., C. Newton, and H. D. Down. 2001. Inheritance and population structure of the white-phased "Kermode" black bear. *Current Biology* 11:1468–1472.

Rose, M. D., and G. A. Polis. 1998. The distribution and abundance of coyotes: the effects of allochthonous food subsidies from the sea. *Ecology* 79:998–1007.

Rowland, M. M., G. C. White, and E. M. Karlen. 1984. Use of pellet-group plots to measure trends in deer and elk populations. *Wildlife Society Bulletin* 12:147–155.

Saunders, D. A., R. J. Hobbs, and C. R. Margules. 1991. Biological consequences of ecosystem fragmentation: a review. *Conservation Biology* 5:18–32.

Schoonmaker, P. K., B. von Hagen, and E. C. Wolf (eds.) 1997. *The Rainforests of Home: Profile of a North American Bioregion.* Washington, DC: Island Press.

Stephens, D. W., and J. R. Krebs. 1986. *Foraging Theory.* Princeton, NJ: Princeton University Press.

Szepanski, M. M., M. Ben-David, and V. Van Ballenberghe. 1999. Assessment of anadromous salmon resources in the diet of the Alexander Archipelago wolf using stable isotope analysis. *Oecologia* 120:327–335.

Taylor, P. D., L. Fahrig, L. Henein, and G. Merriam. 1993. Connectivity is a vital element of landscape structure. *Oikos* 68:571–572.

Terborgh, J., J. A. Estes, P. C. Paquet, *et al.* 1999. The role of top carnivores in regulating terrestrial ecosystems. Pp. 39–64 in M.E. Soulé, and J. Terborgh (eds.) *Continental Conservation.* Washington, DC: Island Press.

Thomson, S., and M. MacDuffee. 2002. Taking stock: assessment of salmon runs on the north and central coasts of British Columbia. Pp. 35–92 in B. Harvey, and M. MacDuffee (eds.) *Ghost Runs: The Future of Wild Salmon on the North and Central Coasts of British Columbia.* Victoria, British Columbia, Canada: Raincoast Conservation Society. Available online at http://www.raincoast.org

Tischendorf, L., and L. Fahrig. 2000. On the usage and measurement of landscape connectivity. *Oikos* **90**:7–19.

Turner, N. J., M. B. Ignace, and R. Ignace. 2000. Traditional ecological knowledge and wisdom of aboriginal peoples in British Columbia. *Ecological Applications* **10**:1275–1287.

Walton, L. R., H. D. Cluff, P. C. Paquet, and M. A. Ramsay. 2001. Movement patterns of Barren-Ground wolves in the central Canadian arctic. *Journal of Mammalogy* **82**:867–876.

Weaver, J. L., P. C. Paquet, and L. F. Ruggiero. 1996. Resilience and conservation of large carnivores in the Rocky Mountains. *Conservation Biology* **10**:964–976.

Weckerly, F. W., and M. A. Ricca. 2000. Using presence of sign to measure habitats used by Roosevelt elk. *Wildlife Society Bulletin* **28**:146–153.

White, G. C. 1992. Do pellet counts index white-tailed deer numbers and population change?: a comment. *Journal of Wildlife Management* **56**:611–612.

Williamson, M. 1981. *Island Populations.* New York: Oxford University Press.

Woodroffe, R., and J. R. Ginsberg. 1998. Edge effects and the extinction of populations inside protected areas. *Science* **280**:2126–2128.

Migratory connectivity

PETER P. MARRA, D. RYAN NORRIS, SUSAN M. HAIG,
MIKE WEBSTER, AND J. ANDREW ROYLE

INTRODUCTION

Migration represents one of the most complex and fascinating behaviors
in nature. Simply defined, migration is the repeated movement of
individuals from one region to another, and can occur over daily, seasonal,
and annual time-frames. Migrations can occur over tens of thousand of
kilometers, as in the case of the Arctic tern (*Sterna paradisaea*) moving
essentially from pole to pole (Hatch 2002), or it can simply involve
movements over distances as small as meters, as is the case of phantom
midge larvae (*Chaoborus* spp.) moving from lake benthos during the day
to the open water at night (Roth 1968). Migratory behavior has enormous
taxonomic breadth including species of anadromous fish that leave natal
rivers to spend several years at sea eventually returning to the same river
to spawn and die (Hodgson and Quinn 2002), to the annual migrations
of black rat snakes (*Elaphe obsoleta*) from winter hibernacula to summer
breeding and foraging areas (Blouin-Demers and Weatherhead 2002),
to the spectacular seasonal movements of long-distance migratory birds
between temperate breeding and tropical winter environments (e.g., Keast
and Morton 1980; Hagan and Johnston 1992). Movements such as these,
by such taxonomically diverse groups of organisms, define migratory
behavior and motivate the need to understand how these movements
interact with and are modified by the physical structure of their
environment.

Connectivity Conservation eds. Kevin R. Crooks and M. Sanjayan. Published by Cambridge
University Press. © Cambridge University Press 2006.

For the majority of migratory birds, the spatial areas traveled are vast and the individuals too difficult to follow throughout the annual cycle. We know basic annual range distributions for most migratory birds but still have almost no information on where specific individuals, age classes, sex classes, or populations disperse during winter or the subsequent breeding season (Alerstam 1990). The geographic linking of individuals or populations between different stages of the annual cycle, including between breeding, migration, and winter stages, is known as *migratory connectivity* (Webster *et al.* 2002). In this chapter, using primarily migratory birds as an example, we compare and contrast migratory connectivity with other forms of ecological connectivity, consider the implications that different strengths of migratory connectivity have for population dynamics, review the latest techniques available to measure migratory connectivity, and finally consider the role of relative abundance in assigning the geographic origin of migratory animals.

MIGRATORY CONNECTIVITY

This book is a tribute to the importance of ecological connectivity, which, as defined in this volume, is the process of movement and how landscape structure can influence the movement of organisms between habitats. Owing to the importance of this concept to basic ecological theory and conservation, connectivity research has a rich history in ecology (Taylor *et al.* 1993; Tischendorf and Fahrig 2000). Examples of connectivity considered in this book, and nicely summarized by Crooks and Sanjayan (Chapter 1), include hydrologic connectivity (Pringle Chapter 10), connectivity in diseases (McCallum and Dobson Chapter 19), and connectivity within marine ecosystems (DiBacco *et al.* Chapter 8; Harrison and Bjorndal Chapter 9).

These and other concepts of ecological connectivity, particularly those in landscape ecology, consider how habitat and landscape structure influence the movement of organisms through space (e.g., Nebel *et al.* 2002; Sanzenbacher and Haig 2002a, 2002b; Taylor *et al.* Chapter 2). Currently, most studies of migratory connectivity are less concerned with such issues because the large spatial scales involved complicate our ability to understand how landscape and/or atmosphere might influence the movement of individuals between breeding areas and wintering areas. For most migratory animals, seasonal movement from point A to point B is a black box (see Webster *et al.* 2002; Webster and Marra 2005). For migratory birds, although the atmosphere likely plays a role in determining movement dynamics and the degree of connectivity (Gauthreaux *et al.* 2005; Marra *et al.* 2005), little is known about how atmospheric factors influence the establishment of migratory routes. Similarly, as birds migrate over land, habitat configuration and structure undoubtedly influence movement dynamics. For example, as birds migrate north in spring or south in autumn, they periodically stop over in habitat patches to replenish fat stores critical for fueling migratory journeys (e.g., Lyons and Haig 1995; Nebel *et al.* 2000). These stop over sites vary in quality and spatial array across the landscape, but the degree to which migratory movements are influenced by the structure, quality, and connectivity of the landscape is poorly known. In general, our knowledge of stopover ecology has been severely hampered by the fact that we can only study birds for short durations; they may migrate in small, undetected groups and they may only land at night to forage. Understanding how bird populations are geographically connected during migration will provide new spatial and temporal resolution that should open unique insights into migration biology and ultimately the protection of important stop over areas.

Currently, migratory connectivity de-emphasizes how landscape structure influences a population's distribution and instead focuses on retention of breeding population structure on the non-breeding grounds (and vice versa), as well as how conditions and events in non-breeding areas affect populations in breeding areas (and vice versa) (see Webster and Marra 2005). In the future, technological advances will likely improve our ability to track migratory individuals (Wikelski *et al.* 2003; Cochran and Wikelski 2005) and thereby allow researchers to directly assess factors such as the effect of landscape structure on migratory movements.

WHY STUDY MIGRATORY CONNECTIVITY?

The fact that individuals spend time each year in two or more widely separated geographic areas has obvious but poorly understood consequences for the biology and conservation of migratory animals. For example, factors and events on the non-breeding wintering grounds (e.g., climate and weather patterns, deforestation, wetland stabilization or drainage) may affect individual condition, subsequent reproductive success, and recruitment on the summer breeding grounds (Heitmeyer and Fredrickson 1981; Kaminski and Gluesing 1987; Marra *et al.* 1998; Norris *et al.* 2004). Subsequent differences in reproductive success in summer can lead to changes in winter population size (Sillett *et al.* 2000). Likewise, events at migratory stop over sites may affect the timing of arrival and the physical condition and survival of birds onto breeding and/or wintering areas (Myers 1983; Ydenberg *et al.* 2002).

Such effects, termed *seasonal interactions* (Fretwell 1972; Myers 1981; Webster *et al.* 2002; Webster and Marra 2005), are likely to be most pronounced at the population level with strong connectivity (i.e., if individuals breeding together in one location also winter near each other; termed *allohiemy* by Salomonsen (1955)), but may be less pronounced if there is weak connectivity (i.e., if individuals breeding in one area spread out over a large geographic range for the winter period; termed *synhiemy* by Salomonsen (1955)). For example, American redstarts (*Setophaga ruticilla*) breeding at the Hubbard Brook Experimental Forest in New Hampshire could either remain clustered in one specific region on the wintering grounds (strong connectivity) or disperse equally across their entire winter range (weak connectivity). The former scenario can have profound implications for population dynamics if, for example, breeding populations are wintering within a country undergoing rapid deforestation. Alternatively, if migratory connectivity is indeed weak and the Hubbard Brook population disperses throughout the wintering grounds, regional land-use practices on the wintering grounds will likely have less of an impact on population abundance. An important point of clarification here is that our definitions of weak versus strong migratory connectivity counter those in the traditional connectivity literature. Specifically, we predict if we find strong migratory connectivity it suggests that populations are tightly linked and have experienced minimal dispersal or mixing. Strong landscape connectivity predicts the exact opposite—high rates of movement and dispersal. Further explanations below will clarify this distinction.

The challenge in studying migratory connectivity is to understand not only the geographic connections among breeding and non-breeding populations (Webster *et al.* 2002) but also how these connections influence the ecology, evolution and conservation of migratory species (Webster and Marra 2005). For example, under strong migratory connectivity, gene flow among subpopulations is limited and we expect greater levels of local adaptation and potential for speciation (Webster and Marra 2005). Migratory connectivity is also critical to conservation efforts as is illustrated with migratory salmonids in which evolutionary and management units are defined by the timing and geography of migratory breeding "runs" (Waples 1991; Neville *et al.* Chapter 13).

An analogous situation exists in the management of North American waterfowl populations which, for management purposes, are defined in terms of geographic subpopulations based on similarity of migratory pathways and wintering grounds (Bellrose 1976). Understanding the connectivity of these populations (referred to as the "derivation of harvest" problem in waterfowl management – see "Statistical approaches for estimating migratory connectivity" section below) is critical in the management of all game species. As a result, considerable effort has been devoted to the assessment of migratory connectivity using band recovery data with waterfowl species (Munro and Kimball 1982) and mourning doves (Nichols and Tomlinson 1993).

Notable examples also exist for non-game species where the ability to link breeding and non-breeding areas has affected critical management decisions. Perhaps most dramatic was the identification of specific Swainson's hawk (*Buteo swainsoni*) wintering sites in Argentina after the population of breeding birds in the western USA and Canada experienced a significant decline (Goldstein *et al.* 1999). Use of satellite transmitters provided locations of fields where excessive monocrotophos, an organophosphate insecticide that farmers apply to alfalfa fields for grasshopper control, was being used and ultimately where tens of thousands of Swainson's hawks were found dead. An international agreement has now been signed and the pesticide will not be used on grasshoppers or alfalfa fields (American Bird Conservancy 1996). With the endangered piping plover (*Charadrius melodus*), over 20 years of banding data indicated birds exhibit strong fidelity to winter and breeding sites, although populations mix in winter sites (Haig *et al.* 2005). This discovery has allowed managers to pay more attention toward protecting specific winter sites. A similar situation has arisen for the buff-breasted sandpiper (*Tryngites subruficollis*), where their small and declining numbers in North America have also been found wintering in specific fields in Argentina leading to further recognition of the importance of their winter sites (Lanctot *et al.* 2002). In the endangered Mariana moorhen (*Gallinula chloropus guami*), a subspecies of less than 300 birds on four islands (Takano and Haig 2004a), identification of movements between wet and dry seasons indicated inter-island movements of birds between the islands of Saipan and Tinian, but only local movements on Guam (Takano and Haig 2004b). Linking islands and identifying habitat preferences during the wet and dry seasons has provided important guidance on subspecies-wide habitat protection. Finally, the Great Basin population of the western willet (*Catoptrophorus semipalmatus*) has shown a surprisingly strong connectivity between breeding sites in western Oregon and California to very specific sites in and around San Francisco Bay (Haig *et al.* 2002), providing further indications for the importance of conservation efforts in the San Francisco Bay to help protect this species.

As shown in the above examples, knowing the migratory connectivity for a given species can often help with vital management decisions. More generally, other factors such as habitat availability on either the wintering or breeding grounds can also strongly affect key aspects of population demographic structure, such as sex ratio dynamics (Runge and Marra 2005). We do not yet know how different strengths of migratory connectivity and population size are influenced by different amounts of habitat loss. In this next section, we develop a simple equilibrium population model to further explore this relationship.

MIGRATORY CONNECTIVITY AND POPULATION DYNAMICS

Here, we explore how habitat loss affects equilibrium population size (E) under different degrees of migratory connectivity in a species with multiple breeding and wintering (non-breeding) populations. We define three general types of connectivity: *strong connectivity*, in which all individuals from one breeding population migrate to a single wintering population and vice versa; *no connectivity*, in which individuals from all breeding populations migrate in equal proportions to all wintering populations; *moderate connectivity*, in which the majority of individuals from each breeding population migrate to one wintering population and a smaller number migrate to the remaining wintering populations. It is important to note that in any scenario that is not strong connectivity, populations can therefore be considered to mix between the breeding and wintering grounds. In the examples provided here, no connectivity equates with "complete mixing" and moderate connectivity equates with "partial" mixing. In all three scenarios, we investigate the effect of habitat loss (population decrease) occurring at one wintering location on (1) E at all breeding and wintering populations, and (2) change in strength of connectivity (i.e., change in number of individuals migrating between each combination of breeding and wintering populations).

The basic model structure for a single breeding and wintering population follows Norris (2005), where the population size (N) at the end of the breeding season in a given year (Ns_t) is represented by N at the end of the previous winter (Nw_t) times per capita breeding output during the summer (b_t):

$$Ns_t = Nw_t b_t \tag{7.1}$$

where b_t is the linear density-dependent function:

$$b_t = b_o - b_1 Nw_t. \tag{7.2}$$

Here, b_o is the intercept (per capita density-dependent breeding rate as density approaches zero) and b_1 is the slope (strength of density dependence: Norris 2005). Similarly, population size at the end of the previous winter (Nw_t) is the product of N at the end of the previous summer in year $t-1$ (Ns_{t-1}) times survival rate ($1 - $ mortality rate $[d_t]$) during the winter:

$$Nw_t = Ns_{t-1}(1 - d_t) \tag{7.3}$$

where d_t is the linear density-dependent function:

$$d_t = d_o + d_1 Ns_{t-1} \tag{7.4}$$

and d_o and d_1 are analogous to b_o and b_1 in Eq. (7.2).

Model output will predict the number of individuals migrating from each of three breeding sites (s_1, s_2, s_3) to each of three wintering sites (w_1, w_2, w_3), yielding a total of nine possible combinations. We express population size in each of these combinations as the number of individuals at the end of the breeding season (prior to fall migration) at breeding location Ns_x that are migrating to wintering location w_x. For example, the number of individuals migrating from breeding location 1 (Ns_1) to wintering location w_1 at time t is expressed as $Ns_1 w_{1(t)}$. Since there are two other wintering sites in the model, the remaining individuals migrating from breeding site 1 at time t are represented as $Ns_1 w_{2(t)}$ and $Ns_1 w_{3(t)}$.

Based on Eqs. (7.1) and (7.3), $Ns_1 w_{1(t)}$, for example, can be written as:

$$Ns_1 w_{1(t)} = Ns_1 w_{1(t-1)}(1 - d_{(t)}^{w_1})b_{(t)}^{s_1} \tag{7.5}$$

where $d_t^{w_1}$ now takes into account individuals arriving to w_1 from all possible breeding locations:

$$d_{(t)}^{w_1} = d_o^{w_1} - d_1^{w_1}(Ns_1 w_{1(t-1)} + Ns_2 w_{1(t-1)} + Ns_3 w_{1(t-1)}) \tag{7.6}$$

and $b_t^{s_1}$ takes into account individuals arriving to s_1 from all possible wintering locations:

$$b_{(t)}^{s_1} = b_o^{s_1} - b_1^{s_1}(Ns_1 w_{1(t-1)} + Ns_1 w_{2(t-1)} + Ns_1 w_{3(t-1)}). \tag{7.7}$$

In other words, the per capita mortality and breeding rates of a given wintering population (d^w: Eq. 7.6) or breeding population (b^s: Eq. 7.7) are dependent upon the total number of individuals arriving into that population from the previous season. The degree of connectivity will determine the proportion of individuals coming from each of the populations the previous season. After successive iterations of the annual cycle, E is reached when

$$Ns_x w_{x(t)} = Ns_x w_{x(t-1)}. \tag{7.8}$$

For all simulations, we used previous published parameters from Eurasian oystercatchers (*Haematopus ostralegus*), a long-distance migratory shorebird. For a population of 2000 individuals, $d_1 = 0.00011$, $b_1 = 0.00005$ (Sutherland 1996). Given values of per capita breeding and mortality at low densities for this species (Goss-Custard *et al.* 1995), Norris (2005) approximated $b_0 = 1.4$ and $d_0 = 0.001$. Using these parameters, $E = 2180$ for a given population.

The three connectivity scenarios are: (1) strong connectivity, the entire population at a breeding site ($E = 2180$) migrates to a single wintering population (Fig. 7.1A), (2) no connectivity, where exactly one-third of each breeding population ($N = 727$) migrates to one of three wintering populations (Fig. 7.1B), (3) moderate connectivity, where 90% of each breeding population ($N = 1962$) migrates to a wintering location and 10% ($N = 218$) migrate to two other wintering sites (Fig. 7.1C). For all three scenarios, we simulated 50% habitat loss at a single wintering population (w_3), where habitat loss is reflected by a change in the strength of the winter density-dependent mortality function (d_1; in this case $d_1^{w_3}$) (Sutherland 1996; Norris 2005).

We assume that all breeding and wintering locations have equal amounts and quality of habitat and that the strength of density dependence is equal between populations. We also assume that individuals cannot change their migratory route once set by one of three connectivity scenarios, an assumption that may or may not be realistic depending on the species (see discussion below). Our aim in this exercise, however, is to examine how migratory connectivity can influence the number of individuals migrating between populations based on population dynamics alone. Other models have considered how migration routes between breeding and wintering areas may develop through evolutionary stable-strategies, assuming individuals have the ability to change migratory routes in response to habitat loss (Dolman and Sutherland 1994; Sutherland and Dolman 1994).

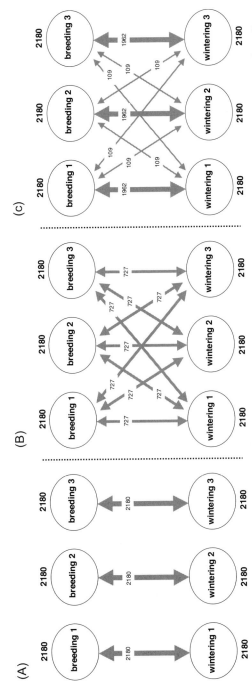

Fig. 7.1. Three connectivity scenarios where $E = 2180$ for each breeding and wintering population. (A) Strong connectivity: breeding populations migrate to a single, mutually exclusive wintering location. (B) No connectivity: an equal number of individuals from a given breeding location migrate to each wintering location. (C) Moderate connectivity: 90% of individuals from a given breeding location migrate to one wintering location and the rest migrate equally to the other two wintering locations (5% each). The numbers embedded in arrows indicate the number of individuals migrating between each of the breeding and wintering sites at equilibrium.

Model output

Under strong connectivity, 50% habitat loss at w_3 resulted in a 45% decrease in E at w_3 and s_3 (Fig. 7.2A). The smaller proportional decrease in E relative to percent habitat loss is due to a "seasonal compensation effect" of density-dependent breeding output the following season (Sutherland 1996; Norris 2005).

With no connectivity (complete mixing), 50% habitat loss at w_3 results in an equal decrease (49%) of individuals migrating from all three breeding locations to w_3 (Fig. 7.2B). Concurrently, E decreases by an equal amount (14%) at all three breeding locations. In contrast, the number of individuals migrating from all three breeding locations to the remaining wintering locations (w_1, w_2) increases by 3% (Fig. 7.2B).

When connectivity is moderate (partial mixing), the number of individuals migrating from s_3 to w_3 decreases by 44% (1962 individuals in Fig. 7.1C to 1091 individuals in Fig. 7.2C) and the number of individuals migrating from s_1 and s_2 to w_3 decreases by 86% (109 individuals in Fig. 7.1C to 15 in Fig. 7.2C). In contrast, the number of individuals migrating from s_3 to the other two wintering locations (w_1 and w_2) increases by 257% from 109 (Fig. 7.1C) to 389 (Fig. 7.2C). Overall, decreases in E at all locations are the same as the no-connectivity scenario (14% decreases at breeding locations).

By simulating habitat loss and altering population size at one wintering location, we have shown that the degree of migratory connectivity can affect population size at multiple breeding and wintering locations, as well as the number and distribution of individuals migrating between these locations. The model generates three predictions:

(1) When habitat is lost at one location, any level of initial mixing (i.e., moderate or no connectivity) between the breeding and wintering populations will result in synchronous population declines and a similar E among populations in the subsequent season.

(2) If individuals mix between the breeding and wintering populations, a decline at one location will result in an increase in E of other populations in the same season.

(3) The proportional distribution of individuals migrating between breeding and wintering populations after habitat loss is strongly affected by the initial degree of connectivity (weak vs. strong). For example, when there is no connectivity (complete mixing), the model predicts that: (A) there will be an equal decline in the number of individuals migrating from all breeding locations (s_1, s_2, s_3) to the site

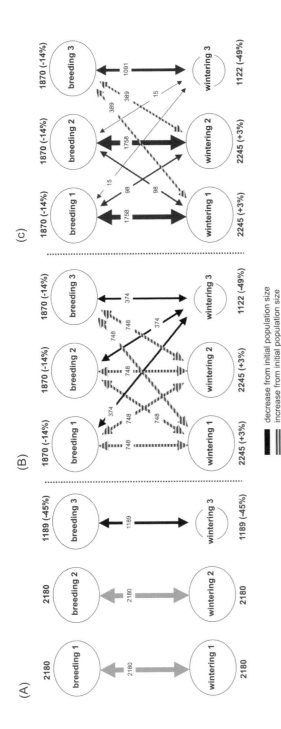

Fig. 7.2. As in Fig. 7.1 but after 50% habitat loss at wintering location 3 (w_3 in model). (A) Strong connectivity, (B) No connectivity, (C) Moderate connectivity. The numbers in parentheses beside E indicate the percent change in E after habitat loss. The pattern of arrows indicates an increase (black stripes) or decrease (solid black) in the number of individuals migrating from a given breeding and wintering location.

experiencing habitat loss (w_3), and (B) individuals migrating to all other wintering locations will increase. Alternatively, when populations are strongly (but not completely) connected, the model predicts that (C) small populations migrating to the site where habitat loss occurs (i.e., s_1w_3 and s_2w_3) will experience a larger proportional decrease in population size compared to large populations. This occurs because of a density-dependent feedback mechanism between breeding and wintering sites. In the model (Fig. 7.2C), the s_3w_3 group experiences a large absolute decrease in population size, thereby increasing the number of individuals migrating from s_3 to other wintering sites (s_3w_1, s_3w_2) by over 250% (through an increase in density-dependent breeding output). The density-dependent feedback occurs as a result of increases in these s_3w_1 and s_3w_2 groups, whereby an increase in the number of individuals in the w_1 and w_2 populations in turn lowers the s_1 and s_2 populations through negative density-dependent effects on reproduction. This, combined with the habitat loss at w_3, acts to decrease the s_1w_3 and s_2w_3 by a large amount relative to its initial size before habitat loss.

Such model predictions can be applied to data from long-term population trends of migratory species. For example, many Neotropical–Nearctic migratory birds exhibit asynchronous population trends across their breeding range (Sauer *et al.* 2004) and recent studies suggest that winter habitat may limit reproductive success on temperate breeding grounds (Marra *et al.* 1998; Norris *et al.* 2004). If a species shows strong connectivity between breeding and wintering locations, severe habitat loss at a single wintering location should only affect a single breeding location and should not affect the population size at other wintering locations. In contrast, our model predicts that species that mix between the breeding and wintering grounds should have relatively *synchronous* population trends across their breeding range or a portion of their range in which mixing occurs (prediction 1). Interestingly, if populations exhibit mixing or weak connectivity, severe habitat loss at one wintering location should *increase* population size at other wintering locations (prediction 2).

In general, our model shows how the degree of migratory connectivity can impact population dynamics. Previous studies that have modeled the effect of habitat loss in multiple breeding and wintering populations have assumed that individuals have the ability to change migratory routes and that individuals redistribute themselves according to an ideal-free distribution (i.e., equal fitness within and between wintering and/or breeding areas: Fretwell and Lucas 1970; Dolman and Sutherland 1994; Sutherland and Dolman 1994). These assumptions may apply for some species (e.g., red-breasted goose, *Branta ruficollis*: Sutherland and Crockford 1993; light-bellied brent goose, *Branta bernicla hrota*: Clausen *et al.* 1998), however, other migratory birds have shown remarkably fixed migration patterns over time, even after major historical events (e.g., Swainson's thrush, *Catharus ustulatus*: Ruegg and Smith 2002). We do not advocate one assumption over the other but rather wish to point out that the strengths of migratory connectivity may change even if migration routes are fixed for some species. Furthermore, many species likely do not distribute themselves between spatially distinct migratory populations according to ideal-free properties. Passerines, in particular, have been shown to be highly territorial on wintering grounds (e.g., Marra *et al.* 1993; Stutchbury 1994), which by definition is not ideal-free and can lead to significant differences in physical condition, timing of migration, and annual survival (Marra *et al.* 1998; Marra and Holmes 2001; Studds and Marra 2005).

The nature of how species are distributed within a season, as well as the degree of flexibility in changing migratory routes, will almost certainly produce different outcomes in relation to habitat loss and migratory connectivity. For example, results of this model show that when there is any degree of mixing between breeding and wintering populations, habitat loss at one wintering location will increase population size of other wintering locations through density-dependent feedback mechanisms. In contrast, models developed by Sutherland and Dolman (1994) and Dolman and Sutherland (1994) suggest that loss of habitat in one wintering area will decrease population size in other wintering areas because individuals are able to change migration routes and settle across wintering areas according to an ideal-free distribution.

One of the most basic requirements for understanding factors that influence population trends of migratory species will be acquiring detailed knowledge of population distributions during breeding and non-breeding seasons. Clearly, we will also need accurate data on dispersal, density dependence, and migratory behavior to be able to accurately predict the consequences of habitat loss across an entire network of migratory populations. Although not incorporated into this model, connectivity between migratory stopover sites and breeding and wintering areas will also likely play a key role in the dynamics of migratory populations and should be incorporated into future theoretical and empirical work.

TOOLS FOR MEASURING MIGRATORY CONNECTIVITY

We are a long way from understanding the importance of migratory connectivity for most migratory birds because of our inability to track individuals between breeding and wintering grounds. However, we are making some progress in the development of tools for measuring geographic connectivity of migratory populations (Webster *et al.* 2002; Rubenstein and Hobson 2004). Below, we briefly summarize techniques currently available to measure geographic origin and also consider their latest applications to measure migratory connectivity in birds.

For decades, devices such as aluminum leg bands, collars, or ear tags have been used to track movements of individually marked animals. Such methods require recapturing or resighting the same individual. Return rates of marked individuals, such as birds marked with aluminum bands, thus far suggests that these methods do not hold much hope for understanding connectivity of migratory bird populations. Monitoring marked waterfowl and shorebirds may be the exceptions to this problem (Alerstam 1990; Bairlein 2001). Satellite telemetry also offers some promise for understanding migratory connectivity, at least for large mammals such as bowhead whales (*Balaena mysticetus*: Heide-Jorgensen *et al.* 2003), sea turtles (Craig *et al.* 2004; Harrison and Bjorndal Chapter 9), and larger migratory birds such as raptors (C. L. McIntyre, pers. comm.), geese (e.g., Green *et al.* 2002; Fox *et al.* 2003), storks (Berthold *et al.* 2001), and seabirds (e.g., Hatch *et al.* 2000). However, the prohibitive costs of satellite transmitters have yet to produce a study with large enough sample sizes over multiple years to assess the range-wide connectivity of any migratory species.

In many ways, molecular genetic approaches have revolutionized our ability to examine migratory connectivity (see also Frankham Chapter. 4; Neville *et al.* Chapter. 13). The classic genetic approach to measuring the movement of individuals among geographically separated populations is to infer gene flow from measures of population structure such as Wright's F_{ST} (Wright 1978), which is a measure of variance in allele frequencies among populations. Recent advances in molecular genetic techniques have offered significant advances on two fronts. First, new classes of molecular markers have been developed, such as microsatellites (Jarne and Lagoda 1996) and amplified fragment length polymorphisms (Mueller and Wolfenbarger 1999), which reveal substantial genetic variation within and across populations. As a consequence of the often substantial variation revealed, these markers can be highly sensitive indicators of genetic differentiation among populations. Second, sophisticated analytical techniques have been developed, such as population assignment tests (Smouse *et al.* 1986) that potentially allow individuals to be assigned to specific populations (Haig *et al.* 1997; Webster *et al.* 2002; Cegelski *et al.* 2003).

Although molecular approaches have the potential to uncover patterns of migratory connectivity, it is currently unclear how useful they will prove in many avian systems. The ease with which birds disperse provides for potentially high levels of gene flow. Thus, developing population-specific markers is difficult when there is little difference among populations. Not surprisingly, studies examining variation at mitochondrial markers have typically uncovered only weak geographic patterns (but see Wenink and Baker 1996; Wennerberg 2001), such as weak east–west genetic variation in North America (e.g., Milot *et al.* 2000; Gorman 2001; Lovette *et al.* 2004). As a consequence, mitochondrial (and other) markers by themselves may be most useful in determining only broad geographic patterns of connectivity (e.g., Kimura *et al.* 2002). However, molecular markers may be combined with other types of markers to increase the precision and resolution of connectivity studies (Clegg *et al.* 2003). It is important to note that a lack of genetic differentiation in mitochondrial DNA could be due not just to gene flow but to historical demographic events such as rapid population expansions, which erase phylogeographic structure. Therefore, the absence of differentiation does not necessarily preclude the existence of connectivity between winter and summer populations.

Stable isotope compositions of animal tissues can also be used to track migratory patterns. Isotope approaches have been applied to migratory animals including birds (Chamberlain *et al.* 1997, 2000; Hobson and Wassenaar 1997; Marra *et al.* 1998, Kelly *et al.* 2002; Rubenstein *et al.* 2002), butterflies (monarch butterflies: Wassenaar and Hobson 1998), fish (salmon: Harrington *et al.* 1998; Kennedy *et al.* 1998), and mammals (African elephants: van der Merwe *et al.* 1990; Vogel *et al.* 1990; Koch *et al.* 1995). The method is based on the fact that natural variations in the stable isotope ratios in animal tissues (bones, muscle, blood, egg shells, feathers, etc.) are incorporated from local climatic conditions, soil type, vegetation, and diet (Mizutani and Wada 1989; Schaffner and Swart 1991; Hobson and Clark 1992). For example, stable isotope ratios of carbon (δ^{13}C: Korner *et al.* 1991; Van Klinken *et al.* 1994) and hydrogen (δD: Epstein *et al.* 1976, Estep and Dabrowski 1980) in plants and animals vary systematically with latitude and climatic conditions, and strontium isotope ratios (δ^{87}Sr) in animal tissues vary as a function of bedrock type (Kennedy *et al.* 1998).

By integrating three isotopes, Chamberlain *et al.* (1997) were able to show that δD and δ^{13}C values of feathers and δ^{87}Sr values in bones of black-throated blue warblers (*Dendroica caerulescens*) varied systematically across their north temperate breeding grounds. Feathers were used because their stable isotopic composition reflects that of the foods eaten during their period of growth (Mizutani *et al.* 1990, 1992; Hobson and Clark 1992) and these ratios remain inert after the feather is grown. Since black-throated blue warblers molt only once per year, between July and September on or near their breeding area (Holmes 1994), the isotopic composition of their feathers reflects that of the food chain at their breeding locality (Chamberlain *et al.* 1997). Studies have also been conducted to assess the utility of using these isotope patterns for identifying the breeding origins of individual birds wintering in the Greater Antilles (Chamberlain *et al.* 1997; Rubenstein *et al.* 2002). These results suggest weak connectivity of individuals between the breeding and wintering grounds, although a greater proportion of individuals wintering in the western Caribbean islands were from northern breeding grounds, whereas those wintering further east were from southern breeding populations.

One study has examined the connectivity of a migratory passerine bird across its entire wintering and breeding range. Combining stable-hydrogen isotopes analyzed from tail feathers and band recovery data, Norris *et al.* (in press) found that American redstarts displayed high levels of regional connectivity between their temperate breeding and tropical wintering grounds. Individuals wintering in Mexico primarily bred in western Canada and the northwestern USA, individuals wintering in Central America bred in the mid-western range, whereas most individuals wintering in the Caribbean bred in the eastern USA and Canada. The latter group showed a unique pattern of chain migration, where northern wintering populations bred at the most northern sites.

Studies to date all indicate that the isotopic composition of bird tissues has great potential for identifying regional and even more localized populations of migratory species. By combining the use of isotopic tracers with molecular genetic approaches, and with other population markers (i.e., parasites, microbial communities, trace elements), we may be able to estimate with even greater precision the migratory connectivity between breeding and wintering populations of migratory birds. Having individual and population markers such as the ones we have described above to assign geographic origin will only solve part of the migratory connectivity riddle. Also required are statistical approaches that incorporate relative abundance when assigning the breeding ground origin for birds on wintering areas (Royle and Rubenstein 2004). Below, we describe the problem and a statistical method for addressing this critical issue.

Statistical approaches for estimating migratory connectivity

A natural way to characterize migratory connectivity between breeding and wintering populations is by defining a set of transition probabilities that describe where birds from each breeding population winter. That is, how breeding birds from any particular population distribute themselves across their wintering range. Formally, let $\psi_{ws} = \Pr(w|s)$ be the probability that a bird from breeding population s winters in region w. We note that these probabilities completely determine where birds winter *given* a particular origin. However, a central question in many studies of migratory connectivity is the "inverse problem": given a sample of birds obtained at some wintering location, from which breeding population did they originate? In probability terms, this is a question about the conditional probability $\gamma_{sw} = \Pr(s|w)$, the "origin probabilities," or the probability that a bird wintering in region w originates from breeding population s. Estimating these conditional probabilities is central to the classical "derivation of harvest" problem in waterfowl management and an analogous problem exists in the management of migratory fish populations. In general, knowledge of ψ_{ws} is insufficient for obtaining estimates of the conditional probabilities γ_{sw} and vice versa.

The relationship between these two sets of conditional probabilities is embodied in Bayes' Rule, which states that

$$\Pr(s|w) = \frac{\Pr(w|s)\,\Pr(s)}{\Pr(w)}. \tag{7.9}$$

Note that this expression involves the marginal probability $\Pr(s)$. This can be thought of as the fraction of the population that originates from s, i.e., it is proportional to the population size of breeding population s. For example, Fig. 7.3 illustrates the relative abundance map of the American redstart (computed from Breeding Bird Survey data) over its breeding range, which can (when suitably scaled) be regarded as an estimate of the probability that a randomly selected bird from the population at large originates from any particular local population.

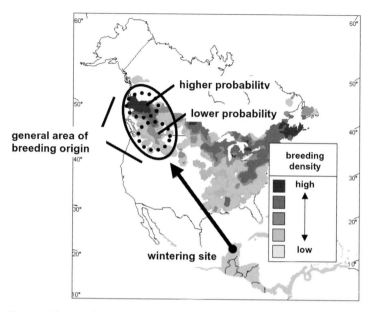

Fig. 7.3. The "probability of origin" problem in estimating migratory connectivity (in this case of the American redstart). A marker (for example, stable isotopes) indicates that an individual sampled at a tropical wintering site bred the previous year within the region in North America indicated by the ellipse. Within this region, there are variable breeding densities, implying that the individual had a higher probability of breeding in the area of high density rather than the area of low density. In this way, the breeding origin of individuals (i.e., migratory connectivity) can be expressed as a set of transition probabilities (in addition to using extrinsic or intrinsic markers).

In the denominator, Pr(*w*) describes the distribution among different wintering populations, and it can be obtained by summation of the numerator of Eq. (7.9) over all possible breeding populations (thus, only knowledge of the distribution of the breeding population is necessary). This expression makes clear that one cannot answer the inverse problem without knowledge of population size, at least relatively, among the potential breeding populations.

It is sometimes possible to estimate the transition probabilities directly, for example by using band recovery data. If R_I birds are banded in s_I and a random sample of birds is selected from populations w_1 and w_2 yielding frequencies r_{11} and r_{21} then estimates of ψ are obtained as the proportions r_{i1}/R_I. Then, in order to obtain estimates of the origin probabilities for any fixed wintering population, it is necessary to, in effect, "weight" these transition probabilities by the size of each breeding population according to Eq. (7.9).

Illustration

Suppose a population exists as two geographically defined breeding populations ($s = 1$, 2) and that there are two wintering areas for the species within which these populations mix. Suppose further that, through intensive banding, the transition probabilities are known to be $\psi_{11} = 0.1$, $\psi_{21} = 0.9$ for breeding population 1 and $\psi_{12} = 0.9$, $\psi_{22} = 0.1$ for breeding population 2. Thus, most (90%) of population 1 birds winter at $w = 2$ and most (90%) of population 2 birds winter at $w = 1$. We would seem to know a lot about this system in the sense that we know the proportion of birds from each breeding population that winter in each of the two wintering populations. However, it is sometimes important (such as in harvest management of North American waterfowl) to estimate the mixture of individuals in some wintering area from the two populations. It is tempting to look at the estimated transition probabilities and declare that birds in $w = 2$ are primarily birds from population 1, and this might conceivably effect management activities directed at this species (e.g., where management objectives or actions for the two breeding populations can be made independently). Put another way, suppose one has a bird in hand, captured at $w = 2$, then one might be tempted to say that that bird most likely originated from population 1. In fact, we really know nothing quantitative about the mix of birds that occurs in the mixed population at $w = 2$. To establish this, suppose that the breeding population sizes are, respectively, $N_1 = 100$ and $N_2 = 1000$. Given the transition probabilities specified previously we expect, on average, 90 birds from population 1 (0.9×100) and 100 birds from population 2 (0.1×1000) in the mixed population at $w = 2$. That is the population of birds at $w = 2$ is made up primarily of birds from population 2. Thus, assessment of origin (the "inverse problem") depends not just on the transition probabilities, but also on the relative size of the two breeding populations.

In many situations it is not possible to measure or estimate transition probabilities directly, but rather an extraneous trait (e.g., an allele or isotope signature) associated with each of the breeding populations is measured. Denote this trait by y, and let $f(y|s)$ denote the probability distribution of this trait for breeding population s. Then y is measured on a sample of individuals in some mixed population (the wintering population, w), yielding $f(y|w)$: the probability distribution for this trait in the wintering population. If we wish to make an inference about the origin of the individuals in the mixed wintering population, this again is a question about the conditional probabilities $\Pr(s|w)$, the probability that a bird wintering in region w originates from breeding population s. To obtain estimates of these conditional probabilities, we note that $f(y|w)$ can be expressed as a mixture of probability distributions according to:

$$f(y|w) = \sum_s f(y|s)\gamma_{sw}. \tag{7.10}$$

Recall that γ_{sw} is the probability that a bird from breeding population s makes a transition to wintering population w. Thus, having obtained estimates of $f(y|s)$ from sampling the breeding populations, one can obtain estimates of the probabilities γ_{sw} simply by rearranging Eq. (7.10) and solving for the unknown values of γ_{sw}. Now, consider estimating the transition probabilities from each breeding population, ψ_{ws}. Applying Bayes' Rule as before we have:

$$\psi_{ws} = \frac{\gamma_{sw}\,\Pr(w)}{\Pr(s)}. \tag{7.11}$$

We see that it is not possible to obtain estimates of $\psi_{ws} = \Pr(w|s)$ unless information about abundance is available, or at least the relative distribution of individuals among the various populations (e.g., Fig. 7.3).

Illustration

Consider two distinct breeding populations composed of 1000 and 100 individuals, and assume that there exists a discrete marker (e.g., an allele) taking on values 1, 2, and 3. Suppose that the frequency of individuals in each population having each value of $y = (1, 2, 3)$ are (800, 150, 50) for population 1, comprising 1000 individuals, and (20, 50, 30) for population 2, comprising 100 individuals. We suppose these two breeding populations migrate to some location (w) with probabilities $\psi_1 = 0.1$ and $\psi_2 = 0.5$, yielding a mixed population with frequency distribution (90, 40, 20). These are computed as, for example, $0.1 \times 800 + 0.5 \times 20 = 90$. In practice, the transition probabilities are unknown and in many problems, neither will be the breeding or wintering population sizes. Rather, the observed data are the allele frequencies in a sample of the mixed population. In this case, the probability distribution of y in the mixed population w is: $f(y = 1|w) = 90/150$, $f(y = 2|w) = 40/150$, and $f(y = 3|w) = 20/150$. To obtain estimates of the conditional probabilities $\gamma_{sw} = \Pr(s|w)$, we substitute these observed quantities into Eq. (7.10) and solve, obtaining $\gamma_1 = 2/3$ and $\gamma_2 = 1/3$. That is, 2/3 of the individuals in this mixed population originate from population 1. To obtain estimates of transition probabilities it is necessary to consider population sizes (Eq. (7.11)), specifically, the relative sizes of the breeding population and also the relative size of the particular mixed (wintering) population in question. Finally, note that the problem of assessing connectivity using isotope markers is analogous to the example considered here, except in this case the "marker" is a continuous trait, defined by the isotope signature measured in bird tissue.

CONCLUSIONS

Phases of the annual cycle, including breeding, migration, and over-wintering, for any migratory animal are inextricably linked yet these connections are poorly understood for most species. Events occurring during winter undoubtedly impact events during migration and the subsequent breeding period. Such *seasonal interactions* can have profound implications for the ecology, evolution, and population dynamics of animals that exhibit migratory behavior (Marra *et al.* 1998; Gill *et al.* 2001; Norris *et al.* 2004; Runge and Marra 2005; Webster and Marra 2005). Unfortunately, for species such as migratory birds, understanding seasonal interactions remain elusive because of the difficulties associated with following the same individuals and/or populations between specific summer, migratory stopover and winter habitats. Such *migratory connectivity* represents a critical area in need of further research and development. However, we are making progress. Technological advances in individual identification and development of genetic and stable isotope signatures to track populations are offering exciting promise to what is a giant gap in our understanding of migratory animals. This, along with more detailed modeling and statistical efforts, are needed to determine how focal breeding populations are affected by large- and small-scale events affecting various wintering populations and vice versa. It will only be through these highly multidisciplinary approaches that we will be able to come to terms with the connectivity of migratory populations and develop effective conservation strategies.

REFERENCES

Alerstam, T. 1990. *Bird Migration*. Cambridge, UK: Cambridge University Press.
American Bird Conservancy. 1996. *Protection for Swainson's Hawk*. Available online at http://www.pmac.net/hawks.htm

Bairlein, F. 2001. Results of bird ringing in the study of migration routes. *Ardea* 89:S7–19.

Bellrose, F. C. 1976. *Ducks, Geese and Swans of North America*, 2nd edn. Harrisburg, PA: Stackpole Books.

Berthold, P., W. van den Bossche, W. Fiedler, *et al.* 2001. Detection of a new important staging and wintering area of the white stork *Ciconia ciconia* by satellite tracking. *Ibis* 143:450–455.

Blouin-Demers, G., and P. Weatherhead. 2002. Implications of movement patterns for gene flow in black rat snakes (*Elaphe obsoleta*). *Canadian Journal of Zoology* 80:1162–1172.

Cegelski, C. C., L. P. Waits, and N. J. Anderson. 2003. Assessing population structure and gene flow in Montana wolverines (*Gulo gulo*) using assignment-based approaches. *Molecular Ecology* 12:2907–2918.

Chamberlain, C. P., J. D. Blum, R. T. Holmes, *et al.* 1997. The use of isotope tracers for identifying populations of migratory birds. *Oecologia* 109:132–141.

Chamberlain, C. P., S. Bensch, X. Feng, S. Akesson, and T. Andersson. 2000. Stable isotopes examined across a migratory divide in Scandinavian willow warblers (*Phylloscopus trochilus trochilus* and *Phylloscopus trochilus acredula*) reflect their African winter quarters. *Proceedings of the Royal Society of London B* 267:43–48.

Clausen, P., J. Madsen, S. M. Percival, D. O'Conner, and D. Q. A. Anderson. 1998. Population development and changes in winter site use by the Svalbard light-bellied brent goose, *Branta bernicla hrota* 1980–1994. *Biological Conservation* 84:157–165.

Clegg, S. M., J. F. Kelly, M. Kimura, and T. B. Smith. 2003. Combining genetic markers and stable isotopes to reveal population connectivity and migration patterns in a Neotropical migrant, Wilson's warbler (*Wilsonia pusilla*). *Molecular Ecology* 12:819–830.

Cochran, W. W., and M. Wikelski. 2005. Individual migratory tactics of New World *Catharus* thrushes. Pp. 274–289 in R. Greenberg, and P. P. Marra (eds.) *Birds of Two Worlds: The Ecology and Evolution of Migration*. Baltimore, MD: Johns Hopkins University Press.

Craig, P., D. Parker, R. Brainard, M. Rice, and G. Balazs. 2004. Migrations of green turtles in the central South Pacific. *Biological Conservation* 116:433–438.

Dolman, P. M., and W. J. Sutherland. 1994. The response of bird populations to habitat loss. *Ibis* 137:S38–S46.

Epstein, S., C. J. Yapp, and J. H. Hall. 1976. The determination of the D/H ratio of non-exchangeable hydrogen in cellulose extracted from aquatic and land plants. *Earth Planetary Science Letters* 30:241–251.

Estep, M. L. F., and H. Dabrowski. 1980. Tracing food webs with stable hydrogen isotopes. *Science* 209:1537–1538.

Fox, A. D., C. M. Glahder, and A. J. Walsh. 2003. Spring migration routes of Greenland white-fronted geese: results from satellite telemetry. *Oikos* 103:415–425.

Fretwell, S. D. 1972. *Populations in a Seasonal Environment*. Princeton, NJ: Princeton University Press.

Fretwell, S. D., and H. L. Lucas, Jr. 1970. On territorial behaviour and other factors influencing habitat distribution in birds. I. Theoretical development. *Acta Biotheoretica* **19**:16–36.

Gauthreaux, S. A., J. E. Michi, and C. G. Belser. 2005. The temporal and spatial structure of the atmosphere and its influence on bird migration strategies. Pp. 182–196 in R. Greenberg, and P. P. Marra (eds.) *Birds of Two Worlds: The Ecology and Evolution of Migration.* Baltimore, MD: Johns Hopkins University Press.

Gill, J. A., K. Norris, P. M. Potts, *et al.* 2001. The buffer effect and large-scale population regulation in migratory birds. *Science* **412**:436–438.

Goldstein, M. I., T. E. Lacher, B. Woodbridge, *et al.* 1999. Monocrotophos-induced mass mortality of Swainson's hawks in Argentina. *Ecotoxicology* **8**:201–214.

Gorman, L. R. 2001. Population differentiation among snowy plovers (*Charadrius alexandrinus*) in North America. M.S. thesis, Oregon State University, Corvallis, OR.

Goss-Custard, J. D., R. T. Clarke, S. E. A. Le V. dit Durrell, and R. W. G. Caldow. 1995. Population consequences of winter habitat loss in a migratory shorebird. I. Estimating model parameters. *Journal of Applied Ecology* **32**:337–351.

Green, M., T. Alerstam, P. Clausen, R. Drent, and B. Ebbinge. 2002. Dark-bellied brent geese *Branta bernicla* as recorded by satellite telemetry do not minimize flight distance during spring migration. *Ibis* **144**:106–121.

Hagan, J. M. III, and D. W. Johnston (eds.) 1992. *Ecology and Conservation of Neotropical Migrant Landbirds.* Washington, DC: Smithsonian Institution Press.

Haig, S. M., C. L. Gratto-Trevor, T. D. Mullins, and M. A. Colwell. 1997. Population identification of western hemisphere shorebirds throughout the annual cycle. *Molecular Ecology* **6**:413–427.

Haig, S. M., L. W. Oring, P. M. Sanzenbacher, and O. W. Taft. 2002. Space use, migratory connectivity, and population segregation among Willets breeding in the western Great Basin. *Condor* **104**:620–630.

Haig, S. M., C. L. Ferland, F. J. Cuthbert, *et al.* 2005. A complete species censuses and evidence for regional declines in Piping Plovers. *Journal of Wildlife Management* **69**:160–733.

Harrington, R. R., B. P. Kennedy, C. P. Chamberlain, J. D. Blum, and C. L. Folt. 1998. ^{15}N enrichment in agricultural catchments: field patterns and applications to tracking Atlantic salmon (*Salmo salar*). *Chemical Geology* **147**:281–294.

Hatch, J. J. 2002. Arctic tern (*Sterna paradisaea*). Pp. 1–40 in A. Poole, and F. Gill (eds.) *Birds of North America*, No. 707. Philadelphia, PA: The Academy of Natural Sciences.

Hatch, S. A., P. L. Meyers, D. M. Mulcahy, and D. C. Douglas. 2000. Seasonal movements and pelagic habitat use of murres and puffins determined by satellite telemetry. *Condor* **102**:145–154.

Heide-Jorgensen, M. P., K. L. Laidre, O. Wing, *et al.* 2003. From Greenland to Canada in ten days: tracks of bowhead whales, *Balaena mysticetus*, across Baffin Bay. *Arctic* **56**:21–31.

Heitmeyer, M. E., and L. H. Fredrickson. 1981. Do wetland conditions in the Mississippi Delta hardwoods influence mallard recruitment? *Transactions of the North American Wildlife and Natural Resources Conference* **46**:44–57.

Hobson, K. A., and R. G. Clark. 1992. Assessing avian diet using stable isotopes. I. Turnover of ^{13}C in tissues. *Condor* **94**:181–188.

Hobson, K. A., and L. I. Wassenaar. 1997. Linking breeding and wintering grounds of neotropical migrant songbirds using stable hydrogen isotopic analysis of feathers. *Oecologia* **109**:142–148.

Hodgson, S., and T. P. Quinn. 2002. The timing of adult sockeye salmon migration into fresh water: adaptations by populations to prevailing thermal regimes. *Canadian Journal of Zoology* **80**:542–555.

Holmes, R. T. 1994. Black-throated blue warbler (*Dendroica caerulescens*). Pp. 1–24 in A. Poole, and F. Gill (eds.) *The Birds of North America*, No. 87. Philadelphia, PA: The Academy of Natural Sciences.

Jarne, P., and J. L Lagoda. 1996. Microsatellites, from molecules to populations and back. *Trends in Ecology and Evolution* **11**:424–429.

Kaminski, R. M., and E. A. Gluesing. 1987. Density and habitat-related recruitment in mallards. *Journal of Wildlife Management* **51**:141–148.

Keast, A., and E. S. Morton (eds.) 1980. *Migrants in the Neotropics: Ecology, Behavior, Distribution and Conservation*. Washington, DC: Smithsonian Institution Press.

Kelly, J. F., V. Atudorei, D. S. Sharp, and D. M. Finch. 2002. Insights into Wilson's warbler migration from analyses of hydrogen stable-isotope ratios. *Oecologia* **130**:216–221.

Kennedy, B. P., C. L. Folt, J. D. Blum, and C. P. Chamberlain. 1998. Natural isotope markers in salmon. *Nature* **387**:766

Kimura, M., S. M. Clegg, I. J. Lovette, *et al.* 2002. Phylogeographical approaches to assessing demographic connectivity between breeding and overwintering regions in a Nearctic–Neotropical warbler (*Wilsonia pusilla*). *Molecular Ecology* **11**:1605–1616.

Koch, P. L., J. Heisinger, C. Moss, *et al.* 1995. Isotopic tracking of change in diet and habitat use in African elephants. *Science* **267**:1340–1343.

Korner, C., G. D. Farquhar, and S. C. Wong. 1991. Carbon isotope discrimination follows latitudinal and altitudinal trends. *Oecologia* **88**:30–40.

Lanctot, R. B., D. E. Blanco, R. A. Dias, *et al.* 2002. Conservation status of the buff-breasted sandpiper: historic and contemporary distribution and abundance in South America. *Wilson Bulletin* **114**:44–72.

Lovette, I. J., S. M. Clegg, and T. B. Smith. 2004. Limited utility of mtDNA markers for determining connectivity among breeding and overwintering locations in three neotropical migrant birds. *Conservation Biology* **18**:156–166.

Lyons, J. E., and S.M. Haig. 1995. Fat content and stopover ecology of spring migrant semipalmated sandpipers in South Carolina. *Condor* **97**:427–437.

Marra, P. P., and R. T. Holmes. 2001. Consequences of dominance-mediated habitat segregation in a migrant passerine bird during the non-breeding season. *Auk* **118**:92–104.

Marra, P. P., T. W. Sherry, and R. T. Holmes. 1993. Territorial exclusion by a Neotropical migrant warbler in Jamaica: a removal experiment with American redstarts (*Setophaga ruticilla*). *Auk* 110:565–572.

Marra, P. P., K. A. Hobson, and R. T. Holmes. 1998. Linking winter and summer events in a migratory bird by using stable-carbon isotopes. *Science* 282:1884–1886.

Marra, P. P., C. M. Francis, R. S. Mulvihill, and F. R. Moore. 2005. The influence of climate on the timing and rate of spring bird migration. *Oecologia* : in press.

Milot, E., H. L. Gibbs, and K. A. Hobson. 2000. Phylogeography and genetic structure of northern populations of the yellow warbler (*Dendroica petechia*). *Molecular Ecology* 9:667–681.

Mizutani, H., and E. Wada. 1989. Nitrogen and carbon isotope ratios in seabird rookeries and their ecological implications. *Ecology* 89:340–349.

Mizutani, H., M. Fukuda, and E. Wada. 1990. Carbon isotope ratio of feathers reveals feeding behavior of cormorants. *Auk* 107:400–437.

Mizutani, H., M. Fukuda, and Y. Kabaya. 1992. ^{13}C and ^{15}N enrichment of feathers of 11 species of adult birds. *Ecology* 73:1391–1395.

Mueller, U., and L. Wolfenbarger. 1999. AFLP genotyping and fingerprinting. *Trends in Ecology and Evolution* 14:389–394.

Munro, R. E., and C. F. Kimball. 1982. *Population Ecology of the Mallard Vol. 7, Distribution and Derivation of the Harvest*, Resource Publication No. 147, Washington, DC: US Fish and Wildlife Service.

Myers, J. P. 1981. Cross-seasonal interactions in the evolution of sandpiper social systems. *Behavioral Ecology and Sociobiology* 8:195–202.

Myers, J. P. 1983. Conservation of migrating shorebirds: staging areas, geographic bottlenecks, and regional movements. *American Birds* 37:23–25.

Nebel, S., T. Piersma, J. van Gils, A. Dekinga, and B. Spaans. 2000. Length of stopover, fuel storage and a sex-bias in the occurrence of red knots *Calidris c. canutus* and *C. c. islandica* in the Wadden Sea during southward migration. *Ardea* 88:165–176.

Nebel, S., D. B. Lank, P. D. O'Hara, et al. 2002. Western sandpipers (*Calidris mauri*) during the nonbreeding season: spatial segregation on a hemispheric scale. *Auk* 119:922–928.

Nichols, J. D., and R. E. Tomlinson. 1993. Analyses of banding data. Pp. 269–280 in T. S. Baskett, M. W. Sayre, R. E. Tomlinson, and R. E. McCabe (eds.) *Ecology and Management of the Mourning Dove*. Harrisburg, PA: Stackpole Books.

Norris, D. R. 2005. Habitat quality and carry-over effects in migratory populations. *Oikos* 109:178–186.

Norris, D. R., P. P. Marra, K. K. Kyser, T. W. Sherry, and L. M. Ratcliffe. 2004. Tropical winter habitat limits reproductive success on the temperate breeding grounds in a migratory bird. *Proceedings of the Royal Society of London B* 271:59–64.

Norris, D. R., P. P. Marra, T. K. Kyser, et al. in press. Migratory connectivity of a widely distributed Neotropical–Nearctic migratory songbird. In M. Boulet and D. R. Norris (eds.) *Migratory Connectivity of Two Species of Neotropical–Nearctic Migratory Songbirds*. Ornithological Monographs No. 61.

Roth, J. C. 1968. Benthic and limnetic distribution of three *Chaoborus* species in a southern Michigan lake (Diptera, Chaoberidae). *Limnology and Oceanography* 13:242–249.

Royle, J. A., and D. R. Rubenstein, 2004. The role of species abundance in determining the breeding origins of migratory birds using stable isotopes. *Ecological Applications* 14:1780–1788.

Rubenstein, D. R., and K. A. Hobson. 2004. From birds to butterflies: animal movement and stable isotopes. *Trends in Ecology and Evolution* 19:256–263.

Rubenstein, D. R., C. P. Chamberlain, R. T. Holmes, *et al.* 2002. Linking breeding and wintering ranges of a migratory songbird using stable isotopes. *Science* 295:1062–1065.

Ruegg, K. C., and T. B. Smith. 2002. Not as the crow flies: a historical explanation for the circuitous migration of Swainson's thrush (*Catharus ustulatus*). *Proceedings of the Royal Society of London B* 269:1375–1381.

Runge, M., and P. P. Marra. 2005. A demographic model for a migratory passerine bird: population dynamics of the American redstart. Pp. 375–389 in R. Greenberg, and P. P. Marra (eds.) *Birds of Two Worlds: The Ecology and Evolution of Migration.* Baltimore, MD: Johns Hopkins University Press.

Salomonsen, F. 1955. The evolutionary significance of bird migration. *Biologiske Meddelelser* 22:1–62.

Sanzenbacher, P. M., and S. M. Haig. 2002a. Residency and movement patterns of wintering dunlin (*Calidris alpina*) in the Willamette Valley, Oregon. *Condor* 104:271–280.

Sanzenbacher, P. M., and S. M. Haig. 2002b. Regional fidelity and movement patterns of wintering Killdeer in an agricultural landscape. *Waterbirds* 25:16–25.

Sauer, J. R., J. E. Hines, and J. Fallon. 2004. *The North American Breeding Bird Survey, Results and Analyses: 1966–2003.* Version 2004.1. Laurel, MD: US Geological Survey Patuxent Wildlife Research Center.

Schaffner, F., and P. Swart. 1991. Influence of diet and environmental water on the carbon and oxygen isotopic signatures of seabird eggshell carbonate. *Bulletin of Marine Science* 48:23–38.

Sillett, S. T., R. T. Holmes, and T. W. Sherry. 2000. Impacts of a global climate cycle on population dynamics of a migratory songbird. *Science* 288:2040–2042.

Smouse, P. E., J. C. Long, and R. R. Sokal. 1986. Multiple regression and correlation extensions of the Mantel test of matrix correspondence. *Systematic Zoology* 28:227–231.

Studds, C. E., and P. P. Marra. 2005. Nonbreeding habitat occupancy and population process: an upgrade experiment with a migratory bird. *Ecology* 86:2380–2385.

Stutchbury, B. J. 1994. Competition for winter territories in a Neotropical migrant: the role of sex and age. *Auk* 111:63–69.

Sutherland, W. J. 1996. Predicting the consequences of habitat loss for migratory populations. *Proceedings of the Royal Society of London B* 263:1325–1327.

Sutherland, W. J., and N. J. Cockford. 1993. Factors affecting the feeding distribution of red-breasted geese *Branta ruficollis* wintering in Romania. *Biological Conservation* **63**:61–65.

Sutherland, W. J., and P. M. Dolman. 1994. Combining behaviour and population dynamics for predicting consequences of habitat loss. *Proceedings of the Royal Society of London B* **255**:133–138.

Takano, L. L., and S. M. Haig. 2004a. Distribution and abundance of the Mariana common moorhen. *Waterbirds* **27**:245–250.

Takano, L. L., and S. M. Haig. 2004b. Inter- and intra-island movement patterns, site fidelity, home range, and core area of the Mariana common moorhen. *Condor* **106**:652–664.

Taylor, P. D., L. Fahrig, K. Henein, and G. Merriam. 1993. Connectivity is a vital element of landscape structure. *Oikos* **68**:571–572.

Tischendorf, L., and L. Fahrig. 2000. On the usage and measurement of landscape connectivity. *Oikos* **90**:7–19.

van der Merwe, N. J., J. A. Lee-Throp, J. F. Thackeray, *et al.* 1990. Source-area determination of elephant ivory by isotopic analysis. *Nature* **346**:744–746.

Van Klinken, G. J., H. van der Plicht, and R. E. M. Hedges. 1994. Bone $^{13}C/^{12}C$ ratios reflect (paleo) climatic variations. *Geophysical Research Letters* **21**:445–448.

Vogel, J. C., Eglinton B., and Auret. J. M. 1990. Isotope fingerprints in elephant bone and ivory. *Nature* **346**:747–749.

Waples, R. S. 1991. Pacific salmon *Onchorynchus* spp., and the definition of "Species" under the Endangered Species Act. *Marine Fisheries Review* **53**:11–22.

Wassenaar, L. I., and Hobson. K. A. 1998. Natal origins of migratory monarch butterflies at wintering colonies in Mexico: new isotopic evidence. *Proceedings of the National Academy of Sciences of the USA* **95**:15436–15439.

Webster, M. S., and P. P. Marra. 2005. The importance of understanding migratory connectivity. Pp. 199–209 in R. Greenberg, and P. P. Marra (eds.) *Birds of Two Worlds: The Ecology and Evolution of Temperate–Tropical Migration Systems.* Baltimore, MD: Johns Hopkins University Press.

Webster, M. S., P. P. Marra, S. M. Haig, S. Bensch, and R. T. Holmes. 2002. Links between worlds: unraveling migratory connectivity. *Trends in Ecology and Evolution* **17**:76–82.

Wenink, P. W., and A. J. Baker. 1996. Mitochondrial DNA lineages in composite flocks of migratory and wintering dunlins (*Calidris alpina*). *Auk* **113**:744–756.

Wennerberg, L. 2001. Breeding origin and migration pattern of dunlin (*Calidris aplina*) revealed by mitochondrial DNA analysis. *Molecular Ecology* **10**:1111–1120.

Wikelski, M., E. Tarlow, A. Raim, *et al.* 2003. Costs of migration in free-flying songbirds. *Nature* **423**:703–704.

Wright, S. 1978. *Evolution and the Genetics of Populations Vol. 2, Variability within and among Natural Populations.* Chicago, IL: University of Chicago Press.

Ydenberg, R. C., R. W. Butler, D. B. Lank, *et al.* 2002. Trade-offs, condition dependence and stopover site selection by migrating sandpipers. *Journal of Avian Biology* **33**:47–55.

(8)

Connectivity in marine ecosystems: the importance of larval and spore dispersal

CLAUDIO DIBACCO, LISA A. LEVIN, AND ENRIC SALA

INTRODUCTION

Connectivity is a concept shared in landscape and metapopulation ecology that is used to describe the movement or exchange of organisms between habitats on various temporal and spatial scales (Gilpin and Hanski 1991; Hanksi and Gilpin 1997; Crooks and Sanjayan Chapter 1; Taylor *et al.* Chapter 2; Moilanen and Hanski Chapter 3) and its population and community consequences. Many marine habitats, such as kelp forests, estuaries, wetlands, seagrass beds, coral and rocky reefs, and deep-sea hydrothermal vents, are naturally fragmented and patchy. As a result, many scientists working with marine populations and associated systems adopt a metapopulation-based interpretation of connectivity where landscapes are viewed as a network of habitat patches or fragments in which species occur as discrete local populations connected by the passive and active migration of individuals. In marine systems, connectivity may be generated by movements of early life stages such as larvae or spores (hereafter referred to as propagules), juveniles, or adults.

The majority of marine organisms, including benthic (living on or in the bottom), demersal (living near and in close association with the bottom), and holoplanktonic (living in the plankton) species, have a complex life cycle characterized by planktonic stages of development (e.g., larvae, spores). In the case of marine invertebrates and fishes,

Connectivity Conservation eds. Kevin R. Crooks and M. Sanjayan. Published by Cambridge University Press. © Cambridge University Press 2006.

propagules exhibit a diversity of nutritional modes, development sites, planktonic durations, and morphological development patterns that can affect patterns of connectivity (Table 8.1). These species with indirect development, involving relatively long-lived, planktonic, feeding larvae, are likely to facilitate dispersal and gene flow between sedentary or sessile adult populations (Thorson 1950; Grahame and Branch 1985;

Table 8.1. *Marine invertebrate and fish larval development patterns and hypothesized relative relationships to connectivity: (H) high connectivity; (M) moderate or variable connectivity; (L) low connectivity*[a]

I. Nutritional mode

(L) Lecithotrophic – pertaining to developmental stages (e.g., larvae) that feed on a yolk reserve
- Characteristic of most invertebrates that brood a small number of large eggs; most lecithotrophs are planktonic
- Most fishes hatch with a yolk reserve; some fishes feed exclusively on yolk reserves until hatching as post-larvae or juveniles (e.g., sharks and skates)

(H) Planktotrophic – pertaining to developmental stages that feed on plankton
- Characteristic of invertebrates and fishes that brood a large number of small eggs
- Most common development mode in marine benthic invertebrates; typically do not feed in early stages of development while feeding organs form
- Most fish larvae begin as lecithotrophs and switch to planktotrophy, a sequence commonly referred to as mixed feeding, as feeding structures develop

II. Duration of planktonic period[b]

(H) Teleplanic – pertaining to larvae that develop in the plankton for many months
- Marine invertebrates with teleplanic larvae (e.g., gastropods) often have amphi-Atlantic distributions and cross-basin gene flow is hypothesized
- Fishes with long-lived larval stages (e.g., eels) can have long migration periods and disperse hundreds to thousands of kilometers

(M) Actaeplanic – pertaining to larvae that develop in the plankton for 1–6 weeks
- Represents the most common larval duration interval observed in marine invertebrates and fishes

(L) Brachyplanic – pertaining to larvae that develop in the plankton from hours to days
- Invertebrate species with brachyplanic larvae (e.g., ascidians) tend to experience limited dispersal

(L) Aplanic – pertaining to larvae with no free-living dispersal stages
- Not common in marine fishes, but examples include species that complete development in egg cases (e.g., sharks, skates), in the mouth and gill cavity of mouth brooders (e.g., cichlids, some catfishes), or live-bearing (viviparous) fishes

(continued)

Table 8.1. (*cont.*)

....

III. Site of development

    (H) Pelagic – pertaining to larvae that develop in the open ocean

    (H) Planktonic – pertaining to larvae that develop in the water column

    (M) Neritic – pertaining to larvae that develop in the nearshore water column

    (L) Benthic/Demersal – pertaining to larvae that develop in close association with the substratum

        • Most common in freshwater fishes, rather than marine fishes

    (L) Aplanktonic – pertaining to larvae that develop on or in the substratum (e.g., brooded in capsules)

IV. Development mode

    (H) Indirect Development – development that includes free-living larval stages or spores

        • Most common mode of development in marine invertebrates and increases their likelihood of dispersal

        • Most fish species have eggs spawned in the plankton, followed by free-living larvae

    (L) Direct development – describes species with no intermediate stage of development; a juvenile emerges and typically crawls away from the parent with limited dispersal

        • More common in marine invertebrates than fish, though mouth-brooding fish may be considered direct developers

....

[a]It is important to recognize that connectivity is not restricted to larval dispersal and exchange among populations. Even if larval developmental stages do not promote connectivity via dispersal, it may be facilitated by juvenile or adult stages. This is particularly true for fish species that are typically more mobile than invertebrates at all stages of development.

[b]See Bradbury and Snelgrove (2001) for a discussion contrasting larval transport in demersal fish and benthic invertebrates.

Source: (Modified from Levin and Bridges (1995).

McConaugha 1992; Levin and Bridges 1995). Most marine propagules are microscopic and exhibit a broad range of morphologies (Figs. 8.1 and 8.2), swimming abilities, and behaviors (Chia *et al.* 1984; Leis and Carson-Ewart 1997; Stobuzki 1998). They typically disperse for weeks in the plankton, making them difficult to track in situ and therefore complicating estimation of population exchange rates (Levin 1990; Thorrold *et al.* 2002). The degree of connectedness of marine populations has become a central issue in marine ecology but remains elusive for most marine taxa (OEUVRE 1998; Cowen *et al.* 2002).

The rate at which populations within a metapopulation exchange individuals and, by inference, the extent to which they self-seed has been termed connectivity (Moilanen and Nieminen 2002; Warner and Cowen 2002; Moilanen and Hanski Chapter 3). If the proportion of recruits that

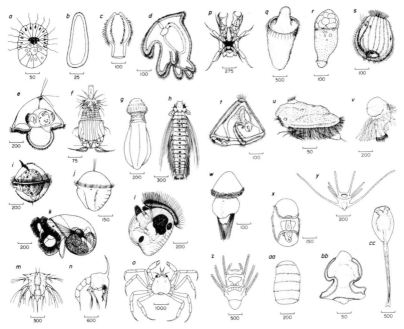

Fig. 8.1. Examples of planktonic marine invertebrate larvae. (a) Sponge amphiblastula; (b) cnidarian planula; (c) ctenophore cydippid; (d) turbellarian Muller's larva; (e) nemertean pilidium; (f) loriciferan Higgins larva; (g) priapulan larva; (h) polychaete larva; (i) polychaete trochophore; (j) polyplacophoran trochophore; (k) gastropod veliger; (l) bivalve veliger; (m) barnacle nauplius; (n) crab zoea; (o) crab megalopa; (p) pycnogonid protonymphon; (q) sipunculan pelagosphera; (r) echiuran trochophore; (s) bryozoan coronate; (t) bryozoan cyphonautes; (u) entoproct trochophore; (v) phoronid actinotroch; (w) articulate brachiopod larva; (x) asteroid bipinnaria; (y) ophiuroid ophiopluteus; (z) echinoid echinopluteus; (aa) holothuroid doliolaria; (bb) enteropneust tornaria; (cc) ascidian tadpole. Scale bars are in micrometers. (Modified from Levin and Bridges 1995.)

immigrate from other populations is high, the population is well connected to the rest of the metapopulation. Processes such as ocean circulation, duration of planktonic development, larval behavior, and settlement success will determine spatial and temporal patterns of population connectivity.

This chapter provides practical definitions of connectivity as it applies to dispersal of planktonic propagules. While planktonic propagules may influence a variety of evolutionary and ecological processes, we focus here on their role in the exchange of individuals between otherwise isolated sessile or sedentary adult populations. In the case of fish dispersal, we focus predominantly on coral reef species that disperse during planktonic

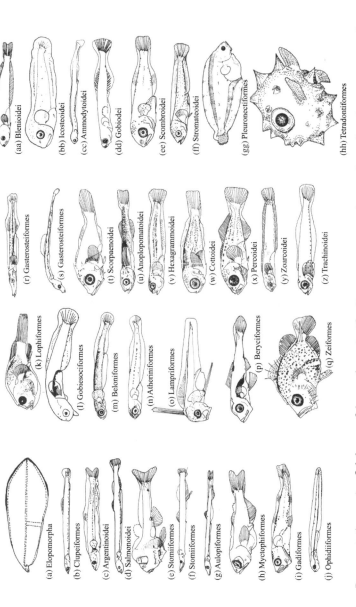

Fig. 8.2. Representative marine fish larvae of the northeast Pacific. Common names and approximate length given in parentheses. (a) Eels (190 mm); (b) herring (19 mm); (c) argentine (22 mm); (d) capelin (16 mm); (e) and (f) hatchetfishes (7–8 mm); (g) barracudina (29 mm); (h) flashlightfish (8 mm); (i) pollock (14 mm); (j) cusk eel (7 mm); (k) dreamer (14 mm); (l) clingfish (7 mm); (m) flying fish (7 mm); (n) topsmelt (8 mm); (o) king-of-the-salmon (9 mm); (p) bigscale (6 mm); (q) oreos (6–23 mm); (r,s) pipefish (6–23 mm); (t) rockfish (8 mm); (u) sablefish (19 mm); (v) greenling (15 mm); (w) irish lords (12 mm); (x) jack Mackerel (10 mm); (y) sandfish (15 mm); (z) pricklebacks (29 mm); (aa) kelpfish (21 mm); (bb) ragfish (21 mm); (cc) sand lance (16 mm); (dd) goby (15 mm); (ee) chub Mackerel (8 mm); (ff) medusa fish (10 mm); (gg) sole (15 mm); (hh) sunfish (5 mm). (Modified from Matarese *et al.* 1989)

larval development, but are more or less sedentary once settled. This perspective should serve to highlight similarities and differences in the application of connectivity measures in marine environments with those of the predominantly terrestrial systems emphasized in this book. The second part of this chapter focuses on the implications of connectivity for understanding population ecology, conservation biology, and biodiversity of marine ecosystems. For a discussion of connectivity issues for wide-ranging marine vertebrates, the reader is directed to Harrison and Bjorndal (Chapter 9).

MEASURING DISPERSAL DISTANCE AND IDENTIFYING SOURCES

Quantification of larval dispersal distances and identification of sources of successful recruits provide much needed information for setting parameters in spatially explicit models that simulate the growth, abundance, age structure, or spread of populations (Eckman 1996; Neubert and Caswell 2000; Moilanen and Nieminen 2002). For example, in the marine realm, patterns and strengths of connectivity and knowledge of recruit sources are important for (1) predicting population responses to environmental change, (2) effectively managing commercial marine resources, (3) conserving populations through the establishment of reserve networks or no-catch zones, and (4) achieving a better understanding and an ability to simulate or model the spread of invasive populations.

Dispersal potential has traditionally been determined by combining estimates of propagule duration in the plankton, typically based on laboratory cultures, with mean current flows characterizing a region. For example, Scheltema (1971, 1988) suggested that gastropod larvae with prolonged planktonic development (i.e., months) could be transported across the Atlantic by exploiting large-scale circulation (e.g., north Atlantic drift, equatorial undercurrent, and equatorial current), though the consequences of such transport on local population dynamics have never been established. This biological–physical coupling approach has evolved with the development of better oceanographic circulation models and greater knowledge of life-history parameters of planktonic organisms (e.g., Werner et al. 1999; Incze and Naimie 2000; Reiss et al. 2002).

The study of propagule dispersal has a relatively long history (Scheltema 1986; Young 1990), yet basic questions of where newly recruited individuals originate and what biological and physical mechanisms facilitate their transport are more recent issues that emphasize themes of connectivity. The tools available to address these are in some cases similar (Table 8.2). They can involve calculations of passive transport based on the distance between populations, population size, or local features of ocean circulation. For example, Graham (2003) showed that *Macrocystis pyrifera* (giant kelp) zoospore abundance sampled from the interior of kelp forests, where dispersal due to tides and currents is expected to be lower, was highly correlated ($R^2 = 76-78\%$) to the relative density and size structure of local reproductive adult sporophytes. In contrast, at sampling sites located near the outer edge of the kelp forest, only 38% of the variability in zoospore supply was explained by local reproductive output at these sites. However, measurements of realized connectivities often involve the use of genetic markers or elemental tags of artificial or natural origin. Gaines and Bertness (1992) were able to distinguish Narragansett Bay larvae from neighboring coastal populations based on mean size and determined that the larger larvae of bay origin recruited to coastal populations primarily during years when freshwater runoff and bay flushing rates were highest. Typically, only for species with unusually large, short-lived, or distinctive larvae have direct observations of larval trajectories been assessable (Olson 1985).

Several recent studies employing genetic techniques (e.g., Pogson *et al.* 2001; Hellberg *et al.* 2002), novel chemical tagging approaches (e.g., Jones *et al.* 1999; Swearer *et al.* 1999, 2002; Thorrold *et al.* 2002; Gillanders *et al.* 2003) and biological–physical models (Cowen *et al.* 2000; Sponaugle *et al.* 2002) have provided unexpected evidence that subpopulations of marine organisms may exhibit less connectivity than previously assumed. Developing improved qualitative and quantitative techniques to character-ize and estimate exchange rates of individuals among marine populations is an active and growing area of research for marine systems (see *Bulletin of Marine Science*, Suppl. Vol. 70; Cowen *et al.* 2002).

Advances in the development and application of artificial and natural tagging techniques (review: Thorrold *et al.* 2002) have significantly increased our ability to track larvae in situ. For example, Jones *et al.* (1999) artificially tagged about 10 million damselfish larvae around Lizard Island on the Great Barrier Reef with tetracycline, a fluorescent stain incorporated in the calcareous matrix of the larval otolith (ear bone) as it accretes. Based on recovery rates, they estimated that 33–66% of tagged larvae were retained near their spawning site after 3 weeks of pelagic development. Other studies have shown that naturally induced, trace ele-mental composition of biogenic hard parts (e.g., otoliths, carbonate shells, and exoskeletons) often reflects differences in the elemental composition of

Table 8.2. *Methodological approaches for the study of larval transport and population connectivities*

....

(A) Visual tracking
  (i) Short-term tracking of large colorful larvae via scuba
  (ii) Aerial photography of buoyant gametes spawned en masse (e.g., as on coral reefs)

(B) Isolated or point source larval release sites
  (i) Islands, patch reefs, seamounts, vents, and submarine canyons
  (ii) Genetically marked populations or colonies within a metapopulation

(C) Modeling approaches
  (i) Eularian approaches: hydrodynamic numerical simulations quantifying distributions relative to:(a) known point sources of egg or larval release, (b) physical oceanographic features
  (ii) Lagrangian tracking:
    • GPSn tracked drifters and drogues
    • Site-specific release of larval mimics (e.g., drift tubes, bottles, and cards)
    • Site-specific dye release studies
  (iii) Distance between populations: estimating physiological or energetic constraints on dispersal.
  (iv) Biological—physical coupled model: coupling biological processes (e.g., migration behavior, development and mortality rates) with physical features (e.g., tides, upwelling) to estimate larval dispersal and transport

(D) Geographic surveys of genetic variation
  (i) Allozymes
  (ii) Mitochondrial DNA (mtDNA)
  (iii) Microsatellites
  (iv) Nuclear sequences

(E) Tagging larvae — to determine dispersal trajectory and origin of recruits
  (i) Artificial tags (i.e., tag—release—recapture methods)
    • Vital and fluorescent stains: (a) calcium replacements (e.g., tetracycline, calcein), (b) Dyes (e.g., Alizarin red)
    • Radiotracers and radioisotopes
    • Genetic markers (e.g., artificial breeding, transgenic mutations)
  (ii) Natural tags (specific to the site of larval origin)
    • Color
    • Developmental markers: (a) morphometric characters, (b) meristic characters
    • Geochemical signatures (typically in calcified structures): (a) elemental and isotopic pollutant markers, (b) water temperature and salinity indicators
    • Genetic markers (e.g., natural cohorts)

....

seawater between sites of larval origin. These naturally induced tags, often referred to as elemental "fingerprints", have been employed to track the dispersal potential of fish (e.g., Swearer *et al.* 1999; Thorrold *et al.* 2001) and invertebrates (DiBacco and Levin 2000; Zacherl *et al.* 2003). Swearer *et al.* (1999) reported that estimates of local retention of fish (wrasse) larvae in the US Virgin Islands was higher than previously thought; in excess of 50% of larvae had trace elemental signatures indicative of retention in nearshore waters.

Refinement of both tagging approaches should allow scientists to examine exchange of larvae and juveniles on timescales (i.e., seasonal, annual) conducive to species management issues. Remarkably, there are no fish or invertebrate species for which quantitative assessments of larval connectivity have been made over multiple locations or subpopulations and years. One issue that will arise is determining the organism's developmental stage at which marine connectivity is assessed (Fig. 8.3). For example, do newly settled larvae, juveniles, or reproductive adults constitute the critical stage where connectivity between populations is established? This is likely to be methodologically driven, and we should be prepared to find out that connectivities are not identical at all stages.

INTRINSIC AND EXTRINSIC INFLUENCES ON CONNECTIVITY

The potential for and magnitude of exchange of individuals between populations may be influenced by autecological aspects of a species as well as by properties of the environment. In the case of marine benthic invertebrates and demersal fishes, extremely basic features of their life history, such as whether or not a free-living larval stage occurs, when larvae are released, how long they remain in the plankton, and where in the water column they develop, can have a profound influence on dispersal and success of propagules (Havenhand 1995; Levin and Bridges 1995). Some of these life-history characteristics are intrinsic properties of species or populations that may be modified by behavior, but in the case of time of release and planktonic duration, extrinsic factors such as water temperature and food availability may also exert strong influences (Morgan and Christy 1994; Reitzel *et al.* 2004).

Timing

The timing of propagule release can vary among species, within a species among different geographical populations, and interannually within

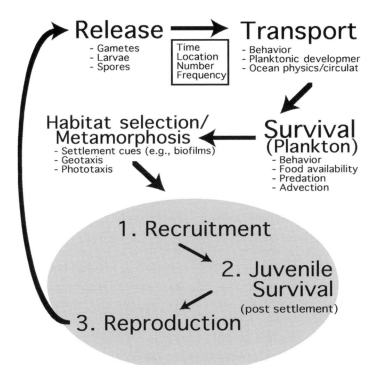

Fig. 8.3. The connectivity journey in the ocean begins with the release of early life-history developmental stages (i.e., gametes, larvae, spores) of marine organisms. Examples of physical and biological factors influencing each step in the journey are listed in the figure. Realized connectivity has been defined a number of ways (see stages 1 to 3 highlighted in gray), including at the time of (1) propagule recruitment (i.e., when planktonic organisms settle out of the plankton into adult populations), (2) juvenile survival (i.e., when propagules have successfully transformed from larval or post-larval stages of development into juveniles, often characterized by metamorphosis), and (3) reproduction (i.e., when individuals recruiting to populations have successfully reproduced).

a population. Strong climate events such as the El Niño/Southern Oscillation (ENSO) cycles or hurricane disturbance can trigger or delay larval release (Barry 1989). In temperate areas, a month or two variation in release date by benthic species can result in larvae experiencing a 3–4 °C difference in water temperature, which may translate into a 100% or more change in development time (Reitzel et al. 2004). Longer larval periods are assumed to strengthen connectivity among populations (Table 8.1), but conversely a longer larval period may increase the likelihood of transport away from suitable habitat while in the plankton. For example, Atlantic cod (*Gadus morhua*) eggs spawned along the northern coast of

Newfoundland are transported south in the Labrador Current towards the Grand Banks, located off the southeastern coast of Newfoundland. DeYoung and Rose (1993) surmised that propagules (eggs and larvae) spawned further south were more likely to be swept off the bank before they were capable of settling. Propagules spawned "upstream" were more likely to complete planktonic development before reaching the Grand Banks and experienced higher recruitment success.

The timing of propagule release by coastal species may also determine the direction and magnitude of transport. On the US west coast, strong winter upwelling induces offshore transport of larvae, which limits larval recruitment to coastal populations (Farrell *et al.* 1991) and subsequently reduces the connectivity between adjacent populations. Relaxation of upwelling can enhance onshore transport and recruitment and strengthen connectivity (Roughgarden *et al.* 1991; Wing *et al.* 1995; Navarrete *et al.* 2002). Extent of onshore transport of larvae by internal waves or tidal bores also varies seasonally (Shanks 1983; Pineda 1991), but the implications for population connectivity have not been explored because the origins of recruiting larvae are not evident. There are few studies that have examined the link between recruitment strength and connectivity. Both genetic and microchemistry time series are being generated for sessile invertebrates, including barnacles (S. Palumbi, pers. comm.) and mussels (J. Fodrie, pers. comm.), to evaluate how changes in larval sources relate to recruit density. A number of coastal or alongshore wind patterns, currents, or eddies reverse during different seasons, affecting the direction of larval transport and ultimately which populations exchange larvae. This is true along the US west coast (Johnson *et al.* 1986; Botsford *et al.* 1994), Chile (Navarrete *et al.* 2002), and the US east coast (Flierl and Wroblewski 1985). ENSO events affect current strength in the eastern Pacific and may alter connectivity among tropical, subtropical, and temperate populations (see *Southern California Academy of Science Bulletin*, Vol. 100).

Swimming behavior

Typical swimming speeds for marine invertebrate larvae (1–10 mm s^{-1}: Chia *et al.* 1984) are generally one to two orders of magnitude slower than ocean or tidal currents while those of fish larvae are not (1–30 cm s^{-1}: Leis and Carson-Ewart 1997; Stobuzki 1998). While most invertebrate propagules cannot swim fast enough horizontally to make headway against typical ocean or tidal currents, vertical migration behavior by larvae has been shown to have strong effects on transport. This is because water parcels in different portions of the water column may move in different directions or at different speeds. Larval retention within estuaries (Cronin 1982) and larval transport out of (DiBacco and Chadwick 2001) or into estuaries (Wooldridge 1991) may be strongly enhanced by tidally timed vertical migration behavior. Such behavior is often seen in crustacean (Epifanio 1988) and fish larvae (Rowe and Epifanio 1994). Vertical migration into or out of surface waters may also influence the likelihood of onshore transport by wind currents, internal waves, or tidal bores (review: Shanks 1995). Migration to the seabed, where net water motion is zero, will inhibit transport during periods of unfavorable flow (DiBacco *et al.* 2001). Any behavior that retains larvae in or near natal habitat will reduce connectivity between populations. Behaviors that increase dispersal distance will increase connectivity if the larvae locate suitable habitat once they have completed planktonic development and are competent to settle.

Species with higher fecundities, more frequent reproductive events, or producing larvae that are less selective of settlement substrates may enhance connectivity. Very often the habitat properties required by a species for settlement will set boundaries for the extent of connectivity possible. Islands, isolated bays and reefs, fragmented habitats, and geographic barriers (e.g., points like Cape Hatteras, NC or Point Concepcion, CA) separate populations and limit connectivity. More contiguous habitats such as long stretches of sandy beach, rocky shoreline, or salt marsh may lead to more continuous gene flow over greater distances. Circulation features, such as gyres and eddies, may act to concentrate, disperse, or return larvae to natal sites (Sammarco 1994). Hydrographic barriers, associated with temperature or salinity fronts or oxygen minima, can isolate propagules in distinct water masses, preventing connections across such features. In contrast, tidal currents, alongshore currents, and internal waves, if predictable, help maintain patterns of connectivity between locations by directly moving propagules or by transporting rafts. These can create the marine equivalent of highways, corridors, or migration routes found on land, but these connecting paths are often variable in size and rate of flow from year to year, subject to the variability in wind and climate regimes.

Post-larval mobility

A number of marine invertebrate taxa and some fishes lack any free-living larval stage. In these taxa, development takes place on or in the adult or within a brood structure such as an egg capsule or jelly mass. For such species, exchange between populations is influenced mainly by adult mobility, by rafting of individuals or developing embryos on floating objects (e.g., logs, marine debris) and by other transport agents such as seaweed wracks, boat hulls, ballast water, or even birds' feet (Highsmith 1985; Hobday 2000; Ruiz and Crooks 2001). In isolated settings where many species lack planktonic larvae, rafting may be the primary means of colonization (Levin and Talley 2002; Moseman *et al.* 2004). Species lacking free-living larvae, including many invertebrates inhabiting tidal flats (e.g., polychaetes), will emerge into the water column as juveniles or adults, often at night, facilitating short-distance transport (Palmer 1988). Other invertebrate and fish species with planktonic larvae may also have dispersal mechanisms during juvenile and adult stages (Martel and Diefenbach 1993). Notably, the consequences of post-larval dispersal for population connectivity have not been quantified in marine invertebrates.

IMPLICATIONS OF CONNECTIVITY BETWEEN MARINE POPULATIONS

Ecological considerations

The exchange of individuals between populations, including propagules, juveniles, or adults, drives much of the dynamics observed on ecological timescales in marine systems and will improve our understanding of life-history and recruitment dynamics, genetic structure, and metapopulation dynamics. The potential for different life stages to disperse successfully forms a key element of an organism's life history. Marine invertebrates and reef fishes exhibit a dispersal dichotomy among species, with long-lived, broadly dispersing larval stages or poorly or non-dispersing early life stages (Levin and Bridges 1995; Cowen and Sponaugle 1997). Whether this dichotomy corresponds to strong and weak connectivities among populations remains uncertain because the links among dispersal potential, realized transport, larval sources and sinks, and successful recruitment are rarely quantified for marine invertebrates (Eckman 1996) or fish (Swearer *et al.* 2002). They have probably been best characterized for reef fishes (Jones *et al.* 1999; Swearer *et al.* 1999; Thorrold *et al.* 2001). However, Victor and Wellington (2000) showed that pelagic larval duration does not appear to determine species range in damselfishes and wrasses in the eastern Pacific Ocean. This suggests that larval life history alone may not be a good proxy for connectivity.

While the dynamics of recruitment have been of interest to marine ecologists for decades (review: Young 1990), the ability to identify sources of recruits is relatively new. Knowledge of population linkages will fundamentally advance understanding of how different life histories evolve and their consequences (Havenhand 1995). It will improve our basic conceptual understanding of what drives recruitment dynamics (i.e., the influx of new members into a population by reproduction or immigration) and the extent to which variability in the exchange and recruitment of individuals between populations is forced intrinsically by source populations or extrinsically by environmental variation.

Population genetic structure is a consequence of gene flow, drift, and selection, and it provides one "script" with which to interpret connectivity. Sessile or sedentary adult populations of marine organisms with lengthy planktonic stages of development typically show few genetic differences over large spatial scales, suggesting that gene flow via larval dispersal is high (Doherty et al. 1995; Palumbi 2001). Palumbi (1994) points out that these conclusions are best addressed on an evolutionary timescale; genetically homogeneous populations result from genetic exchange over relatively long periods of time. Genetic data are less likely to inform the questions of short-term demographic exchange necessary for fisheries stock or reserve management issues (see "Management and conservation" section below). Techniques that can pinpoint the origins and trajectories of new recruits or immigrants, such as artificial and natural tagging techniques discussed above (see "Measuring dispersal distance and identifying sources"), are more likely to decipher exchange and connections on ecological timescales (Thorrold et al. 2002). The relationships between genetic structure and intragenerational population connectivity on ecological or generational timescales are not clear (Hellberg et al. 2002), but elucidation of both ecological and evolutionary linkages is necessary for a comprehensive understanding of connectivity.

Questions have evolved about the dynamics of metapopulations, with different linkage models creating predictions about population size, viability, stability, and consequences of disturbance, invasion, and mutation (innovation). The degree to which specific populations are sources of recruits, or sinks for propagules from other populations, has clear consequences for their sustainability and their significance within the metapopulation. Considerations of the extent to which populations are self-seeding or self-replenishing have cycled over time. Investigations of intertidal organisms in the 1960s to 1970s focused on adult dynamics as key factors controlling population success (e.g., Connell 1961; Dayton 1971; Paine 1974). Recent evidence of limited dispersal and larval retention in several coral reef fish populations previously inferred to disperse more widely (review: Warner and Cowen 2002) may bring local conditions into focus as critical agents for many species and the metapopulation viewpoint may fade. In reality, connectivities are temporally variable, and one perception may not apply readily to all species, populations, or times (Crooks and Sanjayan Chapter 1). Nowhere is this more evident than in the variable distributions of species under conditions of climate change. For example, El Niño events move water masses carrying subtropical species and larvae northward into temperate waters off California (Davis 2001; *Southern California Academy of Science Bulletin*, Vol. 100).

Management and conservation
Fisheries management

Understanding how populations are connected is essential for the successful management of marine fisheries. The traditional fisheries management approach is based on the premise that the productive potential of a fishery is a function, among other factors, of stock size, namely the part of a population managed as a discrete unit (Quinn and Deriso 1999; National Research Council 2001). Besides problems in estimating stock size from inherently biased catch data, defining what constitutes a stock or a population is not straightforward. Ironically, most fisheries management plans do not define the population and may consider them closed systems when they are not. Comparing genetic differentiation or migratory patterns may identify populations. A good example is salmon, which spend their adult life at sea and return to natal rivers to spawn. Salmon populations could be managed as isolated stocks on the basis of abundance estimates in natal streams, but some individuals (up to 27% in Chinook salmon) stray to non-natal streams (Quinn 1993), creating a metapopulation structure with significant dispersal between local populations on an ecological timescale (Cooper and Mangel 1998). The nature (source vs. sink) and the magnitude of the connectivity between populations must be determined for accurate assessment of conservation status of individual populations and more successful fisheries management (Cooper and Mangel 1998). Although models incorporating migration and dispersal have been developed, they are used infrequently because the necessary empirical data are unavailable (Quinn and Deriso 1999).

Interannual variations in recruitment produce large variations in fish and invertebrate stocks, which creates difficulties for fisheries management. Traditionally, variation in the magnitude of recruitment was attributed to physical loss from a system or mortality due to predation in the plankton (Hjort 1916; Thorson 1946; Cushing 1996). Others have suggested that post-settlement, density-dependent mortality may ultimately determine recruitment (e.g., Doherty 2002). All these factors likely play a role, but the degree to which they affect recruitment will depend on their relative impact on different developmental stages and on timescales. However, shifts in connectivity patterns can also produce variability in recruitment of species with passively dispersed larvae that have been largely ignored. For instance, the abundance of the American lobster (*Homarus americanus*) in Maine is presently much larger than decades ago (Steneck and Wilson 2001). Shifts in currents may have modified larval dispersal pathways and connectivity patterns, carrying lobster larvae to nursery grounds where post-settlement mortality is lower (Steneck and Wilson 2001). Identifying sources of larvae that sustain sink populations will help determine source populations that should be exempted from fishing pressure or exploited at much lower rates. Without such a change in approach, fisheries management is unlikely to be effective in dealing with species that have dynamic distributions that depend on recruitment pulses.

Variation in the oceanographic climate can produce changes in the magnitude of recruitment and connectivity between populations. In the southern California bight, recruitment of sheephead (*Semicossyphus pulcher*) in areas surrounded by larval sources occurs consistently from year to year, whereas in areas without larval sources, namely upstream of the distribution of adult populations, supply is highly variable and dependent on anomalous recruitment associated with El Niño or La Niña events (Cowen 1985). Predicted increases in the frequency and the magnitude of El Niño events will modify recruitment patterns and consequently available catch of sheephead. The lack of larval recruits has been exacerbated by recreational fisheries, which typically target the largest and most fecund individuals (Dayton *et al.* 1998).

Understanding variability in connectivity between populations is essential for the conservation of commercially and recreationally fished populations. The above examples demonstrate how connectivity patterns may cause variability in the size of commercial stocks and the structure of metapopulations. If the timescale of change is short, traditional fisheries management and policy decisions clearly will not be adequate. However, at this time it is difficult to predict when the development of a connectivity science will allow fisheries scientists to account for variability in stock size or rates of replenishment and allow managers to reevaluate policy on a continuous basis.

Marine reserves

Marine reserves are an effective conservation tool that operates at the ecosystem level (Palumbi 2001). Reserves increase the abundance and size of target species by providing a refuge from fishing pressure and a means for ecosystem restoration (Roberts 1995; Roberts *et al.* 2001; Russ *et al.* 2003). Reserves can also replenish adjacent areas and enhance local fisheries by spillover of adults or the dispersal of larvae (Roberts *et al.* 2001; Palumbi 2002; Gell and Roberts 2003; Russ *et al.* 2003). Reserve advocates suggest that larger reserves and reserves placed in strategic locations (e.g., where they serve as larval sources) will affect larger areas beyond their boundaries (Planes *et al.* 2000). The potential for self-recruitment may be diminished if reserves are too small, which could compromise the long-term viability of recognized source populations as well as sink areas dependent on source populations for maintaining viable subpopulations. In addition, if the larvae of the species of interest travel long distances, the dilution potential of dispersal processes could make the contribution from a small reserve negligible (Palumbi 2001). Understanding connectivity patterns on ecological timescales for selected species (e.g., adult migration, larval dispersal pathways) will be key in determining the number and size of reserves needed and where they should be placed (Dayton *et al.* 2000; Palumbi 2001).

Networks of marine reserves should enhance metapopulation persistence by increasing connectivity between reserves. Ideally, if a number of source populations were protected we would expect to observe an increase in viable recruits and ultimately replenishment of sinks via larval dispersal. There are several contentious issues among reserve managers and conservation ecologists, including optimal configurations of reserve networks and how to obtain a measure of connectivity between patches or populations. There are no general rules to determine the optimal distance between reserves because connectivity patterns are specific to species and regions and different in insular vs. continental environments. Also, site-specific patterns of connectivity can shift over time due to variations in oceanographic climate (Cowen 2002).

Despite recent advancements in the development of techniques to estimate connectivity rates, it seems unrealistic to expect to determine connectivity rates between two or more populations for all taxonomic groups inhabiting those areas (Sala *et al.* 2002). Given this, how can reserve networks be designed to optimize and ensure connectivity between reserves considering the community rather than individual species? Perhaps a combination of data for representative or fragile species (i.e., model organisms) can be used to provide more general rules to help design effective reserve networks. Sala *et al.* (2002) incorporated connectivity into a general model for designing networks of marine reserves and applied it in the Gulf of California. Because not all species are equally threatened by anthropogenic activities, they focused on vulnerable species, namely commercially fished groupers, and found that average dispersal distances in the Gulf were on the order of 150 km. In order to ensure connectivity via fish larval dispersal and because marine fishes have a mean dispersal distance of about 100 km (Kinlan and Gaines 2003), Sala et al. (2002) assumed that the largest gap between adjacent reserves should not exceed a conservative estimate of 100 km.

Connectivity is achieved de facto when marine reserves are established, even when design criteria ignore propagule dispersal (National Research Council 2001), but it is unlikely that connectivity will be optimal. Natural history, oceanographic data, biogeographic patterns, and common sense could be used to counter a lack of detailed connectivity knowledge to prevent large connectivity gaps in reserve system design. By evaluating species with the highest dispersal capacities, patterns of genetic homogeneity could be used to establish upper limits of propagule exchange and connectivity at evolutionary timescales. However, the absence of genetic difference between two populations does not necessarily mean strong connectivity on shorter, ecological timescales (i.e., seasonal, annual) since genetic homogeneity can be sustained by exchanging a limited number of individuals, orders of magnitude lower than what is required to sustain a metapopulation over ecological timescales (Cowen 2002).

Patterns of connectivity based on adult migration are also critical considerations when designing marine reserves (for a review of long-distance migration of marine vertebrates see Harrison and Bjorndal Chapter 9). Many commercial fishes (e.g., groupers, snappers) aggregate in large numbers at specific locations and times in order to reproduce (Johannes 1978; Domeier and Colin 1997). Fishers often know the locations and periodicity of these aggregations and consequently have eliminated many of them with negative effects, including local ecological extinctions (Sadovy and Eklund 1999). Finally, protecting aggregation sites without protecting migration routes may not be sufficient to prevent the depletion of fish populations. In the Mexican Caribbean a Nassau grouper (*Epinephelus striatus*) spawning site was closed to fishing, but fishers placed gillnets across the reef near the spawning site and caught most groupers migrating towards it (Aguilar-Perera and Aguilar-Davila 1996). A better appreciation of the importance of adult migratory patterns could have prevented fishers from targeting migratory (i.e., connectivity) routes between reefs (Sala et al. 2001). Closing the fishery during the spawning season would also have allowed connectivity mediated by adult migration.

Prevention and control of invasive species

Natural movements of species between habitats occur on very long timescales. Humans have facilitated and accelerated the transport of exotic and native species in marine settings (Ruiz and Crooks 2001; Crooks and Suarez Chapter 18) via (1) transport of fouling organisms on the hulls of ships (Carlton 1996) and in ballast waters (Carlton 1985), (2) intentional introduction of species to stabilize shorelines (e.g., Callaway and Josselyn (1992) for the estuarine cordgrass, *Spartina alterniflora*) or support commercial or recreational fisheries (Moyle 1986), (3) transport of tag-along organisms associated with intentionally introduced species (Carlton 1979), (4) transport of species on long-lived anthropogenic debris that is exchanged between distant ecosystems (Barnes 2002), (5) engineering projects connecting otherwise isolated bodies of water, such as the construction of the Suez canal (Por 1978), and (6) release of pet, bait, or live seafood species (Carlton 1979; Meinesz 1999). Ruiz *et al.* (2000) identified 298 non-indigenous species of invertebrates and algae that have been established in marine and estuarine waters of western North America. Bays, estuaries, and harbors represent the most heavily invaded habitats because many mechanisms of exchange mentioned above are associated with these systems.

Once established, biological invaders have a combination of effects that might threaten native populations and metapopulations and thereby alter patterns of connectivity (Ruiz and Crooks 2001; Crooks and Suarez Chapter 18). Effects include (1) direct consumption of native species via predation, (2) destruction or alteration of native habitat required by native species for reproduction and survival (Cox 1999; Crooks and Khim 1999), (3) exploitative or interference competition with native species for resources (e.g., Alpine and Cloern 1991; Kimmerer *et al.* 1994), and (4) loss of genetic integrity through hybridization (e.g., cordgrasses, mussels). Such effects have been shown to drastically reduce the size of local populations (Race 1982), which in turn potentially affects connectivity between isolated habitat patches.

Invasion prevention has to deal with adequately managing human activities to reduce human-mediated connectivity, and it is basically a political and enforcement problem. Controlling the spread concerns the biology of the invaders, and it is helpful to understand the dispersal patterns of the invasive species in their new environment. For instance, the tropical green alga *Caulerpa taxifolia*, which invaded the Mediterranean in 1984, disperses mostly by vegetative growth. A few years after the first colony appeared in Monaco, smaller colonies appeared as far away as southern France, the Balearic Islands, and Croatia (Meinesz 1999). Experimental work showed that small fragments of *C. taxifolia* could be transported in the anchor chest of pleasure boats and remain viable for more than a week (Sant *et al.* 1996). Boat owners have been alerted to check for *Caulerpa* debris on anchors every time they visit an infested zone to help prevent the spread of the alga.

Biodiversity and connectivity

Connectivity between populations and habitat patches is considered essential to maintain the biodiversity of marine communities. But we have yet to ascertain the subtle relationships between connectivity and biodiversity. On ecological timescales, larval exchange and recruitment promote population persistence, but on evolutionary timescales isolation promotes differentiation and speciation, potentially enhancing biodiversity. In a rare study, Goodsell and Connell (2002) tested the effects of the proximity of habitats on the diversity of marine species inhabiting kelp holdfasts. Their approach involved manipulating the number of kelp holdfasts in 1-m^2 study plots. Patterns of diversity, namely species composition and relative abundance, showed a marked decrease with reductions in the number of habitats (i.e., holdfasts). The effect was reduced when remnant habitats were proximate rather than distant, and rare rather than common taxa primarily showed significant responses. Rare species are expected to be particularly susceptible to habitat loss since they occupy a small number of habitats and are more easily isolated as a result. The responses detected most likely resulted from changes in rates of movement of individuals among patches, rather than demographic changes (Goodsell and Connell 2002).

The temporal and spatial scales considered by Goodsell and Connell (2002) were admittedly small, yet their empirical observations of reduced biodiversity as a result of reduced connectivity have implications for temporal–spatial scales of marine population dynamics. In the ocean, loss of stepping-stone habitats, be they kelp patches, subtidal reefs, rocky outcrops, or coastal wetlands, are more damaging because they exacerbate fragmentation and increase between-habitat distances. Rates of larval exchange and colonization between more distant habitats may be expected to decrease as the time required for larval transport between patches approaches or exceeds mean development times of planktonic larvae. Rates of exchange will also decrease due to increased risks (e.g., predation, physiological tolerance) associated with extended dispersal time in the plankton. Selection promotes loss of connectivity in highly isolated settings, such as Pacific wetlands (Levin 1984), by favoring species with limited dispersal and high probabilities of self seeding.

Limited exchange between marine populations has been reported in numerous genetic studies that measured genetic differentiation among neighboring habitat patches (Burton 1983, 1997; Planes *et al.* 1996). These findings suggest that local patches may not require high connectivity to maintain biodiversity. Planes *et al.* (1996) characterized a surgeonfish population in French Polynesia as local patches occupied and reproductively active with few successful migrations between neighboring patches. Recruitment from other patches would contribute to the maintenance of a diverse assemblage, especially for organisms that may not experience local recruitment. However, we are not aware of any studies that have analyzed entire communities to assess the proportion of species likely to depend on local (or self) recruitment versus immigration from neighboring populations. Until this is done, the relative importance of local recruitment vs. high connectivity on the biodiversity of select communities cannot be accurately evaluated. More careful consideration of this question will require better techniques to study the dispersal of planktonic stages of development or to determine the origin of new recruits (discussed in Mora and Sale 2002).

CONCLUSION

The patchy nature of many marine habitats and the dispersal of free-living propagules via a swiftly moving watery medium make connectivity a central issue for understanding population and community dynamics of marine systems. Among marine scientists, connectivity is defined as the rate at which subpopulations within a metapopulation exchange individuals. Connectivity research is in its infancy in marine systems. There are few species for which there is a synoptic understanding of exchange among populations and the mechanisms responsible. The broad array of developmental patterns and behaviors exhibited by marine fishes and invertebrates, and their interaction with complex bathymetry, water masses, ocean currents, and density structures, create a rich tapestry of connectivity patterns in space and time. Unraveling the linkages among these is one of our greatest challenges. Novel approaches for tracking larvae in situ, including artificial (e.g., fluorescent tags) and natural tagging methods (e.g., genetic markers, geochemical signatures) offer promising means to quantify connectivity and to characterize its variability in space and time. Single measures of connectivity will seldom be sufficient to describe dynamic interactions between populations of a single species, or the connectivity of multi-species patches. Continued exploration of the drivers and dynamics of ecological connectivity is essential for the conscientious stewardship of marine habitats and resources. Effective fisheries management, selection and design of marine reserves, and detection and control of invasive species are among the growing number of issues in the marine environment that require a quantitative understanding of connectivity.

ACKNOWLEDGEMENTS

CDB acknowledges NSF Grant OCE-0326734 and NSERC-Discovery Grant (261480-03), LAL acknowledges NSF Grant OCE 0327209 and ONR grants N00014-00-1-0174 and N00014-01-1-0473, and ES acknowledges the Moore Family Foundation and the Wildlife Conservation Society for support of connectivity research. We acknowledge the conveners and participants of "Population Connectivity in Marine Systems" who have contributed greatly to the development of ideas in this field.

REFERENCES

Aguilar-Perera, A., and W. Aguilar-Davila. 1996. A spawning aggregation of Nassau grouper *Epinephelus striatus* (Pisces: Serranidae) in the Mexican Caribbean. *Environmental Biology of Fishes* 45:351–361.

Alpine, A. E., and J. E. Cloern. 1991. Trophic interactions and direct physical effects control phytoplankton biomass and production in an estuary. *Limnology and Oceanography* 37:946–955.

Barnes, D. K. A. 2002. Biodiversity: invasions by marine life on plastic debris. *Nature* 416:808–809.

Barry, J. P. 1989. Reproductive response of a marine annelid to winter storms: an analog to fire adaptation in plants? *Marine Ecology – Progress Series* 54:99–107.

Botsford, L. W., C. L. Moloney, A. Hastings, *et al.* 1994. The influence of spatially and temporally varying oceanographic conditions on meroplanktonic metapopulations. *Deep Sea Research II* 41:107–145.

Burton, R. S. 1983. Protein polymorphisms and genetic differentiation of marine invertebrate populations. *Marine Biology Letters* 4:193–206.

Burton, R. S. 1997. Genetic evidence for long-term persistence of marine invertebrate populations in an ephemeral environment. *Evolution* 51:993–998.

Callaway, J. C., and M. N. Josselyn. 1992. The introduction and spread of smooth cordgrass (*Spartina alterniflora*) in south San Francisco Bay. *Estuaries* 15:218–226.

Carlton, J. T. 1979. History, biogeography, and ecology of the introduced marine and estuarine invertebrates of the Pacific coast of North America. Ph.D. dissertation, Davis, California, University of California.

Carlton, J. T. 1985. Transoceanic and interoceanic dispersal of coastal marine organisms: the biology of ballast water. *Oceanography and Marine Biology Annual Review* 23:313–371.

Carlton, J. T. 1996. Biological invasions and cryptogenic species. *Ecology* 77:1653–1655.

Chia, F. S., J. Buckland-Nicks, and C. Young. 1984. Locomotion in marine invertebrate larvae: a review. *Canadian Journal of Zoology* 62:1205–1222.

Connell, J. H. 1961. The influence of interspecific competition and other factors on the distribution of the barnacle *Chthamalus stellatus*. *Ecology* 42:710–723.

Cooper, A. B., and M. Mangel. 1998. The dangers of ignoring metapopulation structure for the conservation of salmonids. *Fishery Bulletin* 97:213–226.

Cowen, R. K. 1985. Large-scale pattern of recruitment by the labrid, *Semicossyphus pulcher*: causes and implications. *Journal of Marine Research* 43:719–742.

Cowen, R. K. 2002. Larval dispersal and retention and consequences for population connectivity. Pp. 149–170 in P. F. Sale (ed.) *Coral Reef Fishes: Dynamics and Diversity in a Complex Ecosystem*. San Diego, CA: Academic Press.

Cowen, R., and S. Sponaugle. 1997. Relationships between early life history traits and recruitment among coral reef fishes. Pp. 423–449 in R. C. Chambers, and E. A. Trippel (eds.) *Early Life History and Recruitment in Fish Populations*. London: Chapman and Hall.

Cowen, R. K., K. M. M. Lwiza, S. Sponaugle, C. B. Paris, and D. B. Olson. 2000. Connectivity of marine populations: open or closed? *Science* 287:857–859.

Cowen, R. K., G. Gawarkiewicz, J. Pineda, S. Thorrold, and F. Werner. 2002. *Population Connectivity in Marine Systems*, Report of a workshop to develop science recommendations for the National Science Foundation, November 4–6, 2002, Durango, CO.

Cox, G. W. 1999. *Alien Species in North America and Hawaii*. Washington, DC: Island Press.

Cronin, T. W. 1982. Estuarine retention of larvae of the crab *Rhithropanopeus harrisii*. *Estuarine, Coastal and Shelf Science* 15:207–220.

Crooks, J. A., and H. S. Khim. 1999. Architectural vs. biological effects of a habitat-altering, exotic mussel, *Musculista senhousia*. *Journal of Experimental Marine Biology and Ecology* 240:53–75.

Cushing, D. H. 1996. *Towards a Science of Recruitment in Fish Populations, Excellence in Ecology* Vol. 7. Oldendorf/Luhe, Germany: Ecology Institute.

Davis, J. L. D. 2001. Variability in spot pattern of *Girella nigricans*, the California opaleye: variation among cohorts and among climate periods. *Bulletin of the Southern California Academy of Sciences* 100:24–35.

Dayton, P. K. 1971. Competition, disturbance and community organization: the provision and subsequent utilization of space in a rocky intertidal community. *Ecological Monographs* 41:351–389.

Dayton, P. K., M. J. Tegner, P. B. Edwards, and K. L. Riser. 1998. Sliding baselines, ghosts, and reduced expectations in kelp forest communities. *Ecological Applications* 8:309–322.

Dayton, P. K., E. Sala, M. J. Tegner, and S. Thrush. 2000. Marine reserves: parks, baselines, and fishery enhancement. *Bulletin of Marine Science* **66**:617–634.

deYoung, B., and G. A. Rose. 1993. On recruitment and the distribution of Atlantic cod (*Gadus morhua*) off Newfoundland. *Canadian Journal of Fisheries and Aquatic Sciences* **50**:2729–2741.

DiBacco, C., and D. B. Chadwick. 2001. Use of elemental fingerprinting to assess net flux and exchange of brachyuran larvae between regions of San Diego Bay, California and nearshore coastal habitats. *Journal of Marine Research* **59**:1–27.

DiBacco, C., and L. A. Levin. 2000. Development and application of elemental fingerprinting to track marine invertebrate larvae. *Limnology and Oceanography* **45**:871–880.

DiBacco, C., D. Sutton, and L. McConnico. 2001. Vertical migration behavior and horizontal distribution of Brachyuran larvae in a low inflow estuary: implications for bay–ocean exchange. *Marine Ecology – Progress Series* **217**:191–206.

Doherty, P. 2002. Variable replenishment and the dynamics of reef fish populations. Pp. 327–355 in P. F. Sale (ed.) *Coral Reef Fishes: Dynamics and Diversity in a Complex Ecosystem.* San Diego, CA: Academic Press.

Doherty, P. J., S. Planes, and P. Mather. 1995. Gene flow and larval duration in seven species of fish from the Great Barrier Reef. *Ecology* **76**:2373–2391.

Domeier, M. L., and P. L. Colin. 1997. Tropical reef fish spawning aggregations: defined and reviewed. *Bulletin of Marine Science* **60**:698–726.

Eckman, J. E. 1996. Closing the larval loop: linking larval ecology to the population dynamics of marine benthic invertebrates. *Journal of Experimental and Marine Biology and Ecology* **200**:207–237.

Epifanio, C. E. 1988. Transport of invertebrate larvae between estuaries and the continental shelf. *American Fisheries Society Symposium* **3**:104–114.

Farrell, T. M., D. Bracher, and J. Roughgarden. 1991. Cross-shelf transport causes recruitment to intertidal populations in central California. *Limnology and Oceanography* **36**:279–288.

Flierl, G. R., and J. S. Wroblewski. 1985. The possible influence of warm core Gulf Stream rings upon shelf water larval fish distribution. *Fisheries Bulletin* **38**:313–330.

Gaines, S. D., and M. D. Bertness. 1992. Dispersal of juveniles and variable recruitment in sessile marine species. *Nature* **360**:579–580.

Gell, F.R., and C. M. Roberts. 2003. Benefits beyond boundaries: the fishery effects of marine reserves. *Trends in Ecology and Evolution* **18**:448–455.

Gillanders, B. M., K. W. Able, J. A. Brown, D. B. Eggleston, and P. F. Sheridan. 2003. Evidence of connectivity between juvenile and adult habitats for mobile marine fauna: an important component of nurseries. *Marine Ecology – Progress Series* **247**:281–295.

Gilpin, M. E., and I. Hanski. 1991. *Metapopulation Dynamics.* London: Academic Press.

Goodsell, P. J., and S. D. Connell. 2002. Can habitat loss be treated independently of heliostat configuration? Implications for rare and common taxa in fragmented landscapes. *Marine Ecology – Progress Series* **239**:37–44.

Graham, M. H.. 2003. Coupling propagule output to supply at the edge and interior of a giant kelp forest. *Ecology* **85**:1250–1264.

Grahame, J., and G. M. Branch. 1985. Reproductive patterns of marine invertebrates. *Oceanography and Marine Biology, Annual Review* **23**:373–398.

Hanski, I., and M. Gilpin (eds.) 1997. *Metapopulation Biology: Ecology, Genetics, and Evolution*. London: Academic Press.

Havenhand, J. N. 1995. Evolutionary ecology of larval types. Pp. 79–122 in L. McEdward (ed.) *Ecology of Marine Invertebrate Larvae*. Boca Raton, FL: CRC Press.

Hellberg, M. E., R. S. Burton, J. E. Neigel, and S. R. Palumbi. 2002. Genetic assessment of connectivity among marine populations. *Bulletin of Marine Science* **70**:273–290.

Highsmith, R. C. 1985. Floating and algal rafting as potential dispersal mechanisms in brooding invertebrates. *Marine Ecology – Progress Series* **25**:169–179.

Hjort, J. 1914. Fluctuations in the great fisheries of northern Europe. *Rapports et Procès-verbaux des Réunions, Counseil international pour l'Exploration de la Mer* **20**:1–228.

Hobday, A. J. 2000. Abundance and dispersal of drifting kelp *Macrocystis pyrifera* rafts in the southern California Bight. *Marine Ecology – Progress Series* **195**:101–116.

Incze, L. S., and C. E. Naimie. 2000. Modelling the transport of lobster (*Homarus americanus*) larvae and postlarvae in the Gulf of Maine. *Fisheries Oceanography* **9**:99–113.

Johannes, R. E. 1978. Reproductive strategies of coastal marine fishes in the tropics. *Environmental Biology of Fishes* **3**:65–84.

Johnson, D. F., L. W. Botsford, R. D. Methot, Jr., and T. C. Wainwright. 1986. Wind stress and cycles in Dungeness crab (*Cancer magister*) catch off California, Oregon, and Washington. *Canadian Journal of Fisheries and Aquatic Sciences* **43**:838–845.

Jones, G. P., M. J. Milicich, M. J. Emslie, and C. Lunow. 1999. Self-recruitment in a coral reef fish population. *Nature* **402**:802–804.

Kimmerer, W. J., E. Garstide, and J. J. Orsi. 1994. Predation by an introduced clam as the likely cause of substantial declines in zooplankton of San Francisco Bay. *Marine Ecology – Progress Series* **113**:81–93.

Kinlan, B. P., and S. D. Gaines. 2003. Propagule dispersal in marine and terrestrial environments: a community perspective. *Ecology* **84**:2007–2020.

Leis, J. M., and B. M. Carson-Ewart. 1997. *In situ* swimming speeds of the pelagic larvae of some Indo-Pacific coral-reef fishes. *Marine Ecology – Progress Series* **159**:165–174.

Levin, L. A. 1984. Life history and dispersal patterns in a dense infaunal polychaete assemblage: community structure and response to disturbance. *Ecology* **65**:1185–1200.

Levin, L. A. 1990. A review of methods for labeling and tracking marine invertebrate larvae. *Ophelia* **32**:115–144.

Levin, L. A., and T. Bridges. 1995. Pattern and diversity in reproduction and development. Pp. 1–48 in L. McEdward (ed.) *Ecology of Marine Invertebrate Larvae*. Boca Raton, FL: CRC Press.

Levin, L. A., and T. S. Talley. 2002. Natural and manipulated sources of heterogeneity controlling early faunal development of a salt marsh. *Ecological Applications* **12**:1785–1802.

Martel, A., and T. Diefenbach. 1993. Effects of body size, water current, and microhabitat on mucous-thread drifting in post-metamorphic gastropods *Lacuna* spp. *Marine – Ecology Progress Series* **99**:215–220.

Matarese, A. C., A. W. Kendall Jr., D. M. Blood, and B. M. Vinter. 1989. *Laboratory Guide to Early Life History Stages of Northeast Pacific Fishes*, NOAA Technical Report No. NMFS 80 Seattle, WA: US Department of Commerce.

McConaugha, J. R. 1992. Decapod larvae: dispersal, mortality, and ecology – a working hypothesis. *American Zoology* **32**:512–523.

Meinesz, A. 1999. *Killer Algae: The True Tale of a Biological Invasion*. Chicago, IL: University of Chicago Press.

Moilanen, A., and M. Nieminen. 2002. Simple connectivity measures in spatial ecology. *Ecology* **83**:1131–1145.

Mora, C., and P. Sale. 2002. Are populations of coral reef fish opened or closed? *Trends in Ecology and Evolution* **17**:422–428.

Morgan, S. G., and J. H. Christy. 1994. Plasticity, constraint, and optimality in reproductive timing. *Ecology* **75**:2185–2203.

Moseman, S. M., L. A. Levin, C. Currin, and C. Forder. 2004. Colonization, succession, and nutrition of macrobenthic assemblages in a restored wetland at Tijuana Estuary, California. *Estuarine Coastal and Shelf Science* **60**:755–770.

Moyle, P. B. 1986. Fish introductions into North America: patterns and ecological impact. Pp. 27–43 in H. A. Mooney, and J. A. Drake (eds.) *Ecology of Biological Invasions of North America and Hawaii*. New York: Springer-Verlag.

National Research Council. 2001. *Marine Protected Areas: Tools for Sustaining Ocean Ecosystems*. Washington, DC: National Academy Press.

Navarrete, S. A., B. Broitman, E. A. Wieters, *et al.* 2002. Recruitment of intertidal invertebrates in the southeast Pacific: interannual variability and the 1997–1998 El Niño. *Limnology and Oceanography* **47**:791–802.

Neubert, M. G., and H. Caswell. 2000. Demography and dispersal: calculation and sensitivity analysis of invasion speeds for structured populations. *Ecology* **81**:1613–1628.

OEUVRE 1998. *Ocean Ecology: Understanding and Vision for Research*. NSF Workshop.

Olson, R. R. 1985. The consequences of short-distance larval dispersal in a sessile marine invertebrate. *Ecology* **66**:30–39.

Paine, R. T. 1974. Intertidal community structure: experimental studies on the relationship between a dominant competitor and its principal predator. *Oecologia* **15**:93–120.

Palmer, M. S. 1988. Dispersal of marine meiofauna: a review and conceptual model explaining passive transport and active emergence with implications for recruitment. *Marine Ecology – Progress Series* **48**:81–91.

Palumbi, S. R. 1994. Genetic divergence, reproductive isolation, and marine speciation. *Annual Review of Ecology and Systematics* **25**:547–572.

Palumbi, S. R. 2001. The ecology of marine protected areas. Pp. 509–530 in M.D. Bertness, S.D. Gaines, and M.E. Hay (eds.) *Marine Community Ecology.* Sunderland, MA: Sinauer Associates.

Palumbi, S. R. 2002. *Marine Reserves: A Tool for Ecosystem Management and Conservation.* Arlington, VA: Pew Oceans Commission.

Pineda, J. 1991. Predictable upwelling and shoreward transport of planktonic larvae by internal tidal bores. *Science* 253:548–551.

Planes, S., R. Galzin, and F. Bonhomme. 1996. A genetic metapopulation model for reef fishes in oceanic islands: the case of the sturgeonfish, *Acanthurus triostegus. Journal of Evolutionary Biology* 9:103–117.

Planes, S., R. Galzin, A. Garcia Rubies, *et al.* 2000. Effects of marine protected areas on recruitment processes with special reference to Mediterranean littoral ecosystems. *Environmental Conservation* 27:126–143.

Pogson, G. H., C. T. Taggart, K. A. Mesa, and R. G. Boutilier. 2001. Isolation by distance in the Atlantic cod, *Gadus morhua*, at large and small geographic scales. *Evolution* 55:131–146.

Por, F. D. 1978. *Lessepian Migration: The Influx of Red Sea Biota into the Mediterranean by Way of the Suez Canal.* Heidelberg, Germany: Springer-Verlag.

Quinn, T. P. 1993. A review of homing and straying of wild and hatchery-produced salmon. *Fisheries Research* 18:29–44.

Quinn, T. J., and R. B. Deriso. 1999. *Quantitative Fish Dynamics* New York: Oxford University Press.

Race, M. S. 1982. Competitive displacement and predation between introduced and native mud snails. *Oecologia* 54:337–347.

Reiss, C. S., A. Anis, C. T. Taggart, J. F. Dower, and B. Ruddick. 2002. Relationships among vertically structured *in situ* measures of turbulence, larval fish abundance and feeding success and copepods on Western Bank, Scotian Shelf. *Fisheries Oceanography* 11:156–174.

Reitzel, A. M., B. G. Miner, and L. R. McEdward. 2004. Relationships between spawning date and larval development time for benthic marine invertebrates: a modeling approach. *Marine Ecology – Progress Series* 280 13–23.

Roberts, C. M. 1995. Rapid build up of fish biomass in a Caribbean marine reserve. *Conservation Biology* 9:815–826.

Roberts, C. M., J. A. Bohnsack, F. Gell, J. P. Hawkins, and R. Goodridge. 2001. Effects of marine reserves on adjacent fisheries. *Science* 294:1920–1923.

Roughgarden, J., J. T. Pennington, D. Stoner, S. Alexander, and K. Miller. 1991. Collisions of upwelling fronts with the intertidal zone: the cause of recruitment pulses in barnacle populations of central California. *Acta Oecologia* 12:35–51.

Rowe, P. M., and C. E. Epifanio. 1994. Flux and transport of larval weakfish in Delaware Bay, USA. *Marine Ecology – Progress Series* 110:115–120.

Ruiz, G. M., and J. A. Crooks. 2001. Biological invasions of marine ecosystems: patterns, effects and management. Pp. 3–17 in P. Gallagher, and L. Bendell-Young (eds.) *Marine Invaders: Patterns, Effects, and Management of Non-Indigenous Species.* New York: Kluwer Academic Publishers.

Ruiz, G. M., P. W. Fofonoff, J. T., Carlton, M. J. Wonham and A. H. Hines. 2000. Invasion of coastal marine communities in North America: apparent patterns, processes, and biases. *Annual Reviews of Ecology and Systematics* 31:481–531.

Russ, G. R., A. C. Alcala, and A.P. Maypa. 2003. Spillover from marine reserves: the case of *Naso vlamingii* at Apo Island, the Philippines. *Marine Ecology – Progress Series* 264:15–20.

Sadovy, Y., and A.-M. Eklund. 1999. *Synopsis of Biological Data on the Nassau Grouper*, Epinephelus striatus (*Bloch, 1792), and the jewfish*, E. itajara (*Lichtenstein, 1822*). NOAA Technical Report No. NMFS 146. Seattle, WA: US Department of Commerce.

Sala, E., E. Ballesteros, and R. M. Starr. 2001. Rapid decline of Nassau grouper spawning aggregations in Belize: fishery management and conservation needs. *Fisheries* 26:23–30.

Sala, E., O. Aburto-Oropeza, G. Paredes, *et al.* 2002. A general model for designing networks of marine reserves. *Science* 298:1991–1993.

Sammarco, P. W. 1994. Larval dispersal and recruitment processes in Great Barrier Reef corals: analysis and synthesis. Pp. 35–72 in P. W. Sammarco, and M. L. Heron (eds.) *The Bio-Physics of Marine Larval Dispersal*. Washington, DC: American Geophysical Union.

Sant, N., O. Delgado, C. Rodriguez-Prieto, and E. Ballesteros. 1996. The spreading of the introduced seaweed *Caulerpa taxifolia* (Vahl) C. Agardh in the Mediterranean Sea: testing the boat transportation hypothesis. *Botanica Marina* 3:427–430.

Scheltema, R. S. 1971. The dispersal of the larvae of shoal-water benthic invertebrate species over long distances by ocean currents. In *4th European Marine Biology Symposium*, pp. 7–28.

Scheltema, R. S. 1986. On dispersal and planktonic larvae of benthic invertebrates: an eclectic overview and summary of problems. *Bulletin of Marine Science* 39:290–322.

Scheltema, R. S. 1988. Initial evidence for the transport of teleplanic larvae of benthic invertebrates across the east Pacific barrier. *Biological Bulletin* 174:145–152.

Shanks, A. L. 1983. Surface slicks associated with tidally forced internal waves may transport pelagic larvae of benthic invertebrates and fishes shoreward. *Marine Ecology – Progress Series* 13:311–315.

Shanks, A. L. 1995. Mechanisms of cross-shelf dispersal of larval invertebrates and fish. Pp. 323–367 in L. McEdward (ed.) *Ecology of Marine Invertebrate Larvae*. Boca Raton, FL: CRC Press.

Sponaugle, S., R. K. Cowen, A. Shanks, *et al.* 2002. Predicting self-recruitment in marine populations: biophysical correlates and mechanisms. *Bulletin of Marine Science* 70:341–376.

Steneck, R. S., and C. J. Wilson. 2001. Large-scale and long-term, spatial and temporal patterns in demography and landings of the American lobster, *Homarus americanus*, in Maine. *Marine and Freshwater Research* 52:1303–1319.

Stobuzki, I. C. 1998. Interspecific variation in sustained swimming ability of late pelagic stage reef fish from two families (Pomacentridae and Chaetodontidae). *Coral Reefs* 17:111–119.

Swearer, S. E., J. E. Caselle, D. W. Lea, and R. R. Warner. 1999. Larval retention and recruitment in an island populations of coral-reef fish. *Nature* **402**:799–802.

Swearer, S. E., J. S. Shima, M. E. Hellberg, *et al.* 2002. Evidence of self-recruitment in demersal marine populations. *Bulletin of Marine Science* **70**:251–271.

Thorrold, S. R., C. Latkoczy, P. K. Swart, and C. M. Jones. 2001. Natal homing in a marine fish metapopulation. *Science* **291**:297–299.

Thorrold, S. R., G. P. Jones, M. E. Hellberg, *et al.* 2002. Quantifying larval retention and connectivity in marine populations with artificial and natural markers. *Bulletin of Marine Science* **70**:291–308.

Thorson, G. 1946. Reproduction and larval development of Danish marine bottom invertebrates. *Meddelelser fra Kommissionen for Danmarks Fiskeri- Og Havundersogelser, Serie Plankton* **4**:1–523.

Thorson G. 1950. Reproductive and larval ecology of marine bottom invertebrates. *Biological Reviews* **25**:1–45.

Victor, B. C., and G. M. Wellington. 2000. Endemism and the pelagic larval duration of reef fishes in the eastern Pacific Ocean. *Marine Ecology – Progress Series* **205**:241–248.

Warner, R. R., and R. K. Cowen. 2002. Local retention of production in marine populations: evidence, mechanisms, and consequences. *Bulletin of Marine Science* **70**:245–249.

Werner, F. E., B. O. Blanton, J. A. Quinlan, and R. A. Luettich, Jr. 1999. Physical oceanography of the North Carolina continental shelf during the fall and winter seasons: implications for the transport of larval menhaden. *Fisheries Oceanography* **8** (Suppl. 2): 7–21.

Wing, S. R., J. L. Largier, L. W. Botsford, and J. F. Quinn. 1995. Settlement and transport of benthic invertebrates in an intermittent upwelling region. *Limnology and Oceanography* **40**:316–329.

Wooldridge, T. H. 1991. Exchange of two species of decapod larvae across an estuarine mouth inlet and implications of anthropogenic changes in the frequency and duration of mouth closure. *South African Journal of Marine Science* **87**:519–525.

Young, C. 1990. Larval ecology of marine invertebrates: a sesquicentennial history. *Ophelia* **32**:1–48.

Zacherl, D. C., P. H. Manriquez, G. Paradis, *et al.* 2003. Trace elemental fingerprinting of gastropod statoliths to study larval dispersal trajectories. *Marine Ecology – Progress Series* **248**:297–303.

Connectivity and wide-ranging species in the ocean

AUTUMN-LYNN HARRISON AND KAREN A. BJORNDAL

INTRODUCTION

On land, the study of connectivity grew chiefly from the desire to conserve migratory routes and extensive home ranges of large wide-ranging of mammals. These charismatic animals served as the basis for an expanding and influential branch of conservation biology. The marine analogues of the wildebeest of the plains and the wolves of Yellowstone National Park are the wide-ranging creatures of the sea – cetaceans (whales and dolphins), elasmobranchs (sharks and rays), pinnipeds (seals and sea lions), sea turtles, and large migratory teleost fish such as tuna. Single migration events of these species range from hundreds to tens of thousands of kilometers (Alerstam *et al.* 2003) and patterns of connectivity between populations and between habitats of these organisms provide a distinct interpretation of connectivity among large animal populations.

Connectivity is dependent upon the lens with which the landscape (or seascape) is viewed and upon how the seascape is functionally treated by the species in question (Taylor 1993; Tischendorf and Fahrig 2000; Crooks and Suarez Chapter 18). The view from the eye of a tuna, whale, or sea turtle differs substantially from that of their terrestrial counterparts. Likewise, patterns of connectivity for wide-ranging species in the ocean can differ from those of other marine species such as algae, coral, and reef fish. Larval transport and reef fish connectivity are the subject of an

Connectivity Conservation eds. Kevin R. Crooks and M. Sanjayan. Published by Cambridge University Press. © Cambridge University Press 2006.

extensive body of research and are thus treated separately in this volume (DiBacco *et al.* Chapter 8). This chapter reviews the connections linking populations and habitats of large oceanic animals and presents familiar concepts of terrestrial connectivity within this context. We describe research approaches to connectivity of wide-ranging marine species, including a case study of connectivity in sea turtle populations, and discuss the use of such research in marine management and conservation projects.

CONNECTIVITY IN THE OCEAN

The concept of "patchiness" as it applies to conservation biology was depicted by the famous quartet of maps in MacArthur and Wilson's (1967) seminal treatise on island biogeography. These maps (reproduced from Curtis 1956) starkly demonstrated a pattern of land-use change in a US forest from 1831 to 1950. The pattern of advancing fragmentation is now familiar, and analogous cartographic time series are available from across the globe. MacArthur and Wilson (1967) perceived insular patches of nature to exist within an ocean of development. Forty years ago, a challenge was identified: measuring and conserving connectivity across a landscape (or seascape) were of high priority to conserving the species that depend on such continuity. From its initial roots in landscape ecology, connectivity as an ecological concept is now embraced by many disciplines of conservation biology including metapopulation ecology (Moilanen and Hanski Chapter 3), conservation genetics (Frankham Chapter 4; Neville *et al.* Chapter 13), and the study of ecological functions and processes (Talley *et al.* Chapter 5; Ricketts *et al.* Chapter 11; Crooks and Suarez Chapter 18; McCallum and Dobson Chapter 19).

Connectivity, on land or in sea, requires movement of genes, individuals, or processes. In terrestrial environments, connectivity is highly dependent upon a physically contiguous landscape or habitat type. The permeability of the matrix through which an animal moves is an integral component of landscape-level connectivity. In contrast, highly migratory species in the ocean (and air) move through a fluid environment – water provides a natural conduit or "corridor" for movement. Carr *et al.* (2003) describe connectivity as one of the primary ecological differences between marine and terrestrial ecosystems. In general, marine systems have greater rates of import and export (they are more "open"), have far greater levels of genetic exchange and thus lower genetic differentiation among populations, and greater dispersal between life stages than complementary terrestrial systems (Carr *et al.* 2003).

In contrast with the fragmented landscapes we see on land, ocean systems may "represent a vast and featureless landscape to our untrained eye" (Hyrenbach *et al.* 2000). Oceans are, however, highly dynamic, patchy environments. Wide-ranging migratory species are not distributed at random — their patchy distributions are largely determined by patterns of productivity and prey abundance that are in turn influenced by physical forcing at many scales (Hyrenbach *et al.* 2000). Convergence zones, where opposing currents and/or winds meet and collect floating materials, provide structure and form centers of biodiversity, like "English hedgerows" in the ocean (Carr 1986). Continents, coral reefs, seamounts, hydrothermal vents, submerged valleys and plains, temperature gradients, and currents are just some of the agents contributing to a patchy environment. Some marine patches are equivalent to features of terrestrial landscapes (island "forests" of kelp or coral), while others are physical or chemical in nature (hydrothermal vents, currents). Additionally, anthropogenic activities such as shipping channels, noise pollution, oil exploration, and coastal development can influence the connectivity of marine populations.

CONNECTIVITY RESEARCH AND WIDE-RANGING MARINE ANIMALS

The response of migratory marine megafauna to their patchy and dynamic environment is the subject of an ever-growing field of study. Major objectives of connectivity research for migratory marine animals include assessing foraging and breeding habitat selection (Burton and Koch 1999), elucidating temporal and spatial movement and behavioral patterns in relation to associated physical and oceanographic features (Block *et al.* 2003), revealing genetic connectivity and demography (Bowen *et al.* 1995; Lahanas *et al.* 1998; Laurent *et al.* 1998), and identifying anthropogenic activities that may disrupt connectivity (Schick and Urban 2000).

Conducting such research requires creative technologies. The activities of wide-ranging marine animals are largely invisible to human observers, particularly in comparison to those of many large terrestrial species. Animals that remain immersed for large periods of time elude observation, except during surfacing events. Characteristics of migratory marine megafauna and the nature of the marine environment itself present challenges for the study of their movement patterns and habitat use. These factors include the relative inaccessibility of the oceans, the vast scope of pelagic (open ocean) migrations and ecosystems, the inability to easily and inexpensively capture pelagic species (sea turtles and pinnipeds are notable exceptions due to their use of shore and near-shore habitats), and the difficulty of designing transmitters to withstand corrosive salt water and great pressure when attached to deep-diving animals (Holland *et al.* 2001; Block 2005). Historic ship logbook data from whaling and fishing ventures (Townsend 1935; Reeves *et al.* 2004), non-electronic flipper tag reports (Carr *et al.* 1978), and white papers and agency reports describing coastal, aerial, and marine counting surveys provided our primary insight into the coarse migratory patterns of these species (for example, the US National Marine Mammal Laboratory (2006) issues quarterly research reports including aerial, coastal, and marine surveys for cetaceans and pinnipeds).

Observing detailed movements and oceanographic preferences of individual marine animals is now possible due to rapid advances in electronic tagging technology during the last decade. Smaller size, lower cost, and greater data storage capacity (Holland *et al.* 2001) allow a variety of data to be recorded by electronic tags including three-dimensional location, swimming velocity, water temperature, heart rate, internal body temperature, and pressure (Block *et al.* 2003). Holland (2001) and Block (2005) provide detailed reviews of the evolution and nature of electronic tags for use in acquiring data about marine species in their natural environments.

Two primary tag types and a hybrid of the two are employed on highly mobile marine species: archival storage tags, satellite transmitting tags, and pop-up satellite archival tags. Archival storage tags have been known to store 5 years of movement, physiological, and oceanographic data for Atlantic bluefin tuna (Block 2005). These tags must be recaptured for data to be retrieved — low return rates are common for highly exploited fish species (Holland *et al.* 2001). Satellite tags consist of an antenna which must be out of water to transmit data and are thus most appropriately used on regularly surfacing marine animals (Block 2005). These tags collect data on fine-scale movements, behavior, physiology, and the surrounding environment which are then transmitted back through the Argos satellite system (a number of modules attached to the US National Oceanic and Atmospheric Administration weather satellite system) during a surfacing event (or at regular intervals if the animal remains on the surface for a length of time) (Block 2005). The pop-up satellite archival tag is a hybrid of archival and satellite tag technology. These tags operate and attach in a manner similar to traditional archival tags, but release from the individual at a predetermined time. The tag then floats to the surface and transmits the collected data to the Argos satellite system. Pop-up satellite archival tags do not require retrieval and are thus independent of fishing pressure (a major limitation in the use of strict archival tags) (Block 2005).

Coupled with tagging techniques is the increasing use of biogeo-chemical and genetic approaches to describe the genetic structure and migratory connectivity among populations of wide-ranging pelagic species. Stable isotopes have proven useful in tracking seasonal move-ment of bowhead whales (Schell *et al.* 1989) and in tracking migration and identifying foraging areas of pinnipeds (Burton and Koch 1999). Microsatellite and mitochondrial DNA markers have been widely used to demonstrate how gene flow affects subdivision and conservation of sea turtle species (Bowen *et al.* 1992; FitzSimmons *et al.* 1997; Roberts *et al.* 2004).

CONNECTIVITY RESEARCH: A CASE STUDY

Many highly migratory species do not demonstrate a strong affiliation for the seabed (e.g., tuna, whales, sharks) — ease of movement or choice of travel pathway is largely determined from an oceanographic or ecolog-ical perspective, for example by current, temperature, or prey availability. Other migratory species, such as sea turtles and seals, encounter patchy environments on many levels, from coastal patchiness of available nesting beaches, to oceanographic determinants during vast ocean migrations. The study of connectivity in the ocean can thus be approached from a number of different but complementary perspectives, including genetic, migratory, landscape, and metapopulation connectivity. An understand-ing of these perspectives is critical to conservation issues unique to the marine realm.

Sea turtle research provides an excellent case study to illustrate the relevance of each perspective to connectivity of migratory populations. Individuals cross ocean basins, move from subpolar to tropical waters, and inhabit terrestrial, oceanic, and neritic (nearshore) habitats. Immature sea turtles migrate among a number of nearshore foraging grounds, and adult turtles undertake vast ocean migrations to nest on beaches that may be thousands of kilometers from the neritic habitats in which they feed. The sea turtle life cycle enables researchers to directly observe individuals at accessible nesting and foraging habitats, and tagging techniques allow for long-distance oceanic tracking. These large-scale movements and requirements for non-degraded coastal habitats create significant challenges for the conservation of these species. It is therefore critical to determine the strength and extent of connections between populations and between habitats, the threats to these connections, and the conservation alternatives.

Techniques for measuring population connectivity among sea turtles

Tagging programs initiated in the 1950s began to map the distribution of turtle populations; major tagging programs were undertaken by Harrisson (1956) in Borneo and by Carr (Carr and Giovannoli 1957) at Tortuguero, on the Caribbean coast of Costa Rica. In both programs, non-corrosive metal tags were applied to the front flippers of female green turtles when they came ashore at night to dig a nest and deposit a clutch of eggs. Each tag bore an identification number, return address, and, at Tortuguero, the offer of a reward for the return of the tag. These programs were models for the many other tagging programs that have been established for all species of sea turtles throughout the world.

Tagging programs that targeted sea turtles on foraging grounds were not launched until the 1970s. The simple technology of flipper tags has yielded tremendous information about the extent of connectivity for sea turtles. Flipper tags continue to serve as the basis of many research programs; more than 10 000 tags are applied to sea turtles in the Atlantic each year (Cooperative Marine Turtle Tagging Program, unpublished data). Limiting factors for the utility of this method include tag loss and the requirement of a willing and able individual to recapture the tagged turtle and return the tag. Distribution patterns of turtles based on tag returns can be distorted by distribution of capture effort, such as the presence of commercial fisheries that either target sea turtles or have high incidental capture rates of sea turtles. Some of these limitations have been addressed by more advanced technologies that employ genetic "tags" (mitochondrial DNA sequences have been the most common in sea turtle studies: e.g., Bolten *et al.* 1998), satellite tags (e.g., Ferraroli *et al.* 2004; Hays *et al.* 2004), and stable isotopes (e.g., Hatase *et al.* 2002).

A primer on sea turtle biology

A basic understanding of the sea turtle life cycle will help explain the relevance of connectivity research to their conservation; connectivity concerns differ according to life stage. Three types of life cycles are defined (Bolten 2003a) and have relevance to the extent of connectivity exhibited by the species (Table 9.1). All sea turtles except the flatback have complex

Table 9.1. *The three categories of life cycles exhibited by sea turtles*

Type 1 No oceanic stage	Flatback sea turtle (*Natator depressus*)
Type 2 Exhibit an early oceanic	Green sea turtle (*Chelonia mydas*)
foraging stage after which	Loggerhead sea turtle (*Caretta caretta*)
they recruit to neritic foraging	Hawksbill sea turtle (*Eretmochelys imbricata*)
grounds while still immature	Kemp's ridley sea turtle (*Lepidochelys kempii*)
Type 3 Forage primarily in oceanic	Leatherback sea turtle (*Dermochelys coriacea*)
habitats throughout their lives	Olive ridley sea turtle (*Lepidochelys olivacea*)

Source: Bolten (2003a).

life cycles during which they use terrestrial, oceanic, and neritic habitats. Flatbacks, unlike other species of sea turtle, do not have an oceanic dispersal phase of hatchlings and juveniles (Walker & Parmenter 1990). Their migrations (ranging to over 1000 km) are restricted to the continental shelf of northern Australia and Papua New Guinea (Limpus *et al.* 1983). All species except the flatback and Kemp's ridley have circumglobal distributions.

Considerable variation exists within and among species in the timing and size at which individuals advance to successive life stages. We describe a generalized life cycle of species with an oceanic stage (Type 2) using the life history of the North Atlantic loggerhead (Fig. 9.1) as a frame of reference (reviewed by Bolten 2003b). More detailed accounts, with references, are available in review chapters (Musick and Limpus 1997; Bolten 2003a, 2003b; Limpus and Limpus 2003; Plotkin 2003).

Embryonic development and hatching: coastal habitats

Eggs, in clutches of about 100, are buried in the sand of discrete nesting beaches in tropical and subtropical latitudes. Embryonic development requires approximately 2 months, after which hatchlings emerge at night and move rapidly to the ocean. Hatchling dispersal begins with a period (∼24 h) of constant swimming that carries them beyond nearshore waters. At this point, it is believed that all Type 2 and 3 species enter oceanic habitats.

Surface foraging: open ocean

The location of this early oceanic life stage has been identified only for loggerheads (Bolten 2003a). For all other species, this time at sea (which might span several years) is termed the "lost year" (Carr 1967). In North

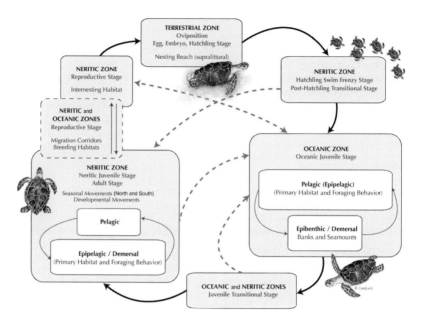

Fig. 9.1. Life cycle of the Atlantic loggerhead sea turtle. Boxes represent life stages and habitat zones; solid lines are primary movements between life stages and habitat zones; dotted lines are secondary or speculative movements. (Modified from Bolten 2003b.)

Atlantic loggerheads, hatchlings emerge from major nesting beaches in the southeastern USA and swim offshore; many become incorporated in the North Atlantic gyre. Waves, wind, and currents primarily determine the movements of post-hatchling loggerheads resulting in variable duration and paths of the journey undertaken by hatchlings. This process is analogous to natal dispersal of fisheries populations and greatly contributes to connectivity − hatchlings incorporated in the North Atlantic gyre are carried from western Atlantic nesting beaches to eastern Atlantic foraging areas.

While crossing the Atlantic and in the first years of life in eastern Atlantic and western Mediterranean waters, loggerheads are primarily surface-living "float-and-wait" predators. As they grow, they gain greater control over their positioning, but still spend 75% of their time within 5 m of the surface. Based on analyses of mitochondrial DNA haplotypes (Bolten *et al.* 1998; Laurent *et al.* 1998), oceanic-stage loggerheads in the eastern Atlantic and western Mediterranean compose a mixed stock representing multiple nesting locations.

Foraging: neritic habitats

In the loggerhead, unknown cues after 7 to 12 years (Bjorndal *et al.* 2003a) prompt the still immature turtles to leave oceanic habitats and recruit to neritic foraging grounds in the western Atlantic where they feed primarily on benthic invertebrates. This stage of growth may last decades, and juveniles of all Type 2 species often undertake extensive developmental migrations among neritic foraging grounds during this life stage.

For example, juvenile green turtles tagged on foraging grounds in the southern Bahamas were later recaptured on foraging grounds in Bahamas, Colombia, Costa Rica, Cuba, Dominican Republic, Haiti, Honduras, Nicaragua, Panama, and Venezuela (Bjorndal *et al.* 2003b). Foraging aggregations of immature sea turtles on neritic foraging grounds are usually mixed stocks, based on analyses of mitochondrial DNA sequences, with source rookeries often covering large geographic ranges (Broderick *et al.* 1994; Bowen *et al.* 1996; Lahanas *et al.* 1998; Bass and Witzell 2000). For example, an aggregation of immature green turtles in Barbados had probable contributions from up to seven rookeries ranging from Costa Rica to West Africa (Luke *et al.* 2004).

Reproductive migration: oceanic and/or neritic habitats

When sexual maturity is attained (best estimates are 15 to 40 years old for Type 2 species), all species begin to make periodic reproductive migrations to nesting beaches that may be thousands of kilometers from their foraging areas. Studies based on genetic markers and flipper tags reveal that females return to their natal beach to nest and exhibit strong site-fixity to that beach or region throughout their reproductive lives (Carr *et al.* 1978; Bowen 1995). Although "mistakes" in natal homing must have occurred for new rookeries to be established, genetic evidence indicates that such mistakes are rare. As a result of natal homing, rookeries can be distinguished by differences in mitochondrial DNA haplotype frequencies (Bowen and Karl 1997).

For all species, these reproductive migrations represent the movement of adults from widely dispersed foraging grounds to a relatively small number of discrete rookeries. Flipper tag returns have revealed that green turtles nesting at Tortuguero, Costa Rica, return to foraging grounds throughout the Caribbean (Carr *et al.* 1978). Adult female green turtles from different rookeries mix on foraging grounds. Green turtles captured on Nicaraguan foraging grounds had been tagged on the nesting beaches at Tortuguero and Aves Island, Venezuela

(Carr *et al.* 1978), whereas females tagged while nesting on Ascension Island and in Suriname were shown to share foraging grounds along the coast of Brazil (Carr 1975). Satellite telemetry has shown that leatherback females from different rookeries move throughout the Atlantic basin on overlapping courses after the reproductive season (Ferraroli *et al.* 2004; Hays *et al.* 2004). Behavior of males is poorly known; research is needed on the timing of reproductive migrations, movement patterns, and mating behavior.

Nesting: coastal habitats

During the nesting season, females usually lay several clutches at about 2-week intervals. During this time, females remain in neritic habitats in close proximity to the nesting beach. Because most sea turtles do not feed during their breeding period (food is often not available near nesting beaches and developing eggs occupy substantial volume in the shell: Bjorndal 1997), these migrations also represent a transport of nutrients from nutrient-rich foraging grounds to nutrient-poor nesting beaches (see Talley *et al.* Chapter 5 for other examples of material flow between marine and terrestrial systems). Only about 30% of the nutrients contained in loggerhead eggs deposited in Florida beaches leave the beach in the form of live hatchlings, whereas approximately 60% is retained in the beach and about 10% is lost through metabolic heat and gases (Bouchard and Bjorndal 2000). Adult Type 2 turtles apparently establish a resident foraging ground before sexual maturity and return to their foraging area after the reproductive season ends.

Connectivity, sea turtles, and conservation

The complex life cycle of sea turtles allows us to examine connectivity through different lenses. Webster *et al.* (2002) described three forms of connectivity: (1) migratory connectivity, the seasonal movements between breeding and foraging sites; (2) landscape connectivity, the regional movement of individuals among habitat patches; and (3) natal dispersal ("connectivity" in the fisheries literature) which in migratory species would be equivalent to the first migration from breeding to foraging sites. Sea turtles exhibit all three forms: initial natal dispersal, landscape connectivity in which individuals move among foraging habitats, and, finally, migratory connectivity when sexually mature individuals undertake reproductive migrations from widely distributed foraging habitats to discrete natal rookeries.

As described above, adult females from a single rookery forage in widely distributed foraging grounds where they mix with females from other rookeries. Immature sea turtles also form mixed foraging aggregations. This distribution and mixing pattern characterize sea turtles as having relatively weak migratory connectivity – that is, the connection between a single foraging ground and single breeding area is weak (Webster *et al.* 2002; Marra *et al.* Chapter 7). Weak migratory connectivity results in individuals from the same breeding population experiencing a broader range of selective pressures on their foraging grounds, and experiencing a greater overlap of selective pressures with other breeding populations.

Patterns of connectivity in sea turtle populations generate many challenges to the conservation of these wide-ranging animals. Among the nesting rookeries in a region, turtles may have very different survival probabilities and different levels of protection. Away from the nesting beach, turtles killed by either directed take or commercial fishery bycatch are mixed stocks from a number of rookeries (Bowen *et al.* 1995; Bass *et al.* 1998; Laurent *et al.* 1998). Endangered nesting populations cannot be targeted for greater protection on foraging grounds – the lack of concentrated individuals from more endangered rookeries precludes the establishment of regulations providing differential protection.

Migratory corridors are used by at least some sea turtles during their reproductive migrations, although the extent of such use is not yet known (Plotkin 2003). Leatherbacks leaving the nesting beach on the Pacific coast of Costa Rica appear to follow a migratory corridor along bathymetric features (Morreale *et al.* 1996), but tracks of leatherbacks departing from French Guiana, Suriname, and Grenada reveal no distinct corridors (Ferraroli *et al.* 2004; Hays *et al.* 2004). The edge of the Great Bahama Bank between Florida and Cuba apparently serves as a migratory corridor for some female loggerheads leaving the nesting beach at Cay Sal, Bahamas (A. B. Bolten, unpublished data).

Identification of migratory corridors is critical to protect these areas from habitat degradation and to protect sea turtles concentrated in these corridors, where they are particularly vulnerable to human-induced mortality. For example, green turtles migrating to Tortuguero are targeted by fishers along migratory corridors off the coast of Panama (Meylan *et al.* 1992) and off the coast of southeast Nicaragua (Mortimer 1981). Although the migratory routes are not physically fragmented in the terrestrial sense, fisheries, oceanographic disruptions (climate change, pollution), and anthropogenic disturbances (beach lighting, open-ocean oil exploration) can essentially "fragment" migratory routes.

To advance our understanding of the extent and effects of connectivity in sea turtles, efforts should be invested in integration of data within and among data types. That is, records of turtle movements from flipper tags should be compiled on a regional basis from the many individual tagging projects in each region; similar integration is needed for satellite tagging tracks and for genetic sequences. For example, a foraging aggregation of immature green turtles off the island of Great Inagua, Bahamas has been the focus of an ongoing study for 30 years. Mixed stock analyses based on mitochondrial DNA haplotypes has revealed that the aggregation has probable contributions from up to six rookeries in the Greater Caribbean and South Atlantic, and flipper tags and satellite telemetry have demonstrated that when turtles leave Inagua, they travel throughout the western Caribbean (Bolten and Bjorndal 2003). Scute samples are now being analyzed for stable isotopes to determine the location and diet of green turtles during the oceanic juvenile stage, before they arrive at Inagua. Integration among data types is essential, as the different forms are often complementary and allow combination of data from different life stages. Metapopulation models, such as that designed during a regional workshop held in the Cayman Islands (Bolten and Chaloupka 2004), should be developed to incorporate these data.

APPLYING PELAGIC CONNECTIVITY RESEARCH TO CONSERVATION

The conservation of endangered sea turtle populations is heavily dependent on connectivity research — as is the conservation of other migratory marine animals. The fluid nature of connectivity in the pelagic realm makes conserving connectivity a particularly complicated task.

Dynamic fragmentation and connectivity

With a medium (water) of physically contiguous connectivity, one might assume that conserving connectivity in the sea is not a problem. Marine connectivity for migratory species is not, however, dependent solely upon contiguous physical habitat (i.e., water) or pristine origin and destination habitats such as nesting beaches, although both of these are also necessary. Distribution, aggregations, and routes of movements are determined largely by patterns of productivity and prey abundance that are in turn influenced by physical forcing at many scales. Hyrenbach *et al.* (2000) describe these factors as foraging "hotspots." Some are static (topographic features such as reefs and canyons), some are persistent but not stationary (currents and frontal systems), and others are ephemeral (wind- and current-driven upwellings).

Threats to connectivity fluctuate in a manner similar to the connections themselves. Some threats to connectivity are static (coastal development and well-established shipping channels), but others manifest themselves in a process of dynamic fragmentation: direct fishing effects (bycatch and target catch), indirect fishing effects on prey species, noise pollution, and whale-watching disturbances. In cases where prey species or endangered stocks are rendered locally extinct, dynamic fragmentation gives way to permanence and genetic/migratory connectivity is broken.

Conservation status of wide-ranging marine species

Predator, nutrient transporter, food for humans and other marine species, ecotourism attraction, and economic asset are some of the important roles played by migratory marine species. Unfortunately, the status of many species is precarious. As in terrestrial environments, the list of over-hunting/overfishing effects is long, sparing no family of wide-ranging species. Early marine naturalists (e.g., explorers and whalers) documented many migratory species in copious abundance. Christopher Columbus in 1503 named the Cayman Islands "Las Tortugas" because the islands were "full of tortoises as was all the sea about, insomuch as they looked like little rocks" (Williams 1970). The turtle fishery was the main industry of settlers on the Caymans; by the late eighteenth century, local turtle stocks previously numbering in the millions had collapsed (Parsons 1962) and in 2002 included only 51 individuals (Aiken *et al.* 2001).

In marine mammals alone, 11 species are in direct peril of extinction and 17 are of significant concern (Van Blaricom *et al.* 2000). The drastic decline of marine megafauna arouses strong emotions worldwide in support of efforts to conserve these organisms. Charisma notwithstanding, these megafauna are often key species (*sensu* Piraino *et al.* 2002) in the function of marine ecosystems. Many top predators are functionally extinct from coastal communities where they were formerly present, leading to the degradation of associated fisheries (Jackson *et al.* 2001) and marine mammal (Estes *et al.* 1998; Springer *et al.* 2003) communities.

With already low numbers of many species, environmental stochasticity and anthropogenic threats to connectivity have a proportionally greater effect than they would on more abundant populations. For example, the 2004 tsunami catastrophe in south Asia may prove devastating for some populations of sea turtle, reducing habitat availability and genetic diversity. Anthropogenic factors also alter connectivity; annual migrations of the highly endangered gray whales are fragmented by a number of activities including shipping lanes, military test ranges, and oil exploration (Moore and Clarke 2002). Management decisions are often based on assumptions of connectivity that may or may not be scientifically sound. This is changing. Our increasing body of knowledge about connectivity in the marine realm and our ability to apply such research plays an increasing role in a wide variety of conservation initiatives.

Conservation of wide-ranging marine species informed by connectivity research

Fishing policy

Tagging studies of tuna in the Atlantic revealed patterns of migratory connectivity with great implications for conservation policy (Block et al. 2001, 2005). Recognized as two management units by the International Commission for the Conservation of Atlantic Tunas (ICCAT), the Atlantic bluefin tuna fishery has been overexploited since 1982 (Magnuson et al. 1994) and the stocks were assumed to be mixing at only low levels — assumptions that were incorporated into base stock assessment models by the ICCAT (Block et al. 2001). Due to a decline in the western breeding population, fishing quotas were reduced on this management unit only. Block and her colleagues demonstrated that tuna from the western Atlantic move to the eastern Mediterranean Sea during the breeding season and are vulnerable to fishing mortality from all Atlantic bluefin tuna fisheries (Block et al. 2001). The connectivity between the management units was much higher than previously assumed, and this new information proved critical to reassessing management strategies applied to the Atlantic bluefin tuna. Similar tagging studies are under way on a suite of wide-ranging species as a part of the Tagging of Pacific Pelagics project and will provide invaluable information on the location of migration corridors, the topographic and oceanographic features by which organisms navigate and/or locate prey, and the regions of conflict between these species and humans (Block et al. 2003).

Coastal and nearshore habitat conservation

Turtle conservation research provides the largest number of examples of the use of tagging and genetic studies to inform management decisions in coastal habitats. As described above, many studies have contributed to our understanding of the connectivity exhibited by sea turtles among nations of the western hemisphere. Recognition of this connectivity has resulted in many regional congresses and in the first international treaty dedicated exclusively to sea turtles. The Inter-American Convention for the Protection and Conservation of Sea Turtles sets standards for the conservation of sea turtles and their habitats (Hykle 2002).

Marine reserve networks

Marine connectivity should be an explicit factor in selection and design of marine reserves (National Research Council 2001). The first marine conservation area dedicated to a migratory marine species, the Pribiloff Islands in 1869, focused on the regulation of northern fur seal hunting (Reeves 2000). Since then, many additional conservation areas have been established to conserve single charismatic marine species, particularly cetaceans and pinnipeds (Reeves 2000).

Connectivity in the traditional sense as used in landscape ecology has until recently seemed irrelevant to the conservation of marine seascapes and large migratory marine vertebrates. The shifting nature of marine migratory corridors makes them fairly unsusceptible to static conservation management techniques. Ideally, protected areas would span the entire range of the species (Reeves 2000) — an impossible feat for species distributed across ocean basins. Boundaries of marine protected areas should be flexible and adaptable to changing anthropogenic pressures and oceanographic forcing. The first seascape-level marine corridor was recently created (Orellana 2004) with the primary goal of protecting the migratory routes of wide-ranging marine species including the leatherback turtle and the blue whale (*Balaenoptera musculus*), as well as to provide the more traditional functions of marine reserves such as fishing zones and spawning area protection. This ambitious project spearheaded by Conservation International extends from Costa Rica to Ecuador, includes four governments, spans over 211 million hectares, and links existing conservation areas into an "Eastern Tropical Pacific Seascape" marine corridor.

CONCLUSION

Although research on connectivity in marine environments has lagged behind that in terrestrial habitats, connectivity has a critical role in conservation of marine organisms and habitats. Charismatic marine megafauna, from Atlantic bluefin tuna to green turtles to blue whales, have been the focus of substantial management efforts. Yet, even with so much attention, wide-ranging ocean animals remain highly vulnerable to extinction. The implications of connectivity must be recognized and incorporated in management plans for marine species and in design of marine protected areas. Studies of connectivity in terrestrial ecosystems will provide valuable insights because some marine connectivity issues are analogous to those experienced on land and in air. However many issues presented in this chapter are unique to the marine realm. We still have much to learn.

REFERENCES

Aiken, J. J., B. J. Godley, A. C. Broderick, *et al.* 2001. Two hundred years after a commercial marine turtle fishery: the current status of marine turtles nesting in the Cayman Islands. *Oryx* **35**:145–151.

Alerstam, T., A. Hedenström, and S. Åkesson. 2003. Long-distance migration: evolution and determinants. *Oikos* **103**:247–260.

Bass, L. A., and W. N. Witzell. 2000. Demographic composition of immature green turtles (*Chelonia mydas*) from the east central Florida coast: evidence from mtDNA markers. *Herpetologica* **56**:357–367.

Bass, A. L., C. J. Lagueux, and B. W. Bowen. 1998. Origin of green turtles, *Chelonia mydas*, at "sleeping rocks" off the northeast coast of Nicaragua. *Copeia* **1998**:1064–1069.

Bjorndal, K. A. 1997. Foraging ecology and nutrition of sea turtles. Pp. 199–231 in P. L. Lutz, and J. A. Musick (eds.) *The Biology of Sea Turtles.* Boca Raton, FL: CRC Press.

Bjorndal, K. A., A. B. Bolten, T. Dellinger, C. Delgado, and H. R. Martins. 2003a. Compensatory growth in oceanic loggerhead sea turtles: response to a stochastic environment. *Ecology* **84**:1237–1249.

Bjorndal, K. A., A. B. Bolten, and M. Y. Chaloupka. 2003b. Survival probability estimates for immature green turtles, *Chelonia mydas*, in the Bahamas. *Marine Ecology – Progress Series* **252**:273–281.

Block, B. A. 2005. Physiological ecology in the 21st century: advancements in biologging science. *Integrative and Comparative Biology* **45**:305–320.

Block, B. A., H. Dewar, S. B. Blackwell, *et al.* 2001. Migratory movements, depth preferences, and thermal biology of Atlantic bluefin tuna. *Science.* **293**:1310–1314.

Block, B. A., D. P. Costa, G. W. Boehlert, and R. E. Kochevar. 2003. Revealing pelagic habitat use: the Tagging of Pacific Pelagics program. *Oceanologica Acta* **25**:255–266.

Block, B. A., S. L. Teo, A. Walli, *et al.* 2005. Electronic tagging and population structure of Atlantic bluefin tuna. *Nature* **434**:1121–1127.

Bolten, A. B. 2003a. Variation in sea turtle life history patterns: neritic vs. oceanic developmental stages. Pp. 243–257 in P. L. Lutz, J. Musick, and J. Wyneken (eds.) *The Biology of Sea Turtles*, vol. 2. Boca Raton, FL: CRC Press.

Bolten, A. B. 2003b. Active swimmers – passive drifters: the oceanic juvenile
stage of loggerheads in the Atlantic system. Pp. 63–78 in A. B. Bolten, and
B. E. Witherington (eds.) *Loggerhead Sea Turtles*. Washington, DC: Smithsonian
Institution Press.

Bolten, A. B., and K. A. Bjorndal. 2003. Green turtles in the Bahamas:
a shared resource. Pp. 225–226 in G. C. Ray, and J. McCormick-Ray
(eds.) *Coastal–Marine Conservation: Science and Policy*. London: Blackwell
Scientific.

Bolten, A. B., and M. Chaloupka. 2004. West Atlantic Green Turtle Population
Modeling Workshop. *Marine Turtle Newsletter* **103**:17.

Bolten, A. B., K. A. Bjorndal, H. R. Martins, *et al.* 1998. Transatlantic developmental
migrations of loggerhead sea turtles demonstrated by mtDNA
sequence analysis. *Ecological Applications* **8**:1–7.

Bouchard, S. S., and K. A. Bjorndal. 2000. Sea turtles as biological transporters
of nutrients and energy from marine to terrestrial ecosystems. *Ecology*
81:2305–2313.

Bowen, B. W. 1995. Molecular genetic studies of marine turtles. Pp. 585–587 in
K. A. Bjorndal (ed.) *Biology and Conservation of Sea Turtles, rev. edn.* Washington,
DC: Smithsonian Institution Press.

Bowen, B. W., and S. A. Karl. 1997. Population genetics, phylogeography,
and molecular evolution. Pp. 29–50 in (P. L. Lutz, and J. A. Musick (eds.)
The Biology of Sea Turtles. Boca Raton, FL: CRC Press.

Bowen, B. W., A. B. Meylan, J. P. Ross, *et al.* 1992. Global population structure and
natural history of the green turtle (*Chelonia mydas*) in terms of matriarchal
phylogeny. *Evolution* **46**:865–881.

Bowen, B. W., F. A. Abreu-Grobois, G. H. Balazs, *et al.* 1995. Trans-Pacific
migrations of the loggerhead turtle (*Caretta caretta*) demonstrated with
mitochondrial DNA markers. *Proceedings of the National Academy of Sciences
of the USA* **92**:3731–3734.

Bowen, B. W., A. L. Bass, A. Garcia-Rodriguez, *et al.* 1996. Origin of hawksbill
turtles in a Caribbean feeding area as indicated by genetic markers. *Ecological
Applications* **6**:566–572.

Broderick, D., C. Moritz, J. D. Miller, *et al.* 1994. Genetic studies of the hawksbill
turtle (*Eretmochelys imbricata*): evidence for multiple stocks in Australian
waters. *Pacific Conservation Biology* **1**:123–131.

Burton, R. K., and P. L. Koch. 1999. Isotope tracking of foraging and
long distance migration in northeast Pacific pinnipeds. *Oecologia*
119:578–585.

Carr, A. 1967. *So Excellent a Fishe: A Natural History Of Sea Turtles*. New York:
Natural History Press.

Carr, A. 1975. The Ascension Island green turtle colony. *Copeia* **1975**:547–555.

Carr, A. 1986. Rips, FADS, and little loggerheads. *BioScience* **36**:92–100.

Carr, A., and L. Giovannoli. 1957. The ecology and migrations of sea turtles. II.
Results of field work in Costa Rica, 1955. *American Museum Novitates*
1835:1–32.

Carr, A. F., M. H. Carr, and A. B. Meylan. 1978. The ecology and migrations of
sea turtles. VII. The West Caribbean green turtle colony. *Bulletin of the
American Museum of Natural History* **162**:1–146.

Carr, M. H., J. E. Neigel, J. A. Estes, *et al.* 2003. Comparing marine and terrestrial ecosystems: implications for the design of coastal marine reserves. *Ecological Applications* **13**:S90—S107.

Curtis, J. T. 1956. The modification of mid-latitude grasslands and forests by man. Pp. 721—736 in W. L. Thomas (ed.) *Man's Role in Changing the Face of the Earth.* Chicago, IL: University of Chicago Press.

Estes, J. A., M. T. Tinker, T. M., Williams, and D. F. Doak. 1998. Killer whale predation on sea otters linking oceanic and nearshore ecosystems. *Science* **282**:473—476.

Ferraroli, S., J. Y. Georges, P. Gaspar, and Y. Le Maho. 2004. Where leatherback turtles meet fisheries. *Nature* **429**:521—522.

FitzSimmons, N. N., A. R. Goldizen, J. A. Norman, *et al.* 1997. Philopatry of male marine turtles inferred from mitochondrial markers. *Proceedings of the National Academy of Sciences of the USA* **94**:8912—8917.

Harrisson, T. 1956. The edible turtle (*Chelonia mydas*) in Borneo. V. Tagging turtles (and why). *Sarawak Museum Journal* **7**:504—515.

Hatase, H., N. Takai, Y. Matsuzawa, *et al.* 2002. Size-related differences in feeding habitat use of adult female loggerhead turtles *Caretta caretta* around Japan determined by stable isotope analyses and satellite telemetry. *Marine Ecology — Progress Series* **233**:273—281.

Hays, G. C., J. D. R. Houghton, and A. E. Myers. 2004. Pan-Atlantic leatherback turtle movements. *Nature* **429**:522

Holland, K. N., S. M. Kajiura, D. G. Itano, and J. Sibert. 2001. Tagging techniques can elucidate the biology and exploitation of aggregated pelagic fish species. Pp. 211—218 in G. R. Sedberry (ed.) *Island in the Stream: Oceanography and Fisheries of the Charleston Bump.* Bethesda, MD: American Fisheries Society.

Hykle, D. 2002. The Convention on Migratory Species and other international instruments relevant to marine turtle conservation: pros and cons. *Journal of International Wildlife Law and Policy* **5**:105—119.

Hyrenbach, K. D., K. A. Forney, and P. K. Dayton. 2000. Marine protected areas and ocean basin management. *Aquatic Conservation: Marine and Freshwater Ecosystems* **10**:437—458.

Jackson, J. B. C., M. X. Kirby, W. H. Berger, *et al.* 2001. Historical overfishing and the recent collapse of coastal ecosystems. *Science* **293**:629—638.

Lahanas, P. N., K. A. Bjorndal, A. B. Bolten, *et al.* 1998. Genetic composition of a green turtle (*Chelonia mydas*) feeding ground population: evidence for multiple origins. *Marine Biology* **130**:345—352.

Laurent, L., P. Casale, M. N. Bradai, *et al.* 1998. Molecular resolution of marine turtle stock composition in fishery bycatch: a case study in the Mediterranean. *Molecular Ecology* **7**:1529—1542.

Limpus, C. J., and D. J. Limpus. 2003. Biology of the loggerhead turtle in western South Pacific Ocean foraging areas. Pp. 93—113 in A. B. Bolten, and B. E. Witherington (eds.) *Loggerhead Sea Turtles.* Washington, DC: Smithsonian Institution Press.

Limpus, C. J., C. J. Parmenter, V. Baker, and A. Fleay. 1983. The flatback turtle, *Chelonia depressa,* in Queensland: post-nesting migration and feeding ground distribution. *Australian Wildlife Research* **10**:557—561.

Luke, K., J. A. Horrocks, R. A. LeRoux, and P. H. Dutton. 2004. Origins of green turtle (*Chelonia mydas*) feeding aggregations around Barbados, West Indies. *Marine Biology* **144**:799–805.

MacArthur, R. H., and E. O. Wilson. 1967. *The Theory of Island Biogeography*. Princeton, NJ: Princeton University Press.

Magnuson, J. J., B. A. Block, R. B. Deriso, *et al.* 1994. *An Assessment of Atlantic Bluefin Tuna*, Committee to Review Atlantic Bluefin Tuna, Ocean Studies Board, Commission on Geosciences, Environment, and Resources, National Research Council. Washington, DC: National Academy Press.

Meylan, P. A., A. B. Meylan, and R. Yeomans. 1992. Interception of Tortuguero-bound green turtles at Bocas Del Toro Province, Panama. *Proceedings of the 11th Annual Workshop on Sea Turtle Biology and Conservation*, Miami, FL, NOAA Technical Memorandum NMFS-SEFC-302, p. 74.

Moore, S. E., and J. T. Clarke. 2002. Potential impact of offshore human activities on gray whales (*Eschrichtius robustus*). *Journal of Cetacean Research Management* **4**:19–25.

Morreale, S. J., E. A. Standora, J. R. Spotila, and F. V. Paladino. 1996. Migration corridor for sea turtles. *Nature* **384**:319–320.

Mortimer, J. A. 1981. Feeding ecology of the West Caribbean green turtle (*Chelonia mydas*) in Nicaragua. *Biotropica* **13**:49–58.

Musick, J. A., and C. J. Limpus. 1997. Habitat utilization and migration in juvenile sea turtles. Pp. 137–163 in P. L. Lutz, and J. A. Musick (eds.) *Biology of Sea Turtles*. Boca Raton, FL: CRC Press.

National Marine Mammal Laboratory. 2006. *Quarterly Research Report*. Avaliable online at http://www.afsc.noaa.gov/Quarterly/CurrentIssue/default.htm

National Research Council. 2001 *Marine Protected Areas: Tools for Sustaining Ocean Ecosystems*, NRC Committee on the Evaluation, Design, and Monitoring of Marine Reserves and Protected Areas in the United States; Ocean Studies Board; Commission on Geosciences, Environment, and Resources. Washington, DC: National Academy Press.

Orellana, C. 2004. Tropical seascape corridor planned. *Frontiers in Ecology and the Environment* **3**:121

Parsons, J. J. 1962. *The Green Turtle and Man*. Gainesville, FL: University of Florida Press.

Piraino, S., S. Fanelli, and F. Boero. 2002. Variability of species' roles in marine communities: change of paradigms for conservation priorities. *Marine Biology* **140**:1067–1074.

Plotkin, P. 2003. Adult migrations and habitat use. Pp. 225–241 in P. L. Lutz, J. Musick, and J. Wyneken (eds.) *The Biology of Sea Turtles*, vol. 2. Boca Raton, FL: CRC Press.

Reeves, R. R. 2000. *The Value of Sanctuaries, Parks, and Reserves (Protected Areas) as Tools for Conserving Marine Mammals*. Hudson, Ontario, Canada: Okapi Wildlife Associates. Available online at http://www.mmc.gov/reports/contract/pdf/mpareport.pdf

Reeves, R., T. D. Smith, E. A. Josephson, P. J. Clapham, and G. Woolmer. 2004. Historical observations of humpback and blue whales in the North Atlantic Ocean: clues to migratory routes and possibly additional feeding grounds. *Society for Marine Mammal Science* **20**:774–786.

Roberts, M. A., T. S. Schwartz, and S. A. Karl. 2004. Global population genetic structure and male-mediated gene flow in the green sea turtle (*Chelonia mydas*): analysis of microsatellite loci. *Genetics* **166**:1857–1870.

Schell, D. M., S. M. Saupe, and N. Haubenstock. 1989. Natural isotope abundances in bowhead whale (*Balaena mysticetus*) baleen: markers of aging and habitat use. *Ecological Studies* **68**:260–269.

Schick, R. S., and D. L. Urban. 2000. Spatial components of bowhead whale (*Balaena mysticetus*) distribution in the Alaskan Beaufort Sea. *Canadian Journal of Fisheries and Aquatic Sciences* **57**:2193–2200.

Springer, A. M., J. A. Estes, G. B. van Vliet, *et al*. 2003. Sequential megafaunal collapse in the North Pacific Ocean: an ongoing legacy of industrial whaling? *Proceedings of the National Academy of Sciences of the USA* **100**:12223–12228.

Taylor, P. D., L. Fahrig, K. Henein, and G. Merriam. 1993. Connectivity is a vital element of landscape structure. *Oikos* **68**:571–573.

Tischendorf, L., and L. Fahrig. 2000. On the usage and measurement of landscape connectivity. *Oikos* **90**:7–19.

Townsend, C. H. 1935. The distribution of certain whales as shown by logbook records of American whaleships. *Zoologica* **19**:1–50.

Van Blaricom, G. R., L. R. Gerber, and R. L. Brownell, Jr. 2000. Extinctions of marine mammals. Pp. 37–69 in S. A. Levin (ed.) *Encyclopedia of Biodiversity*, vol. 4. San Diego, CA: Academic Press.

Walker, T. A., and C. J. Parmenter. 1990. Absence of a pelagic phase in the life cycle of the flatback turtle, *Natator depressa* (Garman). *Journal of Biogeography* **17**:275–278.

Webster, M. S., P. P. Marra, S. M. Haig, S. Bensch, and R. T. Holmes. 2002. Links between worlds: unraveling migratory connectivity. *Trends in Ecology and Evolution* **17**:76–83.

Williams, N. 1970. *A History of the Cayman Islands*. Grand Cayman: The Government of the Cayman Islands.

Hydrologic connectivity: a neglected dimension of conservation biology

CATHERINE PRINGLE

INTRODUCTION

Hydrologic connectivity refers to water-mediated transfer of matter, energy, and/or organisms within or between elements of the hydrologic cycle (*sensu* Pringle 2001). While this property is essential to maintaining the biological integrity of ecosystems, it also serves to perpetuate the flow of exotic species, human-derived nutrients, and toxic wastes in the landscape. All too often, we have acknowledged the importance of hydrologic connectivity in hindsight – as a result of environmental crises. Examples range from: the transport of exotic species that disrupt the food webs of rivers and lakes (e.g., Stokstad 2003); to the occurrence of extremely high levels of persistent organic pollutants (e.g., polychlorinated biphenyls PCBs) that bioaccumulate in global "hotspots" such as the Arctic (e.g., Aguilar *et al.* 2002); to effects of dams which impede riverine transport of essential elements such as silicon to coastal regions – which has been implicated in coastal eutrophication and the creation of coastal dead zones (e.g., Humborg *et al.* 2000).

Management and policy decisions regarding land-use activities are often made in the absence of adequate information on hydrologic connectivity in the landscape. An important area of research is to understand how human alterations of this property (e.g., dams, stream flow regulation, water diversion, inter-basin water transfers, water extraction) influence ecological patterns on local, regional, and global scales. Half of the accessible global freshwater runoff has already been appropriated

Connectivity Conservation eds. Kevin R. Crooks and M. Sanjayan. Published by Cambridge University Press. © Cambridge University Press 2006.

by humans and this could climb to 70% by 2025 (Postel *et al.* 1996). Freshwater ecosystems are being fragmented at a rate unprecedented in geologic history, contributing to dramatic losses in global aquatic bio-diversity and associated ecosystem integrity (e.g., Dudgeon 2000; Pringle *et al.* 2000; Rosenberg *et al.* 2000). Fewer than 42 free-flowing rivers of over 200 km in length exist in the USA; the remaining 98% of US streams have been fragmented by dams and water diversion projects (Benke 1990). Wetlands across the USA and throughout the world have also suffered dramatic declines. Accordingly, the World Wildlife Fund's species population index, measuring the average change over time in populations of almost 200 species of freshwater birds, mammals, reptiles, amphibians, and fishes, has declined by 50% globally over the 30-year period from 1970 to 1999. Current rates of extinction of many freshwater taxa are more than 1000 times the normal "background" rate and, as a whole in the USA, freshwater species are more imperiled than terrestrial species (Master *et al.* 1998).

In this chapter I discuss: (1) a brief history of research on *river connectivity*; (2) the definition of *hydrologic connectivity* and the case for incorporating a better understanding of this property into the field of conservation biology, with emphasis on the vulnerability of "protected" areas to hydrologic disturbances within and outside of their boundaries; (3) a "protected" area case study on the Caribbean National Forest and downstream disruptions in hydrologic connectivity; (4) genetic- to ecosystem-level consequences of both reductions and enhancement of hydrologic connectivity; and (5) emerging environmental challenges involving hydrologic connectivity and the transport of contaminants on regional and global scales.

BRIEF HISTORY OF CONNECTIVITY RESEARCH IN RIVER ECOSYSTEMS

Longitudinal connections within riverine ecosystems have long been recognized by both aquatic and terrestrial ecologists, as illustrated by the widespread use of the term *river corridor* in the literature. However, the term *connectivity* did not really creep into the freshwater literature until the early 1990s (but see Amoros and Roux 1988). Following Crooks and Sanjayan's lead (Chapter 1), I conducted a review of 20 major journals in freshwater ecology and management from 1945 to 2003 (Fig. 10.1) which indicates that *connectivity* surpassed the use of the term *corridor* by the late 1990s — with the trend continuing into the 2000s. In contrast, the term *connectivity* was widely used a decade earlier in journals

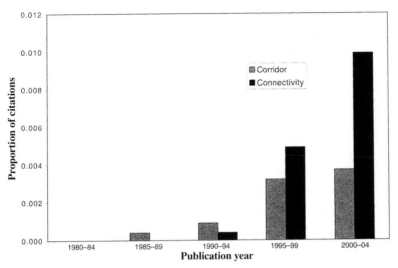

Fig. 10.1. Scientific papers published from 1980 to 2003 in 20 major journals in freshwater ecology and management, with the terms "connectivity" or "corridor" in their titles or keywords. The annual numbers of "connectivity" or "corridor" citations are standardized by the total number of citations in the 20 journals for each time period. The 20 journals examined were: *River Research and Applications, Regulated Rivers Research and Management, Lake and Reservoir Management, Freshwater Biology, Aquatic Conservation, Oikos, Aquatic Sciences, Journal of the North American Benthological Society, Transactions of the American Fisheries Society, Archives für Hydrobiologie, Journal of the American Water Resource Association, Aquatic Botany, Canadian Journal of Fisheries and Aquatic Sciences, Hydrobiologia, Environmental Biology of Fishes, Journal of Great Lakes Research, Copeia, Limnology and Oceanography, Aquatic Ecology,* and *Australian Journal of Marine and Freshwater Resources.*

in the fields of landscape ecology and conservation biology (Crooks and Sanjayan Chapter 1). Connectivity is also a fundamental concept of metapopulation ecology (Moilanen and Hanski Chapter 3). While original metapopulation models were designed and tested on terrestrial biota (typically insects and small mammals), metapopulation theory has been more recently applied to riverine biota such as fishes and mussels (e.g., Stoeckel *et al.* 1997; Policansky and Magnuson 1998; Gotelli and Taylor 1999; Fagan 2002; Neville *et al.* Chapter 13).

Freshwater ecologists frequently use the term connectivity to describe spatial connections within rivers (e.g., Stanford and Ward 1992, 1993; Ward 1997; Amoros and Bornette 1999; Wiens 2002). Ward (1997) defines *riverine* connectivity as energy transfer across the riverine landscape. Ward and Stanford (1989a) define rivers as having interactive pathways along one temporal dimension (timescales) and three spatial dimensions, i.e., longitudinal (headwater–estuarine), lateral (riverine–riparian/floodplain), and vertical (riverine–groundwater). Consideration of dynamic interactions along these four dimensions has proven to be a very effective conceptual framework to understand human impacts on river ecosystems (e.g., Ward and Stanford 1989b; Boon *et al.* 1992; Pringle 1997, 2000; Tockner and Stanford 2002).

I will now move to a discussion of hydrologic connectivity. In contrast to riverine connectivity, it encompasses broader hydrologic connections beyond the watershed (or river basin) on regional and global scales.

HYDROLOGIC CONNECTIVITY AND ITS IMPORTANCE TO THE FIELD OF CONSERVATION BIOLOGY

Hydrologic connectivity (i.e., the water-mediated transfer of matter, energy, and/or organisms within or between elements of the hydrologic cycle) (Pringle 2001, 2003a, 2003b) (Fig. 10.2) plays an important role in maintaining the biological integrity of "natural" landscapes. As stressed earlier, this property also serves to perpetuate the flow of exotic species, human-derived nutrients, and toxic wastes in the landscape. Hydrologic connectivity is complex and often difficult to understand due to: (1) the inherent complexity of water movement within and between the atmosphere and surface–subsurface systems; and (2) the extent and magnitude of human alterations that alter this property (Pringle and Triska 2000). Cumulative effects of these alterations have resulted in regional and global environmental challenges.

Despite its importance, hydrologic connectivity remains a largely neglected dimension of conservation biology. The words "stream" and "river" do not even appear in the indices of major books on the subject of habitat fragmentation (e.g., Shafer 1990; Schelhas and Greenberg 1996; Laurance and Bierregaard 1997; Soulé and Terborgh 1999). Likewise, the field of conservation biology has placed much emphasis on the size, shape and configuration of biological reserves (e.g., the Single Large or Several Small (SLOSS) debate; see Frankham Chapter 4), without any consideration of reserve location with respect to watersheds, underlying regional aquifers, and atmospheric deposition patterns. As just one example, the Biological Dynamics of Forest Fragments Project (also known as the Minimum Critical Size Ecosystem Project) did not use the presence or absence of surface water as a criterion in its experimental

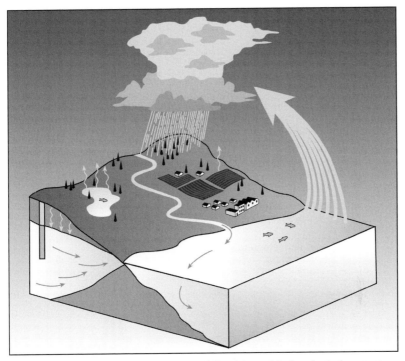

Fig. 10.2. Hydrologic connectivity is defined in a global ecological context as the water-mediated transfer of matter, energy, and organisms between or within elements of the hydrologic cycle (*sensu* Pringle 2001).

design − which involved the creation of forest fragments of different sizes and the monitoring of biodiversity in these fragments through time (R. O. Bierregaard, pers. comm.). This project would be strengthened by incorporation of a hydrologic perspective. The key question then becomes: How does the size, shape, and configuration of a given forest patch (or reserve), with respect to its location in the watershed, affect biota and overall biodiversity?

Worldwide, less than 7% of land area is either strictly or partially protected, yet many biological reserves are in danger of becoming population "sinks" for wildlife if we do not develop a more predictive understanding of how they are affected by alterations in hydrologic connectivity within the greater landscape (Pringle 2001). The location of a biological reserve, relative to watershed boundaries, regional aquifers, and wind and precipitation patterns, can play a key role in affecting biota and biodiversity and in determining the response of biota to disturbance transmitted through the hydrologic cycle. This is illustrated by an examination of the

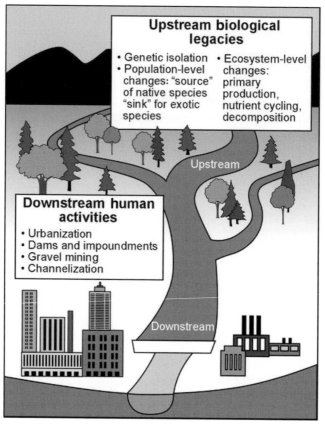

Fig. 10.3. Potential downstream influences on upstream communities. (Modified from Pringle 1997.)

vulnerability of biological reserves, in the USA (Pringle 2000) and throughout the world (Pringle 2001), to cumulative alterations in hydrologic connectivity within and outside of reserve boundaries. Reserves are vulnerable to different types of hydrologic alterations depending on where they are located within the watershed. In lower and middle watershed locations, reserves often experience direct effects of upstream alterations in hydrologic connectivity that cause habitat modifications and exacerbate effects of pollution (Fig. 10.3). In contrast, reserves in upper watersheds may have intact physical habitat and contain important source populations of some native biota, yet hydrologic alterations downstream may cause genetic isolation, extirpation of migratory species, invasion of exotic species, and cascading ecological effects (Pringle 2001). Although many parks in the world that are located in headwater areas of "rock and ice" are esthetically quite pleasing, their biotic integrity has been compromised by hydrologic disturbances that have been transmitted upstream (Pringle 1997).

As human populations grow in regions surrounding biological reserves, there is mounting pressure on water resources associated with these remnant areas as other sources of water are diminished (e.g., Pringle 2001; Dudley and Stolton 2003). Water development proposals focusing on diversion or withdrawal of water for human use have increasingly begun to focus on surface and groundwaters that flow into, under, adjacent to, or out of parks and other public lands (Pringle 2000). Of greatest concern are situations where hydrologic alterations outside of reserve boundaries end up controlling the biology of the reserve. For example, competition for water resources between national wildlife refuges and adjacent human populations is particularly acute in arid western states of the USA, where water supplies for many refuges are controlled by other agencies or organizations. Some refuges pay annual fees to irrigation companies for water delivery and/or pay pumping fees. Wildlife refuges in the USA were established specifically to protect wildlife (primarily migratory birds), yet many refuges report that their existing water rights are not adequate to support wildlife needs during an average year (Pringle 2000).

In the USA and other countries, rates of groundwater pumping and water withdrawals from rivers are so great that there is not enough surface water left in streams and wetlands to meet environmental needs. Groundwaters and surface waters are integrally connected (Winter et al. 1998) and in arid regions of the world it is not uncommon for groundwater pumping to proceed at maximum rates with decreases in river flow and, in some cases, the river drying up completely. In these cases, human water use ends up controlling river levels and ultimately fish, wildlife, and streamside ecosystems.

Cumulative effects of dams and other hydrologic modifications have increasingly altered the biotic integrity of public lands in the USA, from disruption of migratory fish passage and terrestrial wildlife that use riparian corridors (Nehlsen et al. 1991; Pringle et al. 2000), to alteration of riparian vegetation (e.g., Smith et al. 1991; Stromberg and Patten 1992) and the creation of conditions that allow for the proliferation of exotic species (e.g., Johnson and Carothers 1987).

"PROTECTED" AREA CASE STUDY: THE CARIBBEAN NATIONAL FOREST, PUERTO RICO, AND ITS VULNERABILITY TO DOWNSTREAM DISRUPTIONS IN HYDROLOGIC CONNECTIVITY

The ecological consequences of river fragmentation downstream of a "protected" area are illustrated by the Caribbean National Forest of Puerto Rico. The forest (113 km^2) is located in the highlands of northeastern Puerto Rico and is drained by nine major rivers that are characterized by a simple food chain (typical of oceanic islands), dominated by migratory freshwater shrimps and fishes which are food sources for both aquatic and terrestrial organisms, including humans. The migrations of fishes and shrimps form a dynamic link between stream headwaters and estuaries. In the case of shrimps, newly hatched larvae migrate downstream and complete their larval stage in the estuary. Upon metamorphosis, juveniles migrate upstream where they live as adults. When water is withdrawn from rivers for human use (drinking water, irrigation, etc.), this results in direct mortality of shrimp larvae migrating to the ocean. Water diversions for municipal water use at a typical low-head ($2-3$ m) dam cause direct mortality of drifting shrimp larvae. The concrete dam barrier, and low water flows behind the dam, can also negatively affect the upstream migration of juvenile shrimps (Benstead et al. 1999).

At present, all except one of the nine stream drainages within the Caribbean National Forest have low-head dams and associated water intakes on their main channels. The extent of water abstractions is so extreme that, on an average day, an estimated 70% of riverine water draining the forest is withdrawn into municipal water supplies before it reaches the ocean (Crook 2005). Massive larval mortality of shrimp resulting from these water withdrawals could potentially affect upstream recruitment of adults and other ecosystem processes. If migratory shrimps and fishes were to be extirpated above dams and water intakes, as has occurred above high dams without water spillways in other regions of Puerto Rico (e.g., Holmquist et al. 1998; Greathouse 2004; Greathouse et al. 2006), concomitant changes in ecosystem structure and function are predicted to occur (Pringle 1997).

Experimental studies provide evidence of the importance of these migratory species to the structure and function of stream communities. For example, shrimps: (1) significantly reduce algal standing crop and alter algal community composition (e.g., Pringle 1996); (2) reduce fine particulate organic matter and alter its nutritional value for other benthic consumers (e.g., Pringle *et al.* 1999); and (3) increase leaf litter processing rates (e.g., March *et al.* 2002). Consequently, when shrimps are extirpated from above large dams, ecosystem processes such as primary production and nutrient cycling are altered. For example, the standing crop of algae increased by nine-fold, and the carbon and nitrogen content of depositional organic material increased by 27- and 14-fold, respectively, in the absence of shrimps upstream from dams in Puerto Rico (Greathouse 2004).

In conclusion, the Caribbean National Forest in Puerto Rico provides just one example, among many "protected" areas throughout the world (Pringle 2001), where hydrologic alterations outside of its boundaries are affecting ecosystem integrity. This particular case study underlines the importance of understanding how hydrologic modifications occurring downstream are transmitted upstream (Fig. 10.3), resulting in cascading ecological effects. It illustrates the pressing need for innovative new strategies to manage hydrologic connectivity across the boundaries of biological reserves as they become remnant natural areas in human-dominated landscapes. Reserves in biomes ranging from arid deserts to tropical rainforests are vulnerable, regardless of their size and watershed location (e.g., Pringle 2001).

ECOLOGICAL CONSEQUENCES OF ALTERATIONS (i.e., REDUCTION AND ENHANCEMENT) IN HYDROLOGIC CONNECTIVITY

Here, I will briefly examine the ecological consequences of both reducing and enhancing the property of hydrologic connectivity on levels from genes to ecosystems. While some hydrologic alterations result in fragmentation of freshwater ecosystems, others result in enhanced connectivity. In landscapes exposed to increasing levels of fragmentation, consideration and management of hydrologic connectivity is becoming an important strategy.

Genetic and species-level changes

As rivers become increasingly fragmented, populations of aquatic organisms are subjected to reduced gene flow and loss of genetic variation. For obligate riverine species with large home ranges, impoundments may fragment the range of a species, causing losses in genetic diversity and local extinctions of populations. On genetic and species levels it is clearly important to locate and protect upstream areas and river tributaries that are acting as source populations for native fishes (Howe *et al.* 1991) (Fig. 10.3). Isolated populations of native species need to be identified, genetically analyzed, and monitored (e.g., Meffe 1987).

Genetic and species-level effects of dams on economically important migratory fishes such as salmonids have received much attention (Pacific Rivers Council 1993; Neville *et al.* Chapter 13). Over 100 major salmon and steelhead populations or stocks have been extirpated on the west coast of the USA and Canada, while at least 214 more are at risk of extinction (Nehlsen *et al.* 1991). Less is known about the effects of stream fragmentation on North American biota of less economic importance (e.g., populations of freshwater shrimps, other invertebrates, and non-game fishes). Moreover, in tropical areas such as the Amazon basin, fish migratory patterns are so complex (covering huge drainage areas) that effects of stream fragmentation are unknown even for economically important fish species (e.g., Goulding *et al.* 1996).

Many fish species have been extirpated from river drainages throughout the world as a result of disruptions in hydrologic connectivity caused by dams. In a recent synthesis paper (Pringle *et al.* 2000), we summarized the regional effects of dams on fish and mollusk species in the New World. Effects include: (1) extirpation or imperilment of many migratory fish species; (2) faunal range fragmentation and population isolation; (3) extinction or imperilment of geographically restricted species dependent on uniquely riverine habitats; (4) reduction in abundances of flood-dependent species as well as species dependent on freshwater inflows to estuarine habitats; and (5) increases in lentic (i.e., lake/reservoir adapted) and exotic species.

Establishment of new hydrologic connections in the landscape (e.g., inter-basin transfers) can result in dispersal of exotic species into new habitats and hybridization with native species (see also Crooks and Suarez Chapter 18). For example, many exotic species have been introduced into the Laurentian Great Lakes of North America as a result of navigation canals constructed to connect the Great Lakes with the Atlantic Ocean. This complex network of canals has acted to dissolve barriers that had previously kept ships and animals from passing from one watershed to another (see review by Mills *et al.* 1994). Large-scale ballast water releases from ships using these canals have introduced organisms from around the world into the Great Lakes. For example, an allozyme survey showed that patterns of genetic variation in *Daphnia galeata* populations in the Great Lakes indicated hybridization with a previously undiscovered introduction of a European form of the species in the late 1970s or early 1980s (Taylor and Hebert 1993). The introduction most probably occurred when the European form of *D. galeata* was introduced to the Great Lakes in the ballast water of ships.

Population- and community-level changes

Alterations of hydrologic connectivity have consequences that extend from single populations to multiple species within ecological communities. For example, many North American mussel species have been extirpated or are declining, in part because dams have resulted in the extirpation of their migratory fish host species which are necessary for their parasitic glochidial larval state. Just one example is the dwarf wedge mussel (*Alasmidonta heterodon*) of the US Atlantic coast. In the past 20 years, known populations of the mussel have dropped from 70 to 19 (Middleton and Liitschwager 1994). A leading theory for the cause of its demise is that the fish species that serves as its host during the critical glochidial stage of development may also be in decline. Although the identity of key host species for the mussel remains uncertain, it is suspected that the host fish is anadromous, and that dam construction has blocked its access to upstream mussel populations. The mussel is also extremely sensitive to the siltation that results from dams and riverbank erosion, and to toxic chemicals in agricultural and industrial effluents.

One management strategy to enhance longitudinal connectivity of rivers for fish communities is to employ passage devices that allow the upstream passage of fishes that would have been otherwise blocked by dams. Protection of instream flows (e.g., Gillilan and Brown 1997) is of key importance in enhancing hydrologic connectivity on longitudinal, lateral, and vertical dimensions – with benefits to populations of riverine fishes and the flora and fauna of wetlands and riparian areas. There is a growing awareness of the importance of modifiying flow regimes beneath dams to be more reflective of natural flow regimes (e.g., Poff *et al.* 1997).

In some instances, managers have opted to purposely decrease hydrologic connectivity to protect populations of endangered species. The greenback cutthroat trout *(Oncorhynchus clarki stomias)* provides a case in point. It is one of four native species of cutthroat trout found in Colorado. It is vulnerable to displacement by exotic fishes, such as brook trout *(Salvelinus fontinalis)*, which have been introduced into many river drainages of the western USA. Aggressive juvenile brook trout can displace juvenile greenback cutthroats from optimal habitat and make them vulnerable to predation, while other introduced trout species hybridize with them. In order to protect greenback cutthroat trout, permanent physical barriers are maintained at the downstream end of headwater drainages where this endangered species has established populations. These barriers prohibit upstream passage of non-native exotic species. While this strategy has proven to be effective in the short term, the long-term success of this species recovery program is unclear.

Establishment of new hydrologic connections in the landscape, such as inter-basin transfers, can also have community-level effects. Tropical rivers are particularly vulnerable to inter-basin transfers, which can result in the diffusion of diverse faunal communities that were previously isolated. Hazards of these transfers include competition for resources, predation, and the spread of parasitic diseases among geographic isolates (Pringle *et al.* 2000).

Ecosystem- and landscape-level changes

When dominant faunal components of an ecosystem are excluded from upper portions of the watershed as a result of downstream human activities, a cascade of ecosystem-level effects may occur, particularly when the extirpated component was an important food source, predator, host species, or habitat modifier. We are just beginning to acknowledge the magnitude of ecosystem-level consequences of migratory faunal depletion caused by dams (see review by Freeman *et al.* 2003). As just one example, populations of bald eagles and grizzly bears that depend on salmonids as a food source in the Pacific northwest of North America may decrease if this food source is eliminated. It is also well documented that migratory salmonids can provide major inputs of nutrients and energy to freshwater systems when spawning adults return from the sea (Ben-David *et al.* 1998; Gresh *et al.* 2000). Consequently, when dams block salmonid migration routes, patterns of nutrient cycling in entire riverine ecosystems can be altered.

Ecosystem-level effects of the loss of mussel species from streams where they were once abundant provide yet another legacy of river frag-mentation. Given that mussels filter an enormous amount of water, and that they were once plentiful across the landscape, consequences of their elimination likely include substantial losses in system productivity, decreased local retention of nutrients, and alterations in the structure and stability of the benthic stream environment (Strayer *et al.* 1999). Ninety percent of the world's freshwater mussel species are found in North America, and 73% of all mussel species in the USA are at risk of extinction or are already extinct. The prognosis is not good: in 1990, 90% of the listed mussels were still declining, and only 3% were increasing

(Master 1990). Declines have been linked to hydrologic connectivity including the loss of migratory hosts (as theorized for the dwarf wedge mussel above), susceptibility to water-transported contaminants, and stream-flow depletion (including groundwater pumping).

Groundwater exploitation in stream watersheds can sever lateral connections between stream channels and adjacent springs and wetlands, resulting in landscape-level changes in the drainage network and the distribution of biota (e.g., Winter *et al.* 1998; Pringle and Triska 2000). The increasing exploitation of groundwater reserves for municipal, industrial, and agricultural use is having profound effects on riverine ecosystems as groundwater tables are lowered. For example, populations of the anadromous striped bass (*Morone saxatilis*) are dependent upon coldwater refuges within riverine systems during hot summer periods because of their high oxygen requirement. In the southeastern USA, spring-fed stream systems are home to healthy and productive populations of striped bass. These streams have a high thermal diversity, and the fish can actively search out and use spring-fed areas as refuges (Van Den Avyle and Evans 1990). Extensive groundwater withdrawals threaten the springs, and thus the survival of biota such as striped bass dependent on coldwater refuges.

Restoration of hydrologic connectivity in highly modified human-dominated landscapes can also have ecosystem-level effects. For example, dam removal (or provision of fish passage devices around hydroelectric dams) in tributaries of the Laurentian Great Lakes can result in the transport of bioaccumulated toxic chemicals and also non-native species into upstream habitats (summarized by Freeman *et al.* 2002). Consequent cascading ecological effects throughout the food chain include impaired reproduction of bald eagles feeding on fishes contaminated with PCBs and other persistent organic chemicals (Giesy *et al.* 1995).

Ecosystem-level consequences of interbasin transfers can be dramatic. Drawing again upon the Laurentian Great Lakes for an example: construction of the Erie Canal and the St. Lawrence Seaway (for navigation by boats traveling from the Atlantic Ocean into the Great Lakes) have played a major role in the introduction of over 170 non-indigenous species to the Great Lakes. Several of these exotic species have played key roles in destabilizing the native flora and fauna and contributing to cascading trophic changes and ecosystem-level effects through time (Mills *et al.* 1994; Ricciardi 2001; Grigorovich *et al.* 2003). For example, an early aquatic invader was the sea lamprey, *Petromyzon marinus*, which is believed to have invaded the Great Lakes by attaching to the hulls of ships or by migrating through newly constructed canals. Declines of lake trout *(Salvelinus namacush)*, the top predator in the Great Lakes, have been attributed to parasitism by the sea lamprey; in turn, reduction in lake trout populations have contributed to increases in other exotic species (e.g., the alewife, *Alosa pseudoharengus*) and further ecological instability. An early terrestrial invader in the Great Lakes region was purple loosestrife *(Lythrum salicaria)*, which was introduced from Europe in solid ship ballast, dispersing along canals and other "corridors" to eventually displace the native cattail *(Typha* sp.) in wetlands throughout the eastern and midwestern USA, thus impacting wildlife that utilized cattail habitat. More recent invaders include the zebra mussel, *Dreissena polymorpha*, which has spread throughout the Great Lakes, resulting in dramatic ecosystem-level changes in nutrient cycling, primary production, and food-web structure and function. Moreover, the zebra mussel has facilitated the invasion of two co-evolved exotic species – an amphipod *(Echinogammarus ischnus)* and a predatory fish (the round goby, *Neogobius melanostomus*), which are further destabilizing the ecosystem and leading some scientists to suggest that the Great Lakes ecosystem has entered an "invasional meltdown" phase (Ricciardi 2001).

Future invaders of the Laurentian Great Lakes may include the bighead carp *(Aristichthys nobilis)* and the silver carp *(Hypophthalmichthys molitrix)*, which are threatening to invade Lake Michigan through the Chicago Sanitary and Ship Canal, a canal constructed from a river that once naturally drained into Lake Michigan (Stokstad 2003). The canal was designed to reverse the flow of the river and to divert wastes from the city of Chicago into the Mississippi River drainage. Several million dollars have recently been implemented to construct two electric barriers on the Chicago Sanitary Canal to stop the upstream migration of bighead and silver carp into the Great Lakes. Scientists believe that these large omnivorous invaders (aquaculture "escapees," which can reach up to 45 kg and are considered by many fisherman to be "trash" fish) will reach Lake Michigan within the next decade regardless of the barriers, with dramatic ecosystem-level effects and economic losses in fisheries.

EMERGING ENVIRONMENTAL CHALLENGES: HYDROLOGIC CONNECTIVITY AND THE TRANSPORT OF CONTAMINANTS IN THE LANDSCAPE

It is perhaps ironic that, just as we begin to understand how hydrologic connectivity works locally, cumulative anthropogenic alterations in hydrologic connectivity are now exerting effects on regional and global scales

(Rosenberg *et al.* 2000; Pringle 2003c). Perhaps the most compelling global example of how we have underestimated the role of hydrologic connectivity concerns the transport and distribution of persistent organic pollutants such as PCBs to "hotspots" such as the Arctic (e.g., Aguilar *et al.* 2002). The high concentrations found here are partly due to volatilization of PCBs in warmer climates and condensation and deposition in colder regions such as the Arctic. Ocean currents also transport PCBs, and biota that have sequestered PCBs, into the Arctic food web, where they undergo further biological magnification within long-lived animals such as seals and predatory polar bears, potentially affecting the long-term reproductive capacity of these animals and the native peoples that consume them (e.g., Skaare *et al.* 2002; Derocher *et al.* 2003). Atmospheric and hydrologic pathways by which PCBs enter the Arctic food chain were not immediately predicted; as pointed out by Colburn *et al.* (1997), one of the sad ironies is that researchers discovered the high levels of contamination among people in remote Inuit villages while looking for a less-exposed control group.

We also lack an understanding of the role of hydrologic connectivity in the transport of many pathogens (see also McCallum and Dobson Chapter 19). Again, pathways in the hydrologic cycle that play a key role in pathogen transport are often identified in hindsight. As just one example, a particular strain of cholera can live inside algal cells, resting encysted in a dormant state for long periods of time. The emergence of cholera in Peru in 1991 was traced to bilge water drawn from Asian seas by a freighter and discharged into the harbor on the coast of Peru. Cholera-contaminated shellfish subsequently ingested by the human population caused massive outbreaks of cholera (Anderson 1991; Colwell and Spira 1992). The water environment can also play a key role in transporting bacterial genes for antibiotic resistance, as well as providing pathways for resistant bacteria themselves to spread (e.g., McArthur and Tuckfield 2000; Iwane *et al.* 2001). A specific example of our lack of understanding of the role of hydrologic connectivity in pathogen transport concerns the spread of a chytrid fungus, which has been implicated in die-offs of stream-dwelling frogs in high-gradient streams throughout Central America and other regions of the world. It is not clear how this fungal pathogen is transported from river to river and what role, if any, that hydrologic connectivity plays. What is clear is that once the chytrid fungus begins infecting and killing stream-dwelling frogs in a given drainage, they are soon extirpated within that drainage. Accordingly, stream-dwelling frogs throughout Costa Rica (and much of Panama) have been extirpated above 500 m above sea level. These extirpations appeared to occur as a "wave from north to south," moving from valley to valley and leaving a wake of dying frogs infected with chytrid in its path (Young *et al.* 2001; Lips *et al.* 2003).

In a recent synthesis book chapter (Pringle 2003c), I examine three emerging ecological problems that pertain to how alterations of hydrologic connectivity can change the way contaminants are transported and distributed in the landscape: (1) regional declines in migratory birds and wildlife resulting from wetland drainage and contaminated irrigation drainage; (2) bioaccumulation of methylmercury in fish and wildlife in newly created reservoirs; and (3) deterioration of estuarine and coastal ecosystems that receive the discharge of highly regulated (i.e., dammed), silicon-depleted, and nutrient-rich rivers. All three of the problems have surfaced within the last two decades and indicate the degree to which our ability to identify negative environmental impacts of technological change lags behind the application of those advances. They also illustrate the complexity of interactions between hydrologic change and biogeochemistry and the confounding effects of time lags between cause and environmental consequence.

To exemplify, I will briefly discuss the first of these three issues: effects of wetlands loss and contaminated irrigation drainage on wildlife. Wetland drainage and river dewatering for irrigation have resulted in migratory waterfowl and other wildlife being even more dependent on dwindling surface-water supplies. The biological integrity of many of the remaining surface water ecosystems in arid regions of the world has been seriously compromised because of reduced freshwater inflows and the accumulation of pesticides and contaminated subsurface irrigation drainage. Thus, cumulative impacts of wetlands destruction, and contamination of the remaining wetlands, are particularly severe, given that remaining wetlands often comprise only a small portion of historic wetlands and are important to fish and wildlife on regional scales (e.g., as stopovers for migratory waterfowl). California, in the western USA, provides a compelling example of this disturbing trend. Approximately 95% of the interior wetlands in California have been lost (primarily to irrigated agriculture), and this has resulted in over 60% of the Pacific migratory flyway waterfowl population being channeled into available wetlands (Frayer *et al.* 1989). One such wetland is the Sonny Bono Salton Sea National Wildlife Refuge – one of the few wetlands in southern California, attracting more than 380 species of migratory birds to its contaminated waters. Fish and bird die-offs are so common and massive that the US Fish and Wildlife Service operates an incinerator to dispose of dead birds. The National Audubon Society uses the Sonny Bono National Wildlife Refuge as an example of the state of crisis of wildlife refuges in the USA. In 1997, over 14 000 birds (representing 66 species) perished as a result of an outbreak of avian botulism, including over 1400 endangered brown pelicans. The Sonny Bono Salton Sea National Wildlife Refuge is by no means an isolated example: several National Wildlife Refuges in the western USA have experienced die-offs of birds and fishes including Stillwater, Bowdoin, Kesterson, and Ourag National Wildlife Refuges, (Lemly *et al.* 1993). This phenomenon is not just confined to the western USA, but is occurring in other arid regions of the world, such as the Macquarie Marshes of Australia and the Aral Sea in central Asia (see reviews by Lemly *et al.* 1993, 2000). Solutions to the regional environmental problems caused by toxic irrigation drainage are not clear. Even when confined to toxic evaporation ponds, toxic irrigation drainage continues to attract wildlife, given the lack of remaining wetland habitat in many developed landscapes that have been drained for agriculture. It is clear that we need to proactively re-evaluate unsustainable agricultural practices.

CONCLUSION

In summary, it is critical that an understanding of hydrologic connectivity be incorporated into the field of conservation biology − into both theoretical constructs and practical applications. Hydrologic connectivity should be a basis of many cumulative effects and secondary impacts analyses as required under the US National Environmental Protection Act for federal actions, such as permitting for new reservoirs. We also must develop a more predictive understanding of how hydrologic connectivity, and alterations of this property, influence ecological patterns on regional and global scales − before environmental crises occur.

ACKNOWLEDGEMENTS

Many thanks to Mary Freeman and Sandra Tartowski for their suggestions which greatly improved this chapter. I would also like to thank my current graduate students who have provided feedback on material presented in this chapter: Elizabeth Anderson, Marcelo Ardon, Scott Connelly, Kelly Crook, Susan Dye, Effie Greathouse, John Kominoski, Chip Small, and Katherine Smith. I also acknowledge National Science Foundation (NSF)

Grant DEB-0234386 and NSF Grants DEB-0218039 and DEB-00805238 which support the Luquillo Long-term Ecological Research Program.

REFERENCES

Aguilar, A., A. Borrell, and P. J. H. Reijnders. 2002. Geographical and temporal variation in levels of organochlorine contaminants in marine mammals. *Marine Environmental Research* **53**:425–452.

Amoros, C., and G. Bornette. 1999. Antagonistic and cumulative effects of connectivity: a predictive model based on aquatic vegetation in riverine wetlands. *Large Rivers* **11**:311–327.

Amoros, C., and A. L. Roux. 1988. Interaction between water bodies within the floodplain of large rivers: function and development of connectivity. *Munstersche Geographische Arbeiten* **29**:125–130.

Anderson, C. 1991. Cholera epidemic traced to risk miscalculation. *Nature* **354**:255.

Ben-David, M., T. A. Hanley, and D. M. Schell. 1998. Fertilization of terrestrial vegetation by spawning pacific salmon: the role of flooding and predator activity. *Oikos* **83**:47–55.

Benke, A. C. 1990. A perspective on America's vanishing streams. *Journal of the North American Benthological Society* **9**:77–88.

Benstead, J. P., J. G. March, C. M. Pringle, and F. N. Scatena. 1999. Effects of a low-head dam and water abstraction on migratory tropical stream biota. *Ecological Applications* **9**:656–668.

Boon, P. J., P. Callow, and G. E. Petts (eds.) 1992. *River Conservation and Management*. Chichester, UK: John Wiley.

Colburn, T., D. Dumanoski, and P. Peterson-Myers. 1997. *Our Stolen Future*. New York: Penguin Books.

Colwell, R. R., and W. M. Spira. 1992. The ecology of *Vibrio cholerae*. Pp. 107–127 in D. Barua, and W. B. Greenough, III (eds.) *Cholera*. New York: Plenum Press.

Crook, K. E. 2005. Quantifying the effects of water withdrawal on streams draining the Caribbean National Forest, Puerto Rico. M.Sc. thesis, University of Georgia, Athens, GA.

Derocher, A. E., H. Wolkers, T. Colburn, *et al.* 2003. Contaminants in Svalbard polar bear samples archived since 1967 and possible population level effects. *Science of the Total Environment* **301**:163–174.

Dudgeon, D. 2000. Going with the flow: large-scale hydrological changes and prospects for riverine biodiversity in tropical Asia. *BioScience* **50**:793–806.

Dudley, N., and S. Stolton. 2003. Can protected areas quench our thirst? *Conservation in Practice* **4**:30–31.

Fagan, W. F. 2002. Connectivity, fragmentation, and extinction risk in dendritic metapopulations. *Ecology* **83**:3243–3249.

Frayer, W. E., D. D. Peters, and H. R. Pywell. 1989. *Wetlands of California Central Valley: Status and Trends — 1939 to the mid-1980s*. Portland, OR: US Fish and Wildlife Service.

Freeman, M. C., C. M. Pringle, E. A. Greathouse, and B. J. Freeman. 2003. Ecosystem-level consequences of migratory faunal depletion caused by dams. *American Fisheries Society Symposium* **35**:255–266.

Freeman, R., W. Wowerman, T. Grubb, *et al.* 2002. Opening rivers to Trojan fish: the ecological dilemma of dam removal in the Great Lakes. *Conservation in Practice* **3**:35–40.

Giesy, J. P., D. A. Verbrugge, R. A. Othout, *et al.* 1995. Contaminants of fishes from Great Lakes-influenced sections and above dams of three Michigan rivers. III. Implications for health of bald eagles. *Archives of Environmental Contamination and Toxicology* **29**:309–321.

Gillilan, D. M., and T. C. Brown. 1997. *Instream Flow Protection: Seeking a Balance in Western Water Use.* Washington, DC: Island Press.

Gotelli, N. J., and C. M. Taylor. 1999. Testing metapopulation models with stream-fish assemblages. *Evolutionary Ecology Research* **1**:835–845.

Goulding, M., N. J. H. Smith, and D. J. Mahar. 1996. *Floods of Fortune: Ecology and Economy along the Amazon.* New York: Columbia University Press.

Greathouse, E. 2004. Consequences of migratory faunal extirpation upstream from large dams in Puerto Rico. Ph.D. dissertation, University of Georgia, Athens, GA.

Greathouse, E. A., C. M. Pringle, W. H. McDowell, and J. G. Holmquist. 2006. Indirect upstream effects of dams: consequences of migratory consumer extirpation in Puerto Rico. *Ecological Applications* **16**:339–352.

Gresh, T., J. Lichatowich, and P. Schoonmaker. 2000. An estimation of historic and current levels of salmon production in the northeast Pacific ecosystem. *Fisheries* **25**:15–21.

Grigorovich, J. A., R. I. Colautti, E. L. Mills, *et al.* 2003. Ballast-mediated animal introduction in the Laurentian Great Lakes: retrospective and prospective analyses. *Canadian Journal of Fisheries and Aquatic Sciences* **60**:740–756.

Holmquist, J. G., J. M. Schmidt-Gengenbach, and B. B. Yoshioka. 1998. High dams and marine-freshwater linkages: effects on native and introduced fauna in the Caribbean. *Conservation Biology* **12**:621–630.

Howe, R. W., G. J. Davis, and V. Mosca. 1991. The demographic significance of "sink" populations. *Biological Conservation* **57**:239–255.

Humborg, C. D., J. Conley, L. Rahm, *et al.* 2000. Silicate retention in river basins: far-reaching effects on biogeochemistry and aquatic food webs. *Ambio* **29**:45–50.

Iwane, T., T. Urase, and K. Yamamoto. 2001. Possible impact of treated wastewater discharge on incidence of antibiotic resistant bacteria in river water. *Water Science and Technology* **43**:91–99.

Johnson, S. W., and S. W. Carothers. 1987. External threats: the dilemma of resource management on the Colorado River in Grand Canyon National Park, USA. *Environmental Management* **11**:99–107.

Laurance, W. F., and R. O. Bierregaard, Jr. 1997. *Tropical Forest Remnants: Ecology, Management, and Conservation of Fragmented Communities.* Chicago, IL: University of Chicago Press.

Lemly, A. D., S. E. Finger, and M. K. Nelson. 1993. Ecological implications of subsurface irrigation drainage. *Journal of Arid Environments* **28**:85–94.

Lemly, A. D., R. T. Kingsford, and J. R. Thompson. 2000. Irrigated agriculture and wildlife conservation: conflict on a global scale. *Environmental Management* 25:485–512.

Lips, K. R., J. D. Reeve, and L. R. Witters. 2003. Ecological traits predicting amphibian population declines in Central America. *Conservation Biology* 17:1078–1088.

March, J. G., C. M. Pringle, M. J. Townsend, and A. I. Wilson. 2002. Effects of freshwater shrimp assemblages on benthic communities along an altitude gradient of a tropical island stream. *Freshwater Biology* 47:377–390.

Master, L. 1990. The imperiled status of North American aquatic animals. *Biodiversity Network News* 3:1–8.

Master, L. L., S. R. Flack, and B. A. Stein (eds.) 1998. *Rivers of Life: Critical Watersheds for Protecting Freshwater Biodiversity, Special Publication of the Nature Conservancy.* Arlington, VA: NatureServe.

McArthur J. V., and R. C. Tuckfield. 2000. Spatial patterns in antibiotic resistance among stream bacteria: effects of industrial pollution. *Applied and Environmental Microbiology* 66:3722–3726.

Meffe, G. K. 1987. Conserving fish genomes: philosophies and practices. *Environmental Biology of Fishes* 18:3–9.

Middleton, S., and D. Liittschwager. 1994. *Witness: Endangered Species of North America.* San Francisco, CA: Chronicle Books.

Mills, E. L., J. H. Leach, J. T. Carlton, and C. L. Secor. 1994. Exotic species and the integrity of the Great Lakes: lessons from the past. *BioScience* 44:666–676.

Nehlsen, W., J. E Williams, and J. A. Jicharowich. 1991. Pacific salmon at the crossroads: Stocks at risk from California, Oregon, Idaho, and Washington. *Fisheries* 16:4–21.

Pacific Rivers Council. 1993. *The Decline of Coho Salmon and the Need for Protection under the Endangered Species Act.* Eugene, OR: Pacific Rivers Council.

Poff, M. L., J. D. Allan, M. B. Bain, *et al.* 1997. The natural flow paradigm. *BioScience* 47: 769–784.

Policansky, D., and J. J. Magnuson. 1998. Genetics, metapopulations, and ecosystem management of fisheries. *Ecological Applications* 8:119–123.

Postel, S. L., G. C. Daily, and P. R. Ehrlich. 1996. Human appropriation of renewable freshwater. *Science* 271:785–788.

Pringle, C. M. 1996. Atyid shrimp (Decapoda: Atyidae) influence spatial heterogeneity of algal communities over different scales in tropical montane streams, Puerto Rico. *Freshwater Biology* 35:125–140.

Pringle, C. M. 1997. Exploring how disturbance is transmitted upstream: going against the flow. *Journal of the North American Benthological Society* 16:425–438.

Pringle, C. M. 2000. Threats to U.S. public lands from cumulative hydrologic alterations outside of their boundaries. *Ecological Applications* 10:971–989.

Pringle, C. M. 2001. Hydrologic connectivity and the management of biological reserves: a global perspective. *Ecological Applications* 11:981–998.

Pringle, C. M. 2003a. The need for a more predictive understanding of hydrologic connectivity. *Aquatic Conservation* 13:467–471.

Pringle, C. M. 2003b. What is hydrologic connectivity and why is it ecologically important? *Hydrological Processes* 17:2685–2689.

Pringle, C. M. 2003c. Interacting effects of altered hydrology and contaminant transport: Emerging ecological patterns of global concern. Pp. 85–107 in M. Holland, E. Blood, and L. Shaffer (eds.) *Achieving Sustainable Freshwater Systems: A Web of Connections*. Washington, DC: Island Press.

Pringle, C. M., and F. J. Triska. 2000. Emergent biological patterns and surface-subsurface interactions at landscape scales. Pp. 167–193 in J. B. Jones, and P. J. Mulholland (eds.) *Stream and Groundwaters*. San Diego, CA: Academic Press.

Pringle, C. M., N. H. Hemphill, W. McDowell, A. Bednarek, and J. March. 1999. Linking species and ecosystems: different biotic assemblages cause interstream differences in organic matter. *Ecology* 80:1860–1872.

Pringle, C. M., M. C. Freeman, and B. J. Freeman. 2000. Regional effects of hydrologic alterations on riverine macrobiota in the New World: Tropical-temperate comparisons. *BioScience* 50:807–823.

Ricciardi, A. 2001. Facilitative interactions among invaders: is an "invasional meltdown" occurring in the Great Lakes? *Canadian Journal of Fisheries and Aquatic Sciences* 58:2513–2525.

Rosenberg, D. M., P. McCully, and C. M. Pringle. 2000. Global-scale environmental effects of hydrological alterations: introduction. *BioScience* 50:746–751.

Schelhas, J., and R. Greenberg. 1996. *Forest Patches in Tropical Landscapes*. Washington, DC: Island Press.

Shafer, C. L. 1990. *Nature Reserves*. Washington, DC: Smithsonian Institution Press.

Skaare, J. U., H. J. Larsen, E. Lie, *et al.* 2002. Ecological risk assessment of persistent organic pollutants in the arctic. *Toxicology* 181:193–197.

Smith, S. D., A. B. Wellington, J. L. Nachlinger, and C. A. Fox. 1991. Functional responses of riparian vegetation to stream flow diversion in the eastern Sierra Nevada. *Ecological Applications* 1:89–97.

Soulé, M. E., and J. Terborgh (eds.) 1999. *Continental Conservation: Scientific Foundations for Regional Reserve Networks*. Washington, DC: Island Press.

Stanford, J. A., and J. V. Ward. 1992. Management of aquatic resources in large catchments: recognizing interactions between ecosystem connectivity and environmental disturbance. Pp. 91–124 in R. J. Naiman (ed.) *Watershed Management*. New York: Springer-Verlag.

Stanford, J. A., and J. V. Ward. 1993. An ecosystem perspective of alluvial rivers: connectivity and the hyporheic corridor. *Journal of the North American Benthological Society* 12:48–60.

Stoeckel, J. A., D. W. Schneider, L. A. Soeken, K. D. Blodgett, and R. E. Sparks. 1997. Larval dynamics of a riverine metapopulation: implications for zebra mussel recruitment, dispersal, and control in a large-river system. *Journal of the North American Benthological Society* 16:586–601.

Stokstad, E. 2003. Can well-timed jolts keep out unwanted exotic fish? *Science* 301:157–158.

Strayer, D. L., N. F. Caraco, J. J. Cole, S. Findlay, and M. L. Pace. 1999. Transformation of freshwater ecosystems by bivalves. *BioScience* 49:19–27.

Stromberg, J. C., and D. T. Patten. 1992. Mortality and age of black cottonwood stands along diverted and undiverted streams in the eastern Sierra Nevada, California. *Madrono* **39**:205–223.

Taylor, D. J., and P. D. N. Hebert. 1993. Cryptic intercontinental hybridization in *Daphnia* (Crustacea): the ghost of introductions past. *Proceedings of the Royal Society of London B* **254**:163–168.

Tockner, K., and J. A. Stanford. 2002. Riverine flood plains; present state and future trends. *Environmental Conservation* **29**:308–330.

Van Den Avyle, M. J., and J. W. Evans. 1990. Temperature selection by striped bass in a Gulf of Mexico coastal river system. *North American Journal of Fisheries Management* **10**:58–66.

Ward, J. V. 1997. An expansive perspective of riverine landscapes: pattern and process across scales. *River Ecosystems* **6**:52–60.

Ward, J. V., and J. A. Stanford. 1989a. The four-dimensional nature of lotic ecosystems. *Journal of the North American Benthological Society* **8**:2–8.

Ward, J. V., and J. A. Stanford. 1989b. Riverine ecosystems: the influence of man on catchment dynamics and fish ecology. Pp. 56–64 in D P. Dodge (ed.) *Proceedings of the International Large River Symposium*, Canadian Special Publication of Fisheries and Aquatic Sciences No. 160. Ottawa, Ontario, Canada: Department of Fisheries and Oceans.

Wiens, J. A. 2002. Riverine landscapes: taking landscape ecology into the water. *Freshwater Biology* **47**:501–515.

Winter, T. C., J. W. Harvey, O. Lehn Franke, and W. M. Alley. 1998. *Ground Water and Surface Water: A Single Resource*, US Geological Survey Circular (USA) No. 1139. Denver, CO: US Geological Survey.

Young, B. E., K. R. Lips, J. K. Reaser, *et al.* 2001. Population declines and priorities for amphibian conservation in Latin America. *Conservation Biology* **15**:1213–1223.

11

Connectivity and ecosystem services: crop pollination in agricultural landscapes

TAYLOR H. RICKETTS, NEAL M. WILLIAMS
AND MARGARET M. MAYFIELD

INTRODUCTION

As the focus of ecological research has broadened from site-level observations and experiments to landscape-scale studies, interest among ecologists in connectivity among habitat patches has increased (Turner *et al.* 1995; Wiens 1995; Gustafson and Gardner 1996; Hanski and Gilpin 1997; Haddad and Baum 1999). From a conservation perspective, concern with the maintenance of connectivity has grown as habitat loss and fragmentation continue worldwide (Rosenberg *et al.* 1997; de Lima and Gascon 1999; Haddad and Baum 1999). Typically, both basic and applied studies consider connectivity among patches of the same habitat type, such as a set of forest fragments embedded in a matrix of agricultural land uses. These studies, reviewed throughout this volume, focus on measuring patch isolation and examining the utility of corridors to restore connectivity among patches.

However, another type of connectivity, although less well studied and understood, is of equal importance in these complex landscapes: the connectivity among patches of *different* habitat types (Daily *et al.* 2001; Ricketts *et al.* 2001; Talley *et al.* Chapter 5). Many species require different habitat types for different resources or life-history stages. The proximity and availability of different habitat types, therefore, can affect the population dynamics and persistence of individual species and the diversity of communities. For example, bees often nest in one habitat type

Connectivity Conservation eds. Kevin R. Crooks and M. Sanjayan. Published by Cambridge University Press. © Cambridge University Press 2006.

(e.g, tree cavities in forest) but require other types for forage (e.g., wildflower meadows). Connectivity among different habitats in a landscape can also affect important ecosystem processes, many of which confer economic benefits on human populations and are therefore termed "ecosystem services."

Ecosystem services are those natural processes through which ecosystems sustain and fulfill human life (Daily 1997). Examples include water purification and flood control by wetlands, food provision from coastal oceans, and carbon sequestration by forests. These ecosystem services have measurable and often enormous economic value (Daily *et al.* 2000; Heal 2000), which is lost to humanity as the systems that provide them are destroyed (Balmford *et al.* 2002). The scales over which ecosystem services are conferred vary widely. Some, such as carbon sequestration, confer benefits globally by regulating greenhouse gases and moderating climate change. Others, such as water purification, act more locally; only people in the same watershed as functioning wetlands receive the benefits.

For these locally conferred ecosystem services, landscape connectivity between the "source" ecosystem and the "recipient" system or human population center is crucial (Balvanera *et al.* 2001; Talley *et al.* Chapter 5). Source and recipient systems must be sufficiently proximate or otherwise functionally connected to interact ecologically. As mentioned above, water purification and flood control services provided by wetlands only accrue to human populations that are nearby and downstream. Another example is crop pollination. Bees, the most common crop pollinators, generally nest in specific habitats and forage to certain distances from their nests (Saville *et al.* 1997; Michener 2000; Gathmann and Tscharntke 2002). Pollination services by wild bees, therefore, only benefit crops within the foraging distances of key pollinator species. Maintaining sufficient flows of these and other vital ecosystem services requires an understanding of the functional connectivity required between different habitats in the landscape. For most ecosystem services, however, these connectivity requirements are too poorly understood to inform land-use policy in a meaningful way.

In this chapter we explore the importance of landscape connectivity to ecosystem services. We use crop pollination, a locally conferred and enormously valuable ecosystem service, as a model to understand these issues (Kevan 1991; Allen-Wardell *et al.* 1998; Kearns *et al.* 1998; Stubbs and Drummond 2001; Kevan *et al.* 2002; Stickler and Cane 2003). Agricultural landscapes are often highly fragmented, forming complex mosaics of natural habitats and various crop types. Recent studies have shown the importance of connectivity between natural habitats and agricultural fields for enhancing pollinator activity and pollination services to crops (Kremen et al. 2002b; Klein et al. 2003b; Ricketts 2004). We first discuss briefly some general points regarding pollinators and their persistence in agricultural landscapes. We then review three case studies of crop pollination and landscape connectivity: coffee in Costa Rica, watermelon in California (USA), and kiwifruit in New Zealand. Finally, we discuss some general lessons from the three case studies and suggest some future research directions.

POLLINATORS IN AGRICULTURAL LANDSCAPES

Although crops are pollinated by a variety of animal species (e.g., birds, flies, bats, moths), the dominant crop pollinators worldwide are bees (Free 1993). As a result, most information about pollinators and their nesting and foraging behavior exists for bees, but many of the points discussed in this chapter relate to other pollinator groups as well. Even in natural landscapes, bees often rely on landscape connectivity among different habitat types. Many bee species have very specific nesting requirements (Stephen et al. 1969), and these habitats often are spatially disjunct from areas that provide floral resources (e.g., Rust 1990; Williams and Tepedino 2003). The ability of individuals to move among these "partial habitats" (Westrich 1996) is crucial to the persistence of bee populations and the diversity of bee communities in nature. Habitat fragmentation in agricultural landscapes, therefore, can be seen as simply exacerbating this common problem of separation between floral and nesting habitats (Scott-Dupree and Winston 1987; Saville et al. 1997; Walther-Hellwig and Frankl 2000).

Although cultivated areas may represent an inhospitable matrix for many pollinator species (Aizen and Feinsinger 1994), they provide valuable resources for other species and thus help support populations of pollinators (Ricketts 2004). Flowering crops are potential resource bonanzas when they are in bloom and their proximity to nesting sites in natural areas may help bolster bee populations in otherwise degraded landscapes (Westphal et al. 2003). In addition, many solitary bees prefer to nest in exposed soil; therefore, the disturbance created in many agricultural fields may actually create preferred nesting substrate for some of these ground-nesting species (Westrich 1996).

Managed colonies of European honeybees (*Apis mellifera*) are used throughout the world to ensure adequate crop pollination. A large and growing number of studies, however, clearly show that most crops are visited by a diversity of insect species, including bees and various other insect orders (e.g., Cane and Payne 1988; Kremen *et al.* 2002a; Klein *et al.* 2003a). Often, several species can pollinate flowers effectively, and in many cases honeybees are not the most efficient pollinators (Westerkamp 1991; Wilson and Thomson 1991; Freitas and Paxton 1998; Javorek *et al.* 2002). In many regions managed honeybees have escaped into the wild. These feral populations have joined the local bee communities and, like native species, require nesting and foraging habitats and connectivity between them. It is therefore important to distinguish "native" pollinators (i.e., those indigenous to the region) from "wild" pollinators (all local wild bees, including introduced species). While conservation concern clearly focuses on native bees, all wild species can provide crop pollination services.

Crop pollination is therefore an ecosystem service provided by a community, rather than a single species. The members of these communities often differ markedly in their energetics, flight ranges, and nesting and foraging needs. Therefore, the overall effect of landscape connectivity on pollination services results from these species-specific responses, occurring at different scales, and involving different behavioral changes (see "Discussion" section for fuller treatment and references).

CASE STUDIES

The following three case studies illustrate some of the important issues involved in landscape connectivity for pollination services. The studies involve three continents, three crops, and three different pollinator communities. While they highlight important differences and idiosyncrasies among pollination systems, they also reveal surprising commonalities. Most fundamentally, they underscore the value of native habitat as sources of pollinators for neighboring crops and elucidate the scales over which pollination services are conferred.

Case 1: Coffee in Costa Rica

Ricketts (2004) investigated the role of tropical forest fragments in Costa Rica as sources of pollinators to surrounding coffee. Coffee (*Coffea arabica* and *C. canephora*) is one of the developing world's most valuable export commodities. Eleven million hectares are planted in coffee, and the industry employs over 25 million people worldwide (O'Brien and

Kinnaird 2003). Unfortunately, coffee is best cultivated in low- and middle-elevation tropical moist climates and therefore competes for space with some of the most diverse forests on earth (Olson and Dinerstein 1998; Myers *et al.* 2000; Roubik 2002). Coffee-growing regions are often characterized by highly fragmented landscapes, with remnants of native forest surrounded by an agricultural matrix dominated by coffee and a few other crops.

Coffea arabica, the species considered to produce the highest-quality coffee, dominates Central American coffee production. Although it is self-compatible (i.e., able to self-pollinate), several field experiments have shown that yields increase 15% to 50% when bees are allowed to visit flowers, compared to controls with bees excluded (e.g., Free 1993; Roubik 2002). This yield increase is likely a result of higher rates of outcrossing, leading to larger and more robust fruit (Free 1993). Despite these research findings, few coffee farmers recognize pollination as an issue for their crops (T. H. Ricketts *et al.*, unpublished data).

Study system and design

The study was designed to determine whether coffee plants near forest receive higher levels of pollinator (i.e., bee) activity than those farther away. The study site, Finca Santa Fe, was located in the Valle del General, one of the major agriculture regions of Costa Rica. The region is dominated by coffee farms, sugar cane, and cattle pasture, all of which surround hundreds of remnants of tropical/premontane moist forests (Janzen 1983). In this region, coffee flowers between January and April, typically in three or four intense and widespread flushes ("floreas") lasting three days each. No honeybees are currently managed in the area, but feral, Africanized honeybees (*Apis mellifera*) are abundant (Butz Huryn 1997; Roubik 2002).

The landscape is dominated by two large forest patches (111 ha and 46 ha, respectively) and a large coffee farm (approx. 1100 ha) that extends between them (Fig. 11.1). All common pollinators of coffee prefer forest habitats both for nesting sites (i.e., tree cavities) and year-round floral resources (Wille and Michener 1973; Roubik 1989; Griswold *et al.* 1995; Kevan 1999). At the center of the farm, distances to the nearest forest (1600 m) are beyond the typical foraging range of most native bees (Heard 1999). Throughout its extent, Finca Santa Fe uses similar planting and harvest practices, weed and pest control methods, and shade-tree species and density (Ricketts 2004). This landscape therefore allows long,

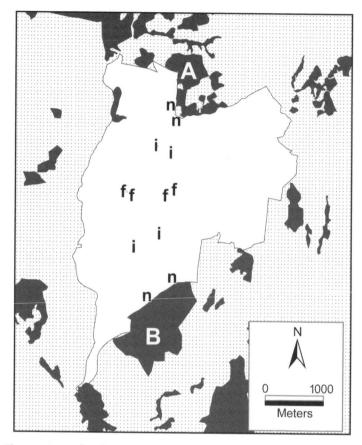

Fig. 11.1. Map of study area and study sites for coffee study. White area is Finca Santa Fe, planted homogeneously in coffee; stippled area is a mix of coffee, pasture, and sugar cane; and black areas are forest. The two focal forest patches are labeled "A" and "B." Study sites are labeled with "n" for near sites, "i" for intermediate sites, and "f" for far sites. The four sites resampled in 2002 are the southern "n" and "i" sites. (Modified from Ricketts (2004).)

replicable distance gradients from forest into coffee fields, with many other important variables held constant.

During the flowering season of 2001, bee activity was observed in 12 sites, arranged in three distance classes along transects from both the north and south forest patches (Fig. 11.1). The three distance classes were "near" (approximately 50 m from forest), "intermediate" (approximately 800 m), and "far" (approximately 1600 m). In 2002, bee activity was observed in an identical manner in four of these same sites (Fig. 11.1).

Pollinator activity and pollination

Pollinator activity was measured daily in each site during each florea. Every visitor and the number of flowers visited were recorded at standard-sized plots during a 20-min interval. Pollen deposition rates were measured as the average number of grains per stigma 24 h after flowers had opened.

Over the two years, 40 bee morphospecies (i.e., field-determined species) were observed visiting coffee flowers. Of these, the 11 most common visitors were eusocial species (i.e., living in colonies with a single queen and many workers (Michener 2000)), including 10 stingless bees in the Meliponini (Apidae) and *A. mellifera* (Table 11.1).

Bee morphospecies richness decreased significantly with increasing distance from forest patches (Fig. 11.2A). Near sites were significantly richer than intermediate and far sites, which did not differ from each other (ANOVA: $F_{2,9} = 13.36$, $P = 0.002$). Rates of bee visitation to coffee

Table 11.1. *Common bee species observed at coffee (Coffea arabica) flowers and their observed total abundances (all observations from both years pooled)*

Species[a]	Abundance
Melipona fasciata	5
Nannotrigona mellaria	33
Meliponini spp.[b]	15
Plebeia jatiformis	209
Plebeia frontalis	93
Trigona (Tetragona) clavipes	3
Trigona (Tetragonisca) angustula	34
Trigona dorsalis	17
Trigona fulviventris	47
Trigonisca sp.	76
Apis mellifera	423
Miscellaneous native species[c]	83

[a]All species are in the tribe Meliponini except for *Apis mellifera* (tribe Apini) and "miscellaneous native species." (mixed tribes).

[b]Found, when later identified, to be composed of three species: *Partamona cupira*, *Trigona fuscipennis*, and *T. corvina*.

[c]Includes 29 rare morphospecies and individuals that eluded morphospecies assignment in the field.

Modified from Ricketts (2004).

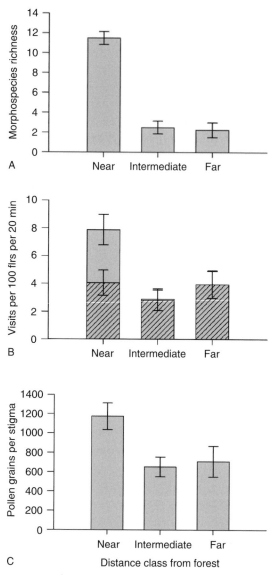

Fig. 11.2. Measures of bee activity and pollen deposition at coffee flowers (mean ± SE) along distance gradients from the forest patches. (A) Accumulated bee richness (i.e., all samples per site pooled); (B) rate of bee visitation to coffee flowers, with visitation rate from honeybees represented by hatched area of bars; (C) pollen grains deposited per stigma.

flowers also significantly decreased with increasing isolation from forest (Fig. 11.2B). Visitation rates in near sites were roughly double that of the intermediate and far sites, which again did not differ from each other (ANOVA: $F_{2,119} = 15.24$, $P<0.001$). Rates of pollen deposition, which are likely the most informative measure of pollinator activity, declined as well, showing a similar pattern to overall visitation rates (Fig. 11.2C); (ANOVA: $F_{2,116} = 7.10$, $P = 0.001$). By every measure, therefore, coffee plants near forest experienced enhanced pollinator activity: they received more visits by more bee species, with higher pollination deposition rates than did plants at greater distances. Although this categorical treatment of isolation is relatively simple, the results here do not change with more sophisticated, continuous isolation measures (e.g., Hanski 1998).

Although overall visitation rates were markedly lower in intermediate and far sites, visitation rates by honeybees were similar across the distance gradient (Fig. 11.2B, hatched portions of bars; ANOVAs: both years $P > 0.65$). The increase in total visitation rate near the two forest patches was therefore caused by native species. In fact, honeybees accounted for almost all floral visits in sites far from forest (Fig. 11.2B).

Stability of pollination service over time

In the second year of the study (2002), honeybee visitation rates throughout the farm were 72% lower than in 2001 ($t_{116} = 3.88$, $P < 0.001$). At sites near the forest, however, several native species increased in visitation rate over the same period, such that overall visitation rates declined only 9% (Fig. 11.3). In contrast, at the intermediate sites, where honeybees were almost the sole pollinators, overall visitation rates dropped 54%. (To ensure the closest comparison between years, these results are based only on data from the four sites sampled in both years; Fig. 11.1.)

These results suggest that the diversity of bees at near sites helped stabilize rates of pollinator visitation over time, with fluctuations in the abundance of one population compensating for asynchronous dynamics of others. Large fluctuations in insect populations are common, including those of important pollinators (Wolda 1978; Roubik 2001). This "averaging," or "portfolio" effect, therefore, may stabilize pollination services if pollinator diversity is maintained. These effects have been identified in other systems and attracted wide interest as potential benefit of diverse ecosystems (Doak et al. 1998; McCann 2000).

In sum, forest fragments in this region of Costa Rica appear to increase both the level and temporal stability of pollination services in nearby coffee fields. This effect was due mainly to native bees, which were

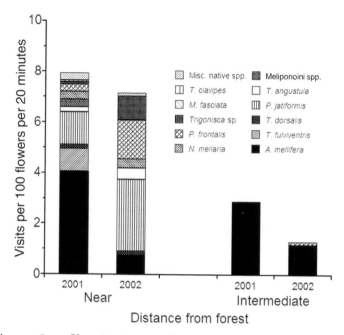

Fig. 11.3. Rate of bee visitation to coffee flowers at near and intermediate distance classes from the south patch in 2001 and 2002. Each bar segment represents the mean visitation rate for the corresponding species (error bars are omitted for clarity). Species appear in the stacked bars in the same order as in the legend. Full names provided in Table 11.1. (Modified from Ricketts (2004).)

restricted almost entirely to the nearest sites to forest, while honeybees visited with equal frequency at all distance classes.

Case 2: Watermelon in California

In a study similar to that described for coffee, Kremen *et al.* (2002b) measured the role of natural habitat in providing pollination service to watermelon (*Citrullus lanatus*) growing in Yolo County in California's central valley. As part of this work, N. M. Williams *et al.* (unpublished data) also assessed the use of natural habitats by these same species, a first step in directly linking pollinator's use of the natural and agricultural habitat types.

This region is a center for fruit and vegetable production in North America, as a state exceeding $16 billion annually (> $165 million in Yolo County alone (California Agricultural Statistics Service 2002)). Watermelon was chosen as an indicator of crop pollination because it has separate-sex flowers and high pollination requirements (500–1000 pollen grains deposited on stigmas to produce marketable fruits), relying on multiple insect visits for successful pollination (Stanghellini *et al.* 1997). Individual flowers are also open for less than one day so total pollination can be assessed within a short period. In this area, pollination is provided by managed honeybees and a diverse set of native bee species (Kremen *et al.* 1993); however, the contributions of native bees have been unmeasured and underappreciated by most growers.

Study system and sites

The landscape comprises a mosaic of row and orchard crops adjacent to and intermixed with a variety of natural habitats (Fig. 11.4). Natural habitats include: chaparral, oak woodland, mixed-oak grassland, and riparian areas, all of which are currently open to cattle or have been grazed

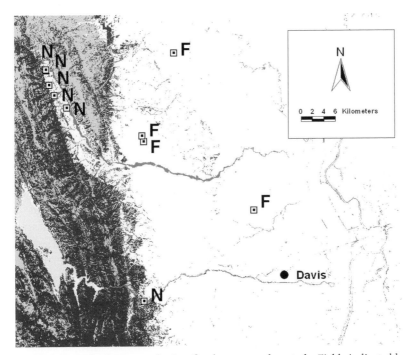

Fig. 11.4. Yolo County study sites for the watermelon study. Fields indicated by boxed dots, "N" sites near to natural habitat, "F" far from natural habitat. Land types were oak and mixed-oak woodland (black), riparian (dark gray), chaparral (gray), open water (Lake Berryessa, very light gray), and urban and agriculture, including some pasture (white).

within the past 10 years. Natural and agricultural habitats form a west to east gradient with primarily natural areas in the foothills of the coastal range in the west and an increasing amount of agricultural land eastward into the Central Valley. Watermelon study fields were located along this gradient to achieve different levels of connectivity with natural habitats.

Pollination was measured at 14 sites that differed in degree of isolation from natural habitat (near vs. far) and in management at the site (certified organic vs. high-pesticide conventional). For this chapter, we focus on isolation from natural habitat and limit comparison to organic farms near ($n = 6$) or far ($n = 4$) from natural habitats. Near sites were selected as those surrounded by >30% natural habitat within 1.0 km of the farm and far sites those surrounded by <1.0% natural habitat within the same radius. Distances between far farms and any substantial patch of natural habitat (i.e., >1 ha) exceeded the foraging distances reported for even the largest bees in this system (Walther-Hellwig and Frankl 2000; Gathmann and Tscharntke 2002).

Watermelon pollination

Visitation rates and pollination for all native bee species and for the honeybee were measured at each site. The contribution of each species was calculated as the number of visits per flower made by that species at the site multiplied by the median number of pollen grains deposited on the stigma of a female flower during a single visit by that species (see Kremen *et al.* 2002b). Total pollination per flower was the sum of deposition by each species.

Isolation from natural habitat significantly affected the pollination services provided by native bees. At farms near natural habitat, the native bee community was sufficient by itself to provide adequate pollination for watermelon (Fig. 11.5C). Pollen deposition by native bees was significantly lower at far farms (Kruskal–Wallis test: $Z = 1.60$, $P = 0.05$) and on average was not enough to ensure adequate pollination without the contribution by managed honeybees.

The decline in pollination service with increasing isolation resulted from a substantial drop in pollinator richness (Fig. 11.5A) (Kruskal–Wallis test: $Z = -1.88$, $P = 0.06$) and a two-fold decrease in overall visitation by native species (Fig. 11.5B) (Kruskal–Wallis test: $Z = -1.10$, $P = 0.13$). In addition, the best pollinators (those that deposited the most pollen per visit) were absent from far sites. Five of the six bee species that deposited the most pollen per visit were absent from all far sites (Table 11.2). Their combined per visit deposition was nearly three times that of the

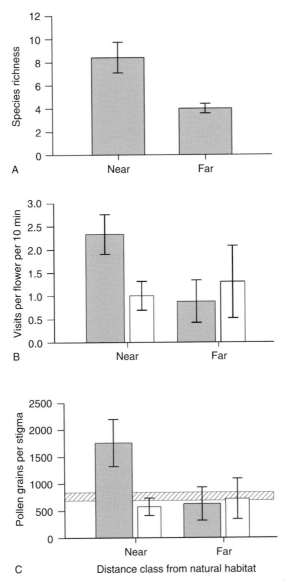

Fig. 11.5. Visitation and pollination of watermelon (mean ± SE) by native bee species on organic farms near ($n = 6$) and far ($n = 4$) from natural habitat (A) Accumulated bee richness; (B) rate of bee visitation by all native bee species (dark bars) and by honeybees (light bars); (C) total pollen deposition per flower by native bees and by honeybees. Hatched line indicates number of pollen grains needed to set marketable fruit.

Table 11.2. *Watermelon (*Citrullus lanatus*) pollinators with median per-visit pollen deposition and presence at FAR farm sites*

Species	Median deposition per visit	Present at FAR sites
Halictus farinosus	196	No
Lasioglossum spp. ($n = 2$)	91	No
Bombus californica	85	No
Peponapis pruinosa	55	Yes
Bombus vasnesenskii	52	No
Melissodes spp. ($n = 4$)	49	No
Agapostemon texanus	43	Yes
Halictus ligatus	21	Yes
Apis mellifera	21	Yes
Halictus tripartitus	17	Yes
Dialictus spp. ($n = 2$)	5	Yes
Evylaeus spp. ($n = 4$)	Too few data	Yes
Hylaeus spp. ($n = 3$)	Too few data	Yes
Sphecodes spp.	Too few data	No

remaining bee species, including the contribution of honeybees. This non-random loss magnified the effect of reduced species richness at far sites (Kremen *et al.* 2002b).

In contrast to native species, visitation and pollen deposition by honeybees did not differ between near and far sites. Despite the fact that there were no managed honeybee hives on these study farms, the high number of honeybees found on the farms likely can be attributed to the managed colonies throughout the surrounding landscape.

Use of natural habitats by watermelon pollinators

The role of natural habitats in supporting the native bee species involved in crop pollination was also explored. During 2002 bees were surveyed in natural habitats within the study area from April through August. Upland habitat types were pooled and upland and riparian were considered as two categories. At nine upland and five riparian sites bees were netted at flowers within ~5 ha plots during two sampling periods at 3–4 week intervals.

Twenty-five of the 26 bee species that visited watermelon during 2001 were collected at upland, riparian, or both types of natural habitat. All 25 were collected at one or more upland sites and 23 were collected at one or more riparian sites. *Peponapis pruinosa*, the one absentee, is a trophic specialist that collects pollen only from the Cucurbita (certain squash and gourd species). It was found visiting watermelon (presumably to collect nectar) at all farms that also grew squash. This bee nests in the ground in areas surrounding squash plantings. Although it is not known where the bees collected from these native plants nested, the near-perfect overlap with species pollinating watermelon supports the importance of connectivity between natural areas and cropland for maintaining native pollinators of crops, whether such areas serve as nesting sites or as areas of floral resources when crops are not flowering.

In sum, the proximity of natural habitats to farms was critical for providing watermelon pollination by native bees. Indeed most farmers whose fields are located near large areas of natural habitat (e.g., N sites Fig. 11.4) do not currently rent managed honeybees for pollination. Whether or not they explicitly understand the roles of native bees, they recognize the general importance of native pollinators and a possible connection to natural habitats. Without such a connection, farms relied on managed honeybees to achieve sufficient pollination of their watermelon crop.

Case 3: Kiwifruit in New Zealand

Since its introduction from China in the 1950s, kiwifruit (*Actinidia deliciosa*) has become one of New Zealand's most important crops, currently representing about 50% of New Zealand's fruit and nut exports (New Zealand 1999). M. M. Mayfield *et al.* (unpublished data) examined the importance of wild and pastoral vegetation in providing pollinators to kiwifruit flowers in the North Island of New Zealand. This kiwifruit-growing region is a matrix of fruit orchards, sheep and cattle pastures, and small patches of woodland. Very little of the region resembles the native forests and scrublands that once covered the area, but small forest patches, hedgerows, windbreaks (required for kiwifruit and other fruit production), and pasture all provide habitat of varying quality for native and exotic insects.

There are two similar species of commercially grown kiwifruit in New Zealand: *Actinidia deliciosa* (Hayward variety) and the new *A. chinensis* (Horti6A variety). Both species have separate male and female vines and have the same pollination system (Pan *et al.* 1997). Few studies have examined the pollination of kiwifruit in China (within its native range), but anecdotal evidence suggests numerous bee species including *Xylocopa*, *Osmia*, *Habropoda*, and possibly some hoverflies (Diptera: Syrphidae) are major pollinators (Steven 1988). Despite extensive research, there is no consensus on the best way to maximize pollination services to kiwifruit in New Zealand, although evidence suggests honeybees and bumblebees improve fruit set and size over wind alone (Craig and Stewart 1988; Costa *et al.* 1993; Goodwin *et al.* 1996; Pomeroy and Fisher 2002). Most growers use the recommended eight honeybee hives per hectare of kiwifruit orchard during the blooming season (Palmer-Jones *et al.* 1976; Craig and Stewart 1988). In addition to honeybees, farmers often supplement bee pollination with pollen collected from male vines and then machine-sprayed throughout the fields (Goodwin *et al.* 1993). Both techniques add considerable cost to this crop's production. Few insect species other than introduced honeybees and bumblebees have been investigated as pollinators.

Although no studies have investigated the use of pollinators native to New Zealand for kiwifruit pollination, windbreaks could be used for the dual purpose of increasing pollinator numbers and wind protection, potentially proving beneficial both biologically and financially. The specific questions addressed were: Does the proximity of wild-lands to kiwifruit orchards influence the richness and abundance of insects visiting kiwifruit flowers? And does increased pollinator richness translate into improved pollination and/or increased fruit quality?

Study system and design

The study took place in three orchards in the central region of the North Island of New Zealand separated from each other by at least 30 km. Two orchards grew *A. deliciosa* Hayward variety fruit and the other grew *A. chinensis* Hort16A variety. The Hayward variety orchards each had eight honeybee hives per hectare of orchard and the Hort16A orchard had approximately two honeybee hives per hectare. One of the Hayward orchards was organic and the others were conventional although none of the orchards received chemical sprays during the blooming season.

In each orchard, two kiwifruit blocks were selected, each measuring at least 100 × 100 m and containing 20–30 rows of vines spaced at ~5 m. One block was separated from natural or pastoral vegetation by at least one other block on all sides (hereafter, "interior blocks"), and the other bordered wild vegetation on at least one side ("edge blocks"). Wild vegetation in this study was any non-agricultural vegetation, in all cases including trees and shrubby native and exotic species. Despite differences in varieties and overall orchard management, data from all orchards were pooled for all analyses because kiwifruit variety and orchard management were not significant correlates in any of the tests discussed in this chapter.

Pollinator activity

In each of the six blocks, six female vines were randomly selected (36 vines total). On each vine flower visitors were observed in 10-min intervals at all times of day throughout the 17-day blooming season. To confirm that insects visiting female flowers were potential pollinators, insect visitation on nearby male vines was also recorded.

Almost all visitors in all sites were honeybees. The non-honeybee visitors included: syrphid flies (Syrphidae: *Melanostoma fasciatum*), beetles (Curculionidae, Cerabycidae and *Conodurus exsul*), bumblebees (*Bombus* spp.), and a variety of rare (fewer than three visits) native bees, flies, and wasps.

More species of pollinators visited in edge blocks than in interior blocks, although this pattern was marginally significant (Fig. 11.6A) (Kruskal–Wallis test: $Z = -1.74574$, $P = 0.0809$). Visitation of all non-honeybee visitors per flower, however, was significantly higher for edge blocks than interior blocks (Fig. 11.6B) (Kruskal–Wallis test: $Z = -5.17295$, $P < 0.001$). This pattern was mirrored on a finer scale with analysis of visitation on a continuous distance gradient from 0 to 111 m from forest edge. Using a split window analysis (Ludwig and Cornelius 1987), a marked decrease in non-*Apis* visits was seen at ~21 meters from the boundary with wild vegetation in two of the three farms. For honeybees, there was no significant difference in visitation between interior and edge blocks (Kruskal–Wallis test: $Z = 0.43869$, $P = 0.6609$).

Fruit set and pollen limitation

In addition to observing pollinator visitation rates, fruit weight and pollen limitation were quantified for the same vines. Pollen limitation is a way of measuring the effectiveness of natural pollination conditions. Pollen limitation is calculated as the ratio of fruit set or fruit weight (used in this case) between two treatments: ambient pollination (i.e., flowers left open to measure production under existing natural pollination) and augmented pollination (i.e., flowers hand-pollinated to ensure sufficient cross pollen). Pollen limitation ratios near 1.0 indicate little pollen limitation, and values close to zero indicate extreme pollen limitation. In this study, 12–18 flowers per observed vine were marked, with half assigned to each treatment (Kearns and Inouye 1993). Staff of the Te Puke Hort Research Station collected mature fruit and weighed them once they had reached full maturity. Fruit in the conventional orchard were sprayed with chemicals to enhance fruit size so they were excluded from analyses.

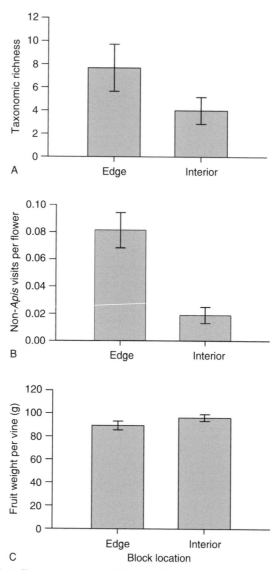

Fig. 11.6. Pollinator activity and fruit weight in kiwifruit fields at edge and interior blocks. (A) Mean number of taxa observed visiting female kiwifruit flowers; (B) visitation rates of non-honeybee insects on kiwifruit flowers (mean number of total insect visitors to female flowers per 10-min observation period); (C) mean fruit weight per vine for the organic and Hort16A orchards. Mean fruit weight per vine is based on between 2 and 10 fruits.

Mean fruit set per site varied from 78% to 100% and there was no significant difference in average fruit weight between edge and interior blocks (Fig. 11.6C) (one-way ANOVA: df $= 201$, $t = 1.650$, $P = 0.1005$). There was also no significant difference in pollen limitation levels in edge verses interior blocks (one-way ANOVA: df $= 181$, $t = 1.750$, $P = 0.0817$).

In sum, while honeybees did not appear to respond to the proximity of wild-lands, richness and visitation rates of non-honeybee pollinators were significantly greater at blocks near wild vegetation, with a dramatic drop-off in visitation beyond 21 m from the edge. These differences, however, did not appear to affect fruit mass or pollen limitation, perhaps because of oversaturation of managed honeybees in these orchards.

Counter-examples

These three studies show consistent effects of connectivity on pollinator behavior, suggesting that connectivity is often crucial for pollination services. Two studies of other crop systems, however, have found little or no effects of landscape configuration on crop pollination, illustrating that the importance of connectivity can vary among systems.

Almond: a temporal disconnect

In a preliminary study of pollination services to almond (*Prunus dolcis*) in California, N. M. Williams *et al.* (unpublished data) examined visitation of native bees and managed honeybees in orchards located along the Sacramento River. All study orchards practiced conventional management, which includes mowing of understory vegetation and application of herbicide, fungicide and other pesticide sprays when needed (Strand 2002). Orchards were within 100 m of mixed riparian forest or grassland. The surrounding landscape contained between 10% and 60% natural or restored habitat within 500 m of the study sites. At each site visitation by all insects to almond flowers were measured throughout the day within standard plots in much the same manner as in previous case studies. Native bees and other potential pollinators were also surveyed at riparian habitat from February (during almond bloom) through August.

Despite the proximity of orchards to natural habitats, there was no substantial visitation by any native insect species (<1 visit per hour per 5 m^3 of almond bloom). In contrast, honeybees from managed hives visited with great intensity (11.2 ± 0.26 bees per 5 m^3 of bloom per 2-min scan, mean $\pm$ SE, $n = 7$ sites, 60 scans per site).

The bee surveys from surrounding natural habitat during and after almond bloom illustrated the importance of a temporal dimension to connectivity between natural habitat and crops for pollination service. During almond bloom, almost no native bee species were found in natural habitats. However their numbers increased substantially later in the season, after almonds had finished flowering (N. M. Williams, in prep). The overall abundance and diversity at these sites along the Sacramento River was low compared to upland habitats in Yolo County (see watermelon case study above), nonetheless bee species known to pollinate almond and other orchard crops (Torchio 1991; Free 1993) were present in the landscape, albeit after the almond bloom. Later in the season bees in the area may well have traveled between agricultural habitats and natural habitats (as was the case for watermelon); however, the commodity requiring service (almond) was temporally disconnected from the areas and the species that might have provided it.

Oil palm: persistence in the matrix?

In a preliminary study of the importance of tropical forest fragments to oil palm (*Elaeis guineensis*) pollination, Mayfield (2005) examined the diversity and visitation rates of oil palm floral visitors in southern Costa Rica. Oil palm originates from West Africa where its major pollinators are weevils (Kevan and Wahid 1995). Individual oil palms bear either female or male flowers at a given time and bloom year round. Male flowers emit an anise-like smell to attract insects and female flowers produce a similar, weaker smell, but provide no pollinator rewards. The West African weevil *Elaeidobius kamerunicus* has been introduced into most regions of the world, including Costa Rica in the 1980s, to ensure adequate pollination (Syed 1979; Dhileepan 1994). While non-weevil insects have been found to be important pollinators in other parts of the world (Syed 1979), they have rarely been studied in Costa Rica.

This study took place in July and August of 2001 in two plantations (Coto47 and PJ) separated by more than 50 km. Both plantations were pygmy palm monocultures with a legume understory and interspersed forest fragments (no more than 50 m diameter) and riparian strips. To determine if female flower visitors were significantly more diverse and frequent near forest fragments, 10-min observations were made at two distance classes from forest (near: 0−15 m, and far: 100−200 m) on both plantations in a similar manner as in the studies described above (Mayfield 2005).

Over the course of the study, 265 visitors were observed over 68 observation periods, representing 17 insect morphospecies. All morpho-species were observed on both male and female flowers. By far the most common visitor was *E. kamerunicus*, the imported weevil. Non-weevil morphospecies included six fly, two bee, three beetle, and five ant morphospecies. Non-weevil visitors made up 28–39% of all visits to female flowers (14–22% excluding ants, which are rarely effective pollinators).

There was no significant difference between distance classes in species diversity (Shannon–Weiner index; Van der Waerden Test: $Z = -0.81973$, $P = 0.4124$) or visitation rates of all visitors (Kruskal–Wallis test: $Z = 0.59503$, $P = 0.5518$). When only non-weevil visitors were considered, there was still no significant difference in visitor abundance between distance classes in Coto47 (Kruskal–Wallis test: $Z = 1.63981$, $P = 0.1010$), but far sites in PJ had significantly more non-weevil visits than near sites (Kruskal–Wallis test: $Z = -1.92432$, $P = 0.0543$). This pattern was due to fly visits, which increased significantly in the far sites of PJ (Kruskal–Wallis test: $Z = 2.60953$, $P = 0.0091$).

Results from this study suggest that connectivity with forest fragments does not increase the diversity or visitation rates of oil palm pollinators in southern Costa Rica. Little is known about foraging and habitat requirements of these fly, beetle, and solitary bee pollinators, but studies of similar species have found that they forage very short distances, a few centimeters to a few meters, not the hundreds of meters seen in social bees (Wratten *et al.* 1995; Opitz 2002). Additionally, many of these pollinators use flowers as oviposition and breeding sites as well as foraging sites (Kevan and Wahid 1995; Wratten *et al.* 1995; Opitz 2002), perhaps removing the need for any other nearby habitats. Kremen *et al.*'s (2002b) study (see above) included a similar finding with the solitary bees that nested in watermelon fields and therefore did not decrease in abundance in more isolated fields. Although this study is preliminary, the patterns found here, along with evidence from other studies, suggest that connec-tivity with forest fragments does not affect pollination services to oil palm.

DISCUSSION

Together, the three major case studies described in this chapter illustrate the importance of connectivity between native habitats and the surround-ing agricultural matrix for crop pollination. Each study examined a different crop on a different continent, yet they all addressed issues of connectivity using similar designs. All investigated the effects of isolation from native habitats on two important variables: pollinator activity (i.e., diversity and visitation rates) and pollination services (pollen deposition and fruit quality). Although results varied somewhat, several common patterns arise from these studies.

Pollinator activity

Isolation from native habitats significantly affected pollinator richness or activity in all three studies (Table 11.3). For coffee and watermelon, pollinator richness and visitation rates were significantly lower at more isolated sites. For kiwifruit, richness did not differ significantly (although a trend was evident), but visitation rates of native species were significantly lower in isolated plots.

In all studies, distances from isolated sites to native habitat appeared to exceed the typical foraging range of most native species (Gathmann and Tscharntke 2002; S. A. Greenleaf *et al.* unpublished data). In the coffee system, for example, effective foraging ranges for most native bee pollinators are typically 100−400 m (Heard 1999), yet the distance to intermediate and far sites were 800 m and 1600 m, respectively. Wild pollinators of kiwifruit (mostly beetles and flies) also seemed to display

Table 11.3. *Summary of results from three case studies and two counter-examples*[a]

Crop	Pollinator visitation	Pollinator richness	Pollination[b]
Case studies			
Coffee	Yes	Yes	Yes
Watermelon	Yes	Yes	Yes
Kiwifruit	Yes	No	No
Counter-examples			
Almond	No	No	—
Oil palm	No	No	—

[a]Elements in the table report whether a significant effect of landscape connectivity was found for the given crop and pollination response. In kiwifruit, for example, landscape connectivity significantly affected pollinator visitation, but not pollinator richness or pollination.

[b]In the coffee and watermelon studies, pollen deposition was used as an index of pollination, while in the kiwifruit study, pollen limitation was examined experimentally; — signifies that no measurement of pollination was attempted.

relatively small foraging ranges; most did not appear to fly more than 21 m into orchards from wild vegetation. In both of these systems, flower visitation in the more isolated sites was dominated by honeybees, which often forage up to several kilometers from their nests (Seeley 1985; Roubik 1989). Connectivity for crop pollination therefore depends on geographic proximity to bee nesting habitats and the foraging ranges of key species. Other important factors may include permeability of surrounding habitats to bee movement (Ricketts 2001) or differences in behavioral response to habitat boundaries (Fagan et al. 1999). This latter factor may also have contributed to the decline of native pollinators beyond the edges of kiwifruit orchards.

In the watermelon study, although limited flight ranges caused a similar loss of connectivity between habitat types, this loss was partially mitigated by the abilities of some bee species to nest and reproduce in the agricultural matrix. Several of the species found at isolated farms readily nest in the ground within agricultural fields or fallow farm borders (Williams and Kim 2006). These small species have relatively low resource demands per individual and therefore may be able to survive and reproduce on the resources provided by flowering weeds when crops are not blooming. Whether these populations actually persist as sink populations (Pulliam 1988), dependent on a continual flow of immigrants from native habitats, remains unclear. The importance of within-field flower resources for supporting pollinator populations, however, is becoming increasingly clear. In some cases year-round bloom is already intentionally maintained to support pollinators (e.g., in some Indian cardamom and coffee plantations: Kuruvilla 1995); in others its impact is incidental (Westphal et al. 2003).

Pollination services

Pollinator richness and visitation rates indicate levels of pollinator activity, but do not measure the process and ecosystem service of pollination directly. For example, bees may be more diverse and active at more connected sites, but this activity may not confer higher rates of cross-fertilization or higher crop yields. Therefore, each of the studies also measured some indicator of pollination service itself. Two of the three studies found that connectivity significantly affected pollination as well as pollinator activity (Table 11.3).

The coffee and watermelon studies each used pollen deposition onto stigmas as a measure of pollination service, and each found dramatic decreases with decreasing connectivity from native habitats (Figs. 11.2C and 11.5C). In coffee, the magnitude of this decrease matched the decline in bee visitation (Fig. 11.2B) almost exactly. In watermelon, the decrease was even greater than that in bee visitation, as the more effective pollinators were largely absent from the far sites. Pollen deposition, however, is still an imperfect measure of the economic value of pollination services, because crops may not respond to higher pollen deposition rates with higher yields or fruit quality. Crops may (1) be self-compatible, (2) require few pollen grains to set marketable fruit, or (3) have a variety of complex post-pollination processes (Marshall 1991; Free 1993; Snow 1994; Richards 2001). One way to avoid many of these problems is to measure pollen limitation using pollen augmentation to estimate the adequacy of ambient pollination (Kearns and Inouye 1993).

The kiwifruit study employed these methods, but found no evidence for increased pollen limitation (i.e., decreased pollination service) with isolation. Despite significantly lower activity by native pollinators at isolated sites, fruit production was unaffected. Lack of an isolation effect is likely because of saturating numbers of managed honeybees placed throughout these kiwifruit orchards. Any service provided by the local native pollinators, therefore, may have been swamped by the superabundant managed honeybees. This raises the possibility, however, that wild pollinators in New Zealand may confer an economic benefit by allowing farms to decrease the number of honeybees they need to employ. In addition, wild pollinators may provide insurance against a decline in this single, managed pollinator from disease, parasites, or economics (Watanabe 1994; Allen-Wardell *et al.* 1998; Kearns *et al.* 1998; Kevan and Phillips 2001; Kevan *et al.* 2002).

Crooks and Sanjayan (Chapter 1) distinguish between "structural connectivity" (i.e., the spatial relationships of habitat patches in a landscape) and "functional connectivity" (i.e., actual movement of individuals among patches) (see also discussion in Taylor *et al.* Chapter 2; Fagan and Calabrese Chapter 12). The results from our two counter-examples illustrate the importance of this distinction; in each case, structural connectivity did not result in functional connectivity (Table 11.3). For oil palm, neither pollinator diversity nor visitation rates differed with proximity of forest. For almond, phenological differences prevented visitation from wild bees, despite close proximity between orchards and native habitats. This result illustrates the importance of measuring structural connectivity along temporal, as well as spatial, dimensions. Furthermore, when considering ecosystem services, perhaps an additional component of connectivity deserves attention: economic connectivity. In our kiwifruit case study, despite structural and functional connectivity, no effects on kiwifruit yield or quality were found. In cases where ecosystem services provide the motivation for conserving connectivity, the ultimate measure of adequate connectivity is whether the services are enhanced.

A community-mediated service

A key lesson from all of our case studies and others is that pollination is a community-level service provided by a diverse set of species (e.g., Kevan 1977; Cane and Payne 1993; Free 1993; Kremen *et al.* 2002a; Klein *et al.* 2003a). As emphasized by several authors in this volume (e.g., Crooks and Sanjayan Chapter 1; Taylor *et al.* Chapter 2; Noss and Daly Chapter 23), each species may respond differently to the same degree of landscape connectivity. The effect of connectivity on pollination services, therefore, is a complex function of the individual responses of the pollinator species involved (Pettersson 1991; Cane 1993; Benedek and Nyéki 1996; Kamler 1997). Species differ in raw flight ability (Heard 1999; Gathmann and Tscharntke 2002), seasonal abundance (Clements and Long 1923; Cane and Payne 1988; Cane 1993; Stern 1996; Wallace *et al.* 1996; Kamler 1997), and behavioral responses to habitat boundaries (Schultz 1998) and matrix type (Ricketts 2001). These differences mean that understanding the factors that drive community-level responses to connectivity requires knowledge of the individual species involved and their responses to the landscape. For example, some species in the community may be influenced primarily by the distance to a fragment of a particular habitat, while others may be more influenced by the quality of resource or availability of nesting sites in adjacent habitats.

The community nature of pollination services argues strongly for conservation of native pollinators. In part because of interspecific differences in life history, diverse pollinator communities may confer stability to pollination services over time. In the coffee study, for example, the diverse native species in sites near forest appeared to stabilize overall visitation rates between the two years (Fig. 11.3). In watermelon as well, populations of many bee species changed significantly between years. Sufficient pollination by native bees required only a subset of species during one season, but the whole community in another (Kremen *et al.* 2002b). In general, an increasing number of studies are suggesting that diversity within communities tends to stabilize them and the important ecological functions they mediate (e.g., McCann 2000; Schwartz *et al.* 2000). Stability may refer to constancy over time, as in these examples, but also can denote resistance to disturbance or resilience after a disturbance has occurred. All of these aspects of stability can also confer economic benefits to people by stabilizing yields from year to year, insuring against population decline in a single pollinator species, or allowing farmers to plan for managed pollinator needs more easily.

Future directions

Although a few studies have identified the importance of connectivity to pollination services, several areas remain unexplored. Additional work in the following areas will enhance basic understanding of pollination across complex landscapes, but also lead to better management of these landscapes by ensuring adequate crop pollination services.

First, it is important to separate the effects of connectivity from the size or type of adjacent habitat. None of the studies included here was designed to separate these factors satisfactorily. The size of a habitat fragment clearly affects its capacity to support pollinator populations. In coffee, for example, 46-ha forest patches seem sufficiently large to provide a diverse pollinator fauna to fields within a few hundred meters. But it remains unclear whether there is a critical size threshold below which important pollinators no longer persist or whether a larger fragment would supply pollinators to greater distances into the plantation. It is likely that such thresholds vary depending on the habitat and pollinator types. The type of habitats adjacent to crop fields may also affect connectivity by altering permeability of the boundary or resistance of the habitat to movement (Bowers 1985; Westerbergh and Saura 1994; Schultz 1998; Haddad 1999; Ricketts 2001). Such effects are particularly relevant for landscape mosaics containing typical agricultural–natural boundaries.

Second, it would be useful to develop aggregate indices of pollination service that take into account species-specific information on pollination effectiveness (Kremen *et al.* 2002b). Simple community measures such as richness or overall visitation rate are convenient, but our studies repeatedly illustrate the important differences among species that such measures mask. Ideally, we would measure each pollinator's effectiveness for a given crop and its behavioral responses to landscape composition. With these data, we could rigorously predict the effect of landscape changes on the pollinator community and the ecosystem service it provides. Measuring these species-specific factors is often prohibitively difficult, but finding easily measured correlates (e.g., nesting habit, body size) may allow us to generalize within species groups. For example, recent reviews have found that body size significantly predicts both foraging distance of bees (Gathmann and Tscharntke 2002) and the scales over which they respond to landscape composition (Steffan-Dewenter *et al.* 2001). Preliminary evidence also suggests that body size may be correlated with per-visit pollinator quality on some crops (Pomeroy and Fisher 2002), though evidence for this is inconsistent (C. Kremen and N. M. Williams, unpublished data). Even if such body size relationships are established, however, the ability of pollinators to transfer pollen is only one of a suite of factors contributing to pollinator quality. If the most effective pollinators are only present in small numbers or vary in abundance among years they are not reliable pollinators (Mayfield *et al.* 2001). Temporal and spatial variability in pollinator communities is likely the rule and can greatly affect the pollination services delivered in a single system (Kremen *et al.* 2002b). A reliable index of pollination services, in this case, would require additional research, including time-intensive surveys of pollinator abundance and community composition over multiple years.

Third, it would be useful to more precisely understand landscape connectivity in terms of individual pollinator movement and pollinator population dynamics (Haddad and Tewksbury Chapter 16). In these case studies and others, connectivity is measured indirectly, using changes in pollinator richness or activity along gradients of increasing isolation (Steffan-Dewenter and Tscharntke 1999; Kremen *et al.* 2002b; Ricketts 2004). While this is a crucial first step in characterizing the role of connectivity in the delivery of pollination services, it leaves several important issues unresolved. We also would like to understand (1) the specific resources that habitat patches supply to pollinators (e.g., nesting sites, floral resources when crops are not in bloom), (2) the potential metapopulation or source–sink dynamics between native and agricultural habitat patches (Pulliam 1988; Hanski and Gilpin 1997), and (3) individual movements of native pollinators between natural habitats and agricultural sites either within or among seasons. All three of these issues will be difficult to address. Following tiny and fast-flying insects over distances more that a few meters has proven a substantial challenge to which we have not yet discovered a reliable solution (Hocking 1953; Saville *et al.* 1997; Osborne *et al.* 1999).

Finally, for this research to eventually gain more direct management relevance, the economic value of pollination services provided by wild pollinators will need to be estimated. Regional planners and decision-makers face difficult trade-offs between economic development and natural resource conservation. Even rough estimates of the value to agriculture of wild pollinators and their habitats will help to inform these decisions. This information would also inform the decisions faced by individual farmers. Maintaining a semi-natural canopy of shade trees on coffee farms could provide nesting sites for important pollinators, as well as habitat for a variety of other native species (Greenberg *et al.* 1997; Perfecto *et al.* 1997). Leaving field boundaries uncultivated could improve on-farm nesting conditions for watermelon pollinators. Making wind-breaks more suitable for pollinators in kiwifruit orchards could increase the number of wild pollinators nesting and foraging in orchards. Each of these management activities may carry real economic benefits to farmers through pollination services. These benefits, however, must be weighed against the cost of forgone production in those natural areas, as well as the costs and benefits of renting managed pollinators during blooming seasons. These sorts of analyses have been conducted for other ecosystem services (e.g., Chichilnisky and Heal 1998) and for the coffee system described here (Ricketts *et al.* 2004).

CONCLUSION

By examining these case studies of crop pollination in complex agricul-tural landscapes, we have illustrated how connectivity among different habitat types can be important not only to individual populations, but also to a key ecosystem service. We found that landscape connectivity generally enhances pollinator diversity and activity in agricultural fields, but this increased activity does not always result in an economic benefit in terms of yields or crop quality. Because pollination is a process mediated by a community of organisms, the overall effect of landscape connectivity is a complex function of the species-specific attributes and responses to the landscape. Nevertheless, crop pollination is a useful example of a highly valuable ecosystem service, conferred at local scales, that often depends on landscape connectivity.

As a result, research in this area has the potential to inform the diffi-cult decisions faced by farmers and landscape managers. Understanding the connectivity requirements of pollinators will help to indicate the amount and distribution of native habitats needed to ensure adequate and stable crop pollination on individual farms and throughout the landscape. Other local ecosystem services (e.g., water purification, flood control, natural pest management) also operate at local scales. Landscape connec-tivity, therefore, is important not only for biodiversity conservation, but also for human health and welfare. Estimating the economic value of ecosystem services such as pollination may help to align the goals of development and conservation in the agricultural landscapes on which we rely.

ACKNOWLEDGEMENTS

We are indebted to many people for help with the studies summarized in this chapter. Field assistance was provided with competence and good humor by J. Florez, A. Kane, M. Hayden, B. Ettinger, B. Reed, T. Oliver, H. McBrydie, N. Nicola, R. Hatfield, L. Krepunske, the Organization for Tropical Studies, the Fallas family, Finca Santa Fe (San Isidro, Costa Rica), and Ruakura and Te Puke Hort Research Stations (Hamilton and Te Puke, New Zealand). C. Kremen, G. Daily, D. Roubik, D. Inouye, and M. Goodwin provided valuable advice. N. Haddad, W. Fagan, K. Crooks, and two anonymous reviewers significantly improved the manuscript with their comments. And the Summit Foundation, NASA, The Smith Fellows Program at the Nature Conservancy, and The Thomas J. Watson Foundation generously supported this work.

REFERENCES

Aizen, M. A., and P. Feinsinger. 1994. Habitat fragmentation, native insect pollinators, and feral honey bees in Argentine "Chaco Serrano". *Ecological Applications* **4**:378–392.

Allen-Wardell G., P. Bernhardt, R. Bitner, *et al.* 1998. The potential consequences of pollinator declines on the conservation of biodiversity and stability of food crop yields. *Conservation Biology* **12**:8–17.

Balmford, A., A. Bruner, P. Cooper, *et al.* 2002. Economic reasons for conserving wild nature. *Science* **297**:950–953.

Balvanera, P., G. C. Daily, P. R. Ehrlich, and T. Ricketts. 2001. Conserving biodiversity and ecosystem services. *Science* **291**:2047

Benedek, B., and J. Nyéki. 1996. Features affecting bee pollination of sweet and sour cherry varieties. *Acta Horticulturae* **410**:21–24.

Bowers, M. A. 1985. Bumble bee colonizatinon, extinction, and reproduction in subalpine meadows in northeastern Utah. *Ecology* **66**:914–927.

Butz Huryn, V. M. 1997. Ecological impacts of introduced honey bees. *Quarterly Review of Biology* **72**:275–297.

Cane, J. H., and J. A. Payne. 1988. Foraging ecology of the bee *Habropoda laboriosa* (Hymenoptera: Anthophoridae), an oligolege of blueberries (Ericaceae: *Vaccinium*) in the southeastern USA. *Annals of the Entomological Society of America* **81**:419–427.

Cane, J. H., and J. A. Payne. 1993. Regional, annual, and seasonal variation in pollinator guilds: intrinsic traits of bees (Hymenoptera: Apoidea) underlie their patterns of abundance at *Vaccinium ashei* (Ericaceae). *Annals of the Entomological Society of America* **86**:577–588.

California Agricultural Statistics Service. 2002. *Summary of County Agricultural Commissioners' Reports 2001–2002*, Pp. 1–5. Sacramento, CA: California Agricultural Statistics Service.

Chichilnisky, G., and G. Heal. 1998. Economic returns from the biosphere. *Nature* **391**:629–630.

Clements, F. E., and F. L. Long. 1923. *Experimental Pollination*. Washington, DC: Carnegie Institution Press.

Costa, G., R. Testolin, and G. Vizzotto. 1993. Kiwifruit pollination: an unbiased estimate of wind and bee contribution. *New Zealand Journal of Crop and Horticultural Science* **21**:189–195.

Craig, J. L., and A. M. Stewart. 1988. A review of kiwifruit pollination: where to next? *New Zealand Journal of Experimental Agriculture* **16**:385–399.

Daily G. C. (ed.) 1997. *Nature's Services*. Washington, DC: Island Press.

Daily, G. C., T. Soderqvist, S. Aniyar, *et al.* 2000. The value of nature and the nature of value. *Science* **289**:395–396.

Daily, G. C., P. R. Ehrlich, and G. A. Sanchez-Azofeifa. 2001. Countryside biogeography: use of human-dominated habitats by the avifauna of southern Costa Rica. *Ecological Applications* **11**:1–13.

de Lima, M. G., and C. Gascon. 1999. The conservation value of linear forest remnants in central Amazonia. *Biological Conservation* **91**:2–3.

Dhileepan, K. 1994. Variation in populations of the introduced pollinating weevil (*Elaeidobius kamerunicus*) (Coleoptera; Curculionidae) and its impact on fruitset of oil palm (*Elaeis guineensis*) in India. *Bulletin of Entomological Research* **84**:477–485.

Doak, D. F., D. Bigger, E. K. Harding, *et al.* 1998. The statistical inevitability of stability–diversity relationships in community ecology. *American Naturalist* **151**:264–276.

Fagan, W., R. S. Cantrell, and C. Cosner. 1999. How habitat edges change species interactions. *American Naturalist* **153**:165–182.

Free, J. B. 1993. *Insect Pollination of Crops*. London: Academic Press.

Freitas, B. M., and R. J. Paxton. 1998. A comparison of two pollinators: the introduced honey bee *Apis mellifera* and an indigenous bee *Centris tarsata* on cashew *Anacardium occidentale* in its native range of NE Brazil. *Journal of Applied Ecology* **35**:109–121.

Gathmann, A., and T. Tscharntke. 2002. Foraging ranges of solitary bees. *Journal of Animal Ecology* **71**:757–764.

Goodwin, R. M., J. H. Perry, H.Haine, P. J. Brown, and A. ten Honten. 1993. *The Apicultural Research Unit Annual Report*, HortResearch Client Report, Pp. 9–59. Hamilton, New Zealand: Horticulture and Food Research Unit of New Zealand Ltd.

Goodwin, R. M., H. M. Haine, and J. H. Todd. 1996. *The Apicultural Research Unit Annual Report*, HortResearch Client Report, Pp. 7–61. Hamilton, New Zealand: Horticulture and Food Research Unit of New Zealand Ltd.

Greenberg, R., P. Bichier, A. C. Angon, and R. Reitsma. 1997. Bird populations in shade and sun coffee plantations in central Guatemala. *Conservation Biology* **11**:448–459.

Griswold T. L., F. D. Parker, and P. E. Hanson. 1995. The bees of Costa Rica. Pp. 650–691 in I. D. Gauld and P. E. Hanson (eds.) *The Hymenoptera of Costa Rica*. New York: Oxford Scientific Press.

Gustafson, E. J., and R. H. Gardner. 1996. The effect of landscape heterogeneity on the probability of patch colonization. *Ecology* 77:94–107.

Haddad, N. M. 1999. Corridor use predicted from behaviors at habitat boundaries. *American Naturalist* 153:215–227.

Haddad, N. M., and K. A. Baum. 1999. An experimental test of corridor effects on butterfly densities. *Ecological Applications* 9:623–633.

Hanski, I. 1998. Metapopulation dynamics. *Nature* 396:41–49.

Hanski I., and M. E. Gilpin (eds.) 1997. *Metapopulation Biology: Ecology, Genetics, and Evolution*. San Diego, CA: Academic Press.

Heal, G. 2000. *Nature and the Marketplace*. Washington, DC: Island Press.

Heard, T. A. 1999. The role of stingless bees in crop pollination. *Annual Review of Entomology* 44:183–206.

Hocking, B. 1953. The intrinsic range and speed of flight in insects. *Transactions of the Royal Entomological Society of London* 104:225–345.

Janzen D. H. (ed.) 1983. *Costa Rican Natural History*. Chicago, IL: University of Chicago Press.

Javorek, S. K., K. E. Mackenzie, and E. G. Reekie. 2002. Comparative pollination effectiveness among bees (Hymenoptera: Apoidea) on lowbush blueberry (Ericaceae: *Vaccinium angustifolium*). *Annals of the Entomological Society of America* 95:345–351.

Kamler, F. 1997. Sunflower pollination in Czech Republic. *Acta Horticulturae* 437:407–411.

Kearns, C. A., and D. W. Inouye. 1993. *Techniques for Pollination Biologists*. Niwot, CO: University Press of Colorado.

Kearns, C. A., D. W. Inouye, and N. M. Waser. 1998. Endangered mutualisms: the conservation of plant-pollinator interactions. *Annual Reviews of Ecology and Systematics* 29:83–112.

Kevan, P. G. 1977. Blueberry crops in Nova Scotia and New Brunswick: pesticides and crop reductions. *Canadian Journal of Agricultural Economics* 25:61–64.

Kevan, P. G. 1991. Pollination: keystone process in sustainable global productivity. *Acta Horticulturae* 288:103–109.

Kevan, P. G. 1999. Pollinators as bioindicators of the state of the environment: species, activity and diversity. *Agriculture Ecosystems and Environment* 74:373–393.

Kevan, P. G., and T. P. Phillips. 2001. The economic impacts of pollinator declines: an approach to assessing the consequences. *Conservation Ecology* 5:80 Available online at http://www.consecol.org/vol5/iss1/art8

Kevan P. G., and M. B. Wahid. 1995. Oil palm pollination with weevils. Pp. 198 in D. W. Roubik (ed.) *Pollination of Cultivated Plants in the Tropics*. Rome, Italy: Food and Agricultural Organization.

Kevan P. G., G. W. Frankie, C. O'Toole, and C. H. Vergara (eds.) 2002. *Pollinating Bees: The Conservation Link between Agriculture and Nature*. Brasilia, Brazil: Ministerio do Meio Ambiente.

Klein, A., I. Steffan-Dewenter, and T. Tscharntke. 2003a. Fruit set of highland coffee increases with the diversity of pollinating bees. *Proceedings of the Royal Society of London B* **270**:955–961.

Klein, A., I. Steffan-Dewenter, and T. Tscharntke. 2003b. Pollination of *Coffea canephora* in relation to local and regional agroforestry management. *Journal of Applied Ecology* **40**:837–845.

Kremen, C., R. Colwell, T. L. Erwin, *et al.* 1993. Terrestrial arthropod assemblages: their use in conservation planning. *Conservation Biology* 7:796–808.

Kremen, C., R. L. Bugg, N. Nicola, *et al.* 2002a. Native bees, native plants, and crop pollination in California. *Fremontia* **41**:41–49.

Kremen C., N. M. Williams, and R. W. Thorp. 2002b. Crop pollination from native bees at risk from agricultural intensification. *Proceedings of the National Academy of Sciences of the USA* **99**:16812–16816.

Kuruvilla, K. M., V. V. Radhakrishnan, and K. J. Madhusoodanan. 1995. Small cardamom plantations: floristic calendar and bee pasturage trees. *Journal of Environmental Resources* 3:32–33.

Ludwig, J. A., and J. M. Cornelins. 1987. Locating discontinuities along ecological gradients. *Ecology* **68**:448–450.

Marshall, D. L. 1991. Nonrandom mating in wild radish: variation in pollen donor success and effects of multiple paternity among one- to six-donor pollinations. *American Journal of Botany* **78**:1401–1418.

Mayfield M. M. 2005. The importance of nearby forest to known and potential pollinators of oil palm (*Elaeis guineensis* Jacq.; Areceaceae) in Southern Costa Rica. *Economic Botany* **59**: 185–196.

Mayfield, M. M., N. M. Wasar, and M. V. Price. 2001. Exploring the "most effective pollinator principle" with complex flowers: bumblebees and *Ipomopsis aggregata. Annals of Botany* **88**:591–596.

McCann, K. S. 2000. The diversity–stability debate. *Nature* **405**:228–233.

Michener, C. D. 2000. *The Bees of the World.* Baltimore, MD: Johns Hopkins University Press.

Myers, N., R. A. Mittermeier, C. G. Mittermeier, G. A. da Fonseca, and J. Kent. 2000. Biodiversity hotspots for conservation priorities. *Nature* **403**:853–858.

New Zealand. 1999. *New Zealand Official Yearbook.* Available online at http://www.stats.govt.nz/domino/external/PASfull/PASfull.nsf/Web/Yearbook+New+Zealand+Official+Yearbook+On+The+Web+1999?OpenDocument#industry

O'Brien T. G., and M. F. Kinnaird. 2003. Caffeine and conservation. *Science* **300**:587.

Olson, D. M., and E. Dinerstein. 1998. The Global 200: a representation approach to conserving the Earth's most biologically valuable ecoregions. *Conservation Biology* **12**:502–515.

Opitz, W. 2002. Flower foraging behavior of the Australian species *Eleale aspera* (Newman) (Coleoptera: Cleridae: Clerinae). *Coleopterists' Bulletin* **56**:241–245.

Osborne, J. L., S. J. Clark, R. J. Morris, *et al.* 1999. A landscape-scale study of bumble bee foraging range and constancy: using harmonic radar. *Journal of Applied Ecology* **36**:519–533.

Palmer-Jones T., P. G. Clinch, and D. A. Briscoe. 1976. Effect of honeybee saturation on the pollination of Chinese gooseberries variety 'Hayward'. *New Zealand Journal of Experimental Agriculture* **4**:255–256.

Pan L. N., B. E. Vaissiere, and C. Z. Chen. 1997. Pollination of *Actinidia chinensis* Planch in Chinese orchards. *Acta horticulturae* **444**:431–437.

Perfecto I., J. Vandermeer, P. Hanson, and V. Cartin. 1997. Arthropod biodiversity loss and the transformation of a tropical agro-ecosystem. *Biodiversity and Conservation* **6**:935–945.

Pettersson, M. W. 1991. Pollination by a guild of fluctuating moth populations: options for unspecialization in *Silene vulgaris*. *Journal of Ecology* **79**:591–604.

Pomeroy, N., and R. M. Fisher. 2002. Pollination of kiwifruit (*Actinidia deliciosa*) by bumble bees (*Bombus terrestris*): effects of bee density and patterns of flower visitation. *New Zealand Entomologist* **25**:41–49.

Pulliam, H. R. 1988. Sources, sinks, and population regulation. *American Naturalist* **132**:652–661.

Richards, A. J. 2001. Does low biodiversity resulting from modern agricultural practice affect crop pollination and yield? *Annals of Botany* **88**:165–172.

Ricketts, T. H. 2001. The matrix matters: effective isolation in fragmented landscapes. *American Naturalist* **158**:87–99.

Ricketts, T. H. 2004. Tropical forest fragments enhance pollinator activity in nearby coffee crops *Conservation Biology* **18**:1262–1271.

Ricketts, T. H, G. C. Daily, P. R. Ehrich, and C. Michener. 2004. Economic value of tropical forest to coffee production. *Proceedings of the National Academy of Sciences of the USA* **101**:12579–12582.

Ricketts, T. H., G. C. Daily, P. R. Ehrich, and J. P. Fay. 2001. Countryside biogeography of moths in a fragmented landscape: biodiversity in native and agricultural habitats. *Conservation Biology* **15**:378–388.

Rosenberg, D. K., B. R. Noon, and E. C. Meslow. 1997. Biological corridors: form, function, and efficacy. *BioScience* **47**:677–687.

Roubik, D. W. 1989. *Ecology and Natural History of Tropical Bees*. New York: Cambridge University Press.

Roubik, D. W. 2001. Ups and downs in pollinator populations: when is there a decline? *Conservation Ecology* **5**:2. Available online at http://www.consecol.org/vol5/iss1/art2

Roubik, D. W. 2002. The value of bees to the coffee harvest. *Nature* **417**:708

Rust, R. W. 1990. Spatial and temporal heterogeneity of pollen foraging in *Osmia lignaria propinqua* (Hymenoptera: Megachilidae). *Environmental Entomology* **19**:332–338.

Saville, N. M., W. E. Dramstad, G. L. A. Fry, and S. A. Corbet. 1997. Bumblebee movement in a fragmented agricultural landscape. *Agriculture Ecosystems and Environment* **61**:145–154.

Schultz, C. B. 1998. Dispersal behavior and its implications for reserve design in a rare Oregon butterfly. *Conservation Biology* **12**:284–292.

Schwartz, M. W., C. A. Brigham, and J. D. Hoeksema. 2000. Linking biodiversity to ecosystem function: Implications for conservation ecology. *Oecologia* **122**:297–305.

Scott-Dupree C. D., and M. L. Winston. 1987. Wild bee population diversity and abundance in orchard and uncultivated habitats in the Okanagan Valley, British Columbia, Canada. *Canadian Entomologist* 119:735–746.

Seeley, T. D. 1985. *Honeybee Ecology*. Princeton, NJ: Princeton University Press.

Snow, A. A. 1994. Postpollination selection and male fitness in plants. *American Naturalist* 144:S69–S83.

Stanghellini, M. S., J. T. Ambrose, and R. Schultheis. 1997. The effect of honey bee and bumble bee pollination on fruit set and abortion of cucumber and watermelon. *American Bee Journal* 137:386–391.

Steffan-Dewenter, I., and T. Tscharntke. 1999. Effects of habitat isolation on pollinator communities and seed set. *Oecologia* 121:432–440.

Steffan-Dewenter, I., U. Munzenberg, and T. Tscharntke. 2001. Pollination, seed set and seed predation on a landscape scale. *Proceedings of the Royal Society of London B* 268:1685–1690.

Stephen, W. P., G. E. Bohart, and P. F. Torchio. 1969. *The Biology and External Morphology of Bees*. Corvallis, OR: Oregon State University Press.

Stern, R. A., and S. Gazit. 1996. Lychee pollination by honeybee. *Journal of the American Society for Horticultural Science* 120:152–157.

Steven, D. 1988. Chinese pollinators identified. *New Zealand Kiwifruit* 51:1.

Stickler, K., and J. H. Cane (eds.) 2003. *For Non-Native Crops, Whence Pollinators of the Future?* Lanham, MD: Entomological Society of America.

Strand, L. 2002. *Integrated Pest Management for Almonds*, 2nd edn. Davis, CA: Division of Agriculture and Natural Resources, University of California–Davis.

Stubbs C. S., and F. A. Drummond (eds.) 2001. *Bees and Crop Pollination: Crisis, Crossroads, Conservation*. Lanham, MD: Entomological Society of America.

Syed, R. A. 1979. Studies on oil palm pollination by insects. *Bulletin of Entomological Research* 69:213–224.

Torchio, P. F. 1991. Bees as crop pollinators and the role of solitary species in changing environments. *Acta Horticulturae* 288:49–61.

Turner, M. G., R. H. Gardner, and R. V. O'Neill. 1995. Ecological dynamics at broad scales. *BioScience* (Suppl.):S29–S35.

Wallace, H. M., V. Vithanage, and E. M. Exley. 1996. The effect of supplementary pollination on nut set of macadamia (Proteaceae). *Annals of Botany* 78:765–773.

Walther-Hellwig, K., and R. Frankl. 2000. Foraging distances of *Bombus muscorum*, *Bombus lapidarius*, and *Bombus terrestris* (Hymenoptera, Apidae). *Journal of Insect Behavior* 13:239–246.

Watanabe, M. E. 1994. Pollination worries rise as honey-bees decline. *Science* 265:1170.

Westerbergh, A., and A. Saura. 1994. Gene flow and pollinator behaviour in *Silene dioica* populations. *Oikos* 71:215–224.

Westerkamp, C. 1991. Honeybees are poor pollinators: why? *Plant Systematics and Evolution* 177:71–75.

Westphal, C., I. Steffan-Dewenter, and T. Tscharntke. 2003. Mass flowering crops enhance pollinator densities at a landscape scale. *Ecology Letters* 6:961–965.

Westrich P. 1996. Habitat requirements of central European bees and the problems of partial habitats. Pp. 1–16 in A. Matheson, S. L. Buchmann, C. O'Toole, P. Westrich, and I. H. Williams (eds.) *The Conservation of Bees.* London: Academic Press.

Wiens, J. A. 1995. Landscape mosaics and ecological theory. Pp. 1–26 in L. Hansson, L. Fahrig, and G. Merriam, (eds.) *Mosaic Landscapes and ecological processes.* London: Chapman and Hall.

Wille, A., and C. D. Michener. 1973. The nest architecture of stingless bees with special reference to those of Costa Rica (Hymenoptera: Apidae). *Revista de Biologia Tropical* 21:278

Williams, N. M., and V. J. Tepedino. 2003. Consistent mixing of near and distant resources in foraging bouts by the solitary mason bee *Osmia lignaria.* *Behavioral Ecology* 14:141–149.

Wilson, P. W., and J. D. Thomson. 1991. Heterogeneity among floral visitors leads to discordance between removal and deposition of pollen. *Ecology* 72:1503–1507.

Wolda, H., 1978. Fluctuations in abundance of tropical insects. *The American Naturalist* 112:1017–1046.

Wratten, S. D., A. J. White, M. H. Bowie, N. A. Berry, and U. Weigmann. 1995. Phenology and ecology of hoverflies (Diptera:Syrphidae) in New Zealand. *Environmental Entomology* 24:595–600.

Part II
Assessing connectivity

Introduction: Evaluating and quantifying the conservation dividends of connectivity

PETER KAREIVA

"Connectivity" is a metaphor and an idea that has captured the imagination of conservation biologists around the world. There remains, however, a big gap between the idea of connectivity and pragmatic insights regarding the on-the-ground actions that should be taken in the name of connectivity if the goal is long-lasting conservation. Like so many ideas in conservation biology, there is often more marketing than critical analysis and more wishful thinking than incisive models and data.

The first question to ask whenever attempting to evaluate connectivity should be: "connectivity for what purpose?" There is an unfortunate tendency to develop measurements of connectivity that are not tightly linked to the reasons for which connectivity is thought to be desirable in any particular setting. If connectivity is being promoted to maintain genetic variability and reduce inbreeding depression, then one would examine spatial variation in neutral genetic variation to assess the degree of evolutionary mixing. One would not track animal movements or even the rate of exchange between populations when maintenance of genetic variability is the issue. This is because dispersal can maintain genetic variability even though it occurs so rarely that it is unlikely to be detected in typical dispersal sampling programs. If connectivity is being promoted because it enhances recolonization in metapopulations, then one should measure colonization events, not actual movements of individuals. If connectivity is thought to be crucial for averaging out the vicissitudes of a fluctuating environment and dampening population oscillations, then it is crucial that population fluctuations are measured with and without connections. Finally, in a world facing severe climate disruption, connectivity might be needed if species are to successfully shift their distributional ranges in response to climatic shifts and habitat displacement. In these cases, one would want to measure movement under the pressure of degraded local conditions in particular, to determine to what extent one would hope for distributional shifts in the face of major environmental changes.

The chapters in this section represent the first generation of methods for quantifying connectivity. They do a tremendous service to conservation because they begin to provide a framework for assessing the benefits different habitat configurations provide in terms of connectivity. The reality is that conservation organizations are investing or prodding governments to invest major resources in maintaining landscape connections. Yet these investments are almost never made in light of any quantification of what is actually being provided. Clearly it is impractical to apply the tools and methods discussed in this section to all "corridor projects." But they should be applied to some of the biggest corridor projects. The organization I work for (The Nature Conservancy) has asserted that a major benefit associated with hundreds of millions of dollars' worth of projects is to maintain landscape connectivity. Rarely, however, are direct measurements made of colonization events or dispersal patterns after these connecting landscapes have been established. The Nature Conservancy is no worse than any other land trust or government agency conducting projects in the name of connectivity. The idea seems so obvious and powerful, it is as though the cost of measuring connectivity is an unnecessary expense. Ideally this book should change that oversight.

It is worth noting what is missing from the quantitative methods developed to date, and discussed in this section. One valuable extension to our theories and models will occur when connectivity is not treated as a separate issue but instead becomes one of the many alternative conservation investments that can be evaluated with a common currency. As we craft a vision for the future that sustains biodiversity, the question is how to manage and deploy land uses, with corridors being only one of hundreds of benefits that land or waterways can provide. Instead of focusing so much on corridors in isolation there is a need to shift our analysis to measuring biodiversity protection in landscapes in which connectivity is but one of many assets to be managed. Because conservation needs to be as cost effective as possible, it is unlikely that biodiversity corridors will ever be pristine habitat; they will instead tend to allow mixed uses, including human economic activity. Connectivity is unarguably important, but I doubt the public or governments will be willing to pay the full price to protect land exclusively as "corridors" − instead there will always be pressure to allow multiple uses in land designated for wildlife as biodiversity corridors. This view may actually favor establishment of corridors in those situations where grazing or logging or low levels of human activity can be tolerated and still allow parcels of land to function as corridors.

Perhaps the most interesting and challenging next generation of models will deal with the value of corridors as escape routes for species faced with untenable climate change. Conservation plans now commonly discuss land use configurations that allow for species to respond to climate change, yet none of the methods developed to date do a good job quantifying this response. The key here is that movement under "normal good conditions" may not be any indication of movement when local conditions are degraded. We need models and measurements that assess the function of potential corridors when the source population is stressed by declining habitat suitability or resource availability. It may well be that habitats or land uses currently considered as inappropriate for corridors will function as corridors if conditions are so bad that a species is forced to move.

The innovations and rigor evident in this section of the book are new and did not exist 10 years ago. When connectivity and corridors first emerged as an idea in conservation, they really represented a hypothesis. Today, the contribution of corridors to biodiversity protection remains largely "just a hypothesis" − but we now have modeling frameworks and empirical metrics that will allow us to critically examine the benefits of corridors in a world short on conservation funding.

12

Quantifying connectivity: balancing metric performance with data requirements

WILLIAM F. FAGAN AND JUSTIN M. CALABRESE

INTRODUCTION

Connectivity is an intuitively appealing and useful notion that is now a central concept in ecology and conservation science. Unfortunately, like many ideas that resonate widely, connectivity is plagued with a variety of definitions, interpretations, and methods of measurement. For example, several recent theoretical reviews have detailed how differences in spatial scale and underlying concepts influence how the issue of connectivity is addressed in landscape ecology and metapopulation ecology (Taylor *et al.* 1993; Tischendorf and Fahrig 2000; Moilanen and Hanski 2001; Tischendorf 2001a; Moilanen and Nieminen 2002). While definitions and measurement might seem boringly technical, conservation scientists must work to identify clear, replicable, and well-understood metrics of connectivity if conservation is to invest funds and efforts wisely and responsibly. Because connectivity is widely believed to facilitate population persistence in a fragmented landscape, on-the-ground applications of connectivity typically involve favoring one landscape configuration or reserve design over alternatives because of improved connectivity (e.g., Siitonen *et al.* 2002, 2003; Singleton *et al.* 2002; Cabeza 2003). The metrics used to make these decisions clearly matter. However, despite the importance of developing the theoretical foundations of connectivity, land managers are more likely to be interested in practical issues surrounding the measurement of connectivity. Most importantly, connectivity metrics must

Connectivity Conservation eds. Kevin R. Crooks and M. Sanjayan. Published by Cambridge University Press. © Cambridge University Press 2006.

be pragmatic and based upon data that might actually be attained on a regular basis. In this chapter we critically review and evaluate the key measures of connectivity, focusing on how different types of empirical data necessitate different metrics, with the aim of giving resource managers and decision-makers a simple guide to these alternatives and their different advantages and limitations.

The general consensus in the literature is that connectivity is defined by the interaction between particular species and the landscape in which they occur (Schumaker 1996; Wiens 1997; Tischendorf and Fahrig 2000; Moilanen and Hanski 2001; this volume). We adopt this species-centered view on the rationale that connectivity, regardless of the spatial scale on which it is defined, is a species-dependent trait. Put another way, we should expect that a single landscape or patch will possess different degrees of connectivity for different species depending on the behaviors, habitat preferences, and dispersal abilities of those species (Johnson and Gaines 1985). The closer a metric comes to estimating this interaction between the species of interest and its landscape, the more useful it will be in understanding the spatial dynamics of the system and providing guidance to managers.

The level of detail a connectivity metric provides depends on the type of connectivity it quantifies. We distinguish three types or classes of connectivity in increasing order of detail: structural, potential, and actual connectivity. Structural connectivity entails indices of spatial pattern that do not incorporate any data on species' dispersal abilities or how likely movement of individual organisms is among patches. Potential connectivity refers to metrics that incorporate some basic (perhaps indirect) knowledge about an organism's dispersal ability together with spatial relationships among landscape elements or habitat patches. Examples of indirect dispersal data include estimates of vagility derived from body size or energy budgets capacity (Cresswell et al. 2000; Porter et al. 2000); examples of generic basic dispersal data include measurements with little spatial detail (such as maximum recapture distances from tagging or banding studies) (Clark et al. 2001). Actual connectivity metrics go a step further, quantifying the movement of individuals through a habitat or landscape and thus providing a direct estimate of the linkages that exist among landscape elements or habitat patches. Previous authors have lumped potential connectivity and actual connectivity together as "functional connectivity" (Crooks and Sanjayan Chapter 1; Taylor et al. Chapter 2). We split the categories, however, because metrics of potential connectivity and actual connectivity require different types of data and provide different levels of detail about spatial dynamics in ecological systems.

Different metrics take various types of information into account to arrive at estimates of connectivity. These differences in data requirements are important because they affect both the degree of detail a connectivity measure can provide and the degree of difficulty associated with obtaining the necessary data. Our approach here is to outline how the three types of connectivity characterized above map onto six different data categories (see next section). In each case, we emphasize the basic data requirements for a given group of connectivity metrics, examine the spatial scales on which the metrics are defined, and highlight which type of connectivity (structural, potential, or actual) each metric estimates. In the discussion, we consider the costs and benefits of performance enhancements and possible extensions to these basic definitions.

Category 1: Patch occupancy data and nearest-neighbor distance

Data on the distance between occupied patches is usually obtainable when standard field surveys have been done to determine patch occupancy patterns for a focal species. The data need not be spatially explicit, provided that the distance between a focal patch and its nearest occupied neighbor is measured. Inter-patch distance is a measure of the isolation of a focal patch, and, in this framework, the structural connectivity of the patch is simply the inverse of its isolation. The inter-patch distance used is usually the Euclidean distance, but could be some other type of distance when the necessary data are available. Though simple to obtain, the nearest occupied neighbor distance is a crude measure of patch-level structural connectivity. Moilanen and Nieminen (2002; see also Moilanen and Hanski Chapter 3) compiled a data set of published studies that used nearest-neighbor distance to quantify connectivity and have shown that it has a much lower rate of detecting a significant effect of connectivity in spatial studies relative to more complex connectivity metrics (see Category 5). Moilanen and Nieminen (2002) also tested the ability of nearest-neighbor metrics to predict colonization events in two detailed empirical butterfly metapopulation data sets and again found that nearest-neighbor measures performed inadequately relative to more complex measures. Furthermore, by randomly deleting patches from the butterfly data sets, Moilanen and Nieminen (2002) demonstrated that the ability of nearest-neighbor measures to detect a significant effect of connectivity was much more influenced by sample size than the other measures were. Bender et al. (2003) obtained similar results using a simulated dispersal process on both real (derived from a geographical information system (GIS)) and artificially generated binary habitat maps. They found that nearest-neighbor distance was consistently the worst or second-worst performer of the four proximity indices they studied, and that it performed especially poorly when patch size and shape were varied (Bender *et al.* 2003).

Despite demonstrably poor performance, the nearest-neighbor distance is one of the most commonly used connectivity metrics (Moilanen and Nieminen 2002; Bender *et al.* 2003). This is most likely due to its simplicity and modest data requirements, but these advantages do not adequately compensate for its limitations. Furthermore, nearest-neighbor metrics are unlikely to benefit substantially from modifications or from further development.

Category 2:   Spatially explicit habitat data

Spatially explicit habitat data are often remotely sensed, cover a large area, and are represented in either raster or vector form in GIS. The increasing availability of this type of data and the inherent geometric capabilities of GIS software make the connectivity metrics in this category relatively easy to calculate. If habitat patches for a particular species or suite of species (e.g., old-growth forest) can be identified from a GIS coverage, then it is possible to calculate several different mathematical indices that quantify aspects of the spatial patterning of elements (e.g., different habitat types) in a landscape. These metrics quantify the number, size, extent, shape, or aspects of the spatial arrangement of landscape elements. Examples of spatial pattern metrics include number of patches, patch area, core area, patch perimeter, contagion, perimeter—area ratio, shape index, fractal dimension, and patch cohesion (Haines-Young and Chopping 1996; Schumaker 1996). Although users often assume that spatial pattern indices provide estimates of actual connectivity, these statistics by themselves only estimate the structural connectivity of a landscape. The link between structural connectivity estimates and actual connectivity has rarely been explored using empirical data. However, several simulation studies have attempted to bridge this gap. Schumaker (1996) combined a correlated random-walk dispersal model with maps generated from real landscapes to show that many commonly used metrics calculated from spatially explicit habitat data correlate poorly with simulated dispersal success. Schumaker (1996) found that shape index and patch cohesion were the best predictors of dispersal success, whereas fractal dimension, number of patches, patch area, core area, patch perimeter, contagion, and perimeter—area ratio were at best weakly correlated with dispersal success. Furthermore, he found that pattern metrics that did not incorporate patch areas performed poorly because small patches that contributed little to connectivity often biased these indices to overpredict connectivity. Tischendorf (2001b) used a similar approach and found that, while some spatial pattern indices were strongly correlated with three measures of simulated dispersal success, 68% of the statistical relationships between 26 metrics and three measures of dispersal success were inconsistent when landscape structure and dispersal behavior were varied. Furthermore, Tischendorf (2001b) found that the scale at which the indices were calculated was crucial. Class-level indices (that are based on particular landcover types) were, in general, more strongly related to the measures of simulated dispersal success than were landscape-level indices (that consider all landcover types) (Tischendorf 2001b).

These simulation analyses demonstrate the potential of spatial pattern indices, but also highlight the fact that more development is required before these metrics can be reliably used to predict connectivity. Currently, only when structural connectivity metrics are combined with knowledge that the movement of focal species through a landscape requires physical habitat connections or close proximity of patches of preferred habitat, do these metrics aptly summarize the connectivity of a landscape. For example, one such case involves so-called "matrix sensitive" species (Ims 1995) that tend not to wander out of their preferred habitat type. Structural connectivity may be more likely to exhibit predictable, direct relationships to actual connectivity for such species.

Metrics based on spatially explicit habitat data have the advantage of being applicable at large spatial scales, which facilitates characterization of the connectivity of entire landscapes. Because of the increasing availability of landscape-scale spatially explicit habitat data, these metrics will be able to quickly characterize large-scale patterns of connectivity in cases where they prove to be good predictors of actual connectivity. However, as Tischendorf's (2001b) analysis demonstrates, the spatial scale at which these metrics are calculated may strongly influence their predictive ability. Clearly, more research is needed to understand the relationship between structural connectivity and actual connectivity before spatial pattern indices can be relied upon to estimate actual connectivity. As several authors have pointed out (Schumaker 1996; Tischendorf 2001b; Fortin et al. 2003), focusing on the relationship between the spatial pattern that these metrics quantify and the underlying ecological processes influencing connectivity, such as demographics, dispersal and behavior, may be the most effective way to develop these metrics into useful predictors of actual connectivity. Unfortunately, there is a dearth of empirical work in this area, which is partially explained by the large spatial scales involved. Nevertheless, the relationships between these metrics and connectivity will ultimately need to be demonstrated with empirical movement data.

Category 3: Point or grid-based occurrence data

This data category involves records of species' spatial occurrences without sufficient detail to delimit actual habitat patches. Data sets fitting in this category include those assembled from museum records or similar compilations arising from long-term surveys of species presence/absence, where patch boundaries are not known or may have changed since the data were collected. Consequently, this category is methodologically distinct because it involves cases where species-level data are used to obtain measures of structural connectivity. In contrast, other approaches to quantifying the structural connectivity of a landscape rely upon habitat or patch-level data.

Both point data, where considerable spatial detail is available regarding the origin of each sample, and grid data, where spatial descriptions are less precise, can be used. In general, though, one's understanding of the system will be constrained by the resolution of the data. Point and grid data can be used to characterize connectivity by applying "scale–area" methods, which use each known occurrence of a species to demarcate, for each of a series of nested scales, what portions of a region are occupied (Kunin 1998) (Fig. 12.1). To derive a species' scale–area curve, one breaks a landscape into a series of equal-sized grid cells at each of several

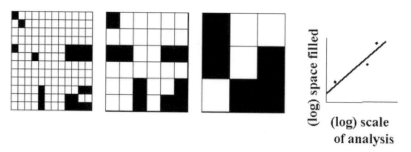

Fig. 12.1. Schematic illustrating the calculation of scale–area metrics. The habitat area occupied by a species is determined for each of several spatial scales (here shown for only three scales). The slope of a power-law regression of area occupied versus scale of analysis yields a statistical measure characterizing how fragmented the spatial distribution is.

resolutions (with a fixed number of fine-scale cells inside each coarser-scale cell). At each resolution, one then determines a species' presence or absence for each cell. Assuming a cell with at least one incidence record is "filled," plotting the total map area filled at a given resolution vs. grid-cell size at each resolution on log–log axes generates a scale–area curve from which one can estimate, via power-law regression, a scale–area slope.

Steep scale–area slopes characterize species that are sparsely distributed over a large area (i.e., species that have fragmented distributions), whereas shallow slopes identify species with occurrences clustered across scales. A scale–area slope statistic provides a scale-independent measure of how strongly species occurrences are clustered across the landscape and allows for consideration of the structural connectivity of a given species distribution. The utility of scale–area slopes in this context hinges on the notion that proximity is the major determinant of the connectivity among occurrences.

Proximity is clearly of overriding importance in some systems. Fagan et al. (2002) demonstrated that for Sonoran Desert fishes, species that under historical conditions exhibited more compactly distributed occurrences were at a distinct advantage when it came to weathering the ensuing decades of anthropogenic alterations to their habitats and landscape. In contrast, species with steep scale–area slopes, whose distributions were more fragmented historically, were at greater risk of local extinction. Fagan et al. (2002) hypothesized that this effect arose because clustered occurrences afforded opportunities for increased recolonization, thereby increasing the likelihood of extant local populations repopulating extirpated ones.

An advantage of this approach is its inherent flexibility for dealing with issues of spatial scale. Because scale–area slopes provide a measure of the spatial clustering of a species' occurrences that is scale independent, the same measure can be used as a predictor variable in analyses of spatial changes taking place at each of several spatial scales. For example, extirpation risks of a species may well vary as a function of spatial scale. For the Sonoran Desert fishes, species with steep scale–area slopes (and presumably lower connectivity among occurrences) have suffered a greater frequency of local extirpations than species with shallow scale–area slopes (Fagan et al. 2002). The same qualitative pattern also holds for the frequency of large-scale losses (e.g., extirpations from whole river basins). However, the relationship between the compactness of a species' historical distribution and its extirpation risk is strongest at intermediate scales (reach units of ~100 km in length), suggesting that it is at this scale where distributional fragmentation and disruption of connectivity have proven most detrimental to persistence for these fishes (Fagan *et al.* 2005).

In the absence of data on dispersal events, scale–area methods cannot provide a direct estimate of actual or potential connectivity. However, dispersal data could be combined with scale–area slope statistics as a way of gauging the dynamic consequences of interspecific differences in fragmentation scores. Such a blending of approaches has not yet been undertaken, and until then, scale–area slopes will remain structural connectivity metrics.

Category 4:   Spatially explicit habitat data with dispersal data

This category combines spatially explicit habitat data derived from GIS coverages with data acquired from independent studies on the dispersal biology of focal species. The requirement for species-specific dispersal data makes it far more difficult to obtain the information necessary to calculate this group of connectivity metrics. However, the advantage of having such data is that one can now address the potential connectivity of a patch or landscape, rather than being restricted to examining structural connectivity (e.g., see Carroll Chapter 15). The data necessary for this approach include, minimally, some measure of a species' dispersal ability through non-habitat "matrix" in the landscape, such as mean or median dispersal distance. More detailed information, in the form of empirically parameterized dispersal functions or data on a species' dispersal ability with respect to particular landscape elements, can be incorporated to improve these measures of connectivity.

With these data, a landscape can be analyzed for a focal species using the mathematical framework of graph theory to identify which patches are potentially connected. Graph-theoretic approaches (Cantwell and Forman 1993; Keitt *et al.* 1997; Bunn *et al.* 2000; Urban and Keitt 2001; Theobald Chapter 17) represent a landscape as a mathematical "graph" (i.e., a network with particular properties) and allow integration of relevant information on habitat patches and potential dispersal routes (Fig. 12.2). The spatial arrangement of patches as well as patch attributes such as areas are obtained from GIS coverages and converted into a graph. Connections (which are termed "edges" in graph theory) among patches (termed "nodes" or "vertices") in the graph are established by using either a fixed critical dispersal distance (Keitt et al. 1997; D'Eon *et al.* 2002) or a dispersal kernel. A fixed critical distance represents the distance after which a species' probability of dispersal is assumed to decline rapidly (van Langevelde 2000) and is similar to using a buffer radius

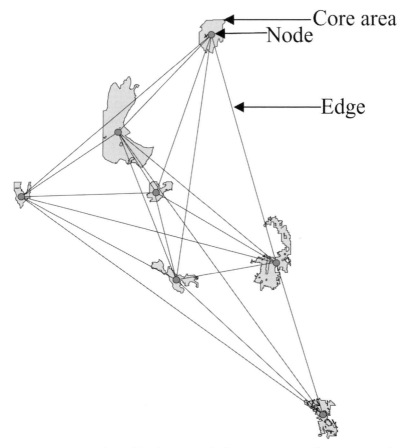

Fig. 12.2. Hypothetical landscape in which a reserve network is represented using techniques of mathematical graph theory. In the figure, each polygon is intended to represent a core habitat unit (in graph theory, a "node") in a reserve network. Edges connecting the nodes may have weights based on a measure of connectivity. (Figure courtesy of Jeff Tracey.)

(see Category 5). A dispersal kernel is a function describing the relationship between dispersal distance and the focal species' probability of dispersal (e.g., Kot *et al.* 1996; Havel *et al.* 2002).

Once the potential connections among patches are established, graph-theoretic approaches then scale up to landscape-level descriptions of connectivity. Various large-scale metrics have been developed within the graph theoretic framework including correlation length, distance to cluster edge, number of graph components, number of nodes in the largest graph component, and diameter of the largest graph component (Keitt *et al.* 1997; Urban and Keitt 2001; D'Eon *et al.* 2002). Such metrics provide an estimate of the potential connectivity of a landscape from the perspective of particular species. Furthermore, various graph operations, such as node and edge deletions, can be used to simulate destruction of habitat patches or dispersal corridors, respectively, allow the prioritization of habitat patches or dispersal routes in terms of their contributions to landscape connectivity (Urban and Keitt 2001). By explicitly acknowledging the scale dependence of connectivity and allowing the examination of the importance of individual patches to overall landscape connectivity, graph approaches provide a bridge between the patch-level and landscape-level views of connectivity (Tischendorf and Fahrig 2000; Moilanen and Hanski 2001; Tischendorf 2001a; Moilanen and Nieminen 2002; Taylor *et al.* Chapter 2; Moilanen and Hanski Chapter 3). If put into practice, lessons from graph-theoretic perspectives on connectivity would allow land managers to identify those specific patches that are most critical to conserve connectivity within a landscape. For example, practical questions such as "How can we harvest *X* hectares of forest without significantly disrupting the connectivity of the landscape?" can be addressed via graph theoretic models that couple spatially explicit habitat coverages with dispersal data.

Category 5:   Spatially explicit patch occupancy, patch, and dispersal data
Spatially explicit patch occupancy data are usually obtained by directly sampling habitat patches for a species of interest and spatially referencing the patch locations. With such data, one can calculate buffer radius or incidence function measures of patch-level potential connectivity, depending on assumptions about the dispersal biology of the focal species (Hanski 1994; Moilanen and Hanski 2001). Moilanen and Hanski (Chapter 3) review these metrics, particularly the incidence function metapopulation model (IFM) (Hanski 1994; Hanski *et al.* 1996) and its extensions, in detail in the context of metapopulation modeling and population persistence. Here we discuss each of these two approaches in terms of data dependence and information content.

For buffer radius measures, patch occupancy data for all patches that lie within a fixed radius of the focal patch are required. The connectivity of a focal patch is a function of the number of occupied patches that lie within that buffer radius and the areas of those patches. The buffer radius can be chosen arbitrarily, but should reflect some aspect of the focal species' dispersal biology. For example, the buffer radius could be the mean or median dispersal distance of the focal species. Moilanen and Nieminen (2002) have shown that buffer measures are sensitive to the choice of buffer radius, suggesting that incorporation of even the most basic dispersal information could substantially improve the predictive performance of this type of potential connectivity metric.

A second set of connectivity measure within this category is derived from the IFM model (Hanski 1994; Hanski et al. 1996). These measures require spatially explicit patch occupancy data for a sufficiently large number of patches in a metapopulation to obtain robust parameter estimates. Furthermore, data on (or an assumption about) the shape of the focal species' dispersal kernel is also required to calculate IFM-type connectivity measures. The negative exponential is generally assumed, but other dispersal kernels (e.g., gamma) could be used. The IFM approach also typically incorporates patch areas as well as a parameter that scales emigration to source patch area (Moilanen and Nieminen 2002). The scaling parameter as well as the parameter of the negative exponential dispersal kernel can be estimated from mark−release−recapture data or by fitting a model to presence/absence data. The basic IFM connectivity measure essentially sums the potential contributions of all occupied patches in a metapopulation to the connectivity of a focal patch. The contributions are weighted by distance and some surrogate for population size (e.g., patch area). As with buffer radius measures, the area of the focal patch and an additional scaling parameter can be added to improve the performance of the metric (Moilanen and Nieminen 2002). If parameter estimation is based on model fit to presence/absence data, erroneous occupancy data or unknown occupied patches can lead to the over-estimation of the focal species' dispersal ability (Moilanen 2002). However, the IFM measure has the advantage of being less sensitive to misestimation of the mean dispersal distance than the buffer measures are to misestimation of the buffer radius (Moilanen and Nieminen 2002).

Incorporating patch occupancy information clearly gives a better estimate of the potential connectivity of a focal patch. However, both buffer radius and IFM connectivities could be calculated without patch occupancy data. This approach, termed "connectivity of landscape elements" (Moilanen and Hanski 2001), is akin to the graph-theoretic method of using minimal dispersal information to establish potential connections in a landscape based on the proximity of patches and some consideration of the dispersal ability of the focal species. Though the mathematical underpinnings of the graph and metapopulation connectivity approaches are very similar, the focus and implementation of the two methods are different. Implementations of graph theory use patch-level connectivities as a basis for developing larger-scale measures of connectivity. In contrast, the metapopulation approach uses patch-level connectivities as a basis for metapopulation modeling and predicting metapopulation persistence (see below).

When the necessary data are available, buffer radius and IFM measures give a very detailed description of patch-level potential connectivity. However, a potential drawback to these metrics is that patch-level connectivities do not necessarily scale up to landscape levels. Presumably, patch-level connectivities could be aggregated or averaged to arrive at a landscape-level connectivity, but no accepted method for developing this linkage currently exists (Moilanen and Hanski 2001). One promising route that moves the patch-level connectivity measures toward applicability on the landscape scale involves the "metapopulation capacity" of a landscape (Hanski and Ovaskainen 2000; Ovaskainen and Hanski 2001; Moilanen and Hanski Chapter 3). If sufficient data are available to parameterize a stochastic patch occupancy model (e.g., the IFM model (Hanski 1994; Hanski *et al.* 1996; Moilanen 1999)), one could then calculate the metapopulation capacity of the study system. Though connectivity is an essential component of metapopulation capacity, it is not directly quantified by this measure. To highlight this distinction, Moilanen and Hanksi (Chapter 3) compare the metapopulation capacity to the simple arithmetic mean of IFM patch-level connectivities, and find that the relationship between these two quantities depends on the spatial configuration of patches. However, when the necessary data are available metapopulation capacity may prove to be a more useful quantity compared to landscape-level connectivity, because it focuses on a landscape's potential to maintain a viable metapopulation over time.

Category 6: Individual movement data

Data on individual movements provide the most direct estimate of actual connectivity. Numerous methods exist for obtaining such data, but generally, these types of studies are too labor-intensive to be conducted at large (landscape) scales. Ims and Yoccoz (1997) review this literature in detail; recent material in this area is reviewed by Tracey (Chapter 14). Here, we will give a brief synopsis and outline major categories dealing with macroscopic evidence for dispersal while excluding detailed coverage of some useful techniques. For example, the use of genetic data to explore the demographic consequences of connectivity (Andreassen and Ims 2001) would fit within this category of approaches. However, because studies involving genetic connections often deal with dispersal over much longer timescales than those of the other approaches we discuss, we refer readers to the applicable chapters (Frankham Chapter 4; Neville *et al.* Chapter 13). Likewise, we omit in-depth coverage of computational approaches for studying individual movement dispersal, even though such studies can provide key insights on how probabilities of patch colonization can depend on landscape characteristics and dispersal behaviors (Gustafson and Gardner 1996).

Tracing movement pathways of individual animals is a very direct method for assessing the actual connectivity of at least a portion of a landscape. For small, relatively sedentary or otherwise easy-to-observe organisms, direct observation of movement pathways may be possible. Otherwise, radiotracking, for animals large enough to carry transmitters, can provide critical long-distance dispersal information for species that may be hard to observe visually (Gillis and Krebs 1999, 2000). Snow tracking, when possible, may provide similar information. Unfortunately, these methods are difficult to apply and may provide insufficient sample sizes for rigorous statistical analysis (Turchin 1998). Despite these diffi- culties, movement pathway studies may be able to identify key dispersal routes that maintain landscape connectivity (Sutcliffe and Thomas 1996). Such detailed studies may be necessary to establish the functional signi- ficance of movement corridors in a landscape (see Tracey Chapter 14).

Mark–release–recapture studies can provide information about how well a focal species can move through a landscape. However, due to logistical constraints, these studies are often restricted to relatively small spatial scales. Careful design of such studies is required to minimize biases introduced by the capturing, marking, releasing, and trapping techniques used. Southwood (1978), Sutherland (1996), and Turchin (1998) provide detailed reviews of many mark–recapture study designs and techniques and also provide taxon-specific methodological recom- mendations. For the purpose of assessing landscape connectivity, mass mark–recapture (MMR) methods, where individuals do not have a unique marking, may be sufficient and may save time and effort. Releasing experimental animals outside of habitat patches and recording the number of recaptures within habitat patches can provide an estimate of how successfully animals can disperse through a landscape (Harrison 1989). Such studies may, however, be confounded by complex interactions between individual behaviors and the landscape (e.g., some individuals may forage instead of dispersing) (Pither and Taylor 1998). Comparisons between different landscape types or different types of intervening matrix can yield insights into species' habitat-specific dispersal abilities. For example, Pither and Taylor (1998) used a comparative design to evaluate how two species of calopterygid damselflies responded to different landscape structure (field vs. forest) when dispersing to their preferred stream habitat.

Measurements of patch-level immigration or colonization rates for unmarked animals can, by themselves, serve as a connectivity metric (van Langevelde 2000). This approach is difficult in practice because immigration or colonization rates must be sufficiently high that useful data can be collected over a reasonable period of time. When occupied natural habitat patches are used, various fencing or directional trapping methods can be employed, though these techniques sometimes cause behavioral changes in animals (Johnson and Gaines 1985, 1987; Valone and Brown 1995). Studies using empty natural habitat patches or artificially created patches are generally easier because immigrants do not need to be separated from patch residents. For example, studies of benthic fouling communities often use artificial substrate such as sponge (Schoener 1974a, 1974b) or glass (Kindt and Small 2002) to measure colonization rates of unoccupied habitat. If artificial patches cannot be made and no empty natural patches are available, patch residents may be experimentally removed in some cases (e.g., Simberloff 1976; Valone and Brown 1995).

Many techniques for the direct measurement of movement may be applied to a variety of taxa. Though landscape-level estimates of actual connectivity are possible in theory, the data-intensive nature of direct measurement methods limits the spatial scales at which they can be applied. Still, in situations where movement data are already available or only a few habitat patches are of interest, measurement of actual connectivity provides a direct, detailed description of how well particular patches are connected in a fragmented landscape.

DISCUSSION

This chapter provides alternative views of quantifying connectivity by focusing on how empirical data define which connectivity metrics can be calculated. It is clear that, across the different connectivity metrics, a trade-off exists between information content and data requirements (Fig. 12.3). For example, the nearest-neighbor measures of Category 1 and

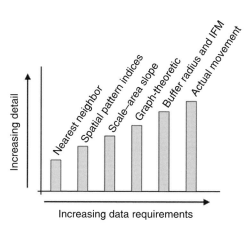

Fig. 12.3. Schematic representation of the trade-off between information content and data requirements among connectivity metrics. Both information content and data requirements increase going from Category 1 (nearest-neighbor distance) to Category 6 (direct observation of individual movement).

spatial pattern indices of Category 2 do not require extensive data to calculate but provide only crude, patch-level (Category 1) or large-scale (Category 2) estimate of structural connectivity. In contrast, the buffer and IFM measures of Category 5 provide very detailed estimates of potential connectivity at the individual patch level, but are extremely data intensive. Likewise, the direct measurements in Category 6 provide the only estimates of actual connectivity, but are, again, applicable mainly to small scales and are extremely data intensive. Given this trade-off, the graph-theoretic approaches of Category 4 may possess the greatest benefit/effort ratio for conservation problems that require characterization of connectivity at relatively large scales. These measures provide a reasonably detailed picture of potential connectivity but have relatively modest data requirements. When habitat patches cannot be reliably delimited, the scale—area approach of Category 3 might be the only option. However, the relationship between scale—area slopes and actual connectivity needs to be developed further.

For many of the metrics discussed here, additional data, not included in the basic definition of the metric, can be incorporated to improve the metric's performance. For example, patch area or other surrogates for population size may be used in specially modified versions of the metrics discussed above. There is an emerging consensus that "area-informed" metrics perform better than their area-less counterparts (Moilanen and Nieminen 2002; Bender *et al.* 2003; Tischendorf *et al.* 2003). This result is to be expected given the general importance of habitat area in many ecological processes. While habitat area is often used as a proxy for population size, other variables may prove to be better surrogates. For example, Moilanen and Hanski (Chapter 3) describe the usage of host plant coverage in patches of a metapopulation of the butterfly *Melitaea cinxia* to improve the performance of the IFM measure. Another factor that may be modified to improve metric performance is the measure of inter-patch distance. In situations where Euclidean distances do not accurately reflect movement distances between patches, "least-cost movement pathways" could be used as an alternative (Bunn *et al.* 2000; Theobald Chapter 17). For example, in metapopulations of stream-dwelling species, dispersal is often constrained to movement pathways very different than Euclidean distances (Dunham and Rieman 1999; Fagan 2002). By focusing on energetic costs, risk/reward ratios, or landscape features that guide dispersal, least-cost pathways may sometimes give a more meaningful measure of inter-patch distances than Euclidean distance. However, to develop meaningful least-cost paths, one must be careful to recognize the scale at which the species of interest perceive landscape elements. Mech and Zollner (2002) describe a method for estimating the perceptual range of forest-dwelling rodents based on body mass, but this method may not be viable for other taxa such as birds and insects. In general, inclusion of extra data to improve metric performance further reinforces the idea of a trade-off between informational yield and data requirements discussed above.

Several "scaling" issues arise when applying connectivity metrics. Calabrese and Fagan (2004) discuss how connectivity depends on spatial scale, while here, we focus on multi-species connectivity. Despite the operational convenience of the standard, species-specific view of connectivity, real conservation problems often involve multiple species. Thus, applying the concept of connectivity to real-world situations will require a broadening of scope to include more than one species at a time. We see this as a key route along which future research on connectivity must develop. Here, we offer a few preliminary suggestions regarding what research on multi-species connectivity should entail.

First, it is worthwhile to consider the broad ways in which species may differ. The simplest scenario to deal with occurs when the species of interest have similar habitat affinities and similar dispersal abilities/behaviors. For example, in their analysis of landscapes in the Mount Lofty

Ranges, South Australia, Westphal *et al.* (2003) showed that many bird species that inhabit native vegetation responded to landscape structure at the same spatial scale, suggesting the landscape was similarly connected for these species. In such cases, the umbrella species approach might be profitably used to characterize connectivity for the suite of species. This strategy could be extended (cautiously) if other species of interest have similar habitat requirements but are better dispersers. Thus for purposes of assessing multi-species connectivity, one would want to choose as an umbrella a species with relatively weak dispersal abilities, because landscape arrangements with high connectivity for the umbrella species should, in principle, afford the other species even greater opportunities for dispersal. More complicated scenarios arise when several species of interest differ in either their habitat affinities or dispersal traits, or in both factors simultaneously. Haddad *et al.* (2003) have shown that the same dispersal corridors can facilitate connectivity for a diverse range of taxa, albeit to different degrees. For species that have similar habitat affinities, if one can characterize the range of their dispersal abilities, then connectivity metrics that consider dispersal ability (such as graph theory and buffer and IFM measures) can be used to place bounds on community-wide connectivity by considering a series of mean or critical dispersal distances (Keitt *et al.* 1997). This "range-bounding" approach can of course be extended to species with different habitat affinities, but the rapidly increasing data requirements will likely limit the applicability of this method. Although large-scale, multi-species empirical studies, such as those by Haddad *et al.* (2003) and Westphal *et al.* (2003), are understandably rare, such efforts are crucial if we are to recognize the extent to which we may generalize the concept of connectivity (Haddad and Tewksbury Chapter 16).

Research on connectivity and the metrics that quantify it is still in an early stage, and much remains to be learned. The categorizations employed here represent frequently encountered types of data and are not intended to cover all possibilities or to be adhered to rigidly. Quite the contrary, combining approaches from more than one category, when possible, can yield substantial insight. For example, van Langevelde (2000) combined graph theory with patch occupancy and patch-specific population size data to relate graph-theoretic connectivity metrics to patch colonization rates. This hybrid approach allowed van Langevelde (2000) to infer the most appropriate critical distance to use in the graph analysis based on how well different distances predicted actual colonization events. Many other types of syntheses are possible and should be fruitful avenues for future research.

It is clear from the literature that scientists and managers do not widely appreciate the differences among alternative connectivity metrics with respect to performance and degree of detail. In this chapter, we have attempted to cut through some of this confusion by outlining the relationships among the metrics in terms of their relative merits and different data requirements. Unfortunately, there is no simple, all-purpose method for maximizing our understanding of connectivity across the range of real-world problems likely to be encountered. It remains up to the end-users to define their questions carefully, assess the data they have or could reasonably obtain, and choose the most appropriate connectivity metrics for their purposes. By understanding the available options and the differences among these alternatives, researchers and practitioners can extract the maximum possible amount of connectivity information from their data.

ACKNOWLEDGEMENTS

We thank K. Crooks, P. Kareiva, and A. Moilanen for comments that substantially improved the manuscript. National Science Foundation support (DEB-0075667) is gratefully acknowledged.

REFERENCES

Andreassen, H. P., and R. A. Ims. 2001. Dispersal in patchy vole populations: role of patch configuration, density dependence, and demography. *Ecology* 82:2911–2926.

Bender, D. J., L. Tischendorf, and L. Fahrig. 2003. Using patch isolation metrics to predict animal movement in binary landscapes. *Landscape Ecology* 18:17–39.

Bunn, A. G., D. L. Urban, and T. H. Keitt. 2000. Landscape connectivity: a conservation application of graph theory. *Journal of Environmental Management* 59:265–278.

Cabeza, M. 2003. Habitat loss and connectivity of reserve networks in probability approaches to reserve design. *Ecology Letters* 6:665–672.

Calabrese, J. M., and W. F. Fagan. 2004. A guide to comparison-shopper's connectivity metrics. *Frontiers in Ecology and the Environment* 2:529–536.

Cantwell, M. D., and R. T. T. Forman. 1993. Landscape graphs: ecological modeling with graph theory to detect configurations common to diverse landscapes. *Landscape Ecology* 8:239–255.

Clark, J. S., M. Lewis, and L. Horvath. 2001. Invasion by extremes: population spread with variation in dispersal and reproduction. *American Naturalist* 157:537–554.

Cresswell, J. E., J. L. Osbourne, and D. Goulson. 2000. An economic model of the limits to foraging range in central place foragers with numerical solutions for bumblebees. *Ecological Entomology* 25:249–255.

D'Eon R. G., S. M. Glenn, I. Parfitt, and M. Fortin. 2002. Landscape connectivity as a function of scale and organism vagility in a real forested landscape. *Conservation Ecology* 6:10. Available online at http://www.conservationecology. org/vol6/iss2/art10/

Dunham, J. B., and B. E. Rieman. 1999. Metapopulation structure of bull trout: influences of physical, biotic, and geometrical landscape characteristics. *Ecological Applications* 9:624–655.

Fagan, W. F. 2002. Connectivity, fragmentation, and extinction risk in dendritic metapopulations. *Ecology* 83:3243–3249.

Fagan, W. F., P. J. Unmack, C. Burgess, and W. L. Minckley. 2002. Rarity, fragmentation, and extinction risk in desert fishes. *Ecology* 83:3250–3256.

Fagan, W. F., C. Aumann, C. M. Kennedy, and P. J. Unmack. 2005. Rarity, fragmentation and the scale-dependence of extinction-risk in desert fishes. *Ecology* 86:34–41.

Fortin, M. J., B. Boots, F. Csillag, and T. K. Remmel. 2003. On the role of spatial stochastic models in understanding landscape indices in ecology. *Oikos* 102:203–212.

Gillis, E. A., and C. J. Krebs. 1999. Natal dispersal of snowshoe hares during a cyclic population increase. *Journal of Mammalogy* 80:933–939.

Gillis, E. A., and C. J. Krebs. 2000. Survival of dispersing versus philopatric juvenile snowshoe hares: do dispersers die? *Oikos* 90:343–346.

Gustafson, E. J., and R. H. Gardner. 1996. The effect of landscape heterogeneity on the probability of patch colonization. *Ecology* 77:94–107.

Haddad, N. M., D. R. Bowne, A. Cunningham, *et al.* 2003. Corridor use by diverse taxa. *Ecology* 84:609–615.

Haines-Young, R., and M. Chopping. 1996. Quantifying landscape structure: a review of landscape indices and their application to forested landscapes. *Progress in Physical Geography* 20:418–445.

Hanski, I. 1994. A practical model of metapopulation dynamics. *Journal of Animal Ecology* 63:151–162.

Hanski, I., and O. Ovaskainen. 2000. The metapopulation capacity of a fragmented landscape. *Nature* 404:755–758.

Hanski, I., A. Moilanen, T. Pakkala, and M. Kuussaari. 1996. The quantitative incidence function model and persistence of an endangered butterfly metapopulation. *Conservation Biology* 10:578–590.

Harrison, S. 1989. Long-distance dispersal and colonization in the bay checkerspot butterfly, *Euphydryas editha bayensis*. *Ecology* 70:1236–1243.

Havel, J. E., J. B. Shurin, and J. R. Jones. 2002. Estimating dispersal from patterns of spread: spatial and local control of lake invasions. *Ecology* 83:3306–3318.

Ims R. A. 1995. Movement patterns related to spatial structures. Pp. 85–109 in L. Hansson, L. Fahrig, and G. Merriam (eds.) *Mosaic Landscapes and Ecological Processes*. London: Chapman and Hall.

Ims R. A., and N. G. Yoccoz. 1997. Studying transfer processes in metapopulations: emigration, migration, and colonization. Pp. 247–265 in I. Hanski and M. E. Gilpin (eds.) *Metapopulation Biology: Ecology, Genetics, and Evolution*. San Diego, CA: Academic Press.

Johnson, M. L., and M. S. Gaines. 1985. Selective basis for emigration of the prairie vole, *Microtus ochrogaster*: open field experiment. *Journal of Animal Ecology* **54**:399–410.

Johnson, M. L., and M. S. Gaines. 1987. The selective basis for dispersal of the prairie vole, *Microtus ochrogaster*. *Ecology* **68**:684–694.

Keitt, T. H., D. L. Urban, and B. T. Milne. 1997. Detecting critical scales in fragmented landscapes. *Conservation Ecology* **1**:4. Available online at http://www.ecologyandsociety.org/vol1/iss1/art4/

Kindt, A. C., and P. F. Small. 2002. Correlation between temperature, colonization rate, and population density of the diatom *Cocconeis placentula* in freshwater streams. *Journal of Freshwater Ecology* **17**:441–445.

Kot, M., M. A. Lewis, and P. van den Dreissche. 1996. Dispersal data and the spread of invading organisms. *Ecology* **77**:2027–2042.

Kunin, W. E. 1998. Extrapolating species abundance across spatial scales. *Science* **281**:1513–1515.

Mech, S. G., and P. A. Zollner, 2002. Using body size to predict perceptual range. *Oikos* **98**:47–52.

Moilanen, A. 1999. Patch occupancy models of metapopulation dynamics: efficient parameter estimation using implicit statistical inference. *Ecology* **80**:1031–1043.

Moilanen, A. 2002. Implications of empirical data quality to metapopulation model parameter estimation and application. *Oikos* **96**:516–530.

Moilanen, A., and I. Hanski. 2001. On the use of connectivity measures in spatial ecology. *Oikos* **95**:147–151.

Moilanen, A., and M. Nieminen. 2002. Simple connectivity measures in spatial ecology. *Ecology* **83**:1131–1145.

Ovaskainen, O., and I. Hanski. 2001. Spatially structured metapopulation models: global and local assessment of metapopulation capacity. *Theoretical Population Biology* **60**:281–302.

Pither, J., and P. D. Taylor. 1998. An experimental assessment of landscape connectivity. *Oikos* **83**:166–174.

Porter, W. P., S. Budaraju, W. E. Stewart, and N. Ramankutty. 2000. Physiology on a landscape scale: applications in ecological theory and conservation practice. *American Zoologist* **40**:1175–1176.

Schoener, A. 1974a. Colonization curves for planar marine islands. *Ecology* **55**:818–827.

Schoener, A. 1974b. Experimental zoogeography: colonization of marine mini-islands. *American Naturalist* **108**:715–738.

Schumaker, N. H. 1996. Using landscape indices to predict habitat connectivity. *Ecology* **77**:1210–1225.

Siitonen, P., A. Tanskanen, and A. Lehtinen. 2002. Method for selection of old-forest reserves. *Conservation Biology* **16**:1398–1408.

Siitonen, P., A. Tanskanen, and A. Lehtinen. 2003. Selecting forest reserves with a multiobjective spatial algorithm. *Environmental Science and Policy* **6**:301–309.

Simberloff, D. 1976. Experimental zoogeography of islands: effects of island size. *Ecology* **57**:629–648.

Singleton, P. H., W. L. Gaines, and J. F. Lehmkuhl. 2002. *Landscape Permeability for Large Carnivores in Washington: A Geographic Information System Weighted-Distance and Least-Cost Corridor Assessment.* US Development of Agriculture Forest Service Pacific Northwest Research Station Research Paper 549. Portland, OR: US Department of Agriculture Forest Service.

Southwood, T. R. E. 1978. *Ecological Methods*, 2nd edn. London: Chapman and Hall.

Sutcliffe, O. L., and C. D. Thomas. 1996. Open corridors appear to facilitate dispersal by ringlet butterflies (*Aphantopus hyperantus*) between woodland clearings. *Conservation Biology* **10**:1359–1365.

Sutherland, W. J. 1996. Predicting the consequences of habitat loss for migratory populations. *Proceedings of the Royal Society of London, B* **263**:1325–1327.

Taylor, P. D., L. Fahrig, K. Henein, and G. Merriam. 1993. Connectivity is a vital element of landscape structure. *Oikos* **68**:571–573.

Tischendorf, L. 2001a. On the use of connectivity measures in spatial ecology: a reply. *Oikos* **95**:152–155.

Tischendorf, L. 2001b. Can landscape indices predict ecological processes consistently? *Landscape Ecology* **16**:235–254.

Tischendorf, L., and L. Fahrig. 2000. On the usage and measurement of landscape connectivity. *Oikos* **90**:7–19.

Tischendorf, L., D. J. Bender, and L. Fahrig. 2003. Evaluation of patch isolation metrics in mosaic landscapes for specialist vs. generalist dispersers. *Landscape Ecology* **18**:41–50.

Turchin, P. 1998. *Quantitative Analysis of Movement: Measuring and Modeling Population Redistribution in Animals and Plants.* Sunderland, MA: Sinauer Associates.

Urban, D., and T. Keitt. 2001. Landscape connectivity: a graph-theoretic perspective. *Ecology* **82**:1205–1218.

Valone, T. J., and J. H. Brown. 1995. Effects of competition, colonization, and extinction on rodent species diversity. *Science* **267**:880–883.

van Langevelde, F. 2000. Scale of habitat connectivity and colonization in fragmented nuthatch populations. *Ecography* **23**:614–622.

Westphal, M. I., S. A. Field, A. J. Tyre, D. Paton, and H. P. Possingham. 2003. Effects of landscape pattern on bird species distribution in the Mt. Lofty Ranges, South Australia. *Landscape Ecology* **18**:413–426.

Wiens, J. A. 1997. The emerging role of patchiness in conservation biology. Pp. 93–107 in S. T. A. Pickett, R. S. Ostfeld, M. Shachak, and G. E. Likens (eds.) *Enhancing the Ecological Basis of Conservation: Heterogeneity, Ecosystem Function and Biodiversity.* New York: Chapman and Hall.

Assessing connectivity in salmonid fishes with DNA microsatellite markers

HELEN NEVILLE, JASON DUNHAM, AND MARY PEACOCK

INTRODUCTION

Connectivity is a key consideration for the management and conservation of any species, but empirical characterizations of connectivity can be extremely challenging. Assessments of connectivity require biologically realistic classifications of landscape structure (Kotliar and Wiens 1990), and an understanding of how landscape structure affects migration, dispersal, and population dynamics (Dunning *et al.* 1992; Rosenberg *et al.* 1997; Hanski 1999; Taylor *et al.* Chapter 2). Empirical assessments of connectivity may be accomplished by studying spatial patterns of habitat occupancy through time (Sjögren-Gulve and Ray 1996; Hanski 1999; Moilanen and Hanski Chapter 3), spatially correlated changes in population demography (Bjornstad *et al.* 1999; Isaak *et al.* 2003; Carroll Chapter 15), and individual movements (Millspaugh and Marzluff 2001; Tracey Chapter 14). These approaches have provided important insights for many species, but they can be difficult to implement for species with slow population dynamics or turnover (extinction and recolonization), complex life histories, and long-distance migrations. For species with these characteristics, molecular genetic markers represent a valuable tool for understanding processes that influence connectivity (Avise 1994; Frankham *et al.* 2004; Frankham Chapter 4). In this chapter, we review applications of molecular genetic markers to assess connectivity in salmonid fishes, a group of relatively well-studied species with

Connectivity Conservation eds. Kevin R. Crooks and M. Sanjayan. Published by Cambridge University Press. © Cambridge University Press 2006.

characteristics that complicate non-genetic approaches to understanding connectivity. Lessons learned from salmonids may apply generally to other species that have received far less attention.

SALMONID ECOLOGY AND CONNECTIVITY

Salmonid fishes are among the most well-studied vertebrates, and for many species, much is known about habitat requirements, life-history diversity, and movement patterns. Most species of salmonids exhibit migratory behaviors. It is generally believed that salmonids have evolved such behaviors to exploit the diverse array of habitats available within the landscape or "riverscape" (Northcote 1992; Fausch *et al.* 2002) (Fig. 13.1). Salmonid fishes often rear in smaller streams and headwater lakes, with some individuals remaining in these natal habitats throughout

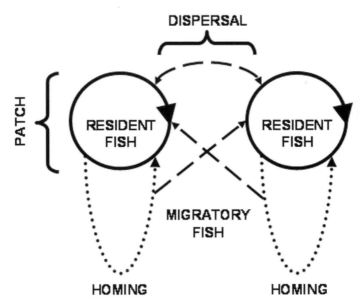

Fig. 13.1. Generalized life history of salmonid fishes, depicting spatial structuring, migratory life histories, and dispersal pathways (from Dunham *et al.* 2002). In this example, non-migratory or resident fish live within natal habitat patches with migratory individuals moving outside of patches and homing to reproduce in their natal habitats (indicated by dotted lines) or dispersing to new patches (indicated by dashed lines). Dispersal by resident fish among patches is also possible.

life. In addition to these non-migratory or "resident" individuals, some individuals may migrate beyond natal habitats to use feeding or refuge habitats, returning to their natal sites to breed (Northcote 1992). Many species or populations exhibit partial migration, with a mixture of individuals with migratory and resident life histories (Jonsson and Jonsson 1993). There are also some with largely "obligate" migratory life histories (e.g., most Pacific salmon: Groot and Margolis 1991), and others with largely resident life histories (Rieman and Dunham 2000). Spatial variation in occurrence of these life histories may have important implications for both connectivity and genetic structure among populations; because migratory fish move away from their natal habitat, they are more likely than resident fish to contribute to connectivity and genetic mixing through dispersal among populations (Hansen and Mensberg 1998; Knutsen *et al.* 2001). Understanding the ecological and evolutionary factors affecting the balance of migration and residency in salmonid fishes is a major area of active research (Hendry *et al.* 2004; Waples 2004).

As a group, salmonid fishes are habitat specialists, with specific requirements for water quality (especially cool temperatures: Elliott 1981), flow regimes (Latterell *et al.* 1998), and a variety of smaller-scale habitat features (Bjornn and Reiser 1991). Due to their specific habitat requirements, salmonids are often distributed discontinuously within or across watersheds (Dunham *et al.* 2002). The resulting landscape geometry of these habitats (e.g., habitat or patch size, degree of isolation) appears to be critical to long-term population persistence (Hanski 1999). As for many species, the general pattern observed for salmonids studied thus far is one of a greater probability of occurrence or persistence in larger and less isolated habitats. Thus, in addition to site-specific habitat features, landscape context (e.g., patch geometry) strongly influences population persistence and connectivity (Rieman and Dunham 2000; Fausch *et al.* 2002; Wiens 2002).

The general influences of landscape geometry on the persistence and occurrence of salmonids are consistent with predictions from metapopulation theory (Hanski 1999), but the specific processes that contribute to these patterns are poorly understood (Dunham and Rieman 1999; Rieman and Dunham 2000; Koizumi and Maekawa 2004). Much of this uncertainty stems from a limited understanding of connectivity. Connectivity can operate in several ways to influence salmonid population persistence (Fig. 13.2). First, connectivity can support or facilitate development of migratory life histories. If a migratory life history is present within

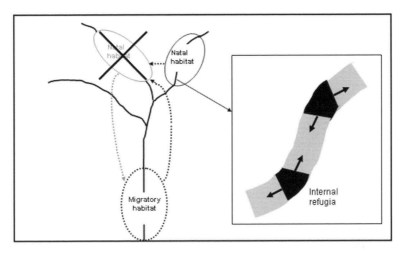

Fig. 13.2. Alternative pathways for repopulation of habitats following extirpation (Rieman *et al.* 1997; Dunham *et al.* 2003). Where individuals use both natal and migratory habitats, repopulation is possible if extirpations are not simultaneous in each habitat. On a smaller scale, repopulation may occur within streams or habitats via internal refugia. This may occur if the scale of a disturbance leading to extirpation is smaller than the size of a habitat.

a population, local extinctions within natal habitats (e.g., headwater streams) may not affect the entire population. For example, if an event causing local extinctions is short-lived, migratory fish outside of the system at the time of disturbance can repopulate natal habitats. This mechanism has been invoked to explain repopulation of habitats following local extirpations caused by wildfires, droughts, and pollution (Dunham *et al.* 1997; Rieman *et al.* 1997; Knutsen *et al.* 2001). If disturbances causing local extirpations are smaller than the natal patch size, smaller-scale or "within-patch" connectivity can be important for repopulation and persistence (Armstrong *et al.* 1994; Rieman *et al.* 1997; Dunham *et al.* 2003). Dispersal among natal patches is yet another mechanism providing gene flow and demographic support that may contribute to population persistence. This form of support is most often invoked in metapopulation theory (Hanski 1999). In summary, an understanding of connectivity in salmonid fishes requires an understanding of landscape structure including patch geometry (Dunham *et al.* 2002), migratory life histories (Northcote 1992), and dispersal processes (Quinn 1993; Rieman and Dunham 2000).

Before moving to discuss molecular markers, it is necessary to clarify how we define two key terms related to connectivity: migration and dispersal. In population genetics, migration is often used to describe gene flow – the transfer of genetic material among populations. In the salmonid literature, and with most other ecological fields, migration is defined as the movement of individuals from natal habitats across landscapes or regions to utilize complementary habitats (Dunning *et al.* 1992) in completing their life cycle. We use this definition of migration whenever possible. The term "dispersal" is also used confusingly across disciplines. In most of the ecological literature, the movement of individuals into non-natal habitats for breeding is often referred to as dispersal. In the salmonid literature this is referred to as "straying," but to avoid confusion we use the term dispersal (Rieman and Dunham 2000). Generally, some dispersing individuals will breed successfully, whereas others may not. Genetic data generally characterize the degree of "effective" dispersal, or dispersal that contributes to gene flow (Slatkin 1987; Peacock and Ray 2001), although new techniques can characterize movement that may or may not lead to gene flow (see below).

MICROSATELLITE MARKERS AND INFERENCES ABOUT CONNECTIVITY

The literature on molecular genetics of salmonid fishes is extensive, and many studies have focused on important systematic and large-scale biogeographic questions (Ryman and Utter 1987; Nielsen and Powers 1995; Hendry and Stearns 2004). Beginning in the late 1980s, the availability of molecular markers to uncover ecological and evolutionary processes operating on smaller spatiotemporal scales has produced a generation of research attempting to quantify connectivity among local populations within and among river basins. Here, we focus on applications of a single class of these higher resolution markers, DNA microsatellites (O'Connell and Wright 1997; Goldstein and Schlotterer 1999; Sunnucks 2000). Microsatellites are specifically targeted regions of neutral DNA, which are co-dominantly inherited. Their fast mutation rate relative to other markers (Hancock 1999) confers high levels of variability and therefore high resolution for distinguishing populations and even individuals. For this reason, these markers are often used to address connectivity among populations for salmonid fishes and other species (Sunnucks 2000; Jehle and Arntzen 2002; Hendry *et al.* 2004).

High-resolution genetic markers such as microsatellites offer many opportunities for understanding connectivity, but careful analysis and interpretation of the data are required to avoid misleading conclusions (Whitlock and McCauley 1999) or confusion of biological and statistical significance (Waples 1998; Hedrick 1999, 2001). A comprehensive overview of methods for statistical analyses of genetic data is well beyond the scope of this chapter (but see Neigel 1997; Balloux and Lugon-Moulin 2001; Rousset 2001 for excellent reviews). We focus instead on selected applications of genetic techniques to assess connectivity in salmonids. We recognize two major classes of analyses for inferring connectivity from patterns of genetic variability revealed by microsatellites: methods based on predefined populations, and methods based on individuals.

Many genetic analyses used to assess connectivity require predefined populations. Populations are most often defined by the researcher as groups of individuals from relatively discrete sampling locations, which are delineated by landscape characteristics thought to restrict gene flow (e.g., rivers and mountain ranges for terrestrial species, waterfalls and dams for aquatic species, or distance for any species). Once populations are defined, population-based analyses involve summarizing information on genetic variability within populations and determining the degree of genetic differentiation among them, which is assumedly influenced by gene flow and levels of connectivity. Common examples include: Wright's F_{ST} (Wright 1951), G_{ST} (Nei 1973), R_{ST} (Slatkin 1995), rare alleles (Slatkin 1985b), and analysis of molecular variance (AMOVA: Weir and Cockerham 1984). Distance measures, such as the chord distance (D_{CE}: Cavalli-Sforza and Edwards 1967) or Nei et al.'s D_A distance (1983), also determine the degree of genetic similarity among populations and are commonly used to build phenograms ("trees") to visualize population relationships. While these measures alone simply characterize the degree of genetic similarity among populations, F_{ST} frequently is used to estimate rates of gene flow empirically from mathematical equations relating F_{ST} to the number of migrants per generation (see Frankham Chapter 4). However, doing so assumes that populations fit the simplistic dynamics of Wright's island model (Wright 1931, 1940). For example, populations are assumed to be in equilibrium between random genetic drift, which causes the loss of alleles over time due to the sampling of individuals each generation, and dispersal, which brings in new alleles. It is also assumed that dispersal rates among populations are symmetrical and that all populations are equal in size (Whitlock and McCauley 1999), defined by the effective population size (N_e). In simple terms, N_e characterizes the rate at which a population loses genetic variability based on the number of individuals actually breeding each generation (see Waples 2002, 2004 for more detail). In addition to these traditional statistical approaches, more complex population-based simulation methods that make fewer assumptions about population dynamics are beginning to be applied (Shrimpton and Heath 2003; Fraser *et al.* 2004; Wilson *et al.* 2004; Neville *et al.* in press). For instance, coalescent-based methods can estimate asymmetrical migration among populations with different N_es, thus characterizing more realistic natural scenarios (Beerli and Felsenstein 1999, 2001; Beerli 2004). Regardless of the method of estimation, population-based measures of genetic variability and differentiation may be tested for correlations with environmental variables, such as geographic distance between populations (i.e., isolation by distance based on Mantel tests) or habitat size and quality, to infer evolutionary processes (Slatkin 1993; Hutchison and Templeton 1999). A major limitation of population-based approaches is the reliance on a priori definitions of population units, which can be highly subjective (Manel *et al.* 2003).

Individual-based analyses do not rely on a priori identification of populations for inferences on gene flow and population genetic structure. These analyses identify the scale at which gene flow among individuals is restricted, which defines the breadth of "genetic neighborhoods" and leads to inferences about dispersal. A variety of spatial statistics (e.g., autocorrelation statistics, Mantel tests, kinship analyses, semivariograms) can examine the degree of structuring of genotypes and connectivity among individuals at different spatial scales (Epperson 2003; Manel *et al.* 2003; Peakall *et al.* 2003). In addition, a recently developed Bayesian clustering approach (STRUCTURE: Pritchard *et al.* 2000) defines population units by iteratively sorting individual genotypes into groups to maximize the fit of the data to theoretical expectations derived from Hardy–Weinberg and linkage equilibrium. When combined with assignment tests (see below), rates of dispersal can be estimated by identifying migrant individuals among the populations it so defined (Pritchard *et al.* 2000). Though the results may be a bit less intuitive for practical use, individual-based methods discern the existence of population structuring in a manner that is less subjective than traditional approaches.

Assignment tests are an important tool for estimating dispersal that can be based either on populations defined a priori (Paetkau *et al.* 1995; Cornuet *et al.* 1999; Banks and Eichert 2000) or by the individual-based clustering method described above (Pritchard *et al.* 2000). Assignment tests compute the probability that an individual's multi-locus genotype belongs to each of a set of reference populations. Individuals with a higher probability of originating in a population other than that in which they were sampled are assumed to be dispersers. A major strength of assignment tests is that they circumvent the drift—migration equilibrium assumption of more traditional analyses (Davies *et al.* 1999) and can be a powerful alternative tool for estimating general dispersal patterns (Rannala and Mountain 1997; Hansen *et al.* 2001). Even when based on predefined populations, their focus on the individual as the sampling unit greatly improves their statistical power to uncover dispersal patterns (e.g., Castric and Bernatchez 2004).

The different analytical approaches outlined above (i.e., individual versus population-based approaches) also yield distinct insights about

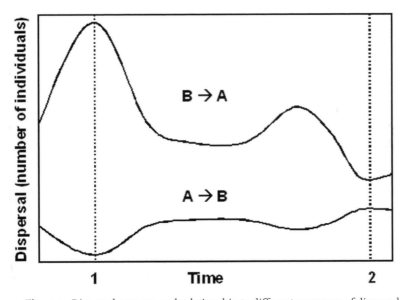

Fig. 13.3. Dispersal patterns and relationship to different measures of dispersal indicated by studies of molecular genetic markers. Arrows (B → A and A → B) represent asymmetrical dispersal, which may be estimated using coalescent approaches. Dashed vertical lines (at times 1 and 2) represent two slices in time where contemporary patterns of movement might be inferred from "instantaneous" methods like the assignment test. Indirect approaches (e.g., F_{ST}, coalescent analyses) estimate long-term rates of effective dispersal (gene flow) averaged across time.

connectivity based on the temporal scale characteristic of each class of analysis (Fig. 13.3). Many methods of estimating dispersal, such as Wright's F_{ST} and coalescent-based methods, estimate long-term, multi-generational average rates of dispersal among populations "indirectly" because they require the use of mathematical equations that relate the observed degree of genetic differentiation to the migration rate (see Slatkin 1985a; Frankham Chapter 4). They also quantify only "effective dispersal," or dispersal that leads to gene flow when an individual reproduces in their new location. Because of their deeper timescale, these methods are influenced by important historical events − such as rare long-distance dispersal − not occurring at the time of study (Peacock 1997). Individual assignment tests, in contrast, quantify connectivity in terms of the number of individuals present in a given population at the time of sampling that likely originated from a different population (Waser and Strobeck 1998). When population differentiation is strong enough to give the test sufficient power (see Cornuet *et al.* 1999; Hansen *et al.* 2001), these "foreign assignments" identify potentially dispersing individuals and thus provide a "direct" measure of current movement patterns (somewhat akin to mark−recapture methods), regardless of whether individuals that moved actually contribute to gene flow (Fig. 13.3). One important caveat regarding the assignment test deserves mentioning: in cases where the test's ability to distinguish potential populations of origin is limited by low levels of differentiation, assignment tests can falsely label individuals as dispersers and more likely characterize historical levels of effective dispersal rather than current movements (Hansen *et al.* 2001; Castric and Bernatchez 2004). Still, where feasible to apply, assignment tests allow us to infer contemporary movement based on the sampling locations of genotypes without actually tracking individual movement. The combination of traditional population-based and individual-based analyses can provide both historical and contemporary perspectives on population dynamics with the same genetic data (Davies *et al.* 1999; Hansen *et al.* 2001).

ASSESSING CONNECTIVITY WITH MICROSATELLITES: CASE STUDIES

A host of recent work on salmonid fishes illustrates the utility of micro-satellites and various genetic analyses for understanding the many factors that influence connectivity. Here, we focus on a selected group of species exhibiting partial migration, including charrs of the genus *Salvelinus* and trout of the genera *Oncorhynchus* and *Salmo*. With these species, microsatellites have been used to reveal patterns of connectivity in relation to influences of river branching patterns, natural and human constructed barriers, life-history variation, historical colonization patterns, and meta-population dynamics.

Influences of stream network structure

One of the most obvious landscape influences on connectivity in aquatic ecosystems is the network structure of stream systems. Transitions in habitat conditions that occur as streams flow together within stream networks (Frissell *et al.* 1986) have been used to identify population bound-aries for stream fishes (Angermeier *et al.* 2002; Dunham *et al.* 2002). Accordingly, many studies of trout and charr have found that genetic population structure can be defined in terms of drainage (e.g., Angers *et al.* 1999; Heath *et al.* 2001; Knudsen *et al.* 2002; Spruell *et al.* 2003; Narum *et al.* 2004) and stream branching patterns (e.g., Spruell *et al.* 1999; Heath *et al.* 2001; Wenburg and Bentzen 2001; Young *et al.* 2004; Neville *et al.* in press). The degree of genetic isolation among populations is also commonly related to stream distance between them, indicated by significant isolation by distance (e.g., Heath *et al.* 2001; Knutsen *et al.* 2001; Taylor *et al.* 2003). However, concordance between habitat structure and genetic structure is not always observed. For instance, F_{ST} estimates in bull charr (*Salvelinus confluentus*) demonstrated significant population differentiation, but genetic relationships among populations did not correspond to their spatial proximity (Spruell *et al.* 1999). This pattern has also been observed in coastal cutthroat trout (*Oncorhynchus clarkii clarkii*: Waples *et al.* 2001). In brook charr (*Salvelinus fontinalis*), Hebert *et al.* (2000) found that the size and hydrographic characteristics (i.e., the complexity of stream branching patterns) of a region were not consistent predictors of the degree of differentiation among populations as indicated by AMOVA and chord distance phenograms. Analysis of molecular variance of brook charr in a different system found low genetic diffe-rentiation among drainages, but differentiation among populations within drainages was relatively high (Castric *et al.* 2001). This pattern was contrary to expectations based on the larger spatial scale of among-drainage comparisons. In addition, Mantel tests showed a slight correla-tion between geographic and genetic distances among populations in one drainage — suggesting that observed genetic patterns are shaped at least in part by dispersal — whereas populations in a nearby drainage showed no such relationship (Castric *et al.* 2001). In a different study of brook charr in tributaries to a large lake, an individual-based clustering method (Pritchard *et al.* 2000) clustered individuals into two major groups combined across streams, contrary to expectations based on stream structure alone of four populations (Fraser *et al.* 2004). Finally, studies of brook charr and Lahontan cutthroat trout (*Oncorhynchus clarkii henshawi*) have found significant differentiation over short distances even without any obvious landscape attributes to influence dispersal (Hebert *et al.* 2000; Neville *et al.* in press).

Various factors may be invoked to explain a lack of correspondence between population genetic structure and patterns of connectivity assumed from stream networks. In several cases where genetic structure was observed within continuous habitats (e.g., within continuous reaches of streams), strong fidelity to natal sites or precise homing behavior was proposed to have created structure on a highly localized scale (Hebert *et al.* 2000; Neville *et al.* in press). A common explanation for lack of isolation by distance in both isolated and/or extremely small populations is that random genetic drift may overwhelm any genetic patterns created by dispersal related to distance or habitat structure (e.g., Spruell *et al.* 1999; Waples *et al.* 2001; Neville *et al.* in press). Alternatively, some populations may not have had sufficient time for drift–migration processes to equilibrate since the historical colonization of habitats (e.g., Castric *et al.* 2001) or due to ongoing metapopulation dynamics, discussed below.

Influences of natural and human-constructed movement barriers

Connectivity in stream networks can be interrupted partially or completely by natural barriers (e.g., waterfalls, desiccated stream segments, thermal barriers) and those constructed by humans (e.g., large dams, water diversions, weirs, and culverts). Barriers to movement have important implications for stream-living salmonids, both by reducing the effective size of populations upstream of the barrier and providing little to no opportunity for upstream dispersal by downstream populations. In this case, genetic drift due to reduced effective size (Waples 2004) should lead to detectable differences in allele frequencies between above- and below-barrier populations. In theory, both the size of the founding population and the time since isolation should be important determinants of differentiation and loss of variation via drift (Nei *et al.* 1975). Isolation of populations upstream of barriers may also lead to selection for resident life histories and loss of migratory behavior (Northcote and

Hartman 1988; Näslund 1993), but in some cases fish continue to disperse downstream over barriers (e.g., Hendricks 2002). If population sizes upstream of barriers are large enough such that genetic drift is reduced, or gene flow from the upstream population to the downstream population is large, then loss of connectivity due to movement barriers may not be evident in genetic data. Prior patterns of genetic differentiation may also be an issue. In cases where differentiation existed prior to formation of a barrier, inferences about post-barrier patterns of differentiation are confounded.

Evidence from selected studies of salmonids indicates both natural and human-caused barriers to upstream fish movement are typically associated with significant impacts on genetic diversity as assessed through a variety of analytical methods (Table 13.1). Common genetic responses to isolation by barriers include increased differentiation from other populations, loss of genetic diversity, a lack of isolation by distance, evidence of genetic bottlenecks, reduced effective population sizes, and asymmetrical patterns of gene flow. However, several studies have not revealed anticipated patterns. Populations of coastal cutthroat trout isolated upstream of impassable waterfalls in two southeast Alaska streams showed no evidence of increased differentiation or loss of diversity. Possible explanations for this pattern included insufficient power of genetic markers to detect true differences, a lack of time for differences to emerge upstream of barriers, or greater than expected dispersal over the presumed barriers (Griswold 2002). Another study (Neville *et al.* in press) employed coalescent methods, assignment tests, and tests for bottlenecks to uncover contrasting dispersal patterns and impacts on genetic diversity in Lahontan cutthroat trout above two barriers. As expected, downstream dispersal was greater than upstream dispersal over the first barrier. Fish above this barrier however, showed no evidence of genetic bottlenecks, possibly due to the large and relatively stable habitat in which these fish lived. At the second barrier, coalescent methods estimated slightly greater (but not significantly so) dispersal rates upstream than downstream, a pattern consistent with the colonization of the previously extirpated above-barrier habitat during a recent period of high flow. Assignment tests indicated no current movement into the above-barrier sample site and tests for bottlenecks demonstrated an extremely strong founder effect at this site (Neville *et al.* in press).

We are not aware of studies designed explicitly to distinguish the relative influences of natural versus human-caused barriers, which may

Table 13.1. Summary of selected studies on salmonid fishes examining the effects of barriers on patterns of genetic diversity within and divergence among populations from DNA microsatellites. Five predicted genetic responses to barriers are considered. Above-barrier populations are expected to have increased divergence and reduced genetic diversity compared to other populations. Above-barrier populations are not expected to fit a pattern of isolation by distance; divergence of populations upstream of barriers should be higher than expected based on geographic distance, due to increased drift and reduced dispersal (Hutchison and Templeton 1999). Bottleneck tests are expected frequently to identify bottlenecks upstream of barriers. Coalescent analyses are expected to reveal smaller estimated effective sizes for isolated populations and lower rates of dispersal from downstream populations

Barrier type	Species	Adherence to predicted genetic response?					References
		Increased divergence	Decreased diversity	Impacted isolation by distance	Bottlenecks and/or reduced N_e	Asymmetrical migration	
Erosion dams	S. leucomaenis	Yes	Yes	—	—	—	Yamamoto et al. (2004)
Natural geomorphic[a]	S. confluentus	Yes	Yes	Yes	—	—	Costello et al. (2003)
	O. clarkii	Yes	Yes	Yes	—	—	Taylor et al. (2003)
	O. clarkii	Yes	Mixed[b]	Yes	Mixed[b]	Mixed[c]	Neville Arsenault (2003)
	O. clarkii	No	No	No	—	—	Griswold (2002)
	S. malma	Yes	Yes	Yes	—	—	Griswold (2002)
Culverts and natural geomorphic	O. clarkii	Yes	Yes	Yes	—	—	Wofford et al. (2004)
Water diversion	O. clarkii	Yes	Yes	Yes	Yes	—	Neville Arsenault (2003)
Natural desiccation	S. confluentus	Yes	Yes	Yes	—	—	Bruce Rieman (pers. comm.)

[a]Natural geomorphic barriers include high waterfalls or steep, high-velocity cascades thought to restrict movement severely.

[b]One above-barrier habitat supported a large, seemingly stable population with relatively high genetic variation, while the other population was severely bottlenecked and had an extremely low N_e.

[c]Contrasting patterns of dispersal were found for two above-barrier populations using coalescent methods; see text.

differ based on time since isolation and influences on effective population size (e.g., Wofford *et al.* 2004; Neville *et al.* in press). However, at least one study has examined these factors in terms of population persistence (Morita and Yamamoto 2002); persistence of white-spotted charr (*Salvelinus leucomaenis*) isolated in streams above erosion barriers was a positive function of habitat size, and negatively related to time since isolation. This result implies patterns of genetic diversity should react similarly, and there is evidence to suggest this for white-spotted charr (Yamamoto *et al.* 2004). Lahontan cutthroat trout above man-made barriers displayed similar genetic impacts to those above natural barriers, suggesting that the influence of isolation on these populations was rapid (given the recent time-frame of man-made barriers) and likely compounded by poor habitat (Neville *et al.* in press).

LIFE HISTORY AND CONNECTIVITY

Dispersal in salmonid fishes can be influenced by a variety of life-history characteristics, including sex-specific dispersal and variable migratory strategies (Hendry *et al.* 2004). Fraser *et al.* (2004) used assignment tests and coalescent-based methods to demonstrate that dispersal of brook charr among streams was sex-biased, with males dispersing more among closer habitats but females being more likely to disperse among distant streams. Such a bias may reflect differences in adaptations to each environment, which dictate the fitness trade-offs of dispersal for each sex (Fraser *et al.* 2004). Assignment tests of native brown trout (*Salmo trutta*) revealed differences between resident and migratory fish in terms of their tendency to interbreed with hatchery fish: the resident component of the population had a much greater hatchery influence from interbreeding than the migratory component, likely due to increased selection against migratory hatchery fish (Hansen *et al.* 2000). In Lahontan cutthroat trout, dispersal is believed to be influenced by spatial segregation of variable life histories, with headwater sites having fish which are less migratory and lower confluence sites being more likely to contain fish with greater migration tendencies (Neville *et al.* in press). Several population-based genetic analyses (F_{ST}, coalescent analyses, assignment tests) demonstrated that fish from sites in the lower reaches of tributary streams and in downstream migratory habitats showed little or no genetic differences. In several cases fish from headwater sites were significantly differentiated from those in geographically close (< 4 km) downstream sites within the same stream even with no barriers to gene flow. These patterns suggest behavioral differences between headwater and downstream fish in terms of their tendency to disperse (Neville *et al.* in press).

HISTORICAL FACTORS AND CONNECTIVITY

A common theme among many studies of salmonid fishes, and one that emphasizes the importance of the temporal depth afforded by molecular genetic markers, is that much of the genetic diversity observed today was shaped not by current dispersal, but by historical colonization patterns (Avise 1994). In such cases, historical demography is thought to override signals from current processes such as ongoing dispersal (Hebert *et al.* 2000; Castric *et al.* 2001). In coastal Maine, for example, AMOVA demonstrated that genetic differentiation among populations of brook charr within watersheds was much greater than differentiation among watersheds, and relationships among watersheds did not fit expectations based on geographical proximity (Castric *et al.* 2001). The authors concluded that insufficient time had elapsed since the initial colonization of this area to allow detectable differences among watersheds to accrue, but that contemporary factors modulating dispersal were important in shaping relationships among populations at smaller spatial scales, such as within watersheds. In a large-scale study of bull charr, all populations had low genetic variability, which is likely attributable to post-glacial founding events (Costello *et al.* 2003). In addition, peripheral populations, which were colonized more recently, exhibited lower levels of genetic diversity compared to those central to the species' distribution. Isolation by distance was weaker in these peripheral populations, indicating that historical colonization effects continue to mask the influence of current dispersal in these populations (Costello *et al.* 2003).

Using canonical correspondence analysis (CCA) to relate various environmental factors to genetic variability, Angers *et al.* (1999) found that altitude contributed significantly to population genetic differentiation in brook charr. Altitude is thought to have mediated historical colonization patterns, i.e., early colonizers of high-altitude areas prevented later colonizers from establishing, creating genetic differences between high- and low-altitude regions (Hamilton *et al.* 1998; Angers *et al.* 1999). Similarly, Castric *et al.* (2001) found that within-population genetic diversity declined with altitude and proposed that this may be due both to the increased isolation and magnified "founder effect" (i.e., fewer colonists reaching these habitats) characterizing high-altitude populations. An alternative explanation for these patterns could involve the influence of contemporary physical disturbances on the likelihood of bottlenecks in headwater streams. Headwater streams are often characterized by disturbance regimes with high magnitude impacts on physical habitat (e.g., debris flows and flooding: Miller *et al.* 2003) that lead to extreme reductions in effective sizes or local extirpations (Dunham *et al.* 2003).

METAPOPULATION DYNAMICS AND CONNECTIVITY

Finally, in addition to the above large-scale changes in connectivity over historical time, it is increasingly recognized that connectivity among populations can also change due to ongoing metapopulation dynamics (Hanski 1999; Moilanen and Hanski Chapter 3). If population turnover (local extinction and recolonization) occurs in salmonids (Rieman and Dunham 2000), we would expect the genetic relationships among populations across a landscape to change on a contemporary timescale (e.g., generations, decades). Though metapopulation dynamics has been discussed extensively in the salmonid literature (Rieman and McIntyre 1995; Cooper and Mangel 1999; Dunham and Rieman 1999; Young 1999; Garant *et al.* 2000; McElhany *et al.* 2000; Rieman and Dunham 2000), relatively few studies have investigated the genetic dynamics of populations over time. Of those that have, many have found marked stability in population structure across timescales ranging from years to decades (Nielsen *et al.* 1999; Tessier and Bernatchez 1999; Hansen *et al.* 2002), suggesting that in many systems, dispersal patterns are relatively stable and extinction−colonization events may not be frequent (Hansen *et al.* 2002). However, several recent studies in systems with volatile environmental conditions, where metapopulation dynamics may be more likely, have suggested contrasting dynamics. Temporal instability in allele frequencies in steelhead (*Oncorhynchus mykiss*) and brown trout has been attributed to changes in connectivity and population structure caused by channel-blocking landslides (Heath *et al.* 2002) and extreme spatiotemporal fluctuations in stream flow and habitat conditions (Ostergaard *et al.* 2003). Extremely low effective population sizes and severe bottlenecks observed in Lahontan cutthroat trout suggest that some populations fluctuate in size and may be vulnerable to extirpation. At least one recolonization event was also well characterized by genetic data (see above, Neville *et al.* in press). These observations may not always confirm the presence of a bona fide "metapopulation" (Harrison and Taylor 1997; Hanski 1999), but they point to the importance of spatial and temporal dynamics in habitat and populations (Smedbol *et al.* 2002), and to cases where connectivity may be especially critical for population persistence (Rieman and Dunham 2000; Dunham *et al.* 2003).

CONCLUDING THOUGHTS

Although salmonid fishes are well studied in comparison to other species, molecular markers such as DNA microsatellites continue to reveal new and important insights about connectivity. In recent years, the availability of a host of new statistical methods has greatly improved the rigor of inferences about gene flow from molecular markers, and many studies now employ multiple methods of analysis. Whether the information gained from each approach is conflicting or complementary in part depends on the nature of the assumptions behind each method, and the questions they were designed to address (e.g., historical gene flow versus contemporary movement). Often, contrasting patterns highlighted by alternative genetic approaches reveal unique insights that improve our understanding of connectivity dramatically.

Equivocal views of connectivity from analyses of molecular genetic markers can also result from incomplete consideration of the intrinsic (e.g., life-history variability) and extrinsic (e.g., stream network patterns, isolation) processes that influence genetic variability. In other words, inferences that are supported statistically with analyses of molecular markers must be also grounded in ecological reality (Hedrick 1999). In most cases, insights from molecular markers will narrow the range of possibilities for likely processes that shape connectivity, rather than pointing to a single influence (e.g., Slatkin 1993; Peacock and Ray 2001). Earlier studies of molecular genetic variation in salmonid fishes were mostly observational or exploratory in nature, and did not always provide clear implications for understanding connectivity. The challenge from this foundation is to develop more rigorous evaluations of alternative hypotheses. Within the literature on salmonid fishes, we see an encouraging trend toward studies that are designed in the context of a priori sets of processes hypothesized to influence connectivity and genetic variation. Insights from these attempts to integrate analyses of molecular markers with ecological processes will produce more useful assessments of connectivity.

ACKNOWLEDGEMENTS

We thank D. Isaak and J. Wenburg for helpful reviews of an earlier version of this manuscript, and L. Bernatchez, S. Jenkins, and S. Bayard de Volo for insightful comments on another manuscript used as a foundation for this chapter.

REFERENCES

Angermeier, P. L., K. L. Krueger, and C. A. Dollof. 2002. Discontinuity in stream-fish distributions: implications for assessing and predicting species occurrences. Pp. 519–528 in J. M. Scott, P. J. Heglund, F. Samson, et al. (eds.) *Predicting Species Occurrences: Issues of Accuracy and Scale*. Covelo, CA: Island Press.

Angers, B., P. Magnan, M. Plante, and L. Bernatchez. 1999. Canonical correspondence analysis for estimating spatial and environmental effects on microsatellite gene diversity in brook charr (*Salvelinus fontinalis*). *Molecular Ecology* 8:1043–1053.

Armstrong, J. D., P. E. Shackley, and R. Gardiner. 1994. Redistribution of juvenile salmonid fishes after localized catastrophic depletion. *Journal of Fish Biology* 45:1027–1039.

Avise, J. C. 1994. *Molecular Markers, Natural History and Evolution*. New York: Chapman and Hall.

Balloux, F., and N. Lugon-Moulin. 2001. The estimation of population differentiation with microsatellite markers. *Molecular Ecology* 11:155–165.

Banks, M. A., and W. Eichert. 2000. WHICHRUN v. 3.2: a computer program for population assignment of individuals based on multilocus genotype data. *Journal of Heredity* 91:87–89.

Beerli, P. 2004. MIGRATE 1.7.6.1 documentation and program, part of LAMARC. Available online at http://evolution.genetics.washington.edu/lamarc.html

Beerli, P., and J. Felsenstein. 1999. Maximum-likelihood estimation of migration rates and effective population numbers in two populations using a coalescent approach. *Genetics* 152:763–773.

Beerli, P., and J. Felsenstein. 2001. Maximum likelihood estimation of a migration matrix and effective population sizes in *n* subpopulations by using a coalescent approach. *Proceedings of the National Academy of Sciences of the USA* 98:4563–4568.

Bjornn, T. C., and D. W. Reiser. 1991. Habitat requirements of salmonids in streams. *American Fisheries Society Special Publication* 19:83–138.

Bjornstad, O. N., R. A. Ims, and X. Lambin. 1999. Spatial population dynamics: analyzing patterns and processes of population synchrony. *Trends in Ecology and Evolution* 14:427–432.

Castric, V., and L. Bernatchez. 2004. Individual assignment test reveals differential restriction to dispersal between two salmonids despite no increase of genetic differences with distance. *Molecular Ecology* 13:1299–1312.

Castric, V., F. Bonney, and L. Bernatchez. 2001. Landscape structure and hierarchical genetic diversity in the brook charr, *Salvelinus fontinalis*. *Evolution* 55:1016–1028.

Cavalli-Sforza, L. L., and A. W. F. Edwards. 1967. Phylogenetic analysis: models and estimation procedures. *Evolution* 32:550–570.

Cooper, A. B., and M. Mangel. 1999. The dangers of ignoring metapopulation structure for the conservation of salmonids. *Fisheries Bulletin* **97**:213–226.

Cornuet, J.-M., S. Piry, G. Luikart, A. Estoup, and M. Solignac. 1999. New methods employing multilocus genotypes to select or exclude populations as origins of individuals. *Genetics* **153**:1989–2000.

Costello, A. B., T. E. Down, S. M. Pollard, C. J. Pacas, and E. B. Taylor. 2003. The influence of history and contemporary stream hydrology on the evolution of genetic diversity within species: an examination of microsatellite DNA variation in bull trout, *Salvelinus confluentus* (Pisces: Salmonidae). *Evolution* **57**:328–344.

Davies, N., F. X. Villablanca, and G. K. Roderick. 1999. Determining the source of individuals: multilocus genotyping in nonequilibrium population genetics. *Trends in Ecology and Evolution* **14**:17–21.

Dunham, J. B., and B. E. Rieman. 1999. Metapopulation structure of bull trout: influences of physical, biotic, and geometrical landscape characteristics. *Ecological Applications* **9**:642–655.

Dunham, J. B., G. L. Vinyard, and B. E. Rieman. 1997. Habitat fragmentation and extinction risk of Lahontan cutthroat trout. *North American Journal of Fisheries Management* **17**:1126–1133.

Dunham, J. B., B. E. Rieman, and J. T. Peterson. 2002. Patch-based models of species occurrence: lessons from salmonid fishes in streams. Pp. 327–334 in J. M. Scott, P. J. Heglund, F. Samson, *et al.* (eds.) *Predicting Species Occurrences: Issues of Scale and Accuracy.* Covelo, CA: Island Press.

Dunham, J. B., K. A. Young, R. E. Gresswell, and B. E. Rieman. 2003. Effects of fire on fish populations: landscape perspectives on persistence of native fishes and nonnative fish invasions. *Forest Ecology and Management* **178**:183–196.

Dunning, J. B., B. J. Danielson, and H. R. Pulliam. 1992. Ecological processes that affect populations in complex landscapes. *Oikos* **65**:169–175.

Elliott, J. M. 1981. Some aspects of thermal stress on freshwater teleosts. Pp. 209–245 in A. D. Pickering (ed.) *Stress and Fish.* London: Academic Press.

Epperson, B. K. 2003. *Geographical Genetics.* Princeton, NJ: Princeton University Press.

Fausch, K. D., C. E. Torgersen, C. V. Baxter, and H. W. Li. 2002. Landscapes to riverscapes: bridging the gap between research and conservation of stream fishes. *BioScience* **52**:483–498.

Frankham, R., J. D. Ballou, and D. A. Briscoe. 2004. *A primer of conservation genetics.* Cambridge, UK: Cambridge University Press.

Fraser, D. J., C. Lippe, and L. Bernatchez. 2004. Consequences of unequal population size, asymmetric gene flow and sex-biased dispersal on population structure in brook charr (*Salvelinus fontinalis*). *Molecular Ecology* **13**:67–80.

Frissell, C. A., W. J. Liss, C. E. Warren, and M. D. Hurley. 1986. A hierarchical framework for stream habitat classification: viewing streams in a watershed context. *Environmental Management* **10**:199–214.

Garant, D., J. J. Dodson, and L. Bernatchez. 2000. Ecological determinants and temporal stability of the within-river population structure in Atlantic salmon (*Salmo salar* L.). *Molecular Ecology* **9**:615–628.

Goldstein, D. B., and C. Schlotterer (eds.) 1999. *Microsatellites: Evolution and Applications*. New York: Oxford University Press.

Griswold, K. E. 2002. Genetic diversity in coastal cutthroat trout and Dolly Varden in Prince William Sound, Alaska. Ph.D. dissertation, Corvallis, OR: Oregon State University.

Groot, C., and L. Margolis (eds.) 1991. *Pacific Salmon Life Histories*. Vancouver, British Columbia, Canada: University of British Columbia Press.

Hamilton, K. E., A. Ferguson, J. B. Taggart, and T. Tomasson. 1998. Post-glacial colonization of brown trout, *Salmo trutta* L.: Ldh-5 as a phylogeographic marker locus. *Journal of Fish Biology* 35:651–664.

Hancock J. M. 1999. Microsatellites, and other simple sequences: genomic context and mutational mechanisms. Pp. 1–9 in D. B. Goldstein, and C. Schlotterer (eds.) *Microsatellites: Evolution and Applications*. New York: Oxford University Press.

Hansen, M. M., and K.-L. Mensberg. 1998. Genetic differentiation and relationship between genetic and geographical distance in Danish sea trout (*Salmo trutta* L.) populations. *Heredity* 81:493–504.

Hansen, M. M., D. E. Ruzzante, E. E. Nielsen, and K.-L. Mensberg. 2000. Microsatellite and mitochondrial DNA polymorphism reveals life-history dependent interbreeding between hatchery and wild brown trout (*Salmo trutta* L.). *Molecular Ecology* 9:583–594.

Hansen, M. M., E. Kenchington, and E. E. Nielsen. 2001. Assigning individual fish to populations using microsatellite DNA markers. *Fish and Fisheries* 2:93–112.

Hansen, M. M., D. E. Ruzzante, E. E. Nielsen, D. Bekkevold, and K.-L. Mensberg. 2002. Long-term effective population sizes, temporal stability of genetic composition and potential for adaptation in anadromous brown trout (*Salmo trutta*) populations. *Molecular Ecology* 11:2523–2535.

Hanski, I. 1999. *Metapopulation Ecology*. New York: Oxford University Press.

Harrison, S., and A. Taylor. 1997. Empirical evidence for metapopulation dynamics. Pp. 27–42 in I. A. Hanski, and M. E. Gilpin (eds.) *Metapopulation Biology: Ecology, Genetics, and Evolution*. San Diego, CA: Academic Press.

Heath, D. D., S. Pollard, and C. Herbinger. 2001. Genetic structure and relationships among steelhead trout (*Oncorhynchus mykiss*) populations in British Columbia. *Heredity* 86:618–627.

Heath, D. D., C. Busch, J. Kelly, and D. Atagi. 2002. Temporal change in genetic structure and effective population size in steelhead trout (*Oncorhynchus mykiss*). *Molecular Ecology* 11:197–214.

Hebert, C., R. G. Danzman, M. W. Jones, and L. Bernatchez. 2000. Hydrography and population genetic structure in brook charr (*Salvelinus fontinalis*, Mitchill) from eastern Canada. *Molecular Ecology* 9:971–982.

Hedrick, P. 1999. Perspective: highly variable loci and their interpretation in evolution and conservation. *Evolution* 53:313–318.

Hedrick, P. W. 2001. Conservation genetics: where are we now? *Trends in Ecology and Evolution* 16:629–636.

Hendrick, S. 2002. Seasonal changes in distribution of coastal cutthroat trout in an isolated watershed. M.Sc. thesis. Corvallis, OR: Oregon State University.

Hendry, A. P., and S. C. Stearns (eds.) 2004. *Evolution Illuminated: Salmon and Their Relatives.* Oxford, UK: Oxford University Press.

Hendry, A. P., V. Castric, M. T. Kinnison, and T. P. Quinn. 2004. The evolution of philopatry and dispersal: homing versus straying in salmonids. Pp. 52–91 in A. P. Hendry, and S. C. Stearns (eds.) *Evolution Illuminated: Salmon and Their Relatives.* Oxford, UK: Oxford University Press.

Hutchison, D. W., and A. R. Templeton. 1999. Correlation of pairwise genetic and geographic measures: inferring the relative influence of gene flow and drift on the distribution of genetic variability. *Evolution* 53:1898–1914.

Isaak, D. J., R. F. Thurow, B. E. Rieman, and J. B. Dunham. 2003. Temporal variation in synchrony among chinook salmon (*Oncorhynchus tshawytscha*) redd counts from a wilderness area in central Idaho. *Canadian Journal of Fisheries and Aquatic Sciences* 60:840–848.

Jehle, R., and J. W. Arntzen. 2002. Microsatellite markers in amphibian conservation genetics. *Herpetological Journal* 12:1–9.

Jonsson, B., and N. Jonsson. 1993. Partial migration: niche shift versus sexual maturation in fishes. *Reviews in Fish Biology and Fisheries* 3:348–365.

Knudsen, K. L., C. C. Muhlfeld, G. K. Sage, and R. F. Leary. 2002. Genetic structure of Columbia River redband trout populations in the Kootenai River drainage, Montana, revealed by microsatellite and allozyme loci. *Transactions of the American Fisheries Society* 131:1093–1105.

Knutsen, H., J. A. Knutsen, and P. E. Jorde. 2001. Genetic evidence for mixed origin of recolonized sea trout populations. *Heredity* 87:207–214.

Koizumi, I., and K. Maekawa. 2004. Metapopulation structure of stream-dwelling Dolly Varden charr inferred from patterns of occurrence in the Sorachi River basin, Hokkaido, Japan. *Freshwater Biology* 49:973–981.

Kotliar, N. B., and J. A. Wiens. 1990. Multiple scales of patchiness and patch structure: a hierarchical framework for the study of heterogeneity. *Oikos* 59:253–260.

Latterell, J. J., K. D. Fausch, C. Gowan, and S. C. Riley. 1998. Relationship of trout recruitment to snowmelt runoff flows and adult trout abundance in six Colorado mountain streams. *Rivers* 6:240–250.

Manel, S., M. K. Schwartz, G. Luikart, and P. Taberlet. 2003. Landscape genetics: combining landscape ecology and population genetics. *Trends in Ecology and Evolution* 18:189–197.

McElhany, P., M. H. Ruckelshaus, M. J. Ford, T. C. Wainwright, and E. P. Bjorkstedt. 2000. *Viable Salmonid Populations and the Recovery of Evolutionarily Significant Units.* NOAA Technical Memorandum NMFS-NWFSC-42. Seattle, WA: US. Department of Commerce.

Miller, D., C. Luce, and L. Benda. 2003. Time, space, and episodicity of physical disturbance in streams. *Forest Ecology and Management* 178:121–140.

Millspaugh, J. J., and J. M. Marzluff. 2001. *Radio Tracking and Animal Populations.* New York: Academic Press.

Morita, K., and S. Yamamoto. 2002. Effects of habitat fragmentation by damming on the persistence of stream-dwelling charr populations. *Conservation Biology* 16:1318–1323.

Narum, S. R., C. Contor, A. Talbot, and M. S. Powell. 2004. Genetic divergence of sympatric resident and anadromous forms of *Oncorhynchus mykiss* in the Walla Walla River, USA. *Journal of Fish Biology* **65**:471–488.

Näslund, I. 1993. Migratory behavior of brown trout, *Salmo trutta* L.: implications of genetic and environmental influences. *Ecology of Freshwater Fish* **2**:51–57.

Nei, M. 1973. Analysis of gene diversity in subdivided populations. *Proceedings of the National Academy of Sciences of the USA* **70**:3321–3323.

Nei, M., T. Maruyama, and R. Chakraborty. 1975. The bottleneck effect and genetic variability in populations. *Evolution* **29**:1–10.

Nei, M., F. Tajima, and Y. Tateno. 1983. Accuracy of genetic distances and phylogenetic trees from molecular data. *Journal of Molecular Evolution* **19**:153–170.

Neigel, J. E. 1997. A comparison of alternative strategies for estimating gene flow from genetic markers. *Annual Review of Ecology and Systematics* **28**:105–128.

Neville, H. M., J. B. Dunham, and M. M. Peacock. In press. Landscape attributes and life history variability shape genetic structure of trout populations in a stream network. *Landscape Ecology*.

Nielsen, E. E., M. M. Hansen, and V. Loeschcke. 1999. Genetic variation in time and space: microsatellite analysis of extinct and extant populations of Atlantic salmon. *Evolution* **53**:261–268.

Nielsen, J. L., and D. A. Powers (eds.) 1995. *Evolution and the Aquatic Ecosystem: Defining Unique Units in Population Conservation*. Bethesda, MD: American Fisheries Society.

Northcote, T. G. 1992. Migration and residency in stream salmonids: some ecological considerations and evolutionary consequences. *Nordic Journal of Freshwater Research* **67**:5–17.

Northcote, T. G., and G. F. Hartman. 1988. The biology and significance of stream trout populations (*Salmo* spp.) living above and below waterfalls. *Polskie Archiuvum Hydrobiologii* **35**:409–422.

O'Connell, M., and J. M. Wright. 1997. Microsatellite DNA in fishes. *Reviews in Fish Biology and Fisheries* **7**:331–363.

Ostergaard, S., M. M. Hansen, V. Loeschcke, and E. E. Nielsen. 2003. Long-term temporal changes of genetic composition in brown trout (*Salmo trutta* L.) populations inhabiting an unstable environment. *Molecular Ecology* **12**:3123–3135.

Paetkau, D., W. Calvert, I. Stirling, and C. Strobeck. 1995. Microsatellite analysis of population structure in Canadian polar bears. *Molecular Ecology* **4**:347–354.

Peacock, M. M. 1997. Determining natal dispersal patterns in a population of North American pikas (*Onchotona princeps*) using direct mark–resight and indirect genetic methods. *Behavioral Ecology* **8**:340–350.

Peacock, M. M., and C. Ray. 2001. Dispersal in pikas (*Ochotona princeps*): combining genetic and demographic approaches to reveal spatial and temporal patterns. Pp. 43–56 in J. Clobert, A. Dhondt, E. Danchin, and J. Nichols (eds.) *The Evolution of Dispersal*. Oxford, UK: Oxford University Press.

Peakall, R., M. Ruibal, and D. B. Lindenmayer. 2003. Spatial autocorrelation analysis offers new insights into gene flow in the Australian bush rat, *Rattus fuscipes. Evolution* **57**:1182−1195.

Pritchard, J. K., M. Stephens, and P. Donnelly. 2000. Inference of population structure using multilocus genotype data. *Genetics* **155**:945−959.

Quinn, T. P. 1993. A review of homing and straying of wild and hatchery-produced salmon. *Fisheries Research* **18**:29−44.

Rannala, B., and J. L. Mountain. 1997. Detecting immigration by using multilocus genotypes. *Proceedings of the National Academy of Sciences of the USA* **94**:9197−9201.

Rieman, B. E., and J. B. Dunham. 2000. Metapopulations of salmonids: a synthesis of life history patterns and empirical observations. *Ecology of Freshwater Fishes* **9**:51−64.

Rieman, B. E., and J. D. McIntyre. 1995. Occurrence of bull trout in naturally fragmented habitat patches of varied sizes. *Transactions of the American Fisheries Society* **124**:285−296.

Rieman, B. E., D. C. Lee, and R. F. Thurow. 1997. Distribution, status, and likely future trends of bull trout within the Columbia River and Klamath Basins. *North American Journal of Fisheries Management* **17**:1111−1125.

Rosenberg, D. K., B. R. Noon, and E. C. Meslow. 1997. Biological corridors: form, function, and efficacy. *BioScience* **47**:677−687.

Rousset, F. 2001. Genetic approaches to the estimation of dispersal rates. Pp. 18−28 in J. Clobert, A. Dhondt, E. Danchin, and J. Nichols (eds.) *The Evolution of Dispersal.* Oxford, UK: Oxford University Press.

Ryman N., and F. Utter (eds.) 1987. *Population Genetics, and Fishery Management*, Washington Sea Grant Program. Seattle, WA: University of Washington Press.

Shrimpton, J. M., and D. D. Heath. 2003. Census vs. effective population size in chinook salmon: large- and small-scale environmental perturbation effects. *Molecular Ecology* **12**:2571−2583.

Sjögren-Gulve, P., and C. Ray. 1996. Large-scale forestry extirpates the pool frog: using logistic regression to model metapopulation dynamics. Pp. 111−137 in D. R. McCullough (ed.) *Metapopulations and Wildlife Conservation and Management.* Washington, DC: Island Press.

Slatkin, M. 1985a. Gene flow in natural populations. *Annual Reviews of Ecology and Systematics* **16**:393−430.

Slatkin, M. 1985b. Rare alleles as indicators of gene flow. *Evolution* **39**:53−65.

Slatkin, M. 1987. Gene flow and the geographic structure of natural populations. *Science* **236**:787−792.

Slatkin, M. 1993. Isolation by distance in equilibrium and non-equlibrium populations. *Evolution* **47**:264−279.

Slatkin, M. 1995. A measure of population subdivision based on microsatellite allele frequencies. *Genetics* **139**:457−462.

Smedbol, R. K., A. McPherson, M. M. Hansen, and E. Kenchington. 2002. Myths and moderation in marine "metapopulations"? *Fish and Fisheries* **3**:20−35.

Spruell, P., B. E. Riemen, K. L. Knudsen, F. M. Utter, and F. W. Allendorf. 1999. Genetic population structure within streams: microsatellite analysis of bull trout populations. *Ecology of Freshwater Fish* **8**:114−121.

Spruell, P., A. R. Hemmingsen, P. J. Howell, N. Kanda, and F. W. Allendorf. 2003. Conservation genetics of bull trout: geographic distribution of variation at microsatellite loci. *Conservation Genetics* 4:17–29.

Sunnucks, P. 2000. Efficient genetic markers for population biology. *Trends in Ecology and Evolution* 15:199–203.

Taylor, E. B., M. D. Stamford, and J. S. Baxter. 2003. Population subdivision in westslope cutthroat trout (*Oncorhynchus clarki lewisi*) at the northern periphery of its range: evolutionary inferences and conservation implications. *Molecular Ecology* 12:2609–2622.

Tessier, N., and L. Bernatchez. 1999. Stability of population structure and genetic diversity across generations assessed by microsatellites among sympatric populations of landlocked Atlantic salmon (*Salmo salar* L.). *Molecular Ecology* 8:169–179.

Waples, R. S. 1998. Separating the wheat from the chaff: patterns of genetic differentiation in high gene flow species. *Journal of Heredity* 89:438–450.

Waples, R. S. 2002. Definition and estimation of effective population sizes in the conservation of endangered species. Pp. 147–168 in D. R. McCullough, and S. R. Beissinger (eds.) *Population Viability Analysis*. Chicago, IL: University of Chicago Press.

Waples, R. S. 2004. Salmonid insights into effective population size. Pp. 295–314 in A. P. Hendry, and S. C. Stearns (eds.) *Evolution Illuminated: Salmon and Their Relatives*. Oxford, UK: Oxford University Press.

Waples, R. S., R. G. Gustafson, L. A. Weitkamp. *et al.* 2001. Characterizing diversity in salmon from the Pacific Northwest. *Journal of Fish Biology* 59:1–41.

Waser, P. M., and C. Strobeck. 1998. Genetic signatures of interpopulational dispersal. *Trends in Ecology and Evolution* 13:43–44.

Weir, B. S., and C. C. Cockerham. 1984. Estimating *F*-statistics for the analysis of population structure. *Evolution* 38:1358–1370.

Wenburg, J. K., and P. Bentzen. 2001. Genetic and behavioral evidence for restricted gene flow among coastal cutthroat trout populations. *Transactions of the American Fisheries Society* 130:1049–1069.

Whitlock, M. C., and D. E. McCauley. 1999. Indirect measures of gene flow and migration: $F_{st}/=1/(4N_m+1)$. *Heredity* 83:117–125.

Wiens, J. A. 2002. Riverine landscapes: taking landscape ecology into the water. *Freshwater Biology* 47:501–515.

Wilson, A. J., J. A. Hutchings, and M. M. Ferguson. 2004. Dispersal in a stream dwelling salmonid: inferences from tagging and microsatellite studies. *Conservation Genetics* 5:25–37.

Wofford, J. E. B., R. E. Gresswell, and M. A. Banks. 2004. Influence of barriers to movement on within-watershed genetic variation of coastal cutthroat trout. *Ecological Applications* 15:628–637.

Wright, S. 1931. Evolution in Mendelian populations. *Genetics* 16:97–159.

Wright, S. 1940. Breeding structure of populations in relation to speciation. *American Naturalist* 74:232–248.

Wright, S. 1951. The genetical structure of populations. *Annals of Eugenics* 15:323–354.

Yamamoto, S., K. Morita, S. Kitano. *et al.* 2004. Phylogeography of white-spotted charr (*Salvelinus leucomaenis*) inferred from mitochondrial DNA sequences. *Zoological Science* 21:229–240.

Young, K. A. 1999. Managing the decline of Pacific salmon: metapopulation theory and artificial recolonization as ecological mitigation. *Canadian Journal of Fisheries and Aquatic Sciences* **56**:1700–1706.

Young, S. F., J. G. McLellan, and J. B. Shaklee. 2004. Genetic integrity and microgeographic population structure of westslope cutthroat trout, *Oncorhynchus clarki lewisi*, in the Pend Oreille Basin in Washington. *Environmental Biology of Fishes* **69**:127–142.

Individual-based modeling as a tool for conserving connectivity

JEFF A. TRACEY

INTRODUCTION

Animal movement

Functional (or *behavioral*) connectivity has been defined as "the degree to which the landscape facilitates or impedes movement among resource patches" (Taylor *et al.* 1993; Forman 1997; Taylor *et al.* Chapter. 2). When an animal moves, it must expend energy and it may take risks such as being more visible to predators. So why should an animal move at all? The reason is that the landscape in immediate proximity to an animal may not satisfy its present or anticipated needs. Therefore, we would expect movement to have some purpose for animals; in other words, it is a goal-oriented and behavior-mediated search. Bell (1990) suggests three factors that determine searching behavior: the characteristics and abilities of the animal, the resources and risks in the external environment, and resource requirements as determined by the internal state of the animal. We could think of movement as an activity that allows an animal to match its internal needs to its external environment: if it is threatened it finds safety, if it is hungry it finds food, if it is cold it finds warmth, and if it is ready to reproduce it finds a mate. An animal, however, must weigh all of these needs simultaneously each time it moves based on which needs are most important and what it knows about the landscape. An ordered set of these decisions, which results in a movement path, determines in large part the success of the individual. And the set of movement paths for all individuals in a landscape at a given point in time determines the functional connectivity of that landscape.

Connectivity Conservation eds. Kevin R. Crooks and M. Sanjayan. Published by Cambridge University Press. © Cambridge University Press 2006.

Background

Functional landscape connectivity is directly related to animal movement responses to landscapes. Most approaches for evaluating functional connectivity focus on the landscape, and only implicitly consider animal movement. I believe that animal movement behavior warrants equal and explicit evaluation in studies of functional connectivity. From this perspective, we consider the interaction of individual animals with the landscape; that is, how it is perceived by and affects movement behavior of individuals, and the costs and benefits the local landscape provides to them.

Before proceeding further, I will define some terminology and present some basic concepts related to landscapes and animal movement. First, we are interested in two kinds of spatial variation: discrete and continuous. We conceptualize discrete landscape variation in terms of distinct entities in space, such as an individual animal or a highway. I will refer to these entities as *objects*, but they are sometimes referred to as landscape features or landscape elements. Objects are often represented via vector-based models in which the geometry of objects are described by points, lines, or polygons. For example, we may represent animal locations as points, rivers or roads as lines, and urban development or lakes as polygons. Several important concepts related to the perception of objects by animals have been developed. Movement in relation to objects in a landscape has been called *object orientation* (Jander 1975). Further, Jander (1975) defined *detection space* as "the area around the searcher within which it detects objects." Later, Lima and Zollner (1996) defined *perceptual range* as the distance from which an animal can perceive a particular landscape element. More recently, Olden and others (2004) discuss concepts pertaining to the relation between the strength of environmental stimulus and components of perceptual range. In the absence of spatial learning, we assume that an object must be within an animal's detection space before it will respond to it through movement. These movement responses to objects (or landscape features) may be qualitatively described as attractive, repulsive, or neutral. Animals also respond to continuous landscape variation, often referred to as *fields*, and represented by raster (e.g., grid-based) models. Examples of landscape variables that might be represented with raster models are elevation and temperature. Movement in response to continuous landscape variation, depending on the mechanism of orientation, has been described by *taxis* and *kinesis* models (Jander 1975; Benhamou and Bovet 1992; Turchin 1998).

The effects of human-made (or built) objects on animal movement are the primary focus of movement-based evaluation of landscape connectivity. For some species, human-made objects may be attractive, such as newly sprouted corn fields to feeding sandhill cranes or trash bins to foraging racoons. In some cases human-made structures can provide shelter; for example, rattlesnakes may overwinter in piles of concrete debris left behind following housing construction. But for many species and in many cases, human-made objects are avoided or hazardous. Some objects, such as urban development and its associated night-time illumination, may have no natural analogue. Objects such as freeways can reduce connectivity by restricting movement due to behavioral avoidance as well as increasing mortality due to vehicle collisions when animals do risk a crossing (Clevenger and Wierzchowski Chapter 20).

The effects of a particular type of object on functional connectivity may not necessarily be inferred from habitat utilization studies. For example, suppose an animal encounters an urban development that it wants to avoid. Such avoidance can be achieved in many ways:

- the animal may avoid crossing the boundary into the urban area (and perhaps have to circumvent it),
- it may immediately leave the urban area if it is entered,
- it may decrease time in the urban area by moving more often, moving at a greater rate, or moving in straighter paths, or
- some combination of the above.

Each of these responses can produce low utilization of the urban development. However, the different kinds of avoidance responses can have different effects on landscape connectivity. In the first two responses, the urban development acts as a barrier to movement, and therefore, may reduce connectivity or alter connectivity by redirecting movement to other areas. The third response, due to the rapid movement through urban development, could actually promote connectivity among core habitat areas if the risk of mortality in the urban area is not too high.

Animal movement is poorly understood (Marsh and Jones 1988; Turchin 1998; Van Vuren 1998), and in spite of some promising conceptual and empirical work (Jander 1975; Lima and Zollner 1996; Haddad 1999; Zollner 2000), there is a notable lack of models that are useful for analyzing movement in relation to objects. In general, it is important not to construct models that are more complicated than necessary for a particular application. But if the model leaves out essential features of the system, it will not be useful. Animal movement models (often with much biological realism) are frequently used in simulations employed for conservation purposes, but they are generally based on rules of movement and parameter estimates that seem reasonable (this has been called a *standard of plausibility*: Lima and Zollner 1996), rather than those that are based on analysis of movement data. On the other hand, very simple models of movement (e.g., simple or correlated random walks: Turchin 1998) are not useful by themselves because they do not include the responses to objects and fields that are central to functional connectivity. An important part of my research is to develop animal movement models that are sufficiently biologically realistic and yet simple enough that their parameters can be estimated from empirical data.

In the models I describe, I consider movement occurring in discrete (but potentially small) time intervals which can be conceptualized as an iterative process in which the animal cycles through three basic steps: (1) collecting information on the animal's needs and the landscape, (2) information processing, and (3) using the processed information to make a movement. Notice here that the animal is the system under consideration: information is input to the system (the animal), it is processed (which is a form of computation), and the output is a movement and/or other actions such as feeding. These processes occur within individual animals, and so I take an individual-based approach (Huston *et al.* 1988) to modeling movement.

Preview

In this section I have presented a background on the relation of animal movement to functional connectivity. In the remainder of this chapter, I give a brief introduction to statistical models of animal movement that my colleagues and I have developed, and an overview of a computer program that I have constructed for simulating animal movement on data layers derived in a geographic information system (GIS). We are applying these models to the evaluation of landscape connectivity for large carnivores, and I provide a case study for pumas (*Puma concolor*) in coastal southern California. I conclude with a brief summary and recommendations for future modeling, empirical, and applied work.

GENERAL APPROACH

The approach I describe involves three general areas of research. First, I develop alternative mathematical models of movement in relation to landscape features and develop statistical procedures for parameter estimation and model selection. Second, I construct programs for simulating animal movement on GIS models of landscapes using the models we have developed. Third, I develop the application of the statistical models and simulation programs so that they can be used to quantify and visualize connectivity from simulation output and to evaluate landscape connectivity in support of reserve design and other conservation efforts. Simulations can be run on current or alternative landscapes. Movement can even be simulated on models of potential future landscapes, such as those produced by urban growth models, providing proactive guidance to conservation planning. Therefore, this approach provides a framework for data supported evaluation of connectivity using models that include the essential features of connectivity. In the remainder of this section I will describe each of these areas of research in more detail, and in the next section I will present a case study that illustrates our approach.

Animal movement data

The models I present are designed for data on the movement paths of individual animals such as the type collected via radiotelemetry, global positioning system (GPS) tracking, snow tracking, direct observation, and other methods in which locations are sampled at discrete time intervals. Ideally, these time intervals should be regular. These data consist of spatial coordinates (x, y) and a time at which the location was sampled (Fig. 14.1). From these data we can calculate the move distance and move angle between each pair of consecutive locations. The *turn angle* can also be calculated as the difference between consecutive move angles. These measurements serve as dependent variables in our statistical models.

Data on objects are required to quantify the relation of the animal locations to the landscape. These data are usually in the form of GIS layers (Fig. 14.1). For polygon data (e.g., landcover), we identify the type of polygon that each animal location falls within and the type of the nearest neighboring polygon. This allows us to identify the type of patch boundary to which the animal may be responding. We combine the movement and landscape data to obtain measurements of the angle and distance from each animal location to the nearest point on the boundary of each type of object in the landscape (Fig. 14.1). I refer to these as animal-to-object angles and distances. For continuous landscape variation, such as elevation, we might calculate the slope (which consists of the *gradient*, or steepness, and the *aspect*, or orientation) of the terrain at the animal's location. These measurements are related to the animal's perception of the landscape, and are predictor variables (or covariates) in our statistical models.

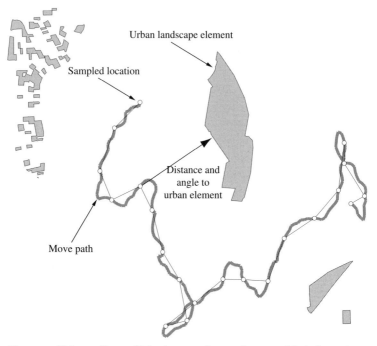

Fig. 14.1. Using radio or GPS telemetry data and geographic information system layers we can quantify animal movement in relation to landscape features (or objects). The gray polygons represent urban landscape elements. The thick dark gray line is the animal's true movement path, and the thin black lines show the approximated movement path formed by joining the sample locations (white-filled circles) with straight line segments. The animal-to-object angle for one of the locations is shown as an arrow. In this case the object is the closest point on the urban boundary to each animal location. This example was produced by simulation.

For each animal location, we also subtract animal-to-object angles and aspects from the move angle to obtain angle of movement relative to each object and field, which I refer to as a *response angle* (Fig. 14.2). For example, if the animal moves directly toward an object, its response angle is 0.0 degrees; alternatively, if it moves directly away from the object its response angle is 180.0 degrees. The response angle is also considered a dependent variable in the statistical models.

Careful consideration should be given to the kinds of movement and landscape data that are analyzed. Large-scale movements might be more important to connectivity than small-scale movements. Ranging,

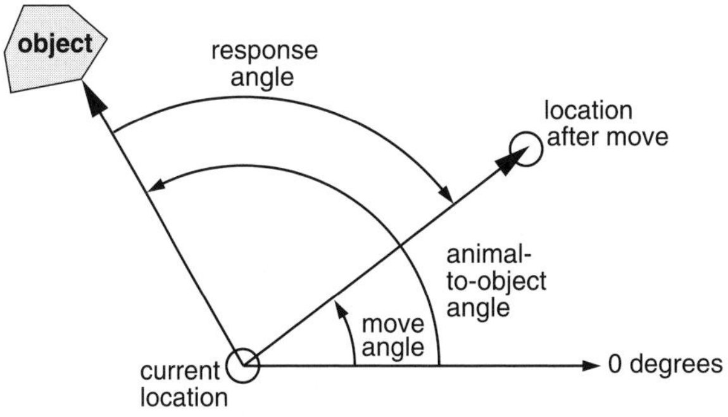

Fig. 14.2. Response angle. We calculate the *animal-to-object angle* from the animal's current location to the nearest point on the object, and the *move angle* from the animal's current location to its location after the move (the next observed location). The *response angle* for a move is the move angle minus the animal-to-object angle constrained to the interval from 0 to 360 degrees (or in radians, 0 to 2π).

topographic orientation (Jander 1975), migration, and dispersal probably contribute more to connectivity among core areas in a reserve network than most other kinds of movements. We may also want to narrow our focus to particular kinds of animals. It may be reasonable to assume that newborn, dispersing, or relocated animals are moving primarily in response to their immediate perceptions rather than spatial memory. Furthermore, not all types of landscape features need to be included. The number of categories in landcover layers might be reduced to a smaller number of meaningful classes based on the particular application and the results of previous habitat utilization studies and analysis of movement data. Some data, such as the distribution of prey, might be important to determine where animals might establish home ranges, but might be less important to understanding movement across a landscape between core areas. One should select data that matches the assumptions of the statistical models used for analysis and that matches the connectivity evaluation objectives.

Statistical models of animal movement
The statistical models of movement I present in this section are a mathematical description of how the dependent variables (e.g., move distance, and move turn, and response angles) are related to the predictor variables

(e.g., animal-to-object distance and angle, slope). As such, they represent information processing and decision-making within individual animals. For this reason, the models are designed to make inferences about individual animals, not a population of animals. Parameters are estimated for each alternative model and the best model is selected separately for each individual. Since the animal movement simulation program I describe (see p. 352) is also individual-based, this is all we need. For some applications, population-level inference may be desirable, and the models I present in this section can be extended to do so, but an explanation is beyond the scope of this chapter. Here, I briefly describe two distinct approaches for modeling animal movement in relation to landscapes: non-linear regression models and finite mixture models. In general, rather than modeling movement on a grid, we allow the animals to draw response angles and distances from continuous distributions.

Our first approach to modeling animal movement in relation to landscape features is a *non-linear regression model* framework (Tracey *et al.* 2005). We first propose a theoretical distribution for both the response angle and the move distance. For the response angle model, we assume that an animal has a fixed mean response angle for movement in relation to a particular type of object. For example, if the animal tends to move away from the object it will always do so "on average." However, we assume that the strength of the response (which depends on a distribution parameter called *concentration* and can be thought of as the inverse of the variance) increases as the animal gets close to the object. In other words, when the animal is far from the object, the response angle distribution is less clustered about the mean response angle, and the animal is more free to choose response angles far from the mean. If the animal is far enough from the object, the response angle will become uniform, and the animal will no longer move in relation to the object at all. But as the animal approaches the object, the distribution becomes more clustered about the mean response angle, and the animal will be more likely to choose a response angle close to the mean. We fit the model by estimating the mean response angle and regressing the *concentration* parameter on the animal-to-object distance using a non-linear function. For the move distance models, we assume that the variance of the theoretical move distance distribution remains constant, but that the mean move length of the distribution may potentially change as the animal approaches the object. For example, if the animal is repulsed by the object then its mean move length may increase, but if the animal is attracted to the object its mean move length may decrease. Again, the mean move length is modeled as a non-linear function of the distance to the object. The assumptions mentioned above may be relaxed by extending the model structure; for example, we can make the mean move distance dependent upon the response angle. We are currently extending these non-linear regression models to account for movement in relation to multiple types of objects and fields in the landscape, and to allow population-level inference.

Our second approach to modeling animal movement in response to landscapes might be called *decision tree models*, but statistically they are called *finite mixture models* (McLachlan and Peel 2000). In these models, an animal is confronted with one or more objects or fields. It also has one or more possible responses to each object or field. For each move, the animal makes a series of decisions that eventually lead to selection of one type of movement response to one of the objects or fields. The probability of making each choice is a non-linear function of the predictor variables such as distance to each object or the slope of a field. For each possible choice, there is a corresponding response angle distribution and move distance distribution with fixed parameters. A movement is made by selecting the response, and then drawing a response angle and move distance from the corresponding distribution.

In both approaches, we must account for cases when the animal does not respond to *any* of the types of objects or fields that we include in the model. When this occurs, we assume that the animal moves according to a *default rule*. The default rule acts as a null model. For example, in the non-linear regression response angle models, the response angle becomes uniform (flat) as the animal-to-object distance becomes large and the animal moves according to a simple random walk where all directions of movement are equally likely. In the finite mixture models, an animal may "decide" not to respond to any of the objects or fields. For the finite mixture models, more types of default rules can be used, including a simple random walk, a correlated random walk, and directional bias. In a correlated random walk, which models the distribution of turn angles described above (p. 347), an animal has a tendency to move in a direction that is related to its previous direction of movement. An animal with directional bias has a higher probability of moving in a fixed compass direction. In both the non-linear regression approach and the finite mixture approach, when certain model parameter estimates are close to 0.0, the entire model reduces to the default rule, which suggests that an individual animal may not be responding to the types of objects or fields we have included in the model.

Above, I have presented two general approaches to modeling animal movement in response to landscape variation. Within each of these approaches, we can propose many specific alternative models for movement that vary in terms of the types of objects and fields included, the functional relation between response and predictor variables, and whether or not to include temporal autocorrelation in movements. When using the finite mixture approach, we can also vary the number of possible responses to each type of field or object, and the type of default rule used. Given sufficient real-world movement data for each individual, parameters can be estimated for each model for each animal using numerical optimization of the corresponding likelihood functions, and then the best alternative model for each animal can be selected using likelihood ratio tests (LRT, for nested models) or Akaike information criterion (AIC, for non-nested models). Then we can use these parameterized, selected models to evaluate connectivity.

Individual-based movement simulations

In order to use these statistical models to evaluate connectivity, we must have computational tools that can extract data from animal locations and GIS data layers and that can simulate animal movement on GIS landscape models. The structure of the individual-based movement simulation program I have developed has four basic parts: (a) a *main* function that can be thought of as the final executable program, (b) a *landscape component*, (c) an *animal component*, and (d) a *simulation control component*. Here I will review these parts of the simulation program.

The landscape component of the simulation program reads GIS data files into spatial data structures that support efficient search and retrieval and provides functions that query the spatial data structures. It can handle both raster-based data for continuous variation and vector-based data for discrete objects. This component is used both in the simulations and in programs that extract covariates that quantify the relation between animal locations and landscape elements from movement data and GIS data layers. The landscape model component performs queries that are used by the individuals in the animal component of the model. Two types of GIS data are managed by the landscape component: landscape data and core habitat data. The landscape data (for landcover, roads, terrain, and so on) represent the landscape variation to which the animals respond through movement. The core habitat specifies where individuals begin their simulated paths and also play a role in the stopping conditions for a movement path (see below). The core habitat data *does not* play a role in how the animal moves across the landscape; rather, we are concerned with predicting connectivity among the core habitat areas.

The animal component keeps track of the state and movement history of each individual and is responsible for simulating movement behavior. Each individual in the simulation corresponds to one of the real individuals in our data set, and its movement behavior is specified by the statistical model that has been selected and parameterized using field data from that individual. For each movement model there is a *perceive* function that calculates animal-to-object distances and angles and the gradient and aspect of the fields using landscape component query functions, and a *move* function that uses this information to generate a move angle and distance according to the individual's statistical movement model. The simulation generates movement paths for each individual by iteratively applying the perceive and move functions.

Interactions between individuals and the landscape are coordinated by the simulation control component. This component also sets the initial state of the simulation, controls the ordering of events within a time step and the number of realizations and time steps that the model will be run, and initiates the writing of output files at the end of the simulation.

Running a movement simulation consists of three basic steps, which are orchestrated by the main function. First, we set up the landscape and individuals. The GIS data layers are read by the landscape component, while the individual component creates a fixed number of each type of individual in each core area (each "type" of individual corresponds to a model that has been selected and parameterized from data from a real animal). Second, we simulate movement by allowing each individual to make one move per time step. An individual's movement path terminates when one of three stopping conditions is met: (a) the simulated animal moves outside the boundary of the landscape, (b) the simulated animal reaches a maximum number of moves (based on a realistic upper bound on dispersal path length), or (c) it successfully reaches another core area. Third, a *path summary file* and a *move summary file* are written. The path summary file contains information on each simulated movement path. Each record in this file includes the individual identification number, initial core, stopping condition, number of moves, total path length, and final core. The move summary file contains information on each location occupied by each individual. Each record in this file includes the individual identification number, path number, move number, landcover type occupied, animal-to-object distances and angles, slope and aspect of fields, move distances and angles, and response angles. Records in the path summary file can be matched to records in the move summary using the identification numbers. We can use this output to evaluate functional connectivity.

Evaluating functional connectivity

We use the movement simulation output to evaluate functional landscape connectivity by comparing alternative landscapes, determining how each alternative landscape meets specified connectivity objectives, and identifying functional corridors (movement routes between core habitat areas). When preparing to evaluate connectivity, we first select the focal species based on conservation status, ecological role, or other factors. We also take data availability into account. We select and parameterize models using movement data from the focal species according to the methods outlined above (pp. 349–352). We must select the alternative landscapes upon which we will run the simulations; for example, we might be interested in how the current landscape compares to landscapes that might result from implementing different reserve network designs or future land use patterns. For each alternative, we develop GIS data layers containing information needed for each movement model used in the simulations. It is useful (if not essential) to develop clear objectives for conserving connectivity for the focal species. Core habitat areas must be identified; for the sake of comparison, the same core habitat areas should be used with all of the alternative landscapes. The role of connectivity between each pair of core areas in species conservation (based on genetics, demographics, and other considerations) should be described and prioritized. We could even use population viability analysis (PVA) models to specify dispersal rates between core areas that will be likely to ensure species persistence. This provides a basis for determining how well each alternative landscape satisfies our conservation objectives. Once we have finished preparing the models and alternative landscapes, we run the movement simulations on each alternative landscape as described above (p. 353).

We use the output produced from the movement simulations to quantify and visualize functional connectivity. We quantify connectivity between pairs of core habitat areas in terms of *success*, *risk*, and *cost* and we may combine these quantities into a single pair-wise connectivity measure. If a simulated animal reaches a core other than the one from which it began, it is considered a *successful* disperser. Otherwise, it is considered *unsuccessful*. We can think of success as the probability that an individual will find its way from one specific core habitat area to another specific core area given that it survives the journey.

The probability of survival is related, in part, to the anthropogenic risks it encounters along its path and how much energy it must expend during dispersal. We quantify risks due to human-caused mortality in terms of numbers of road crossings (perhaps categorized by types of roads) and numbers of encounters with urban areas. We quantify cost as the length of the movement path outside of core habitat areas. All of the information we need to determine these quantities is contained in the path summary files output by the movement simulations. We may combine success, risk, and cost for each simulated movement path into a single probability of dispersal between a pair of core habitat areas:

Pr(successful *and* survives)

= Pr(success *given* survives) Pr(survives risks *and* survives cost)

where the event that an animal successfully moves from one core area to another and survives the journey is a Bernoulli random variable; that is, it takes on a value of either success (1) or failure (0). Here, we assume that mortality from natural causes, other than those related to movement costs, are negligible. For each pair of core areas, we can add up the number of simulated paths between the pair with a path connectivity value of 1, and divide each result by the total number of simulated paths to obtain a final pair-wise connectivity measure.

Methods for visualizing connectivity from simulation output are varied, and the method we should use depends on our objectives. If we want to get a general sense of connectivity across a landscape or visually compare alternative landscapes, we can use connectivity graphs. A graph consists of nodes that are connected by edges (see Fagan and Calabrese Chapter 12; Theobald Chapter 17). In our case, the nodes correspond to core habitat areas, and might be visually represented as circles or polygons depicting the shape of the core area. The edges correspond to connectivity. Connectivity is directional (Gustafson and Gardner 1996); therefore, these edges can be depicted by arrows (one in each direction) between each pair of core areas. We can visually depict the degree of connectivity between pairs of core areas by plotting numerical values (for success, risk, cost, or pair-wise connectivity measure) next to the corresponding edges, or by varying the widths of the arrows according to these values. In order to visualize movement routes through *functional corridors* between a pair of core areas, we can plot the movement paths between them using the information in the move summary files. However, not all of the movements in a successful movement path, such as moving into and back out of cul-de-sacs, contribute equally to connectivity and may detract from identifying functional corridors. Therefore, an important area of future work will be to develop techniques to identify functional corridors from simulation output. As a final example, if we are interested in areas of high risk, we can plot locations where simulated movement paths intersect roads and urban areas (see Clevenger and Wierzchowski (Chapter 20) for an example for roads).

We can directly compare connectivity between pairs of core areas within or among alternative landscapes and how well an alternative landscape satisfies our connectivity conservation objectives using the quantities described above. Further, we can gain an overall sense of patterns of connectivity at broad scales, where land ought to be protected (as predicted from the simulations) in order to preserve connectivity at finer scales, and how differences in the alternative landscapes alter predicted patterns of movement between core habitat areas by visualizing connectivity in different ways. We can use these results as a basis for selecting alternative reserve designs, identifying weakness in existing or future landscapes, and for identifying and prioritizing areas for protection to help ensure the conservation of connectivity.

CARNIVORE MOVEMENT IN SOUTHERN CALIFORNIA

Background

In areas with increasing urbanization, the loss and fragmentation of habitat is virtually inevitable (Soulé 1991; Beier et al. Chapter 22). In coastal southern California, intensive development over the past century has fragmented the landscape and has helped create a "hotspot" of endangerment and extinction in the region (Myers 1990; Dobson et al. 1997). Pumas (*Puma concolor*) are the largest predator remaining in the region and are particularly sensitive to habitat fragmentation (Beier 1993; Crooks 2000, 2002). Pumas are an "area-dependent" species, requiring large areas for home ranges and dispersing long distances. For example, female pumas in the Santa Ana Mountains of southern California occupied minimum convex polygon home ranges that average an estimated 218 km^2 (Beier and Barrett 1993), and juveniles had an average estimated dispersal distance of 63 km (Beier 1995). Because fragmentation of the natural landscape of coastal southern California is continuing at a rapid rate, large-scale assessments of regional connectivity are critical. I conducted a pilot study, which used the methods outlined above, to evaluate landscape connectivity for pumas in relation to an existing landscape, a habitat conservation plan landscape, and a worst-case land-scape in the Southern California Ecoregion (Tracey and Crooks 2004). Here I use results for the existing and worst-case landscapes to illustrate the approach described above.

Alternative landscapes

For the movement simulations, I constructed GIS landcover layers for an existing landscape and a worst-case landscape scenario (Fig. 14.3). The existing landscape layer serves as a baseline for comparison. This layer was constructed from Southern California Association of Governments (SCAG) and San Diego Association of Governments (SanDAG) land-use layers, which were created in approximately 1995. Landcover in the existing landscape layer (and all other alternative landscape layers) was categorized into four types: habitat, disturbed, urban, and water. So-called "vacant" areas, undeveloped local and regional parks, and open space preserves were classified as habitat landcover. Housing, commercial, industrial, developed military, and other such areas were classified as urban landcover. Water consists of lakes, reservoirs, and the Pacific Ocean. All other land-use types were classified as disturbed landcover, and include such areas as roads, rural residential areas, local developed parks, and agricultural lands. I constructed the worst-case landscape layer by regarding all public land as "protected" and assigning it to the habitat landcover type (Fig. 14.3). I used polygons for bodies of water from the existing landcover layer, and all private lands within the Southern California Ecoregion were assumed to be completely converted to the urban landcover type.

I constructed a GIS layer for puma core habitat areas from a protected lands data layer and a mountain lion wildlife habitat relation data layer. Protected lands layers were obtained from the California Spatial Information Library and The Nature Conservancy. I assumed all government land was adequately protected, although this is probably overly inclusive. I selected the two highest-quality habitat categories from layers for mountain lion wildlife habitat suitability (Hunter *et al.* 2003). The protected lands and suitable habitat layers were intersected to produce a layer of potential mountain lion core areas. From this layer, I selected polygons larger than 90 km^2, which is about 20 km^2 less than the smallest female minimum convex polygon home range reported by Beier and Barrett (1993). The result was a layer of 12 core habitat polygons (Table 14.1, Fig. 14.4).

(A)

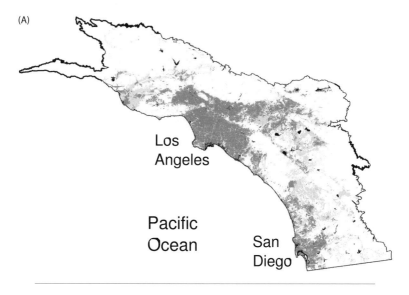

Los
Angeles

Pacific
Ocean

San
Diego

(B)

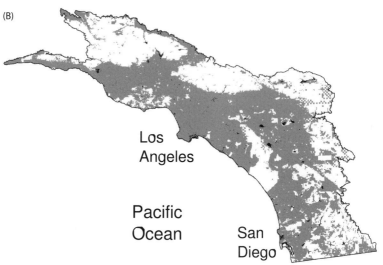

Los
Angeles

Pacific
Ocean

San
Diego

Fig. 14.3. A GIS representation of the existing (A) and worst-case (B) land-cover scenarios used in the simulations. White areas indicate habitat, light gray indicates disturbed, dark gray indicates urban, and black indicates water. The black boundary delineates the Southern California Ecoregion.

Table 14.1. *A list of core areas used in the simulations*

Number	Name	Area (km²)
1	Los Padres NF 1	1115
2	Los Padres NF 2	102
3	San Bernadino NF	213
4	Santa Ana Mountains	508
5	Palomar Mountains	583
6	Los Coyotes	301
7	Cleveland NF	1343
8	Miramar	98
9	Otay Mountain	122
10	Angeles NF	1791
11	Santa Monica Mountains 1	116
12	Santa Monica Mountains 2	273

Data

For the pilot study, I estimated model parameters using radiotelemetry data from two subadult dispersing male pumas and landcover data. The movement data were collected by Paul Beier and his colleagues in the Santa Ana Mountains of coastal southern California (Beier 1995) between October 1990 and September 1992 (animals M8 and M10). In the analysis, I used data from 143 movements made by M8 and 260 movements made by M10. These data were collected during 12-h or 24-h monitoring sessions during which animals were located every 15 min, and the locations were rounded to the nearest 100 meters in a Universal Transverse Mercator (UTM) coordinate system. I used landscape data for the existing landscape to analyze movement in response to landcover polygons. Therefore, we had to test models for six boundary types: habitat–disturbed, habitat–urban, disturbed–habitat, disturbed–urban, urban–habitat, and urban–disturbed. I used a computer program that I constructed, which utilized the landscape component of the simulation program (p. 352), to derive the response (dependent) variables and covariates (independent variables) from the movement and landscape data. The animal-to-object angles and distances were the points on the landcover polygon boundary closest to the animal locations (Fig. 14.1). I assumed that the animal responded to the nearest boundary type.

I proposed a total of 30 alternative finite mixture models for movement in each landcover type for response to each boundary type. Each model varies the kind of default model (simple random walk, correlated random

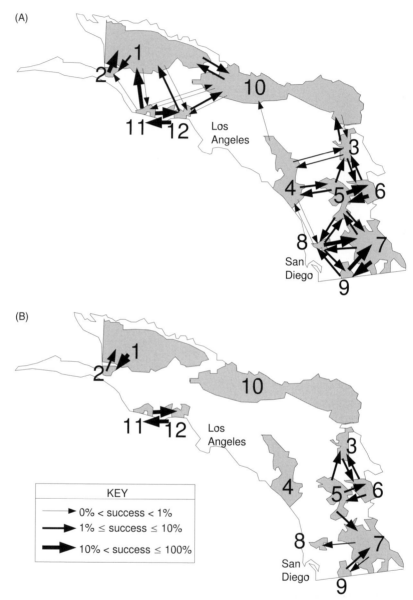

Fig. 14.4. Connectivity of existing (A) and worst-case (B) southern California landscapes. Connectivity, as measured by successful simulated dispersal among pairs of puma core areas (in gray), is depicted by arrows. The width of the arrow indicates the category derived from the number of successful simulated paths. The absence of an arrow between a pair of core areas indicates that no successful simulated dispersal occurred between them.

walk, and directional bias), whether there is no response, a single response, or two responses to a boundary type, and whether the probability of response was a constant, exponential, or logistic function of the distance to the boundary. In the single-response models, the animal has a tendency to move in one mean angle in relation to the boundary type (e.g., toward or away). In the two-response models the animal can move in two mean directions in relation to the boundary type, with each mean direction selected with some probability; for example, a rule might be to move parallel to the boundary to the left with probability Pr(left) and to the right with probability Pr(right). I also included two variations of the two-response models, each with the two responses being either symmetric or asymmetric about a 0.0 degree response angle. I estimated the parameters for each model for each puma using programs that I have written in the R statistical language (R Development Core Team 2004), and then selected the best model for each animal using AIC or LRTs.

The analysis of the data using the models indicated that the two pumas generally moved according to a correlated random walk when they were in habitat. When they were in habitat near the urban edge, they tended to move parallel to it (either left or right), probably in an attempt to circumvent an urban area. They did not seem to avoid the habitat—disturbed boundary, but if they entered a disturbed area they tended to move back toward the habitat. Parameters were estimated and models were selected for two pumas, but since the results were very similar, simulations for only one puma (M10) are presented below. Recall that the statistical models I described are designed to make individual-level inferences; hence, the number of individuals for which we have data is not relevant to making such inferences. When evaluating connectivity, however, we would like to have data from many animals so that we have a sample of movement behaviors representative of the animal population in our region. Our ongoing modeling efforts are incorporating data from more radio-collared animals in the region, including animals fitted with GPS collars.

Simulation results

We started with 250 individuals in each of the 12 core areas, for a total of 3000 simulated individuals. We allowed them to make a maximum of 7200 moves, which (at 15-min time steps) translates to 75 days of continuous movement. However, if a simulated animal left the boundary of the Southern California Ecoregion or successfully reached another core area, that individual's dispersal was terminated.

In the existing landscape, 1471 individuals (49%) left the Southern California Ecoregion, 792 individuals (26.4%) reached the maximum number of moves, and 737 individuals (24.6%) successfully reached another core area. Of the 132 possible linkages between cores ((number of cores)^2 − number of cores), 35 were realized in the simulations on the existing landscape. Of these, nine of the linkages had ≤2 successful paths, seventeen had from 3 to 24 successful paths, and nine of the linkages had ≥25 successful paths. In the worst-case landscape, 1238 individuals (41.3%) left the Southern California Ecoregion, 1518 individuals (50.6%) reached the maximum number of moves, and 244 individuals (8.1%) successfully reached another core area. Of the 132 possible linkages, twelve were realized in the simulations on the worst-case landscape. Of these, two of the linkages had ≤2 successful paths, eight had from 3 to 24 successful paths, and two of the linkages had ≥25 successful paths. A landscape connectivity graph showing the number of successful simulated movement paths between core areas in each landscape is shown in Fig. 14.4. By comparing the connectivity graphs for the existing and worst-case landscapes, we can see that the model predicts a near-complete breakdown in connectivity should the worst-case landscape be realized in coastal southern California.

Objectives for connectivity conservation

It is constructive to interpret the simulation results in light of a clear set of objectives for conserving connectivity. Below, I suggest five connectivity objectives for the Southern California Ecoregion and use the model output to identify vulnerabilities and to provide guidance for improving and conserving connectivity. The connections associated with each objective

Table 14.2. *A list of connections related to each connectivity strategy goal*

Goal number[a]	Linkages[b]
1	1⇔10, 3⇔5, 3⇔6, 3⇔10, 5⇔7, 6⇔7
3	1⇒11, 1⇒12, 3⇒4, 5⇒4, 10⇒11, 10⇒12
4	1⇔2, 5⇔6, 7⇔9, 11⇔12
5	4⇒3, 4⇒5, 4⇒10, 5⇔8, 7⇔8, 11⇒1, 11⇒10, 12⇒1, 12⇒10

[a]Each goal is described in detail in the text. Goal 2 is omitted from this table because it depends on evaluation of connectivity in neighboring regions.
[b]A double arrow (⇔) indicates connectivity in both directions is related to the goal. A single arrow (⇒) indicates connectivity only in one direction is related to the goal. The numbers connected by arrows are the core area numbers in Table 14.1.

are listed in Table 14.2. Connections between a pair of core habitat areas can contribute to more than one objective.

Objective 1: Conserve connectivity among large core areas

Description If we are to ensure that mountain lions have a continued presence in coastal southern California, then it will be essential to maintain connectivity among the larger core areas in the eastern parts of the Southern California Ecoregion. This includes connectivity between pairs of large habitat core areas (cores 1, 3, 5, 6, 7, 10) in both directions.

Results In the existing landscape, the simulations predict low to medium connectivity among large cores. Several critical weaknesses exist, most notably movement in the southern direction from core 10 to core 3 and from core 3 to core 5. In the worst-case landscape, connections among the large core areas in the northern part of the study area were predicted to be lost, essentially fragmenting the landscape at a regional scale.

Objective 2: Connectivity throughout California

Description By maintaining connectivity between large core habitat areas in the Southern California Ecoregion and large core habitat areas in adjacent ecoregions, we can help conserve connectivity at larger spatial scales. Important benefits of this connectivity are demographic and genetic exchange among mountain lion populations in neighboring ecoregions, and preserving opportunities for range shifts for mountain lions and other species in the face of regional or global climate change. Achieving this goal is also dependent upon successfully conserving connectivity among large core habitat areas.

Results This goal can not be addressed directly without conducting connectivity evaluations in neighboring regions. Using information from the move summary files, we can identify areas on the boundary of the Southern California Ecoregion that were encountered by simulated animals. We also note that more individuals left the confines of the Ecoregion in the existing landscape scenario compared to the worst-case scenario, suggesting that exchange of individuals with adjacent ecoregions would be reduced in the worst-case landscape.

Objective 3: Connectivity for important coastal core areas

Description Several coastal core areas, most notably the Santa Ana Mountains (core 4) and the Santa Monica Mountains (cores 11 and 12), can still support mountain lions. However, for these populations to persist,

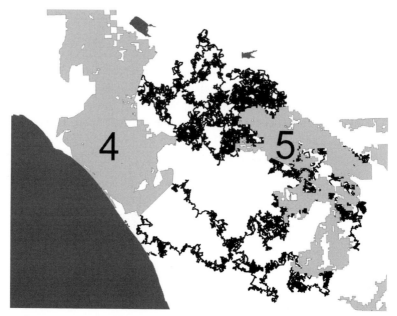

Fig. 14.5. Simulated locations for lowest-cost successful dispersal locations from the Palomar Mountains (core 5) to the Santa Ana Mountains (core 4). The core areas are shown in light gray, water in dark gray, and the simulated lowest-cost successful dispersal paths are shown in black.

connectivity from larger eastern cores *into* the coastal cores must be maintained. On the other hand, the coastal core habitat areas, due to the relatively small puma populations they contain, probably will make little contribution to the long-term viability of the larger core areas (but see Objective 5), so successful dispersal from these areas to the larger eastern cores is less essential.

Results In the existing landscape scenario, connectivity from larger inland cores to the coastal core areas is low. Therefore, it is critical to protect such connectivity immediately. In Fig. 14.5, we show predicted low-cost movement paths from the Palomar Mountains (core 5, an inland core) to the Santa Ana Mountains (core 4, a coastal core). Such predictions can help identify areas that should be immediately protected to conserve connectivity. In the worst-case landscape scenario, all connectivity from inland to coastal core areas is predicted to be lost.

Objective 4: Connectivity for nearby small cores
Description By connecting small nearby core habitat areas to other core habitat areas (large or small) we can increase the effective area of both core

areas. Some smaller core habitat areas, such as the two protected regions of the Santa Monica Mountains (cores 11 and 12), will require connectivity with each other in order to form a single larger core habitat area. Some small core habitat areas can add to the effective area of larger, nearby core habitat areas. For example, the Chino Hills (not listed as a core) are in very close proximity to the Santa Ana Mountains (core 4), and can make additional habitat available to pumas in the Santa Ana Mountains (this particular corridor is discussed further in Beier *et al.* Chapter 22). Other small core areas are strategically important because they help achieve other connectivity objectives; for example, Otay Mountain (core 9) is important in achieving Objectives 1 and 2.

Results In the existing landscape, many of the small core habitat areas show high connectivity to nearby large core habitat areas, and this connectivity changes little in the worst-case landscape.

Objective 5:   Create redundant connections
Description In some cases, connectivity between pairs of core habitat areas may create redundancy in linkages among other core areas. The coastal core habitat areas, such as Santa Ana Mountains, Santa Monica Mountains, and Otay Mountain, can play a role in maintaining alternate, albeit less viable, connections among the large cores.

Results Most of the redundant connections occur through coastal areas (except for connections from core 5 to 8 and from core 7 to 8) so the results for Goal 3 also pertain to this objective.

CONCLUSION

Modeling is by nature an iterative process in which models are proposed, evaluated, and then revised or replaced. Therefore, the existing models need to be applied to data for a range of species and landscapes and then evaluated. We need to develop variations and extensions of existing models, propose new types of models and adapt existing model frameworks (e.g., neural networks) to the study of animal movement. Another important area of work is to develop better methods for visualizing connectivity from the large volumes of output produced by the movement simulation programs.

In field studies, careful consideration must be given to the scale (both grain and extent) of landscape change with which we are concerned. The scale of landscape change and the scale of movement behavior of the focal animals will dictate the frequency at which we should sample movements. Additionally, there must be correspondence between the grain of the landscape data and the frequency of sampling locations. In general, most tracking studies relocate animals too infrequently to give useful information about local movement responses to landscape variation. With new GPS tracking collars, however, it is practical to obtain this information for larger species. Another advantage of GPS methods is that we often obtain many more animal locations, which can permit us to use more detailed movement models. For other species, field sessions involving more frequent relocations can be conducted using traditional radiotelemetry techniques (similar to Beier *et al.* 1995). It is also important to try to collect movement data that occur in a variety of landscape contexts, and over a range of the covariates such as distances to objects. Otherwise, we may not be able to fit models that include responses to all of the landscape variation we judge important to our specific application. Finally, it may be helpful to collect behavioral information associated with movements or segments of individual movement pathways.

Conservation practitioners should prepare to use this approach in several ways. First, important focal species for connectivity studies should be identified. Second, GIS data for the region of interest, including elevation, roads, vegetation and landcover, streams or riparian zones, bridges and road undercrossings, and other relevant information should be acquired, evaluated for accuracy, and prepared for use in movement simulation models. Third, information on projected land-use trends and conservation plans should be acquired and studied. In some cases, GIS layers for projected land use can be obtained from local governments, resource agencies, or universities. This information should be used to create GIS layers for possible future alternative landscapes. It is important that the alternative future landscapes are similar to the GIS data used to analyze the movement data (for example, with respect to spatial resolution, landcover categories, and ranges of continuous spatial variation). Finally, objectives for conserving connectivity for each focal species should be developed to guide interpretation of the simulation results. The use of data-supported, individual-based movement models has the potential to be a useful addition to our conservation toolkit.

ACKNOWLEDGEMENTS

I dedicate this chapter to my wife, Rosana, and my children Andrew and Marissa. I thank them for their love and support. I am grateful to my advisors which include Kevin Crooks, Jun Zhu, and Ted Case and to collaborators that include Paul Beier, Robert Fisher, Lisa Lyren, Seth Riley, Ray Sauvajot, and Walter Boyce. I thank Scott Morrison of The Nature Conservancy, and Colleen Miller and Dave Lawhead of the California Department of Fish and Game. This project was funded by the US Geological Survey, the California Department of Fish and Game, The Nature Conservancy, the University of Wisconsin–Madison, and Colorado State University. This research is based upon work partially supported by the National Science Foundation under IGERT Grant No. DGE-0221595 through the Program for Interdisciplinary Mathematics, Ecology, and Statistics (PRIMES) at Colorado State University.

REFERENCES

Beier, P. 1993. Determining minimum habitat areas and habitat corridors for cougars. *Conservation Biology* 7:94–108.

Beier, P. 1995. Dispersal of juvenile cougars in fragmented habitat. *Journal of Wildlife Management* 59:228–237.

Beier P., and R. H. Barrett. 1993. *The Cougar in the Santa Ana Mountain Range, California*, Final Report. Sacramento, CA: California Department of Fish and Game.

Beier, P., D. Choate, and R. H. Barrett. 1995. Movement patterns of mountain lions during different behaviors. *Journal of Mammalogy* 76:1056–1070.

Bell, W. J. 1990. *Searching Behavior*. New York: Chapman and Hall.

Benhamou, S., and P. Bovet. 1992. Distinguishing between elementary orientation mechanisms by means of path analysis. *Animal Behavior* 43:371–377.

Crooks, K. R. 2000. Mammalian carnivores as target species for conservation in southern California. Pp. 105–112 in J. E. Keeley, M. Baer-Keeley, and C. J. Fotheringham (eds.) *Second Interface between Ecology and Land Development in California*, Open-File Report 00-62. Sacramento, CA: US Geological Survey.

Crooks, K. R. 2002. Relative sensitivities of mammalian carnivores to habitat fragmentation. *Conservation Biology* 16:488–502.

Dobson, A. P., J. P. Rodriguez, W. M. Roberts, and D. S. Wilcove. 1997. Geographic distribution of endangered species in the United States. *Science* 275:550–553.

Forman, R. T. T. 1997. *Land Mosaics*. New York: Cambridge University Press.

Gustafson, E. J., and R. H. Gardner. 1996. The effect of landscape heterogeneity on the probability of patch colonization. *Ecology* 77:94–107.

Haddad, N. M. 1999. Corridor use predicted from behaviors at habitat boundaries. *American Naturalist* 153:215–227.

Hunter, R., R. Fisher, and K. Crooks. 2003. Landscape-level connectivity in coastal southern California as assessed by carnivore habitat suitability. *Natural Areas Journal* 23:302–314.

Huston, M., D. DeAngelis, and W. Post. 1988. New computer models unify ecological theory. *BioScience* 38:682–691.

Jander, R. 1975. Ecological aspects of spatial orientation. *Annual Reviews of Ecology and Systematics* **6**:171–188.

Lima, S. L., and P. A. Zollner. 1996. Towards a behavioral ecology of ecological landscapes. *Trends in Ecology and Evolution* **11**:131–135.

Marsh, L. M., and R. E. Jones. 1988. The form and consequences of random walk models. *Journal of Theoretical Biology* **133**:113–131.

McLachlan, G., and D. Peel. 2000. *Finite Mixture Model*. New York: John Wiley.

Myers, N. 1990. The biodiversity challenge: expanded hot-spots analysis. *Environmentalist* **10**:243–256.

Olden, J. D., R. L. Schooley, J. B. Monroe, and N. L. Poff. 2004. Context-dependent perceptual ranges and their relevance to animal movements in landscapes. *Journal of Animal Ecology* **73**:1190–1194.

R Development Core Team. 2004. *R: A Language and Environment for Statistical Computing*. Vienna, Austria: R Foundation for Statistical Computing. Available online at http://www.r-project.org/

Soulé, M. E. 1991. Land use planning and wildlife maintenance: guidelines for conserving wildlife in an urban landscape. *Journal of the American Planning Association* **57**:313–323.

Taylor, P. D., L. Fahrig, and G. Merriam. 1993. Connectivity is a vital element of landscape structure. *Oikos* **68**:571–573.

Tracey, J. A., and K. R. Crooks. 2004. *Evaluating Landscape Connectivity in Coastal Southern California using Individual-Based Movement Models*, Final Report. San Diego, CA: The Nature Conservancy and the California Department of Fish and Game.

Tracey, J. A., J. Zhu, and K. R. Crooks. 2005. A set of nonlinear regression models for animal movement in response to a single landscape feature. *Journal of Agricultural, Biological, and Environmental Statistics* **10**:1–24.

Turchin, P. 1998. *Quantitative Analysis of Movement*. Sunderland, MA: Sinauer Associates.

Van Vuren, D. 1998. Mammalian dispersal and reserve design. Pp. 369–393 in T. Caro (ed.) *Behavioral Ecology and Conservation Biology*. New York: Oxford University Press.

Zollner, P. A. 2000. Comparing the landscape level perceptual abilities of forest sciurids in fragmented agricultural landscapes. *Landscape Ecology* **15**:523–533.

15

Linking connectivity to viability: insights from spatially explicit population models of large carnivores

CARLOS CARROLL

INTRODUCTION

Increasingly, conservation groups and agencies attempt to create regional reserve designs that move beyond a simple aggregation of important sites to form a biologically functional network. As natural habitats are converted for human uses, remaining natural areas simultaneously become smaller and more isolated, reflecting the twin processes of habitat reduction and fragmentation (Wilcove *et al.* 1986). Maintaining connectivity between these remnant natural habitat patches is important for several reasons (Crooks and Sanjayan Chapter 1); on a timescale of generations, a single reserve or patch of natural habitat is unlikely to be large enough to sustain populations of area-sensitive species that are subject to the processes of demographic and environmental stochasticity (Harrison 1994); on a timescale of tens of generations, a single reserve may not be large enough to sustain a population's genetic diversity and maintain evolutionary processes, or allow the species to shift its range in response to long-term environmental trends such as climate change (Frankel and Soulé 1981).

Early literature on assessing landscape connectivity (e.g., Forman and Godron 1986) focused primarily on classifying landscape structure rather than relating this structure to population dynamics of particular species (Hanski 1994; Tischendorf and Fahrig 2000; Moilanen and Hanski 2001). In contrast, functional connectivity, at the scale discussed

Connectivity Conservation eds. Kevin R. Crooks and M. Sanjayan. Published by Cambridge University Press. © Cambridge University Press 2006.

in this chapter, is a population-level process that implies that individuals of a species successfully disperse between connected patches and survive to breed in the destination patch. Functional connectivity depends not only on the permeability of the linkage habitat, but also upon conditions in the source and destination patches, such as the production of sufficient potential dispersers. Therefore functional connectivity is best addressed in a whole-landscape context by examining the roles of all landscape elements in promoting or hindering effective dispersal. Functional connectivity, as used here, may be further divided into potential or actual connectivity based on whether the metric is based on model results or field data (Fagan and Calabrese Chapter 12). The models described in this chapter evaluate potential connectivity, but ideally they can inform and be informed by data on actual connectivity. For example, the newly emerging field of landscape genetics provides means to test the hypotheses of connectivity models with genetic data on regional population structure and the likelihood of rare long-distance dispersal events (Proctor *et al.* 2002; Manel *et al.* 2003; Frankham Chapter 4; Neville *et al.* Chapter 13).

Large mammalian carnivores, such as the grizzly bear (*Ursus arctos*) and wolf (*Canis lupus*), are often proposed as focal species for evaluating landscape connectivity, especially in areas such as western North America where large areas remain suitable for species that avoid humans (Soulé and Terborgh 1999). Regional-scale connectivity analyses have also been performed for other taxa, including ungulates in Europe (Bruinderink *et al.* 2003). Typically, the carnivore species have large area requirements, with a population of 500 individuals encompassing tens of thousands of square kilometers (Noss *et al.* 1996), which is larger than the size of most protected areas. Human-associated mortality is an important limiting factor for the large carnivores, and their dispersal through a landscape is often limited or blocked by areas of development or high human access (Thiel 1985), due to behavioral avoidance of developed habitat or excessive mortality in those areas (Paquet and Carbyn 2003). For example, roads may be a good predictor of wolf habitat suitability not because they are physical barriers to dispersal but because they alienate habitat by increasing human access and hence wolf mortality (Mladenoff *et al.* 1995). If we assume that these landscape change processes that fragment large carnivore habitat will eventually affect a broader suite of less-sensitive species, large carnivores may be useful indicator species (Lambeck 1997) for landscape connectivity. In order to increase the generality of conservation guidelines while retaining the link to species biology, it is often useful to compare connectivity needs of several carnivore species that differ in their area requirements, dispersal ability, and habitat associations. Carnivores may also function as keystone species (Power *et al.* 1996) in some ecosystems, so their continued presence at ecologically effective densities may be important for maintaining ecosystem processes (Soulé *et al.* 2003).

Conservation biologists have long debated whether resources devoted to corridors might be better spent on other goals (e.g., Noss 1987; Simberloff and Cox 1987; Crooks and Sanjayan Chapter 1). Because assessing the trade-off between connectivity and other design goals is difficult without long-term field data on dispersal and demographics, this may be where mechanistic models are most useful. Spatially explicit population models can help planners decide when to allocate resources to protect relatively secure core areas, to stem the degradation of threatened buffer zones, or to restore linkages that are already degraded but might contribute to long-term persistence of metapopulations. This chapter reviews examples of the use of spatially explicit population models to evaluate connectivity for large carnivores in the Rocky Mountains of the USA and Canada and elsewhere in North America. I examine what distinguishes priorities derived from such complex models from those suggested by simpler connectivity metrics, and whether model results are robust enough to our uncertainty about biological processes such as dispersal to provide reliable insights for conservation planners.

WHY USE SPATIALLY EXPLICIT POPULATION MODELS TO EVALUATE CONNECTIVITY?

Spatially explicit population models (SEPMs), (Table 15.1) can be broadly defined as models that represent population processes in combination with the spatial location of individuals and landscape features (Dunning *et al.* 1995; South *et al.* 2002). SEPMs can be divided into those that map an individual's spatial location onto a lattice of cells and those in which a population's or individual's location is independent of any grid structure. The former type of lattice-based models can then be divided by their resolution, such that each grid cell may hold a population, an individual's home range, or be only a portion of a single home range (South *et al.* 2002). Among the latter type of non-lattice-based models are so-called "pseudo-spatial" models that follow aggregate populations that inhabit patches that have a location and distance from other patches, but no internal landscape structure or shape (e.g., VORTEX: Lacy 1993).

Table 15.1. *Definitions of model acronyms used in this chapter*

Acronym	Term	Explanation
ESLI	Ecologically scaled landscape index	Landscape indices based on patch area and isolation scaled to species-specific data on home range size and dispersal distance.
IFM	Incidence function model	A metapopulation model that predicts patch occupancy based on extinction and colonization dynamics dependent on patch area and isolation as well as more complex factors such as Allee and rescue effects.
LCP	Least-cost path model	A model developed in a geographic information system (GIS) that identifies the path between two patches that carries the least total cost. Total cost is based upon the sum of the cost of every cell in the path, with a cell's cost being based on various environmental factors such as slope that impede or facilitate movement or survival.
SEPM	Spatially explicit population model	A model that represents population processes in combination with the spatial location of individuals and landscape features, often by mapping an individual's spatial location onto a lattice of cells.
SPOM	Stochastic patch occupancy model	Patch-based metapopulation models similar to IFMs that incorporate the effect of stochastic processes on patch occupancy.

The lattice-based SEPM considered in depth in this chapter, PATCH (Schumaker 1998; Schumaker *et al.* 2004), represents each home range as a single hexagonal cell. Designed for studying territorial vertebrates, PATCH links the survival and fecundity of individual territory-holders to geographic information systems (GIS) data on mortality risk and habitat productivity measured at the location of the individual's home range by scaling the values in a demographic matrix (Schumaker 1998). Lower GIS habitat scores translate into lower survival rates or reproductive output. Because SEPMs of this type address species distribution and demography on a spatial scale that encompasses many home ranges, they are not suitable for evaluating conservation issues that depend on within-home-range movements, for example, response of carnivore foraging movements to a small development project or road barrier (Clevenger and Wierzchowski Chapter 20).

Spatially explicit population models may be further categorized based on whether dispersal of individuals is dependent on habitat characteristics of the originating cell, the intervening matrix, or density dependence (triggered by the number of individuals in the originating patch) (South *et al.* 2002). In the PATCH model, users can choose the level of search "intelligence" that dispersers will exhibit. Intelligence options include: (1) a simple random walk with varying degrees of linearity and with increased tendency to settle as the disperser approaches a maximum number of steps; (2) the ability to sense and ascend habitat gradients; or (3) knowledge of habitat beyond immediately adjacent cells, such as the ability to settle on the closest vacant territory or the highest-quality territory within a search radius (Schumaker 1998). Moreover, adult organisms are classified as either territorial or floaters. The movement of territorial individuals is governed by a site fidelity parameter, but floaters must always search for available breeding sites (Schumaker 1998). In contrast to other SEPMs (see Lamberson *et al.* 1992), PATCH has no explicit mortality associated with dispersal beyond that which dispersers experience at the site where they are located at the end of each yearly time step. PATCH was originally developed for study of the northern spotted owl *(Strix occidentalis caurina)*, and this formulation of dispersal mortality may be more realistic for avian dispersers that can fly over habitat, rather than for less vagile terrestrial species.

Spatially explicit population models are often used to study idealized landscapes in order to elucidate general rules governing the impact of landscape structure on species distribution (With and King 2001). Alternately, SEPMs can be used to explore site- and species-specific conservation problems. Although these latter applications may be more relevant to real-world planning questions, they usually require much more biological detail, and it can be more difficult to draw from them general lessons for conservation. Spatially explicit population models that link the movement of individuals to a species dispersal ability and habitat affinity are, by their nature, well suited for analyzing functional connectivity. Besides identifying corridors whose protection has a strong effect on population viability (Carroll *et al.* 2003a), these models can also help identify the location of source and sink habitat in a landscape and evaluate the vulnerability of habitat to landscape change processes such as development (Carroll *et al.* 2004). Spatially explicit population models can help reveal non-linear responses (e.g., population viability thresholds) of species to increasing habitat protection. However, the added complexity of SEPMs increases model sensitivity to poorly known parameters such as maximum dispersal distance (Ruckelshaus *et al.* 1997).

CONNECTIVITY IN CONTEXT: CLASSIFYING LANDSCAPES BASED ON PATCH/MATRIX CONTRAST

The relative level of connectivity between core habitat patches is one of several factors affecting the persistence of area-sensitive species in fragmented landscapes (Carroll *et al.* 2004). The importance of connectivity as compared to other factors, such as patch size and total habitat area, will vary with the dispersal ability of the species and the quality of the landscape matrix (Andrén 1994). This point is illustrated by the status of grizzly bear and wolf populations across the 750 000 km² Yellowstone-to-Yukon region (Chadwick 2000) in the Rocky Mountains of the northern USA and Canada (Carroll *et al.* 2004). The Yellowstone-to-Yukon region shows a strong contrast in the condition of landscape matrix from its more-developed southern end in the northern USA to its northern end in northwestern Canada. In the northern portion of the region, protected areas are embedded within a relatively benign landscape matrix, whereas in the south, they more closely resemble habitat islands (Carroll *et al.* 2001; Noss *et al.* 2002).

The Greater Yellowstone Ecosystem is the strongest example in the western USA of a situation that increasingly typifies protected areas throughout the world: a large core refugium is surrounded by rapidly growing human populations (Noss *et al.* 2002). In such a landscape with "high contrast" between protected areas and the landscape matrix, it may be so challenging to re-establish connectivity for some species that protection of linkage zones represents a lower conservation priority than protection of buffer habitat. In some areas of the Greater Yellowstone Ecosystem, protection of blocks of habitat may help achieve both goals. For example, protection of grizzly bear habitat in the buffer zone of the northwestern Greater Yellowstone Ecosystem (e.g., Madison Valley) might also help re-establish connectivity with the central Idaho core area. Protection of areas at the northern extremity of the Greater Yellowstone Ecosystem, in a proposed corridor to the more distant Northern Continental Divide Ecosystem, may be a less realistic conservation investment, since the area would likely function as sink habitat with little impact on enhancing functional connectivity or grizzly bear population persistence. Habitat restoration could move the overall regional landscape towards a condition that would eventually recreate the historically connected grizzly bear metapopulation. However, the area of newly protected habitat necessary for this to occur is much greater (Carroll *et al.* 2003a) than that identified by analyses focused solely on linear carnivore movement corridors (Boone and Hunter 1996; Walker and Craighead 1997; Singleton *et al.* 2002).

Unlike in these high-contrast landscapes, in "low-contrast" areas such as northern British Columbia, human activities are still at sufficiently low levels that vulnerable species such as large carnivores still use much of the landscape matrix. Because species in these low-contrast landscapes are not restricted to defined corridors, the usual planning paradigm of core and buffer habitat linked by corridors (Noss and Harris 1986) may not be as useful as would be a "reversed paradigm" that maintains wild-lands as the landscape matrix with human settlements linked by "developed corridors" (Noss 1992). Thus what I categorize as "medium-contrast" landscapes remain as the situation in which connectivity may be a critical component of conservation planning. These landscapes are often located on the margin of a species' continuously inhabited range, where this continuous distribution is beginning to break up into isolated populations due to fragmentation, but enough habitat remains in the matrix that core areas are not yet comparable to islands.

Carroll *et al.* (2004) tested the applicability of the patch/matrix paradigm across the Yellowstone-to-Yukon region by predicting the ability of existing park systems to sustain carnivore populations (grizzly bears and wolves) based on both a SEPM and a simpler logistic regression model that used only data on park area and connectedness (or isolation). The patterns of persistence of grizzly bear predicted by the PATCH model for the region's parks agreed with those from the area-isolation logistic regression models for the grizzly bear in developed (northern US Rocky Mountains) and semi-developed (southern Canadian Rocky Mountains) landscapes. The area-isolation logistic regression model for the grizzly bear performed poorly where the landscape matrix contained large amounts of suitable habitat (northern Canadian Rocky Mountains). Moreover, park area and connectedness were poor predictors of gray wolf occurrence due to this species' broader-scale range dynamics and greater ability to inhabit the landscape matrix. Based on the logistic regression results, a doubling of park area corresponded to a 47% and 57% increase in projected grizzly bear population persistence in developed and semi-developed landscapes, respectively. In comparison, a doubling of a park's connectedness index corresponded to an 81% and 350% increase in grizzly bear population persistence in developed and semi-developed landscapes,

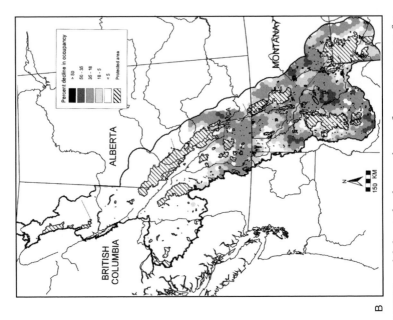

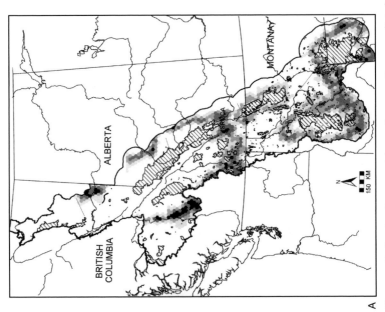

Fig. 15.1. Reduction in potential carrying capacity as predicted by the PATCH model due to landscape change from 2000–2025 for (A) grizzly bear and (B) gray wolf in the Rocky Mountains of Canada and the United States. (Adapted from Carroll *et al.* 2003a.)

respectively, suggesting that conservation planning to enhance connectivity may be most effective in the earliest stages of landscape degradation (Carroll *et al.* 2004). The PATCH results suggest that the role of the landscape matrix for sustaining connectivity varied between the two species, causing parks in the US Northern Rockies to support a functionally connected metapopulation of gray wolves, versus several disjunct populations of grizzly bears. However, in these model scenarios, landscape change trends move the US Northern Rockies landscape towards a condition where the wolf subpopulations would also become increasingly isolated (Carroll *et al.* 2003a) (Fig. 15.1). This is due not only to increasing barriers to movement (e.g., highways), but also to degradation of source habitat and consequent reduction in the numbers of dispersers and the area of sink habitat that can be sustained by this dispersal (Fig. 15.1). Range contraction (decrease in probability of occupancy in Fig. 15.1) occurs primarily on the edges of core habitat for the grizzly bear, but throughout the landscape matrix for the wolf.

CORRIDOR PLANNING IN MEDIUM-CONTRAST LANDSCAPES: AN EXAMPLE

Although SEPM results may be most informative for conservation planning at regional scales, these models may also aid corridor design at the finer subregional scales more commonly considered by planners. For example, Carroll *et al.* (2002) used the PATCH model to compare the effects on connectivity of contrasting conservation proposals in the semi-developed, medium-contrast landscape along the Highway 3 area in southwestern Alberta and southeastern British Columbia (Canada/USA transboundary region), which separates the large parks and undeveloped areas to the north from more isolated southern refugia that range in size from Glacier/Waterton Parks (4500 km^2) in the Northern Continental Divide Ecosystem to smaller areas of around 1000 km^2 in Idaho's Selkirk Mountains, and the Cabinet/Yaak (Montana) and Granby (British Columbia) areas (Proctor *et al.* 2002).

The land-use scenarios considered by Carroll *et al.* (2002) include:

(1) current carrying capacity
(2) current trends to 2025, assuming development on both public and private lands
(3) current trends to 2025, assuming no further road construction on public lands

(4) creation of an absolute barrier or zone of inhospitable habitat along the Highway 3 area

(5) proposed Waterton Park expansion, enlarging the park to encompass areas primarily in the North Fork of Flathead (British Columbia) (Weaver 2001)

(6) proposed Southern Rocky Mountains Conservation Area, connecting the Northern Continental Divide Ecosystem to the Rocky Mountains parks across the Highway 3 area in the area of Fernie, British Columbia (Weaver 2001).

Under current landscape conditions, the SEPM results predicted that the probability of maintaining a continuously distributed population of grizzly bears across the Highway 3 area over the long term is low. Importantly, these simulation results are equilibrium predictions, in that current predictions depict the current capacity for an area to support a carnivore species over the long term (200 years), which may be lower (e.g., grizzly bears in southeastern British Columbia) or higher (grizzly bears in central Idaho) than the number of animals currently inhabiting that area. Field data suggests that the highway area currently functions as a semi-permeable sex-biased "filter," with lack of female grizzly bear dispersal creating genetic divergence between bear populations to the north and south of the highway (Proctor *et al.* 2002). In the SEPM results of Carroll *et al.* (2002), landscape change through 2025 made this connection even more tenuous due to the retreat of grizzly bear range to the south and north. Because of a negative "ripple effect" of habitat loss in other parts of southeastern British Columbia on grizzly bear distribution to the north of Highway 3, connectivity in this area will be more difficult to maintain. Assuming no further road construction on public lands greatly reduced range loss in the model, but range contraction was still extensive in the Rocky Mountain Front, the immediate Highway 3 area, and the Columbia Trench. The effect of an absolute barrier (e.g., expanded multi-lane highway) in the Highway 3 area was noticeable under current conditions, but minor under 2025 conditions, as projected development already effectively excluded bears from the highway zone. For grizzly bears, the Waterton Park expansion was effective at counteracting the effects of landscape change to the south of the highway, but the larger Southern Rocky Mountains Conservation Area proposal was most effective at retaining a level of connectivity at or higher than the current condition, despite increasing development in other parts of the transboundary region (Fig. 15.2A). The positive ripple effect for both

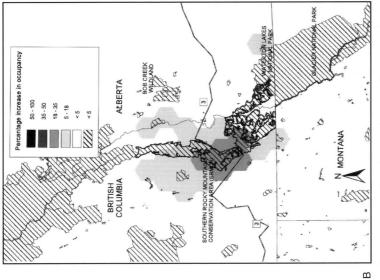

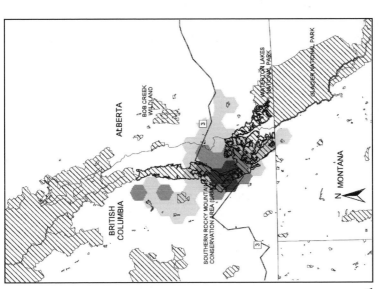

Fig. 15.2. Increase in potential carrying capacity in response to proposed park expansion within the Southern Rocky Mountains Conservation Area as predicted by the PATCH model for (A) grizzly bear and (B) wolf in the Canada/USA transboundary region under future landscape conditions (2025). (Adapted from Carroll *et al.* 2002.)

the Waterton expansion and the Southern Rocky Mountains Conservation Area on bear distribution was extensive beyond the boundaries of the proposed park areas in both Alberta and British Columbia.

For the wolf, the PATCH model predicted a continuous distribution across the Highway 3 area throughout the area under current landscape conditions (Carroll *et al.* 2002). Connectivity was most pronounced on the Alberta side of the border. However, with landscape change, long-term connectivity was effectively lost by 2025. When compared to the situation for grizzly bears, loss in carrying capacity for wolves was more widespread throughout the region because wolves are more tolerant of human impacts and hence use more areas in the landscape matrix that are at risk from development. Loss in demographic potential for wolves was greatest along the Rocky Mountain Front. In the model results, the Southern Rocky Mountains Conservation Area preserved connectivity both within its boundaries and via a ripple effect in the Alberta Highway 3 area (Fig. 15.2B).

In summary, greater dispersal ability and subsequent interlinkage of populations in the wolf versus the grizzly bear makes the effect of the Highway 3 barrier scenario more noticeable, but still of little more impact than predicted future conditions in which development trends alone approximate a barrier there. For the wolf, the positive ripple effect of the Conservation Area is similar, but not as strong as it is for the grizzly bear. The effect of the Waterton expansion is also less pronounced for the wolf than for the bear.

CORRIDOR PLANNING AND PATTERNS OF DISPERSAL

The pattern of range expansion in response to habitat restoration scenarios shown in the above SEPM simulations for the Highway 3 area can help in siting corridors where they are most likely to exhibit functional connectivity. These simulations can similarly be used to rank various restoration proposals by measuring the relative magnitude of their effect on species distribution. The SEPM-based evaluation of the Highway 3 area (Carroll *et al.* 2002) shows similarities with earlier work using static models of habitat suitability (Apps 1997). Similarly, a SEPM-based analysis of wolf viability in the northeastern USA and southeastern Canada (Carroll in press) identified a key linkage area between Adirondack (New York) and Algonquin (Ontario) parks that had been described previously using the least-cost path (LCP) analysis (Quinby *et al.* 2000; for a description of the least-cost path technique see Theobald Chapter 17). The SEPM analysis, in addition to identifying the linkage zone, rated its relative probability as a dispersal route as similar to linkages to the east from Quebec to northern Maine, but highlighted risks of a rapid decline in functional connectivity over time, as well as threats to the viability of the extant source population (Algonquin) and potential destination population (Adirondacks) (Carroll 2003). These results suggest that the level of conservation effort required to protect the narrow Algonquin-to-Adirondacks linkage was much greater than that necessary to protect any intraregional connectivity (e.g., between a reintroduced wolf population in Maine and adjacent New Brunswick). Since a reintroduced Maine wolf population was found to be demographically viable without dispersal from Canada, intra-regional linkages may be a higher priority than inter-regional connectivity in this situation.

In contrast to the Algonquin-to-Adirondacks results, a study using a SEPM to assess wolf viability in the southern Rocky Mountains (USA) failed to identify a likely corridor from Yellowstone south to Colorado (Carroll et al. 2003b). (More recently, wolves have dispersed across this distance but have been killed before establishing territories due to livestock conflicts or road mortality.) In some regional landscapes such as southern Wyoming, the pattern of dispersal as simulated in PATCH is fairly uniform, whereas in other regions, such as the Adirondacks, it is channeled into corridor-like paths. This pattern is due to both the level of habitat contrast in the landscape, and the overall likelihood of effective dispersal between source and destination area. In the Algonquin-to-Adirondacks example, the source and destination areas were relatively close (~250 km) and the intervening landscape was highly modified by agriculture except in one area. In the Colorado example, source and destination area were more distant (~500 km), and the intervening landscape was sparsely settled and characterized by extensive land use such as grazing (Carroll et al. 2003b). The channelization of dispersal paths in the Algonquin-to-Adirondacks region was accompanied by a high sensitivity of model results to variation in dispersal parameters. This sensitivity may occur because this regional landscape is currently at a threshold for effective wolf dispersal (Carroll 2003). The narrow dispersal corridor predicted for Algonquin-to-Adirondacks cause the least-cost path results to mimic more complex SEPM results better than they would match the broad wave of dispersal predicted by SEPMs in many regions.

The regional contrasts in SEPM results may be, in part, an artifact of how dispersal mortality is treated in the PATCH model. Because there is no explicit dispersal mortality except at the end of each yearly time step

(Schumaker 1998), the likelihood of a disperser traversing a short but highly hostile landscape may be overestimated. Sensitivity of results to poorly known parameters, particularly dispersal distance, is an often-cited weakness of SEPMs (Ruckelshaus *et al.* 1997; Coulson *et al.* 2001). Other authors have identified this behavior as most typical of simple SEPMs that lack a demographic context, or that use a dispersal function that is not sensitive to landscape structure (Mooij & DeAngelis 1999; South 1999). In several SEPM-based studies of wolf population dynamics across a range of North American landscapes (Carroll *et al.* 2003a, 2003b, in press), probability of recolonization of distant habitat varied with the parameter used for maximum dispersal distance (Table 15.2). However, model results, and the resulting conservation recommendations, were qualitatively similar across the biologically plausible range of dispersal distances for this species (Table 15.2).

Compared to other SEPM parameters, the importance of dispersal varies with landscape context and appears to be most important at moderate levels of fragmentation (Rushton *et al.* 2000; Carroll *et al.* 2004). Furthermore, dispersal parameterization is rarely a significant factor in determining population persistence in SEPMs (Macdonald and Rushton 2003). The inherent uncertainty attached to predictions of rare dispersal events suggests that SEPMs may not provide useful estimates of the probability of natural recolonization of a patch of potential habitat from a distant source population (Carroll *et al.* 2003b). Of more concern than the dispersal distance parameter itself may be the differences in pattern between actual dispersal paths and the simplified rules, such as correlated

Table 15.2. *Sensitivity of wolf dispersal success in different study regions to different parameters for maximum dispersal distance using the PATCH model*

	Wolf dispersal distance (km)		
	250	500	1500
Recolonization probability (%)			
Maine	0.42	3.54	10.62
Adirondacks	0.30	12.33	43.68
Colorado	0.34	0.95	2.20
Adirondacks (2025)	0	0	0
Colorado (2025)	0	0	0

Source: Unpublished data from studies reported in Carroll *et al.* (2003a) and Carroll (2003).

random walks, used in SEPMs; dispersing animals likely respond to complex environmental cues, including conspecific attraction and fine-scale habitat structure (Lindenmayer *et al.* 2003; Tracey Chapter 14). Heterogeneity between individuals and contrasts between the same individual's behavior in the initial and later stages of a dispersal path (Morales and Ellner 2002) may add additional complexity. Although much of this complex behavior may not result in qualitative contrasts in the distribution of dispersers, they must be considered in SEPMs if they are to provide realistic estimates of dispersal success. Unfortunately, it is the few longest-distance dispersers that most influence model results, and these individuals are often unrecorded by field studies (Shigesada and Kawasaki 2002).

COMPARING SPATIALLY EXPLICIT POPULATION MODEL RESULTS TO THOSE FROM SIMPLER LANDSCAPE INDICES AND MODELS

Least-cost path techniques (Theobald Chapter 17) are one example of methods relating landscape structure to connectivity that are easier to calculate and less "data hungry" than are SEPMs, and that are more capable of being generalized to new situations (Table 15.1). Ecologically scaled landscape indices (ESLIs) (Vos *et al.* 2001) also attempt to make simple landscape indices, such as patch area and isolation, more biologically relevant by scaling them based on species-specific data on home range size and dispersal distance. Although it might be possible for SEPM results to be generalized into rules or indices that approximate simulation results in novel landscapes, comparisons of most landscape metrics against predictions of connectivity (i.e., dispersal success) from SEPMs have shown that the two match poorly (Schumaker 1996). As demonstrated above, representing a landscape mosaic of diverse habitat types as a binary system of discrete habitat patches, embedded in a non-habitat matrix, works best in highly fragmented landscapes (Carroll *et al.* 2004). These types of landscapes are increasingly common in conservation planning for endangered species, especially in regions with high human population density. However, it is important to avoid sacrificing biological relevance in the quest for model generality, and it remains unclear how informative simpler models such as least-cost paths are in the many regions and species contexts with intermediate levels of landscape contrast.

Metapopulation models (Moilanen and Hanski Chapter 3) are structurally more complex and potentially more biologically realistic than least-cost path techniques. However, the populations of large carnivores treated in this chapter often do not strictly conform to a classic metapopulation model. Some use is made of matrix habitat, and thus populations are not confined to island-like patches that experience repeated extinction and recolonization (Carroll *et al.* 2004). Habitat structure within patches is also important to the likelihood of population persistence. For poor dispersers such as grizzly bears, if "matrix" habitat is benign enough to allow dispersal, then it is likely to also occasionally support territorial individuals, and thus should be considered for its demographic role as sink habitat. If it is not benign enough to allow dispersal, then area effects alone, rather than connectivity, will be sufficient to predict viability. For many species, such as grizzly bears, the regional metapopulation is in a non-equilibrium state (Harrison 1994), and extinction events occur in small patches but few colonization events occur. For other species, such as wolves, both extinction and colonization events are common, so in landscapes that are fragmented to the degree that wolves no longer constitute a single patchy population, metapopulation concepts in the strict sense may indeed be relevant.

While the classic division of landscapes into patch and matrix is attractive due to its simplicity, biologists are increasingly moving towards more variegated habitat models that portray landscapes in shades of gray rather than black and white (Fischer *et al.* 2004). Least-cost path and more recent metapopulation models can incorporate varying levels of matrix permeability but cannot expand the patch-matrix model to allow for the demographic effect of poor but suitable (i.e., sink) habitat (Possingham *et al.* 2005). In addition, least-cost path techniques do not explicitly incorporate population viability. Irrespective of a species' dispersal ability, a least-cost path model will always identify a "best" linkage between source and destination patch. A least-cost path model assumes that "source" populations are known and fixed, whereas an SEPM makes no such assumptions, but attempts to specify where sources and sinks are in the landscape. Linkage areas identified in an SEPM-based analysis must meet a biological threshold for effective dispersal (i.e., the path is not too costly to be used), and the core and buffer habitat that anchor the ends of a corridor must provide sufficient dispersers to make the corridor effective. An SEPM-based analysis might lead to placement of a corridor in an area with a more "costly" path but stronger anchor habitat, or in the shifting of priorities away from linkage zones if these appeared as poorer conservation bargains.

CONCLUSION

Choosing between the various models of connectivity depends on an assessment of what level of model complexity might provide better guidance in a particular conservation planning context. There is a need for a typology of landscape and species combinations that can suggest to planners which types of models are most informative for their problem. I have made the case that for wide-ranging species in medium-contrast landscapes, a simple connectivity metric such as patch isolation is unlikely to substitute for SEPM-based mechanistic and context-specific predictions of the probability of functional connectivity and persistence in a patch. Spatially explicit population models may be especially useful in providing information on population vulnerability under novel future scenarios that is hard to extract from other, simpler metrics (Carroll et al. 2004). However, these types of scenarios are a small subset of the situations that confront conservation planners. The match between models and metrics is expected to be better for landscapes and species combinations for which the binary habitat/non-habitat landscape is a reasonable approximation. For example, a good candidate for simpler models might be a species that shows strong associations with a single type of habitat (such as the northern spotted owl with old-growth forest), in a landscape with processes (e.g., clearcut logging) that tend to produce hard edges between patch and matrix at a scale similar to that of the species' home range. Similarly, the vagility of the species should be intermediate in relation to the level of contrast on the landscape, not being so low that all patches are isolated nor so high that the intervening matrix has little effect on dispersal success. A rigorous comparison of the conservation priorities identified by ESLIs, graph theory, SPOMs, and SEPMs (Table 15.1) in such a landscape would help planners assess the strengths and weaknesses of the diverse approaches to modeling connectivity.

The most important contribution of SEPMs to connectivity planning may not be their specific predictions, but rather the way in which they link connectivity tightly to its role in promoting population viability. The separation of connectivity from viability has led to potential misuse of the former concept in conservation planning. For example, conservation organizations increasingly use the term "corridor" to refer to regional landscapes that would, in traditional conservation planning terminology, be instead a planning landscape divided into components of cores, buffers, and corridors, each with distinct management regimes (Noss and Harris 1986). Use of corridors in this broad sense tends to obscure the distinct roles played by the different components, e.g., strictly protected habitat that can sustain sources of species vulnerable to human-induced mortality versus less secure habitat (corridors in the narrow sense) that may sustain movement of these species. In this case, designation of the landscape as a "corridor," which is assumed to require few restrictions on land use, may be a means of avoiding the harder challenges to slowing the loss of both core and connective habitat.

ACKNOWLEDGEMENTS

Research discussed in this chapter was supported by World Wildlife Fund Canada, the Turner Endangered Species Fund, and The Wildlands Project. Kevin Crooks, Brett Dickson, and Hugh Possingham provided helpful reviews of the manuscript.

REFERENCES

Andrén, H. 1994. Effects of habitat fragmentation on birds and mammals in landscapes with different proportions of suitable habitat: a review. *Oikos* 71:355–366.

Apps, C. D. 1997. *Identification of Grizzly Bear Linkage Zones along Highway 3 Corridor of Southeast British Columbia and Southwest Alberta.* Calgary, Alberta, Canada: Aspen Wildlife Research.

Boone, R. B., and M. L. Hunter, Jr. 1996. Using diffusion models to simulate the effects of landuse on grizzly bear dispersal in the Rocky Mountains. *Landscape Ecology* 11:51–64.

Bruinderink, G. G., T. Van Der Sluis, D. Lammertsma, P. Opdam, and R. Pouwels. 2003. Designing a coherent ecological network for large mammals in northwestern Europe. *Conservation Biology* 17:549–557.

Carroll, C., R. F. Noss, and P. C. Paquet. 2001. Carnivores as focal species for conservation planning in the Rocky Mountain region. *Ecological Applications* 11:961–980.

Carroll, C., R. F. Noss, and P. C. Paquet. 2002. *Rocky Mountain Carnivore Project,* final report. Toronto, Ontario, Canada: World Wildlife Fund Canada. Avaliable online at http://www.wwf.ca/en/res_links/rl_resources.asp/

Carroll, C., R. F. Noss, P. C. Paquet, and N. H. Schumaker. 2003a. Use of population viability analysis and reserve selection algorithms in regional conservation plans. *Ecological Applications* 13:1773–1789.

Carroll, C., M. K. Phillips, N. H. Schumaker, and D. W. Smith. 2003b. Impacts of landscape change on wolf restoration success: planning a reintroduction program using dynamic spatial models. *Conservation Biology* 17:536–548.

Carroll, C., R. F. Noss, P. C. Paquet, and N. H. Schumaker. 2004. Extinction debt of protected areas in developing landscapes. *Conservation Biology* 18:1110−1120.

Carroll, C. 2003. Impacts of Landscape Change on Wolf Viability in the Northeastern U.S. and Southeastern Canada: Implications for Wolf Recovery. Wildlands Project Special Paper No. 5. Richmond, VT: Wildlands Project. Available online at http://www.klamathconservation.org/

Chadwick, D. 2000. *Yellowstone to Yukon*. Washington, DC: National Geographic Society.

Coulson, T., G. M. Mace, E. Hudson, and H. P. Possingham. 2001. The use and abuse of population viability analysis. *Trends in Ecology and Evolution* 16:219−221.

Dunning, J. B., Jr., D. J. Stewart, B. J. Danielson, *et al.* 1995. Spatially explicit population models: current forms and future uses. *Ecological Applications* 5:3−11.

Fischer, J., D. B. Lindenmayer, and I. Fazey. 2004. Appreciating ecological complexity: habitat contours as a conceptual landscape model. *Conservation Biology* 18:1245−1253.

Forman, R. T. T., and M. Godron. 1986. *Landscape Ecology*. New York: John Wiley.

Frankel, O. H., and M. E. Soulé. 1981. *Conservation and Evolution*. Cambridge, UK: Cambridge University Press.

Hanski, I. 1994. A practical model of metapopulation dynamics. *Journal of Animal Ecology* 63:151−162.

Harrison S. 1994. Metapopulations and conservation. Pp. 111−128 in P. J. Edwards, R. M. May, and N. R. Webb (eds.) *Large-Scale Ecology and Conservation Biology*. Oxford, UK: Blackwell Scientific Publications.

Lacy, R. C. 1993. VORTEX: A computer simulation model for population viability analysis. *Wildlife Research* 20:45−65.

Lambeck, R. J. 1997. Focal species: a multi-species umbrella for nature conservation. *Conservation Biology* 11:849−856.

Lamberson, R. H., R. McKelvey, B. R. Noon, and C. Voss. 1992. A dynamic analysis of northern spotted owl viability in a fragmented forest landscape. *Conservation Biology* 6:505−512.

Lindenmayer, D. B., H. P. Possingham, R. C. Lacy, M. A. McCarthy, and M. L. Pope. 2003. How accurate are population models? Lessons from landscape-scale population tests in a fragmented system. *Ecology Letters* 6:41−47.

Macdonald, D. W., and S. Rushton. 2003. Modelling space use and dispersal of mammals in real landscapes: a tool for conservation. *Journal of Biogeography* 30:607−620.

Manel, S., M. K. Schwartz, G. Luikart, and P. Taberlet. 2003. Landscape genetics: combining landscape ecology and population genetics. *Trends in Ecology and Evolution* 18:189−197.

Mladenoff, D. J., T. A. Sickley, R. G. Haight, and A. P. Wydeven. 1995. A regional landscape analysis and prediction of favorable gray wolf habitat in the northern Great Lakes region. *Conservation Biology* 9:279−294.

Moilanen, A., and I. Hanski. 2001. On the use of connectivity measures in spatial ecology. *Oikos* 95:147−152.

Mooij, W. M., and D. L. DeAngelis. 1999. Error propagation in spatially explicit population models: a reassessment. *Conservation Biology* 13:930−933.

Morales, J. M., and S. P. Ellner. 2002. Scaling up animal movements in heterogeneous landscapes: the importance of behavior. *Ecology* **83**:2240–2247.

Noss, R. F. 1987. Corridors in real landscapes: a reply to Simberloff and Cox. *Conservation Biology* **1**:159–164.

Noss, R. F. 1992. The Wildlands Project: land conservation strategy. *Wild Earth* (Special Issue):10–25.

Noss, R. F., and L. D. Harris. 1986. Nodes, networks, and MUMs: preserving diversity at all scales. *Environmental Management* **10**:299–309.

Noss, R. F., H. B. Quigley, M. G. Hornocker, T. Merrill, and P. C. Paquet. 1996. Conservation biology and carnivore conservation in the Rocky Mountains. *Conservation Biology* **10**:949–963.

Noss, R. F., C. Carroll, K. Vance-Borland, and G. Wuerthner. 2002. A multicriteria assessment of the irreplaceability and vulnerability of sites in the Greater Yellowstone Ecosystem. *Conservation Biology* **16**:895–908.

Paquet P. C., and L. N. Carbyn. 2003. Gray wolf (*Canis lupus* and allies). Pp. 482–510 in G. A. Feldhamer, B. C. Thompson, and J. A. Chapman (eds.) *Wild Mammals of North America*, 2nd edn. Baltimore, MD: Johns Hopkins University Press.

Possingham H. P., J. Franklin, K. Wilson, and T. J. Regan. 2005. The roles of spatial heterogeneity and ecological processes in conservation planning. Pp. 386–406 in G. M. Lovett, C. G. Jones, M. G. Turner, and K. C. Weathers (eds.) *Ecosystem Function in Heterogeneous Landscapes*. New York: Springer-Verlag.

Power M. E., D. Tilman, J. A. Estes, *et al.* 1996. Challenges in the quest for keystones. *BioScience* **46**:609–620.

Proctor, M. F., B. N. McLellan, and C. Strobeck. 2002. Population fragmentation of grizzly bears in southeastern British Columbia, Canada. *Ursus* **13**:153–160.

Quinby, P., S. Trombulak, T. Lee, R. Long, *et al.* 2000. Opportunities for wildlife habitat connectivity between Algonquin Provincial Park and the Adirondack Park. *Wild Earth* **10**:75–80.

Ruckelshaus, M., C. Hartway, and P. Karieva. 1997. Assessing the data requirements of spatially explicit models. *Conservation Biology* **11**:1298–1306.

Rushton, S. P., G. W. Barreto, R. M. Cormack, D. W. Macdonald, and R. Fuller. 2000. Modelling the effects of mink and habitat fragmentation on the water vole. *Journal of Applied Ecology* **37**:475–490.

Schumaker, N. H. 1996. Using landscape indices to predict habitat connectivity. *Ecology* **77**:1210–1225.

Schumaker, N. H. 1998. *A User's Guide to the PATCH model*, EPA/600/R-98/135. Corvallis, OR: US Environmental Protection Agency, Environmental Research Laboratory. Available online at http://www.cpa.gov/wed/pages/models.htm

Schumaker, N. H., T. Ernst, D. White, J. Baker, and P. Haggerty. 2004. Projecting wildlife responses to alternative future landscapes in Oregon's Willamette Valley. *Ecological Applications* **14**:381–400.

Shigesada N., and K. Kawasaki. 2002. Invasion and the range expansion of species: effects of long-distance dispersal. Pp. 350–373 in J. M. Bullock, R. E. Kenward, and R. S. Hails (eds.) *Dispersal Ecology*. Malden, MA: Blackwell.

Simberloff, D., and J. Cox. 1987. Consequences and costs of conservation corridors. *Conservation Biology* **1**:163–171.

Singleton, P. H., W. L. Gaines, and J. F. Lehmkuhl. 2002. *Landscape Permeability for Large Carnivores in Washington: A Geographic Information System Weighted-Distance and Least-Cost Corridor Assessment*, Research Paper PNW-RP-549. Portland, OR: US Department of Agriculture, Forest Service, Pacific Northwest Research Station.

Soulé M. E., and J. Terborgh. 1999. *Continental Conservation: Scientific Foundations of Regional Reserve Networks*. Covelo, CA: Island Press.

Soulé M. E., J. A. Estes, J. Berger, and C. M. Del Rio. 2003. Ecological effectiveness: conservation goals for interactive species. *Conservation Biology* 17:1238–1250.

South, A. 1999. Dispersal in spatially explicit population models. *Conservation Biology* 13:1039–1046.

South A. B., S. P. Rushton, R. E. Kenward, and D. W. Macdonald. 2002. Modelling vertebrate dispersal and demography in real landscapes: how does uncertainty regarding dispersal behaviour influence predictions of spatial population dynamics? Pp. 327–349 in J. M. Bullock, R. E. Kenward, and R. S. Hails (eds.) *Dispersal Ecology*. Malden, MA: Blackwell.

Thiel, R. P. 1985. Relationship between road densities and wolf habitat suitability in Wisconsin. *American Midland Naturalist* 113:404–407.

Tischendorf, L., and L. Fahrig. 2000. On the usage of landscape connectivity. *Oikos* 90:7–19.

Vos, C. C., J. Verboom, P. F. M. Opdam, and J. F. Ter Braak. 2001. Toward ecologically scaled landscape indices. *American Naturalist* 157: 24–41.

Walker, R., and L. Craighead. 1997. Analyzing wildlife movement corridors in Montana using GIS. *Proceedings of the ESRI User Conference 1997*. Available online at http://gis.esri.com/library/userconf/proc97/proc97/to150/pap116/p116.htm

Weaver, J. L. 2001. *The Transboundary Flathead: A Critical Landscape for Carnivores in the Rocky Mountains*, WCS Working Papers No. 18. Bronx, NY: Wildlife Conservation Society.

Wilcove D. S., C. H. McLellan, and A. P. Dobson. 1986. Habitat fragmentation in the temperate zone. Pp. 237–256 in M. E. Soulé (ed.) *Conservation Biology: The Science of Scarcity and Diversity*. Sunderland, MA: Sinauer Associates.

With, K. A., and A. W. King. 2001. Analysis of landscape sources and sinks: the effect of spatial pattern on avian demography. *Biological Conservation* 100:75–88.

16

Impacts of corridors on populations and communities

NICK M. HADDAD AND JOSH J. TEWKSBURY

INTRODUCTION

This chapter focuses specifically on the most popular approach to maintain connectivity in conservation and management, which is to create or maintain habitat corridors. The popularity of corridors in conservation derives from the direct and intuitive relationship to their purported function: by physically connecting otherwise isolated fragments, corridors should increase the movement of both individuals and genes. In doing so, corridors provide sources of immigrants to offset local extinction, and sources of genetic diversity to reduce harmful effects of inbreeding and drift. The most fundamental spatial models in ecology, including island biogeographic models (MacArthur and Wilson 1967) and metapopulation models (Levins 1969; Hanski 1999), predict that movement between patches will increase population size and persistence and, through the rescue of declining populations (Brown and Kodric-Brown 1977), maintain local species richness. We recognize that studies focusing on corridors represent only a small fraction of studies on connectivity, and the large literature examining effects of patch isolation on colonization and occupancy in metapopulations has been reviewed elsewhere (see Table 9.1 in Hanski 1999; Moilanen and Nieminen 2002; Molainen and Hanski Chapter 3). The goal of this chapter is to assess existing evidence for corridor effects on populations and communities, and to discuss future directions that would permit more rigorous evaluation of their use in conservation.

Connectivity Conservation eds. Kevin R. Crooks and M. Sanjayan. Published by Cambridge University Press. © Cambridge University Press 2006.

We focus on population and community impacts of corridors because evidence for the necessary prerequisite – that corridors increase movement and gene flow – has been growing and has also been reviewed elsewhere. In a review by Beier and Noss (1998) of 32 published studies as of 1997, 21 studied some aspect of animal movement through or within corridors, and many supported the role of corridors in increasing movement. Since that review, a number of other studies have demonstrated that corridors enhance movement rates of plants and animals between otherwise isolated patches (e.g., Coffman et al. 2001; Berggren et al. 2002; Tewksbury et al. 2002; Haddad et al. 2003). Still other recent studies have documented the role of connectivity in enhancing gene flow (Aars and Ims 1999; Hale et al. 2001; Mech and Hallett 2001; Kirchner et al. 2003; Neville et al. Chapter 13). While some studies have not found evidence for corridor effects (Rosenberg et al. 1998; Bowne et al. 1999; Danielson and Hubbard 2000), no studies have shown that corridors decrease movement rates.

The growing number of studies that show how corridors affect movement between patches provide a critical base of support for the idea that corridors should enhance population viability. From this work, it follows that corridors will reduce stochastic temporal variation in local and regional population sizes by increasing the rates of immigration from high-density to lower-density patches. Another possibility that has emerged from theoretical results is that corridors may have a negative effect by synchronizing dynamics and causing simultaneous extinction (Petchey et al. 1997; Earn et al. 2000; Hudgens and Haddad 2003). However, in cases most typical in conservation where small populations have rare dispersal and low growth rates, corridors should reduce local extinctions and allow individual patches to maintain a larger number of species with stable population dynamics (Brown and Kodric-Brown 1977; Gonzalez and Chaneton 2002). Other potential negative effects of corridors that have been discussed extensively in the literature are reviewed elsewhere in this volume (Crooks and Sanjayan Chapter 1; Crooks and Suarez Chapter 18).

The link between corridor effects on movement and their effects on the demography and persistence of populations, and ultimately, the maintenance of local and regional biodiversity, is critical for the appropriate use of corridors in management. Yet, there is currently a paucity of studies addressing population or community effects of corridors. Consistent empirical evidence regarding population and community effects of corridors would support their expanded implementation in conservation.

In this chapter, we first review empirical corridor studies that focus on population and community effects, seeking synthesis across studies. We highlight deficiencies in the existing literature, describe conditions under which corridor effects are expected, and discuss problems with study scale and design. We then go on to detail how new research could more effectively test the case for conservation benefits of corridors on populations and communities.

A REVIEW OF CORRIDOR EFFECTS ON POPULATIONS AND COMMUNITIES

The literature

We reviewed all empirical studies that examined terrestrial and microcosm corridor effects on population size or persistence, or on species diversity. Ideally for conservation, population studies would focus on how corridors affect population viability, and would thus measure persistence. Population growth can also be a strong indicator of persistence, especially when corridors may tip the balance between decreasing and increasing population trends. However, these measures are often difficult to obtain as they require long-term studies for meaningful estimates. Other population-level responses to corridors, such as size and survivorship, are less useful in assessing conservation value, but are still correlated with population viability. Regarding diversity, the most relevant response variables for conservation are often those that describe change in community composition after fragmentation, such as the rate of species loss, particularly with regard to species of management concern. One commonly measured community response, species richness, could be used to assess loss. In our review, we list response variables measured in existing corridor studies.

In our analysis, we included studies of corridor effects within patches relative to similar, isolated areas. Although a number of studies have shown that corridors affect population sizes by providing habitat for plants or animals within corridors (e.g., Machtans et al. 1996; Laurance and Laurance 1999; Perault and Lomolino 2000; Pryke and Samways 2001; Mönkkönen and Mutanen 2003), we are interested in the effects of corridors on populations or communities within patches they connect. We also did not include corridor studies where different numbers and configurations of corridors were added (Holyoak 2000), unless there were also treatments of unconnected fragments.

We searched the following journals using ISI Web of Science: *Biological Conservation, Conservation Biology, Ecological Applications, Ecological Monographs, Ecology, Ecology Letters, Ecography, Journal of Animal Ecology, Journal of Applied Ecology, Journal of Ecology, Nature, Oikos, and Science.* We searched using the following terms: (corridor*) and (population* or communit* or biodiversity). Our search extended from 1977 to 2003, and was conducted on 15 December 2003.

We found 15 studies that tested for corridor effects on populations and five studies that tested for corridor effects on diversity (Table 16.1). Some studies were included in both categories, as they analyzed both population and community responses. Studies covered a variety of species, including population studies on mammals, insects, microorganisms, birds, and a lizard (in order of decreasing frequency), and diversity studies on arthropods and birds. No studies focused on plants. Most studies (15/19) were experimental, as they manipulated and replicated landscape pattern.

At first glance, support for the idea that corridors affect population size or persistence appears strong. Of the 15 studies focusing on population responses to corridors, 13 demonstrated some corridor effect. Yet there were often caveats along with observed effects. Some studies (Fahrig and Merriam 1985; Mansergh and Scotts 1989; Dunning *et al.* 1995) were unreplicated and others (Burkey 1997; Schmiegelow *et al.* 1997; Haddad and Baum 1999; Schmiegelow and Mönkkönen 2002) observed corridor effects that may have been caused by patch shapes, edges, or habitat types that were confounded with corridor effects (see below, section "Designing corridor studies in variable environments"). Regarding corridor effects on diversity, all measured species richness and only two microcosm experiments showed convincing positive effects (Gilbert *et al.* 1998; Gonzalez and Chaneton 2002). The only other study to report evidence for corridor effects on diversity had no replication (MacClintock *et al.* 1977).

We conclude that the empirical literature to date shows ambiguous support for corridor effects on populations or communities. In support of corridors, most studies reported some positive effect. These effects are apparent even above many other local factors that are known to impact populations (like effects of local environments and other landscape-level effects) and that might obscure corridor effects. Despite these results, evidence remains weak because of confounding effects, and because some species performed more poorly in patches connected by corridors (Holyoak and Lawler 1996; Burkey 1997). At this time, current evidence

Table 16.1. *Studies reporting responses of populations or communities within patches that are either connected by corridors or isolated*

Study	Species	Scientific name	Study type	Corridor effect	Measured population response
On populations					
Fahrig & Merriam (1985)	Mammal: white-footed mouse	*Peromyscus leucopis*	Observational	Yes	Size
Mansergh & Scotts (1989)	Mammal: mountain pygmy possum	*Burramys parvus*	Observational	Yes	Survivorship
La Polla & Barrett (1993)	Mammal: meadow vole	*Microtus pennsylvanicus*	Experimental	Yes	Size
Ims & Andreassen (1999)	Mammal: Townsend's vole	*Microtus townsendii*	Experimental	No	Growth
Coffman *et al.* (2001)	Mammal: meadow vole	*M. pennsylvanicus*	Experimental	No	Size
Hannon & Schmiegelow (2002)				Yes	Survivorship
Schmiegelow *et al.* (1997)	Birds		Experimental	Yes for 7 of 23 species	Size
Dunning *et al.* (1995)	Bird: Bachman sparrow	*Aimophila aestivalis*	Observational	Yes	Size

Reference	Taxa	Species	Method	Effect	Response variable
Boudjemadi et al. (1999)	Herpetile: common lizard	*Lacerta vivipara*	Experimental	Yes in rich habitats No in poor habitats	Survivorship and fecundity
Haddad & Baum (1999)	Insects: butterflies		Experimental	Yes for 3 of 4 species	Size
Forney & Gilpin (1989)	Insects: fruit fly	*Drosophila hydei, D. pseudoobscura*	Experimental (Microcosm)	Yes for 1 of 2 species	Persistence
Shirley & Sibly (2001)	Insect: fruit fly	*D. melanogaster*	Experimental (Microcosm)	Yes in polluted areas No in non-polluted areas	Persistence
Gonzalez et al. (1998)	Microarthropods		Experimental (Microcosm)	Yes for 18 of 21 species	Persistence and size
Burkey (1997)	Microorganisms		Experimental (Microcosm)	Yes	Persistence
Holyoak & Lawler (1996)	Microorganisms		Experimental (Microcosm)	Yes for 2 of 2 species	Persistence and size
On diversity					
MacClintock et al. (1977)	Birds		Observational	Yes	Species richness
Schmiegelow et al. (1997)	Birds		Experimental	No	Species richness, log series α, Jaccard similarity
Collinge (2000)	Insects		Experimental	No	Species richness
Gilbert et al. (1998)	Microarthropods		Experimental (Microcosm)	Yes	Species richness
Gonzalez & Chaneton (2002)	Microarthropods		Experimental (Microcosm)	Yes	Species richness

offers tentative support for corridors, and much more work on population and community responses is needed.

When to expect positive corridor effects

In considering why our review did not strongly support corridor effects, it is important to be clear about the mechanisms or conditions under which we expect corridors to impact populations. Certain species in any community will perceive corridor habitat as being of equal or lesser quality than other surrounding habitat. It should be clear that these species, often habitat generalists, will not respond to corridors. Thus even when corridor experiments are conducted at the appropriate scale and are well controlled and replicated, we do not expect all species to respond positively. Corridor research and application therefore should focus on species that are either specialists for the habitats and corridors of interest, or are likely to exhibit reduced survival when traveling through matrix habitat. Even for habitat specialists, patches must be separated by distances large enough to restrict movement to the rate of few or no individuals per generation without corridors. If movement rates between unconnected patches are high, then immigration does not limit population size or diversity, and corridor utility for increasing population viability depends on their capacity to reduce mortality risk relative to matrix habitat (Hudgens and Haddad 2003).

Finally, corridor effects are likely to be highly scale-specific, both in terms of the scale of the landscape relative to an organism's size or movement distances, and in terms of the timescale of study relative to an organism's movement rate and generation time. Our literature review points directly to this issue of scale. Studies focusing on smaller organisms were generally more likely to find corridor effects than studies focusing on larger organisms (Table 16.1). While just over half of the studies focusing on mammals or birds found corridor effects for a majority of species, all studies focusing on insects and on microcosms found such effects (Table 16.1). One reason for more consistent responses with smaller species is that the landscape size can be better matched to the organism's home range or ambit. Although this could be a strength of model systems, we found that researchers tended to adjust their corridor length to the size of the organism, and that the length of organisms in microcosm studies was not significantly longer relative to corridor length (Fig. 16.1A). Though again not significant, microcosm studies were conducted for many more generations (and include the four right-most points on Fig. 16.1B), allowing a greater time for population dynamics to

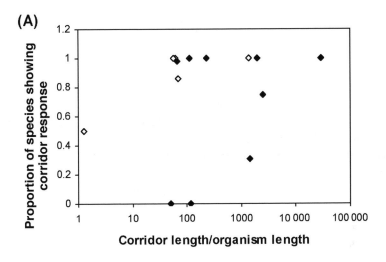

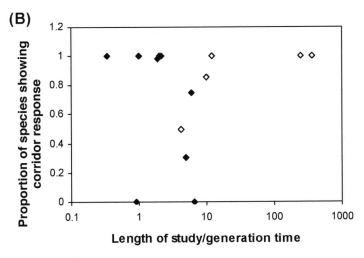

Fig. 16.1. Effects of spatial and temporal scale on population responses to corridors. Each data point represents one population study in Table 16.1, and shows the proportion of species that demonstrated population response to corridors (that is, populations performed significantly differently in connected relative to unconnected patches; $n = 14$ studies), as a function of (A) the ratio of corridor length to average organism length within a study, and (B) the ratio of study duration to the average generation time within a study. Neither relationship was significant in analyses with all studies, or excluding microcosm studies (white diamonds).

respond to corridors. Focus on scale should not become overly restrictive in the study or implementation of corridors, in that any corridor is likely to benefit many organisms, and likewise any organism is likely to respond to corridors at many scales. For example, our work with colleagues on one experiment with 1-ha patches and 150-m long corridors that was initially designed with a focus on butterflies has demonstrated corridor effects on birds, small mammals, insects, and plants (Tewksbury *et al.* 2002; Haddad *et al.* 2003; Levey *et al.* 2005). Still, more attention should be devoted to matching the scales of organisms and landscapes.

Designing corridor studies in variable environments

It is evident from our review that, in addition to issues of scale, issues of landscape variability also need more attention in the design of future studies. Corridor effects are only likely under a certain set of limited conditions, as their role is to increase the likelihood of rare events. Thus, corridors may have no effect on populations in years where dispersal is high (under density-dependent dispersal, these may be at times when populations are growing), but these same corridors may be critical in years when reproduction is lower, and dispersal is limited. Corridors may also have no effects when habitats in the landscape are stable, but be essential in the face of disturbance (Shirley and Sibly 2001). Thus, studies are needed that address the variability of effects rather than simply mean effects. These studies would be conducted for longer time periods and would more explicitly link corridor effects with population stresses.

Our review also makes clear that a critical aspect of study design is to assure that corridor effects are not confounded by other factors. These are not simply factors determined by the local environment, but rather are factors intrinsic to the landscape design. Adding a corridor affects not only connectivity, but patch size and shape, which can also affect population sizes and diversity (Harrison and Bruna 1999; Orrock *et al.* 2003). Haddad and Baum (1999) showed how the addition of a corridor changes edge effects within patches, increasing the area available to edge avoiding butterflies, and thus increasing their abundances. Schmiegelow *et al.* (1997) and Collinge (2000) both discuss how adding corridors affected the size of their experimental patches by adding the area of a corridor, and thus influenced population size and diversity. Only four experiments have controlled for the added area and change in shape caused by corridors in testing for their effects (Gonzalez *et al.* 1998; Boudjemadi *et al.* 1999; Gonzalez and Chaneton 2002;

Tewksbury *et al.* 2002). Perhaps it is because of the difficulty in separating effects of these uncontrolled variables that most studies in our review were experimental rather than observational. Of critical importance in future studies is to account for environmental and landscape variability in the design of experimental and observational studies.

FUTURE DIRECTIONS LINKING THEORY, MODEL SYSTEMS, AND MANAGEMENT

The next decade in corridor research should include an explicit focus on how corridors, and the movement they facilitate, affect populations and communities. Unlike studies of movement alone, which can usually be conducted over short time periods and at a variety of scales, studies of population and community consequences will have to more carefully incorporate into their design an understanding of isolation's impact, both in time and space. Because corridors often have their effects on extinction and on the recolonization that follows, the role of corridors is likely to be observed only after long-term studies or in particularly stressful years.

Theoretical predictions as well as some microcosm studies point to strong impacts of corridors on population and community dynamics, yet studies on macroinvertebrates and vertebrates show weaker and inconsistent effects. Why is this? While corridors may influence movement in many organisms, theory predicts stronger corridor effects on populations linked by rare events — either because the patches are sufficiently distant that migration is rare, or because the organisms are relatively sedentary. Microcosm studies appear to back this claim, with clear population effects in moss microcosm systems (patch area $= 79\,\text{cm}^2$) where the microinvertebrates in the moss are specialists on moss habitat (Gilbert *et al.* 1998; Gonzalez *et al.* 1998; Gonzalez and Chaneton 2002). In those studies, fragmentation created a matrix of completely unsuitable habitat, maximizing the barriers to dispersal between isolated patches and the benefits incurred by corridors that promote exchange. Larger-scale studies are typically leaky systems — corridors may increase movement between patches, but the degree of influence relative to movement through the matrix is often hard to determine and variable between and within species. In the following sections, we outline several research approaches to address corridor effects on populations and communities of species of management concern.

Corridors as conduits for rare events

Most corridor studies to date are small in scale, typically covering centimeters to hundreds of meters. These studies have provided a great deal of insight into how corridors function. Yet their mismatch with scales of landscape conservation is striking. This mismatch is further compounded by typical study species, which are usually common and mobile. These characteristics are convenient for obtaining results in short-term studies. Unlike species of conservation concern, common, mobile species are likely to move through inhospitable matrix, especially when distances between patches are relatively short. Higher movement rates between connected patches may not have population consequences for these species, particularly over short time periods, as movement rates between isolated patches are often sufficient to offset extinction in most years (Hudgens and Haddad 2003).

When movement events are rare and corridors buffer populations against local extinctions during stressful periods, corridors are likely to be most valuable in conservation. Perhaps that is why microcosm studies show such consistently strong effects. Gonzalez and colleagues (1998) examined microarthropod communities on moss patches for only 6 months, and demonstrated some of the strongest effects of corridors to date. While the study appears short in duration, it spanned at least several generations for all species, much greater than typical studies of corridors.

Organisms of management concern that disperse over smaller areas and have generation times of a year or less are much more likely to benefit from empirical studies of corridor effects, because small-scale studies are applicable. There are many species of plants, insects, small mammals, amphibians, and reptiles with relatively short generation times that make up much of total biodiversity and that are likely to benefit from small-scale corridors. Still, full accounting of the spatial and temporal dynamics that mediate corridor effects in these species will require studies lasting multiple generations.

For many rare species with short generation times, local extinctions and colonization dynamics are imposed by natural disturbances, and elucidating the role of corridors requires long-term studies of movement and population sizes to determine population viability. For example, our work with collaborators on an endangered butterfly, the St. Francis satyr (*Neonympha mitchellii francisci*), is designed to determine the role of corridors in facilitating colonization and maintaining viable populations. This sedentary subspecies occurs in small (0.1–0.6 ha) wetland openings

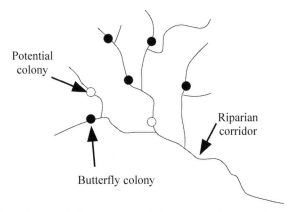

Fig. 16.2. Metapopulation structure of the St. Francis satyr. Riparian corridors may serve as movement corridors to promote colonization of habitats created by disturbance.

along streams (Fig. 16.2) that are maintained by disturbances caused by beavers and by fire, without which vegetation quickly succeeds to unsuitable riparian forest. Butterflies can survive neither the disturbance nor the succession. We believe that riparian habitats serve as corridors to promote colonization of new sites. Yet, in 3 years of research encompassing six butterfly generations, we have marked approximately 750 butterflies and observed just three movements between patches, all between the closest (separated by 300 m), connected patches.

The St. Francis satyr case study highlights an important role of corridors: their effects may be particularly important under stress or disturbance. Successional habitat dynamics for St. Francis satyr occur over many years or decades. The creation of new openings by disturbance generates the opportunity for natural experiments. We have already observed colonization of two sites that had been flooded and then abandoned by beavers (both within a couple of hundred meters of existing sites). It will take many years to observe a sample size of openings that permits conclusions that will affect landscape-level habitat restoration. Although studies of the interaction between corridors and disturbance are rare, Shirley and Sibly (2001) created a microcosm experiment with fruit flies that demonstrated the important interactions between corridors and environmental disturbance. They investigated metapopulation response under unpolluted and polluted conditions, and found that corridors increased population persistence in patches disturbed with pollution (Shirley and Sibly 2001). An important area of future research will be to further understand how the interrelationship between corridors and disturbance affects populations and communities.

Thinking big: large-scale manipulations

Conservation at the landscape scale usually involves corridors that may extend kilometers to hundreds of kilometers. At these scales, there have been only a few successful studies of movement (Beier 1995), gene flow (Hale *et al.* 2001; Mech and Hallett 2001), or population sizes (Dunning *et al.* 1995). One response variable that is not included in the corridor studies we reviewed, but is typical of connectivity studies in metapopulations, is patch occupancy (Moilanen and Nieminen 2002; Moilanen and Hanski Chapter 3). One approach to expand the number of corridor studies at larger scales may be to focus more specifically on patch occupancy in landscapes with and without natural corridors (MacKenzie *et al.* 2006).

The primary constraint on studies at large scales is the difficulty in finding replication and in controlling for variables that may confound corridor effects. In addition to the confounding factors discussed above, connected patches tend to be larger than unconnected patches (Villard *et al.* 1999; Fahrig 2003). Overcoming confounding effects of other variables will likely require a great deal of replication and/or judicious pairing of control and treatment sites, which can be difficult to find and sample at large spatial scales.

Ideally, larger-scale, longer-term studies will involve some level of controlled experimentation. Experimental manipulation of both habitat and disturbance levels allows isolation of mechanisms and greater time-efficiency by eliminating confounding variables. Long-term studies where landscapes are manipulated over large areas are much more likely to yield definitive results regarding corridor effects, and collaborative teams of researchers working at large scales may be much more effective than individual researchers working separately at smaller scales. Evidence from one long-term, large-scale fragmentation experiment, the Biological Dynamics of Forest Fragments Project in Brazil, suggests that responses accumulate over time. This experiment was created starting in 1980 in Manaus, Brazil to test the effects of fragmentation and patch size on tropical ecosystems. It is only after more than a decade of study, that major community and ecosystem impacts of fragmentation have been documented (e.g., Laurance *et al.* 1997, 2001; Bierregaard and Gaston 2001; Ferraz *et al.* 2003; see also Cook *et al.* 2005). Given the critical roles of habitat loss and fragmentation as the most important factors impacting the loss of biodiversity (Wilcove *et al.* 1998), more such studies are needed.

One approach to studying effects of connectivity at large scales is to take advantage of manipulations that occur as part of landscape management. Such manipulations occur every day through forestry, development, agriculture, and other changes in land use. Although many alterations come through habitat loss, habitat restoration should also provide opportunities for experimental assessment of responses to connectivity. Because large-scale habitat modification for scientific research alone can create serious ethical concerns, we recommend coordinating research plans along with planned habitat modifications (destruction or restoration), so that useful information can be gained in the context of adaptive management. Unfortunately, land-use manipulations are typically uncontrolled with respect to landscape factors like connectivity and other important environmental factors that might obscure landscape-level responses, thus limiting their usefulness in guiding future management. As pointed out by Beier and Noss (1998), a good example of how a study can be designed around landscape management was conducted by Mansergh and Scotts (1989). By measuring responses before and after corridor restoration at a ski resort, they demonstrated positive effects of corridors on mountain pygmy possum survivorship.

In lieu of controlled experimentation, new research will have to be creative in identifying opportunities for replicated large-scale manipulations allowing isolation of corridor effects. In our own work, we have found that academic partnerships and close collaboration with land-management agencies are critical for the success of these projects. With investigators from three additional academic institutions, we have been working closely with the US Forest Service at the Savannah River Site in South Carolina to assess large-scale effects of corridors. The Savannah River Site is an 80 000-ha site managed for plantation pine forest and as native habitat for wildlife. The Forest Service employs clearcut forestry, and creates clearings that range in size from 5 to 50 ha. These clearings vary in their connectivity, as some of the clearings are connected to others by long, straight utility rights-of-way and roads (Fig. 16.3). Because these rights-of-way are subjected to frequent, thorough disturbance (by herbicide and mowing), they are unlikely to be long-term sources of butterflies, but instead serve as corridors between suitable habitats.

To test the effects of large-scale, open corridors in landscapes managed for forestry, we have studied butterfly species that thrive in early

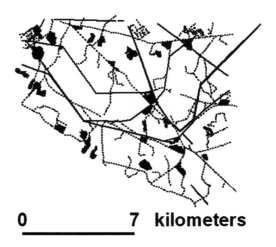

0   7 kilometers

Fig. 16.3. Fragmentation of cleared patches caused by forest management at the Savannah River Site, South Carolina. Black areas are clearings 1–7 years of age. Solid black lines are utility rights-of-way that may serve as corridors for dispersing butterflies and other organisms. Dashed lines are roads, primarily small forest roads, that are less likely to serve as corridors. Cleared patches vary in isolation, both in their distance to other patches and in their connection to other patches via utility rights-of-way.

successional habitats for up to 8 years after forest harvest, when the pine canopy starts to close. After extensive studies of two of these species, the buckeye (*Junonia coenia*) and the variegated fritillary (*Euptoieta claudia*), using large, replicated, experimental landscapes, we have shown that both are more likely to move between connected patches separated by up to 400 m (Haddad 1999a; Tewksbury *et al.* 2002). To test these corridor effects at the much larger scales of operational forestry, we conducted repeated surveys for both species in all (*n* = 137) clearcut openings on the Savannah River Site (some connected by utility right-of-way corridors, others not). After controlling for potentially confounding effects such as stand type, area, and age, very preliminary analyses from the first year of a multi-year study indicate that the presence of a corridor increases population sizes of fritillaries, but not buckeyes (B. Danielson and N. Haddad, unpublished data). Thus, at least for the fritillary, results from small-scale studies appear to "scale up" to larger areas. It is worth noting that the one species whose responses did scale up — the fritillary — is more sedentary than the species whose responses did not show corridor effects at the largest scale, the buckeye (Haddad 1999a). While far from conclusive, this supports the theoretical prediction that corridor effects are more likely in situations in which migration between patches is relatively rare. This work highlights a challenge in ecology and conservation, which is to reconcile often smaller-scale experimental with often larger-scale observational data.

Linkage across life-history and trophic levels

Population- and community-level responses to corridors are potentially caused by many different mechanisms spanning trophic levels and acting on different life-history stages. This diversity of corridor effects could magnify or dampen the observed response for any particular population or community. Most often, researchers consider the effects of corridors on movement or gene flow within an individual species. Considered more deeply, however, there can be multiple stages at which movement can be important, and multiple interactions that can result in positive or negative effects of corridors on individual species. For even a single species, these may include a diversity of interactions with different groups — predators, competitors, mutualists, parasites — all of which may respond to fragmentation and corridors. For many plants, the initial effect of corridors on movement rates will be a direct function of corridor effects on pollinators and seed dispersers, and plant establishment will additionally be influenced by the response of seed-predators, parasites, and herbivores. This diversity of interactions may cause contrasting responses and can make detection of net population responses difficult. The positive effects of corridors may be dampened or reversed by negative effects (Fig. 16.4) (see also Crooks and Sanjayan Chapter 1), and more work is needed to assess the balance of positive and negative effects on population and community structure.

At both small and large scales, assessing net corridor effects on populations will most likely be done through long-term studies. In studies of short duration (i.e., the typical duration of a grant funding cycle or of a dissertation program), approaches that focus on aspects of population demography might provide more rapid assessment of corridor effects on survivorship and reproduction at key life-history stages (see Mansergh and Scotts 1989; Beier and Noss 1998; Coffman et al. 2001). This approach may occasionally allow researchers to model corridor effects on populations, especially if key life-history attributes and developmental stages are easily identified. With structured data, population models can be used to assess population dynamics and viability in the presence or absence of corridors (discussed below). Yet conclusions from demographic models must be approached with caution, as parameters estimated over

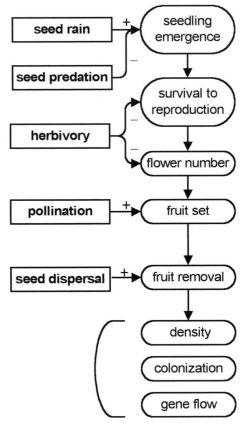

Fig. 16.4. Effects of corridors on various life stages of plants. Multiple plant–animal interactions could cancel or enhance corridor effects on a single plant species. Interactions are in square boxes, with arrows pointing to life-history stage affected (in ovals) with positive and negative signs indicating the most likely effect on plant demography. Density-dependent linkages between interactions (seed rain and seed predation, pollination, and herbivory), and feedbacks onto animal counterparts of the plants (seed predators, herbivores, pollinators, etc.), further increase the complexity of effects that may be influenced by patch isolation and connectivity.

relatively short time intervals may fail to capture corridor impacts in extreme years (catastrophes or bonanzas: Morris and Doak 2002) when corridors may be most important.

We have conducted an experimental study over the past decade where we have had some success at determining corridor effects on aspects of population demography (Tewksbury *et al.* 2002; Haddad *et al.* 2003). These experiments at the Savannah River Site have involved 1-ha experimental patches that are separated by 100 + m. The patches are open, early-successional habitat surrounded by plantation pine forest. The openings are suitable for many species of plants, insects, mammals, and birds, while the pine forest is not. Our work has focused mainly on movement of insects, small mammals, birds, and plants, but we have made some inroads into understanding effects on population demography. For example, we have found that corridors increase plant pollination rates and dispersal of fruiting plants (Tewksbury et al. 2002; Haddad et al. 2003). Both of these factors should lead to higher seed numbers in connected patches, leading to predictions of higher population sizes (Tewksbury et al. 2002). Yet corridor effects on plant demography are complex, and effects at one life stage may be modified or reversed by effects on other life stages. Studies with our collaborators, for example, have shown that the same corridors that increase pollination and seed dispersal (Tewksbury et al. 2002) also lead to increases in seed predation by mammals that use corridors as foraging conduits (Orrock et al. 2003; Brinkerhoff et al. 2005). The net impact of these corridor-related effects on whole plant population demography is unclear (Fig. 16.4). Further research should focus on integrating landscape effects across life-history stages.

Model-directed experimental research

Empirical approaches will benefit from informed models that guide targeted experiments. Three types of modeling exercises may be effective at informing future research: individual-based models, numerical models, such as demographic models linking effects at different life stages, and analytical models that determine the types of organisms most likely to respond to corridors. Models will serve to assess impacts of empirically measured responses at one life-history stage on total population size. They will also be important in determining which demographic characteristics and behaviors should be the focus of further empirical study.

Individual-based models, often spatially explicit, have the advantage of linking some of the most available data on corridor use — movement data — to impacts on populations and communities. To date, such models have had some success at linking local behavior to larger-scale distribution (Tilman et al. 1997; Tischendorf and Wissel 1997; Haddad 1999b; Levey et al. 2005). Individual-based models can incorporate understanding of fine-scale decision rules that link dispersal to landscape structure (Tracey Chapter 14). Habitat-specific data on movement distances and turning can be used to predict corridor effects on patch colonization. Of specific interest is the role of habitat boundaries, as they define corridor structure and function (Cadenasso et al. 2003), and vary in their permeability across species (Haddad 1999b). Other decision rules, such as the effects of density or presence of related individuals (as in Le Galliard et al. 2003), may also be included. One advantage to such models is that landscape characteristics can be easily varied to generate predictions about corridor impacts. For example, three different models have predicted that there is an asymptote to the effect of corridor width on movement rates (Tilman et al. 1997; Tischendorf and Wissel 1997; Haddad 1999b). Although this prediction has received some empirical support (Andreassen et al. 1996), more is needed. A disadvantage of this approach is that adapting such models to population-level questions involves a large and often intractable number of parameters. Such individual-based models may be most effective in predicting corridor effects on species with large ranges and long generation times, for which population trends are very difficult to obtain. In these circumstances, individual-based models of movement in relation to habitat boundaries, coupled with spatially explicit, habitat-specific data on reproduction and survival, may provide much needed insight into the probable effectiveness of proposed changes in land management.

Two other types of models may have greater practical applicability in determining corridor effects on population size and community structure. Numerical models, such as matrix models and structural equation modeling (Grace and Pugesek 1998; Caswell 2001), can be used to project population trends and assess population viability. The primary advantage to these approaches is that they integrate empirical estimates from research targeted at specific life-history stages. Rather than ignore all corridor effects but those on movement and on population sizes, a model-based approach integrates the impacts of corridors on species and interactions across life-history stages. In addition to determining the contribution of each stage to population growth, sensitivity analyses can be used to determine how management targeted at a specific stage might be used to increase population sizes in connected landscapes. To date, the use of these approaches has been limited, primarily because researchers have yet to gather data on the effects of corridors across life-history stages.

Another approach that can be used to predict responses to corridors is analytical modeling. Analytical models can have a closed form solution, such that the equations that describe the effects of corridors on populations can be expressed as a mathematical function. These are often much simpler than the real-world corridors they represent, but they can provide guidance about which characteristics are in need of targeted research, and which species are in need of conservation. Metapopulation models have become increasingly common in assessing effects of connectivity, including corridors (Henein and Merriam 1990; Hess 1996; Anderson and Danielson 1997; McCallum and Dobson 2002; Moilanen and Hanski Chapter 3). These models have provided some of the few analyses of how species interactions in fragmented landscapes with or without corridors may affect host populations in the context of disease dynamics (Hess 1996; McCallum and Dobson 2002; McCallum and Dobson Chapter 19).

Analytical models that assess the role of corridors on population size are less common. In simple population growth models, Earn and colleagues (2000) showed how corridors can synchronize population dynamics and lead to metapopulation extinction. Hudgens and Haddad (2003) used simple logistic population growth models to determine for what types of species and situations corridors were likely to affect populations. They modeled a two-patch system that accounted for corridor effects by including terms for corridor and matrix migration and mortality. Their results generated a number of implications for further research and future conservation. First, their results suggest that corridors are likely to benefit species with high population growth rates in the short term, whereas they are likely to benefit species with low population growth rate in the long term. Since most species of conservation concern have low growth rates, this result further reinforces the need for studies of long-term population responses to corridors. Second, their results suggest that the type of population dynamics exhibited by a species will determine corridor effectiveness. Species that have large population oscillations with years of low population size are likely to be harmed by corridors, as corridors are likely to synchronize population dynamics among patches. Corridors are likely to benefit species of conservation concern that are experiencing sustained population decline. Third, they showed that it is usually not just dispersal through corridor and matrix habitats, but also corridor and matrix mortality that determine corridor benefits. Only when species have extremely low matrix migration are corridors likely to benefit populations through their role in increasing migration. With higher matrix migration, corridors are likely to increase population viability by reducing mortality during dispersal. These results emphasize the need for demographic studies not only in patches, but in corridor and matrix habitats as well.

CONCLUSION

Over the past decade, empirical research has generated broad support for the hypothesis that habitat corridors increase movement through fragmented landscapes for many species. The effects of corridors on population viability, however, are less well studied, and the empirical understanding of corridor effects on community structure and diversity is still in its infancy. In our review of the existing evidence, we find that support for corridor effects on populations is growing, though with many caveats. There is more support for corridor effects in smaller taxa with shorter generation times. This result reflects greater ease in matching temporal and spatial scales of experiments with smaller species. Although corridors are intended to promote movement, their ultimate effectiveness in conservation must be measured by their population- and community-level effects in promoting colonization, reducing extinction, and increasing population viability.

Empirical support for corridors at the population and community level would strengthen arguments to maintain or construct corridors, rather than to allocate resources toward other conservation strategies. Empirical tests of corridor effects on populations and communities will require long-term experiments conducted at large spatial scales, coupled with creative approaches to obtaining difficult to observe events, such as rare, long-distance dispersal. While such studies require careful coordination, we believe there are opportunities for such large-scale studies in conjunction with planned habitat restoration or destruction. For rare species, purely observational studies may be the only ones that are possible to inform conservation decisions. However, they rarely can overcome the many intercorrelated factors confounding the effects of connectivity. Rather than focusing only on population abundance or species numbers, successful studies are likely to integrate research on movement, habitat-specific demography, and density-dependent interactions between species in an assessment of corridor effects on population viability. Empirical studies would be aided by insights from analytical, demographic, and individual-based models that help to focus research on specific life-history characteristics and life-cycle stages that link corridor effects to population dynamics. We are hopeful that such approaches will help stimulate more rapid progress in understanding the impacts of corridors on populations and communities, and, thus, their value in conservation.

ACKNOWLEDGEMENTS

We would like to acknowledge work by our many collaborators who have contributed to our understanding of corridor effects, including Jory Brinkerhoff, Ellen Damschen, Brent Danielson, Brian Hudgens, Daniel Kuefler, Doug Levey, John Orrock, Sarah Sargent, Patricia Townsend, Aimee Weldon, and many others. We would also like to thank our agency collaborators, including John Blake and Ed Olson at the Savannah River Site and Erich Hoffman at Fort Bragg. We thank Becky Bartel, Paul Beier, Kevin Crooks, Ellen Damschen, Bill Fagan, Andrew Gonzalez, Daniel Kuefler, Aimee Weldon, and Neal Williams for comments on this chapter. Our work on corridors has been funded by National Science Foundation DEB-9907365, the US Forest Service, Savannah River, and the Department of Defense at Fort Bragg.

REFERENCES

Aars, J., and R. A. Ims. 1999. The effect of habitat corridors on rates of transfer and interbreeding between vole demes. *Ecology* **80**:1648–1655.

Anderson, G. S., and B. J. Danielson. 1997. The effects of landscape composition and physiognomy on metapopulation size: the role of corridors. *Landscape Ecology* **12**:261–271.

Andreassen, H. P., S. Halle, and R. A. Ims. 1996. Optimal width of movement corridors for root voles: not too narrow and not too wide. *Journal of Applied Ecology* **33**:63–70.

Beier, P. 1995. Dispersal of juvenile cougars in fragmented habitat. *Journal of Wildlife Management* **59**:228–237.

Beier, P., and R. F. Noss. 1998. Do habitat corridors really provide connectivity? *Conservation Biology* **12**:1241–1252.

Berggren, A., B. Birath, and O. Kindvall. 2002. Effects of corridors and habitat edges on dispersal behavior, movement rates, and movement angles in Roesel's bush-cricket (*Metrioptera roeseli*). *Conservation Biology* **16**:1562–1569.

Bierregaard R. O., Jr., and C. Gaston. 2001. The Biological Dynamics of Forest Fragments project: overview and history of a long-term conservation project. Pp. 5–12 in R. O. Bierregaard Jr., C. Gascon, T. E. Lovejoy, and R. C. G. Mesquita (eds.) *Lessons from Amazonia: The Ecology and Conservation of a Fragmented Forest*. New Haven, CT: Yale University Press.

Boudjemadi, K., J. Lecomte, and J. Clobert. 1999. Influence of connectivity on demography and dispersal in two contrasting habitats: an experimental approach. *Journal of Animal Ecology* **68**:1207–1224.

Bowne, D. R., J. D. Peles, and G. W. Barrett. 1999. Effects of landscape spatial structure on movement patterns of the hispid cotton rat (*Sigmodon hispidus*). *Landscape Ecology* 14:53–65.

Brinkerhoff, R. J., N. M. Haddad, and J. L. Orrock. 2005. Corridors and olfactory predator cues affect small mammal behavior. *Journal of Mammalogy* 86:662–669.

Brown, J. H., and A. Kodric-Brown. 1977. Turnover rates in insular biogeography: effect of immigration on extinction. *Ecology* 58:445–449.

Burkey, T. V. 1997. Metapopulation extinction in fragmented landscapes: using bacteria and protozoa communities as model ecosystems. *American Naturalist* 150:568–591.

Cadenasso, M. L., S. T. A. Pickett, K. C. Weathers, *et al.* 2003. An interdisciplinary and synthetic approach to ecological boundaries. *BioScience* 53:717–722.

Caswell, H. 2001. *Matrix Population Models: Construction, Analysis, and Interpretation*, 2nd edn. Sunderland, MA: Sinauer Associates.

Coffman, C. J., J. D. Nichols, and K. H. Pollock. 2001. Population dynamics of *Microtus pennsylvanicus* in corridor-linked patches. *Oikos* 93:3–21.

Collinge, S. K. 2000. Effects of grassland fragmentation on insect species loss, colonization, and movement patterns. *Ecology* 81:2211–2226.

Cook, W. M., J. Yao, B. L. Foster, R. D. Holt, and L. B. Patrick. 2005. Secondary succession in an experimentally fragmented landscape: community patterns across space and time. *Ecology* 86:1267–1279.

Danielson, B. J., and M. W. Hubbard. 2000. The influence of corridors on the movement behavior of individual *Peromyscus polionotus* in experimental landscapes. *Landscape Ecology* 15:323–331.

Dunning, J. B., Jr., J. R. Borgella, K. Clements, and G. K. Meffe. 1995. Patch isolation, corridor effects, and colonization by a resident sparrow in a managed pine woodland. *Conservation Biology* 9:542–550.

Earn, D. J. D., S. A. Levin, and P. Rohani. 2000. Coherence and conservation. *Science* 290:1360–1364.

Fahrig, L. 2003. Effects of habitat fragmentation on biodiversity. *Annual Reviews of Ecology and Systematics* 34:487–515.

Fahrig, L., and G. Merriam. 1985. Habitat patch connectivity and population survival. *Ecology* 66:1762–1768.

Ferraz G., G. J. Russell, P. C. Stouffer, *et al.* 2003. Rates of species loss from Amazonian forest fragments. *Proceedings of the National Academy of Sciences of the USA* 100:14069–14073.

Forney, K. A., and M. E. Gilpin. 1989. Spatial structure and population extinction: a study with *Drosophila* flies. *Conservation Biology* 3:45–51.

Gilbert, F., A. Gonzalez, and I. Evans-Freke. 1998. Corridors maintain species richness in the fragmented landscape of a microecosystem. *Proceedings of the Royal Society of London B* 265:577–582.

Gonzalez, A., and E. J. Chaneton. 2002. Heterotroph species extinction, abundance and biomass dynamics in an experimentally fragmented microecosystem. *Journal of Animal Ecology* 71:594–602.

Gonzalez, A., J. H. Lawton, F. S. Gilbert, T. M. Blackburn, and I. Evans-Freke. 1998. Metapopulation dynamics, abundance, and distribution in a microecosystem. *Science* 281:2045–2047.

Grace, J. B., and B. H. Pugesek. 1998. On the use of path analysis and related procedures for the investigation of ecological problems. *American Naturalist* 152:151–159.

Haddad, N. M. 1999a. Corridor and distance effects on interpatch movements: a landscape experiment with butterflies. *Ecological Applications* 9:612–622.

Haddad, N. M. 1999b. Corridor use predicted from behaviors at habitat boundaries. *American Naturalist* 153:215–227.

Haddad, N. M., and K. A. Baum. 1999. An experimental test of corridor effects on butterfly densities. *Ecological Applications* 9:623–633.

Haddad, N. M., D. R. Bowne, A. Cunningham, *et al.* 2003. Corridor use by diverse taxa. *Ecology* 84:609–615.

Hale, M. L., P. W. W. Lurz, M. D. F. Shirley, *et al.* 2001. Impact of landscape management on the genetic structure of red squirrel populations. *Science* 293:2246–2248.

Hannon, S. J., and F. K. A. Schmiegelow. 2002. Corridors may not improve the conservation value of small reserves for most boreal birds. *Ecological Applications* 12:1457–1468.

Hanski, I. 1999. *Metapopulation Ecology.* Oxford, UK: Oxford University Press.

Harrison, S., and E. Bruna. 1999. Habitat fragmentation and large-scale conservation: what do we know for sure? *Ecography* 22:225–232.

Henein, K., and G. Merriam. 1990. The elements of connectivity where corridor quality is variable. *Landscape Ecology* 4:157–170.

Hess, G. R. 1996. Linking extinction to connectivity and habitat destruction in metapopulation models. *American Naturalist* 148:226–236.

Holyoak, M. 2000. Habitat patch arrangement and metapopulation persistence of predators and prey. *American Naturalist* 156:378–389.

Holyoak, M., and S. P. Lawler. 1996. The role of dispersal in predator–prey metapopulation dynamics. *Journal of Animal Ecology* 65:640–652.

Hudgens, B. R., and N. M. Haddad. 2003. Predicting which species will benefit from corridors in fragmented landscapes from population growth models. *American Naturalist* 161:808–820.

Ims, R. A., and H. P. Andreassen. 1999. Effects of experimental habitat fragmentation and connectivity on root vole demography. *Journal of Animal Ecology* 68:839–852.

Kirchner, F., J. Ferdy, C. Andalo, B. Colas, and J. Moret. 2003. Role of corridors in plant dispersal: an example with the endangered *Ranunculus nodiflorus.* *Conservation Biology* 17:401–410.

La Polla V. N., and G. W. Barrett. 1993. Effects of corridor width and presence on the population dynamics of the meadow vole (*Microtus pennsylvanicus*). *Landscape Ecology* 8:25–37.

Laurance, S. G., and W. F. Laurance. 1999. Tropical wildlife corridors: use of linear rainforest remnants by arboreal mammals. *Biological Conservation* 91:231–239.

Laurance, W. F., S. G. Laurance, L. V. Ferreira, *et al.* 1997. Biomass collapse in Amazonian forest fragments. *Science* 278:1117–1118.

Laurance, W. F., D. Perez-Salicrup, P. Delamonica, *et al.* 2001. Rain forest fragmentation and the structure of Amazonian liana communities. *Ecology* 82:105–116.

Le Galliard J., R. Ferrière, and J. Clobert. 2003. Mother-offspring interactions affect natal dispersal in a lizard. *Proceedings of the Royal Society of London B* 270:1163–1169.

Levey, D. J., B. M. Bolker, J. J. Tewlsbury, S. Sargent, and N. M. Haddad. 2005. Effects of landscape corridors on seed dispersal by birds. *Science* 309:146–148.

Levins, R. 1969. Some demographic and genetic consequences of environmental heterogeneity for biological control. *Bulletin of the Entomological Society of America* 15:237–240.

MacArthur, R. H., and E. O. Wilson 1967. *The Theory of Island Biogeography.* Princeton, NJ: Princeton University Press.

MacClintock, L., R. F. Whitcomb, and B. L. Whitcomb. 1977. Island biogeography and the "habitat islands" of eastern forest. II. Evidence for the value of corridors and minimization of isolation in preservation of biotic diversity. *American Birds* 31:6–12.

Mackenzie, D. I., J. D. Nichols, J. A. Royle, *et al.* 2006. *Occupancy Estimation and Modeling: Inferring Patterns and Dynamics of Species Occurence.* Boston, MA: Academic Press.

Machtans, C. S., M. Villard, and S. J. Hannon. 1996. Use of riparian buffer strips as movement corridors by forest birds. *Conservation Biology* 10:1366–1379.

Mansergh, I. M., and D. J. Scotts. 1989. Habitat continuity and social organization of the mountain pygmy-possum restored by tunnel. *Journal of Wildlife Management* 53:701–707.

McCallum, H., and A. Dobson. 2002. Disease, habitat fragmentation and conservation. *Proceedings of the Royal Society of London B* 269:2041–2049.

Mech, S. G., and J. G. Hallett. 2001. Evaluating the effectiveness of corridors: a genetic approach. *Conservation Biology* 15:467–474.

Moilanen, A., and M. Nieminen. 2002. Simple connectivity measures in spatial ecology. *Ecology* 83:1131–1145.

Mönkkönen M., and M. Mutanen. 2003. Occurrence of moths in boreal forest corridors. *Conservation Biology* 17:468–475.

Morris, W. F., and D. F. Doak. 2002. *Quantitative Conservation Biology: Theory and Practice of Population Viability Analysis.* Sunderland, MA: Sinauer Associates.

Orrock, J. L., B. J. Danielson, M. J. Burns, and D. J. Levey. 2003. Spatial ecology of predator–prey interactions: corridors and patch shape influence seed predation. *Ecology* 84:2589–2599.

Perault, D. R., and M. V. Lomolino. 2000. Corridors and mammal community structure across a fragmented, old-growth forest landscape. *Ecological Monographs* 70:401–422.

Petchey, O. L., A. Gonzalez, and H. B. Wilson. 1997. Effects on population persistence: the interaction between environmental noise colour, intraspecific competition and space. *Proceedings of the Royal Society of London B* 264:1841–1847.

Pryke, S. R., and M. J. Samways. 2001. Width of grassland linkages for the conservation of butterflies in South African afforested areas. *Biological Conservation* 101:85–96.

Rosenberg, D. K., B. R. Noon, J. W. Megahan, and E. C. Meslow. 1998. Compensatory behavior of *Ensatina eschscholtzii* in biological corridors: a field experiment. *Canadian Journal of Zoology* 76:117–133.

Schmiegelow, F. K. A., and M. Mönkkönen. 2002. Habitat loss and fragmentation in dynamic landscapes: avian perspectives from the boreal forest. *Ecological Applications* 12:375–389.

Schmiegelow, F. K. A., C. S. Machtans, and S. J. Hannon. 1997. Are boreal birds resilient to forest fragmentation? An experimental study of short-term community responses. *Ecology* 78:1914–1932.

Shirley, M. D. F., and R. M. Sibly. 2001. Metapopulation dynamics of fruit flies undergoing evolutionary change in patchy environments. *Ecology* 82:3257–3262.

Tewksbury, J. J., D. J. Levey, N. M. Haddad, *et al.* 2002. Corridors affect plants, animals, and their interactions in fragmented landscapes. *Proceedings of the National Academy of Sciences of the USA* 99:12923–12926.

Tilman D., C. L. Lehman, and P. Kareiva. 1997. Population dynamics in spatial habitats. Pp. 3–20 in D. Tilman, and P. Kareiva (eds.) *Spatial Ecology: The Role of Space in Population Dynamics and Interspecific Interactions*. Princeton, NJ: Princeton University Press.

Tischendorf, L., and C. Wissel. 1997. Corridors as conduits for small animals: attainable distances depending on movement pattern, boundary reaction, and corridor width. *Oikos* 79:603–611.

Villard, M., M. K. Trzcinski, and G. Merriam. 1999. Fragmentation effects on forest birds: relative influence of woodland cover and configuration on landscape occupancy. *Conservation Biology* 13:774–783.

Wilcove, D. S., D. Rothstein, J. Dubow, A. Phillips, and E. Losos. 1998. Quantifying threats to imperiled species in the United States. *BioScience* 48:607–615.

17

Exploring the functional connectivity of landscapes using landscape networks

DAVID M. THEOBALD

INTRODUCTION

Understanding landscape connectivity is an important research challenge for conservation science (Taylor *et al.* 1993). A relatively recent development in pursuit of this challenge is the differentiation between functional and structural connectivity of landscapes. Structural connectivity is based on the spatial arrangement of different types of habitat in a landscape, while functional connectivity recognizes the behavioral response of individuals, species, or ecological processes to the physical structure of the landscape (Baudry and Merriam 1988; Bennett 1999; Crooks and Sanjayan Chapter 1; Taylor *et al.* Chapter 2; Fagan and Calabrese Chapter 12). A number of conceptual and practical developments have contributed to the recognition of and ability to conduct analyses of functional landscape connectivity.

The patch—matrix—corridor conceptualization of landscapes (Forman and Godron 1986) that built on the theory of island biogeography (MacArthur and Wilson 1967) has strongly influenced work on landscape connectivity. Many researchers have recognized that island biogeography's assumption of islands of habitat distributed throughout a homogenous matrix of non-habitat (i.e., the so-called "inhospitable sea") is overly simple and does not recognize an organism-centric perspective of the landscape that differentiates perception, mobility, and resource use by individuals and species (Johnson *et al.* 1992; Wiens 1994). For example, Vos and Stumpel (1995) examined the relationship between pond occupancy by tree frogs and distance to the nearest occupied pond, but the heterogeneity of the intervening landscape was ignored. Species movement is influenced not just by inter-patch distances, but by the characteristics of the intervening matrix such as vegetation type, structure, and land use (Wiens *et al.* 1993). Taylor *et al.* (1993) argued that landscape connectivity is distinct from landscape composition and physiognomy (or structure), and more recently Taylor *et al.* (Chapter 2) refined the distinction between structural and landscape connectivity. Increasingly, empirical studies have found that matrix heterogeneity influences movement among patches (e.g., van Langevelde 2000; Roland *et al.* 2000; Ricketts 2001). Also, in reaction to the proliferation of work identifying corridors on landscapes, Puth and Wilson (2001) have argued that patch boundaries and corridors are at the opposite extremes of a continuum based on permeability and directionality of ecological flows. Another influential approach to understanding landscape connectivity has been metapopulation dynamics (Hanski 1998; Moilanen and Hanski Chapter 3). This work has emphasized movement by juvenile dispersal among patches that support populations of a species. Movement here is defined as the process by which individual organisms are displaced in space over time (Turchin 1998). However, work on landscape connectivity needs to recognize that a landscape is organized at a variety of levels by movements that occur at a range of temporal scales (Table 17.1). Typically,

Table 17.1. *Functional landscape connectivity incorporates a range of movements that occur at a variety of scales, from centuries to hours. Although the temporal scale of movement types will vary greatly with different species (and thus the examples below are general guidelines), it is important to recognize that the functional definition of a "patch" depends on the temporal context conditioned by the movement type being considered*

Scale	Movement type	Patch types
Century	Genetic exchange	Population
Decadal	Natal dispersal, genetic exchange	Population
Yearly	Natal dispersal, genetic exchange	Home range (lifetime/annual)
Seasonally	Seasonal migration	Resource (winter to summer range; fish habitat to spawning; butterfly nectar resource, larval resource, roost sites, overwintering sites), home range (seasonal)
Weekly	Foraging	Resource
Daily	Roost to forage	Roost, resource
Hourly	Forage, safety	Resource, refuge, or escape habitat

the number of linkages or corridors at each level increases along the temporal scale from long to short term. That is, there are only a few linkages for broad-scale connectivity that facilitate genetic and natal dispersal, while numerous short-scale linkages facilitate seasonal to weekly to daily movements among resource patches (e.g., Dennis *et al.* 2003). These conceptual developments shift the perspective of a landscape from discrete patches (or even binary landscape) to a gradient of subtle changes in the landscape (Theobald and Hobbs 2001). Progress has been made from simple, structural representations to more complex notions of functional connectivity of landscapes.

Complementing these conceptual developments has been progress in developing ways to implement them, primarily through geographic information system (GIS) approaches. Knaapen *et al.* (1992) were among the first to explicitly recognize that matrix quality affects dispersal success among isolated habitat patches. They quantified habitat isolation using the notion of minimum cumulative resistance, which is computed as the probability of moving through a landscape as the product of the distance and the resistance of the landscape. Numerous researchers have subsequently used the idea of landscape resistance to quantify the response of movement through a heterogeneous landscape.

A common way to understand landscape resistance is through individual-based models (IBMs) that parameterize individual response to landscape structure through measures of landcover resistance, for example Boone and Hunter (1996), Gustafson and Gardner (1996), Schippers *et al.* (1996), Cramer and Portier (2001), Tischendorf *et al.* (2003), Gardner and Gustafson (2004), Kramer-Schadt *et al.* (2004), and Carroll (Chapter 15). A second approach to understanding functional movement through a landscape is to estimate an ecological or effective distance among patches (e.g., Paetkau *et al.* 1997; Ferreras 2001). Typically ecological distance relies on least-cost path analysis, which is based on finding the optimal or least-cost path of cells between two source patches. An early and widely cited conservation science application of the least-cost method was Walker and Craighead's (1997) modeling of grizzly bear connectivity in the Greater Yellowstone Ecosystem. More recent examples of work that has employed least-cost path include Halpin and Bunn (2000), Servheen *et al.* (2001), Ray *et al.* (2002), Singleton *et al.* (2002), Adriaensen *et al.* (2003), Miller *et al.* (2003), Larkin *et al.* (2004), Beier *et al.* (Chapter 22), Clevenger and Wierzchowski (Chapter 20), and Rothley and Rae (2005). Effective distance generated by least-cost path analysis has been found to be more useful than straight-line distances and increasingly effective distance has been extracted and used to weight a graph (or network) representation of a landscape (e.g., Bunn *et al.* 2000). Although most work on effective distance to date has occurred in terrestrial ecosystems, effective distance has begun to be used to account for hydrologic (stream) distance and flow direction in freshwater eco-systems (e.g., Olden *et al.* 2001; Fagan 2002; Shurin and Havel 2002; Cottenie *et al.* 2003) and in marine ecosystems as well (e.g., Halpin and Bunn 2000).

In this chapter I review the biological assumptions and computational issues associated with quantifying functional landscape connectivity through least-cost path analysis. I then offer a refinement to standard least-cost path methodology that couples estimation of effective distance with graph-theoretical techniques. I review basic graph metrics and develop the idea of a *landscape network*, which incorporates key innovations such as functionally defined patches and multiple pathways. The approach presented here is offered in pursuit of challenging questions that conservation scientists commonly face: is loss of connectivity leading to species imperilment? Which habitat patches are the most important for long-term viability? Which are the most important linkages? How much fragmentation of habitat will occur given different scenarios of land-use change? Where should a mitigation structure, such as an over- or under-pass, be placed to restore wildlife movement and maximize connectivity? To address these pressing questions, a useful solution should be con-ceptually elegant and computationally feasible to ensure it will be useful when dealing with complex real-world problems.

COMPUTING EFFECTIVE DISTANCE

The term *effective distance* (or cost-weighted distance) has been used to distinguish straight-line or Euclidean distance from distance that is modified by landscape resistance (Berry 1993; Ferreras 2001; Michels *et al.* 2001). Graphical and mathematical methods to compute least-cost routes existed as early as 1968 (Werner 1968). Berry (1993) noted that spatial analysis has been hindered because typical GIS methods are based on straight-line distance, which ignores important landscape parameters. More recently, Miller and Wentz (2003) broadened this critique and argued for the need to escape the limiting framework of Euclidean distance that typifies most spatial analysis in geographic information science.

Typically, effective distance has been computed using an algorithm that accumulates values as "waves ripple outward" from a "source" location through a matrix of cells or raster data set (Eastman 1989; Berry 1993; Douglas 1994). Commonly this algorithm is called a cost-weighted function because it incorporates some form of cost associated with moving across a location (or cell). The cost distance value at a cell reflects the cost associated with traveling across a surface from a source (or patch) to nearby locations. A number of synonymous terms have been used to describe how movement is affected by landscape heterogeneity. Most commonly, the terms friction, cost, resistance, and impedance have been used — note that permeability is the inverse of these (Singleton *et al.* 2002). Friction is defined as the energy expenditure and mortality risks associated with crossing a cell (Ray *et al.* 2002).

Outputs from cost-weighted methods

Although there are a number of possible outputs from cost-weighted algorithms, five are particularly important to understand functional landscape connectivity. I first review three of these outputs (cost-weighted distance, allocation zones, and least-cost paths), which are fundamental to assess landscape connectivity, and then extend these methods to discuss outputs allowing identification of multiple pathways among source patches (distribution functions of cost distances and *N*th-optimal corridors).

First, the cost-weighted distance method computes the minimum cumulative cost at each *destination* (or non-source) cell back to the nearest *source* cell. That is, beginning at a source cell, each adjacent cell is visited in turn, and the cost-weight associated with the adjacent cell is summed from the source to a destination cell. Source cells represent patches of habitat (computed in a GIS as a contiguous region of like-valued cells). Inter-patch distance is computed from patch edge to edge, not from a single *x*, *y* point that represents a patch such as the centroid (Fig. 17.1). Edge-to-edge distance is more biologically realistic and is more robust to the variety and complexity of real-world patches that are often convoluted and narrow, particularly in locations of rapid environmental gradients (Keitt *et al.* 1997; Crooks 2002; Bender *et al.* 2003).

The accumulated distance value from the nearest source, which is also called the *least-cost distance*, is computed and stored at each destination cell. That is, each destination cell contains the least-cost distance that accumulated from its nearest source cell. Note that if there is more than one patch or contiguous region of source cells in a landscape, the

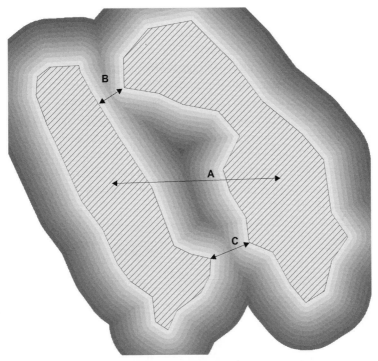

Fig. 17.1. An illustration of straight-line distance between two patches of habitat near the Sangre de Cristo and Wet Mountains, Colorado. The straight-line inter-patch distance (center-to-center, A) is 46.5 km and edge-to-edge is 8.7 km (B) and 11.7 km (C). Pathway B is the minimum least-cost pathway, while pathway C is an additional pathway. The gray tones radiating from the patches are straight-line distance buffers out to 13 km away from the edge of patches.

cost-weighted distance "waves" emanating from each patch will meet oncoming waves from adjacent source patches to create "ridges" in the cost-weighted surface. These ridges demarcate the *allocation boundary*, which forms halfway between the nearest source cells. As a result, the least-cost distance between sources is twice the least-cost distance at the ridgeline or allocation boundary. If a landscape is homogeneous, then a uniform cost-weighted surface is assumed (e.g., all cost values are 1) and the distance waves from different source patches meet at precisely midway between the source cells. The straight-line (or Euclidean) distance between patches is then twice the midpoint distance (Fig. 17.1). A useful way to visualize least-cost distances is to represent the values as a surface, where elevation represents least-cost distance.

Cost-weighting is used to parameterize the behavioral response of an individual organism so that ecological knowledge of a species can be explicitly incorporated into a model. This is accomplished by estimating friction (or its inverse, permeability) values at each location, based on in situ factors such as landcover, housing and road densities, and slope (e.g., Singleton *et al.* 2001) (Fig. 17.2). Usually cost weights are simply estimated by biological experts, and a high priority for additional research is to develop more rigorous methods of estimating resistance parameters from empirical data (e.g., Ferreras 2001; Ricketts 2001).

Different responses of an individual to a heterogeneous landscape can be incorporated during modeling of inter-patch movement by modifying

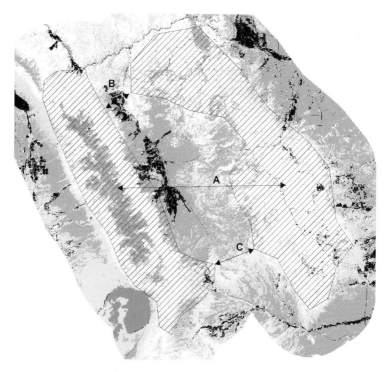

Fig. 17.2. Following from Fig.17.1, here patches are shown with a permeability surface that is used to weight the effective distance. For example, for a forest carnivore species, lighter shades show higher-permeability landcover such as coniferous forest, shrubland, and wetlands, while darker shades show lower-permeability landcovers such as agricultural cropland, urban areas, and highways. The straight-line inter-patch distance (center-to-center, A) is 46.5 km and edge-to-edge is 8.7 km (B) and 11.7 km (C). Pathway B is the minimum least-cost pathway, while pathway C is an additional pathway.

the cost weights or resistance values. However, as with other methods of distance computation such as straight-line distance between patch centroids, the delineation of patches used to compute inter-patch distance is critical. That is, inter-patch effective distance that incorporates matrix heterogeneity is meaningful only if the patches and costs have been defined in a way that reflects an ecological process or the habitat requirements of a given species (With and Crist 1995; Theobald and Hobbs 2001).

A second important output of cost-weighted functions is a map of zones defined by the allocation boundaries. *Allocation zones*, which are also known as Voronoi or Thiessen polygons (Boots 1979), demarcate the zone of influence or service area around a point (e.g., the area serviced around a fire station) or a patch. Recall that each cell contains the least-cost distance back to the nearest patch, and so the boundaries of allocation zones trace the locations (or cells) where the maximum least-cost distance occurs between any two sources (Adriaensen et al. 2003) (Fig. 17.3). The boundaries of the allocation zones then occur along the ridgelines formed between adjacent "plains" or "pits" of the source patches on the least-cost distance surface.

A third important output is the *least-cost path* (LCP), which is the path of least-cost values that link one source patch to another. The location where the least-cost path crosses the allocation boundary—through the "saddle" of the ridge—is the shortest least-cost distance between patches (Fig. 17.3). Recall that because the allocation boundary is midway between patches, the effective distance between patches is twice the least-cost distance value along the allocation boundary. Most landscape connectivity analyses to date have been based on the least-cost path distance, for example Krist and Brown (1994), Walker and Craighead (1997), Bunn et al. (2000), Hoctor et al. (2000), Quinby et al. (2000), Ferreras (2001), Michels et al. (2001), Servheen et al. (2001), Meegan and Maehr (2002), Schadt et al. (2002), Vuilleumier and Prelaz-Droux (2002), Joly et al. (2003), Rouget et al. (2003), Sutcliffe et al. (2003), Coulon et al. (2004), Beier et al. (Chapter 22), and Noss and Daly (Chapter 23).

It is important to note that the LCP is not a prediction of the movement path of an individual species moving across a landscape. Rather, it simply identifies the potential travel route that minimizes the cost of movement (Walker and Craighead 1997). The LCP can be interpreted as the path that provides an individual the greatest likelihood of survival (Russell et al. 2003). Increasingly researchers have recognized that movement is influenced by a species' perceptual ability (e.g., Russell et al. 2003; Schooley and Wiens 2003; Olden et al. 2004; Tracey Chapter 14), and it is

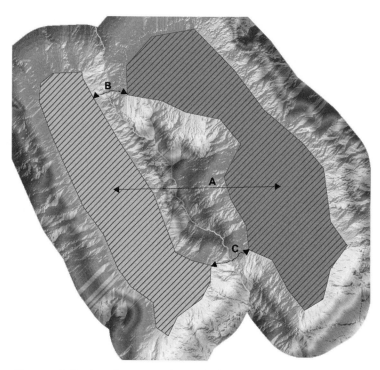

Fig. 17.3. Following Figs. 17.1 and 17.2, the least-cost distances between the two patches are shown below as a surface generated from cost distances (so peaks and ridges are higher cost-weights). Note that the least-cost distance (B) is 12 km, while C is 18 km (although C could have been less than B, given different landcover types in the intervening matrix). Also note the allocation boundary between the patches is illustrated as a gray line tracing the ridge of least-cost values.

unrealistic that an individual can realize a globally optimal path represented by the LCP. As a result, the LCP likely underestimates the effective distance traveled by an individual on average. Also, it is unrealistic to assume that a single cell-width pathway is sufficiently wide for a species to move through a landscape, particularly for larger species (Walker and Craighead 1997; Bunn *et al.* 2000). In practice, the LCP often is buffered some distance (e.g., 200 m) to form a corridor (e.g., Hoctor *et al.* 2000; Singleton *et al.* 2002; Hilty and Merenlender 2004). Even with these limitations, LCP has been the primary way to estimate effective distance because it is readily computed in a variety of GIS software. Moreover, effective distance has been found to be a better estimate of ecological proximity than straight-line distance (e.g., Bunn *et al.* 2000; Michels *et al.*

2001; Arnaud 2003; Chardon *et al.* 2003; Sutcliffe *et al.* 2003; Verbeylen *et al.* 2003; Coulon *et al.* 2004). In summary, LCP is typically used to identify both the most likely movement path and to estimate the effective distance between pairs of patches.

From the least-cost path to multiple pathways

An important extension of LCP is to quantify the probability that an animal can successfully move among patches to better incorporate biological understanding of how a species or process operates. Here I introduce a novel, fourth output from cost-weighted methods to obtain the full distribution of cost-distance values along the allocation boundary. That is, the list of all cost-distance values along the allocation boundary (Fig. 17.4) is ordered from minimum to maximum to generate a cumulative distribution function. Extending the LCP (the minimum cost-distance) to the full distribution provides two important advantages. First, although movement paths of individuals are likely to travel along or close to the LCP, there is some possibility of movement along pathways away from it. Although untested, it is reasonable that the full distribution of effective distances would approximate the distribution of distances generated by models of individual-based movement that follow correlated or biased random-walk paths (e.g., Boone and Hunter 1996; Gustafson and Gardner 1996; Schippers *et al.* 1996; Tracey Chapter 14). Also, a probabilistic or Monte Carlo simulation approach to understanding inter-patch connectivity can be implemented by drawing values from the cumulative distribution function (e.g., Russell *et al.* 2003). A simple, useful alternative is to rely on statistical measures such as quartiles or quintiles. Estimating effective distance using the distance at the 25th percentile Q_{25}, for instance, is a more robust measure than simply the minimum. Moreover, it can distinguish "too narrow" from "wide enough" pathways. For instance, in Fig. 17.5 there are five possible pathways where the cost-distance values located along the allocation boundary are below the Q_{25} threshold. However, if a species of interest requires a pathway to be at least 500 m wide, then only two pathways are wide enough at Q_{25}—from 10 to 20 km (near path B) and from 68 to 74 km (near path C).

Thus, a second advantage of extending LCP is that multiple pathways can be detected and represented, freeing one from the limited and often unrealistic assumption of LCP analysis of a single pathway. Multiple pathways are fundamentally important to distinguish because they allow the redundancy of connectivity to be measured (Baudry and Merriam 1988;

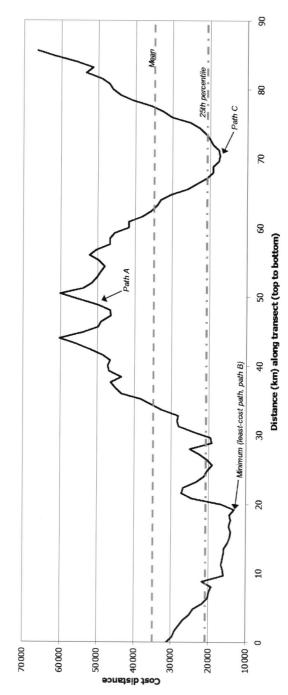

Fig. 17.4. A cross-sectional view of the profile of least-cost distances along the allocation boundary ridge (top shown on left side, bottom shown on right side), between patches in Fig. 17.3. Note that the least-cost path (B) travels through the minimum cost-distance value (~12 000). Note also that the straight-line pathway represented by A (from Fig. 17.3) crosses roughly between 40 to 50 km, while B is the least-cost path at 18 km and C occurs at roughly 70 km. The cost-distance values are reordered to generate a cumulative distribution function. By having the full distribution of distance values, multiple pathways can be identified (e.g., below the 25th percentile line). Also, pathways can be constrained so that they must have a certain width, and so the cost-distance threshold can be adjusted (above the least-cost distance) to include wider pathways (e.g., a threshold of 20 000 would identify two primary pathways, one roughly 10 km wide – near path B – and one roughly 5 km wide – near path C).

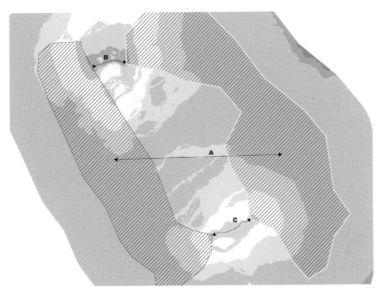

Fig. 17.5. Following Figs. 17.1, 17.2, and 17.3, corridors between patches can be determined by generating a corridor surface calculated by adding the least-cost distance surface generated from one patch to the least-cost distance surface generated from an adjacent patch. Smaller values on this corridor surface depict locations that are near the optimal pathway, while larger values are less optimal. The corridor for the 10th percentile distance is shown as dark gray (under pathway B), while the 25th percentile corridor is shown by light gray surrounding pathway B and under pathway C (bounded by the white area).

Jordan 2000), and they occur when two or more (nearly) equally suitable corridors contribute to the probability of connecting two patches (Knaapen *et al.* 1992). Also, multiple pathways are more likely when the intervening matrix is heterogeneous and/or when the ratio of the length of the allocation boundary formed between nearby patches to the intervening distance between the patches is high. In practice, multiple pathways are frequent when patches are functionally defined (rather than simple centroids).

A fifth output that can be generated from the cost-weighted approach is called the "*N*th-optimal corridor" (J. Berry, pers. comm.) and is especially helpful to visualize the two-dimensional geometry pathway between patches (Fig. 17.5). This analysis identifies an optimal pathway (the least-cost distance) and the corresponding corridor containing the top *N*th (e.g., 10th or 25th percentile) best paths. Essentially, Berry is recognizing that the least-cost distance pathway is the first or most optimal route, and that an optimal corridor surrounding this path will form a set of "nearly optimal" alternative routes. Thus, by relaxing the threshold for identifying "optimal," corridors can be identified that build around the least-cost path. These Nth-optimal corridors can be used to locate the shape of the pathway or "nearly optimal route" between patches, as well as to identify "pinch points" or landscape features that are particularly constraining to movement.

To summarize, multiple pathways can be identified by finding cells along the allocation boundary that have values below a threshold (e.g., Q_{25}), but are not contiguous. This allows a more robust, realistic representation of effective distance among patches that recognizes that there is some (perhaps small) probability of moving across any part of the intervening matrix and that multiple pathways between patches of habitat may exist. Also, a minimum width of a pathway can be specified to filter out possible pathways that are too narrow to be ecologically effective.

GRAPH THEORY

A powerful approach to understand connectivity based on effective distance is to represent landscapes using a graph (or network) data structure, where patches defined from raster data are represented by nodes, which are then connected by effective distance pathways represented by edges (Urban and Keitt 2001; Fagan and Calabrese Chapter 12). As discussed above, when identifying functional patches and computing inter-patch effective distance, landscapes are often represented as continuous or gradient data (as opposed to discrete feature-based data). An advantage of a continuous spatial representation, typically implemented using raster data, is that heterogeneity in the matrix quality can be explicitly examined and possible functional behavior can be easily incorporated. However, a common challenge to all scientists is to include important aspects under study, but to simplify as much as possible. Graphs represent the essential topology of features and have long been used to represent the connectivity of linear features such as transportation and utility corridors (e.g., Haggett and Chorley 1969). Urban and Keitt (2001) provide an excellent introduction to graph theory for ecologists, and below I follow much of their terminology in the brief review of some of the basics of graph theory.

A graph or network is composed of a set of nodes V that may or may not be connected by edges E (Fig. 17.6). In an ecological setting, a node represents a patch and is typically symbolized using a simple 0-dimensional point, but the size of the point can be made proportional to

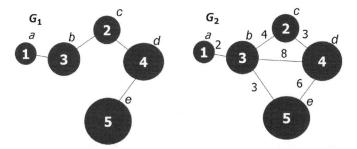

Fig. 17.6. An example of two graphs (or networks), G1 and G2, with similar nodes but different edge configurations. Both graphs have five nodes (a, b, c, d, e) of different sizes (e.g., $a = 1$, $c = 2$, etc.). G1 has four edges (E_{ab}, E_{bc}, E_{cd}, and E_{de}), while G2 has two additional edges (E_{bd} and E_{be}). The edge distances are specified on G2 (e.g., $d_{ab} = 2$, $d_{be} = 3$).

represent an attribute (e.g., area) of a patch. An edge (aka link or linkage) represents the logical connection between a pair of nodes and is typically symbolized by a straight line. An edge can also have attributes, such as the effective distance d_{ij} between a pair of nodes i and j. A graph is completely or "fully" connected if each node can be reached via the edges from any other node in a graph. Graphs provide an elegant and efficient means to represent and analyze landscape connectivity. In fact, the node–edge representation of graph theory is used to implement cost-weighted functions in a raster data set, where each cell is a node linked to its eight neighbors via an edge (Environmental Systems Research Institute 2003).

Landscape networks

Extending the work of traditional graph theory to recognize the landscape context, here I develop the construct of a landscape network (LN), which is distinguished by four key characteristics. First, an LN is represented as a geometric network, which stores not only the topology of the graph, but also the geometry of nodes and edges (edges can be composed of multiple line segments). Although a strict interpretation of a graph (e.g., by mathematicians) ignores the geometry and focuses solely on the topological structure of a graph, the spatial framework or landscape context in which the network is embedded remains important (Cantwell and Forman 1993). The spatial context is important to determine the size and shape of a patch, the angle between edges incident to a node, and the pathway represented by an edge. The location of both nodes and edges are georeferenced (i.e., spatially represented), so that direction and distance can be computed. Also, an LN can be combined with additional layers of geographic information within a GIS to allow additional relationships between the LN and other layers of spatial data. In ESRI's Geodatabase architecture, this type of data structure is called a *geometric network* (Zeiler 1999), and edges are represented as a one-dimensional polyline, which can be a simple straight line between two nodes or may be a complex "wiggly" line (with more than two vertices) like a stream. In a geometric network, the location where two or more edges intersect is represented by a node, which has a spatial location and associated attributes such as area (Zeiler 1999).

Second, nodes are used to represent patches, and these patches should be functionally defined to reflect some behavioral response to landscape structure. Many graph-based metrics (e.g., the degree or number of edges per node) are based solely on the presence (or absence) of nodes. Yet, ecologists are often interested in attributes associated with nodes, such as size, shape, or quality. LNs allow asymmetrical distances or interactions to be represented, such as typically occur among patches of unequal size or quality (i.e., using directed graphs). Third, effective distance is used as an attribute (or weight) of an edge, and multiple pathways between nodes can be represented as unique edges. Finally, LNs recognize "stepping stone" movement from patch to patch and employ planar graph algorithms.

A standard way to represent and analyze graphs is through a series of matrices (Fig. 17.7). The simplest form is *adjacency matrix A*, where a value of 1 in an $m \times m$ matrix represents connectivity (or 0 for unconnected) between node i and j, assuming a graph with m nodes. A *distance matrix D* stores the measured distance d_{ij} of a single edge (not a pathway composed of multiple edges) between nodes i and j in each matrix element. A related distance matrix D' can be computed where all pair-wise distances are computed between nodes i and j, which may or may not require traversal of multiple pathways through the network (Ahuja *et al.* 1993; Urban and Keitt 2001).

Typically, an undirected graph assumes that edge direction is not important and equivalent-distance (symmetry) is assumed between nodes where $d_{ij} = d_{ji}$, so that only one-half of the matrix needs to be computed. A directed graph, where $d_{ij} <> d_{ji}$, can be represented by computing the other half of the matrix. Note that multiple pathways between nodes i and j are difficult to represent in standard adjacency and distance matrices.

Edge distances are typically found for all pairs of nodes (called a fully connected graph), which can require considerable computational

$$A = \begin{bmatrix} - & 1 & 0 & 0 & 0 \\ 1 & - & 1 & 1 & 1 \\ 0 & 1 & - & 1 & 0 \\ 0 & 1 & 1 & - & 1 \\ 0 & 1 & 0 & 1 & - \end{bmatrix}$$

$$D = \begin{bmatrix} - & 2 & 0 & 0 & 0 \\ 2 & - & 4 & 8 & 3 \\ 0 & 4 & - & 3 & 0 \\ 0 & 8 & 3 & - & 6 \\ 0 & 3 & 0 & 6 & - \end{bmatrix}$$

$$D' = \begin{bmatrix} - & 2 & 6 & 9 & 5 \\ 2 & - & 4 & 8 & 3 \\ 6 & 4 & - & 3 & 7 \\ 9 & 8 & 3 & - & 6 \\ 5 & 3 & 7 & 6 & - \end{bmatrix}$$

$$R = \begin{array}{ccc} 1 & 2 & 2 \\ 2 & 1 & 2 \\ 2 & 3 & 4 \\ 2 & 4 & 8 \\ 2 & 5 & 3 \\ 3 & 2 & 4 \\ 3 & 4 & 3 \\ 4 & 2 & 8 \\ 4 & 3 & 3 \\ 4 & 5 & 6 \\ 5 & 2 & 3 \\ 5 & 4 & 5 \end{array}$$

Fig. 17.7. Data structures used to represent graph G_2 from Fig. 17.6. An adjacency (A), edge distance (D), and shortest-path distance (D') matrix for all pairs of nodes. Each matrix element represents a symmetrical relationship between nodes i and j (where columns and rows are represented for nodes a, b, c, d, and e). R (right) is a list of relationships (Theobald 2001) for each edge represented in G_2, denoting the *from node* (column 1), *to node* (column 2), and *distance* (column 3).

resources because algorithms require m^2 computations (aka order of $O(m^2)$). Moreover, estimating the effective distance using LCP for all pairs of nodes requires generating a least-cost distance raster for each pair of nodes, which is quite time-consuming (Bunn et al. 2000). As a consequence, the standard method is restricted to relatively small graphs (e.g., $m < 100$) or currently requires access to supercomputers. Interestingly, many edges in a fully connected graph are eventually removed because of their unrealistically long distances.

In addition to computational limitations, the implicit biological assumption of a fully connected graph is that organisms and processes do not respond to any nodes as they move between patch i and j. A more realistic assumption for most organisms, particularly terrestrial mammals, is that movement across a landscape will proceed in a "stepping stone" fashion as a series of steps from a node to an adjacent node (in graph theory this is termed a path of edges), from patch to patch (Wiens et al. 1993). Situations where movement is constrained between adjacent nodes are represented by a *planar graph*, so called because it is embedded in two-dimensional space and edges do not cross one another. In a planar graph, the number of edges is reduced considerably, to roughly 1–5% of a fully connected graph. Using landscape networks, adjacent nodes in planar graphs are found by developing a list of the boundaries between allocation zones, and an *R* list (Fig. 17.7) is used as an efficient data structure for representing these relationships. In addition to being more ecologically realistic, the computational complexity of algorithms for planar graphs is greatly reduced (Ahuja *et al.* 1993). Moreover, estimation of effective distance in GIS is reduced because only a single cost-distance raster is needed, rather than m^2 for non-planar graphs. One limitation to this planar-based approach is that direction-dependent distances cannot be estimated easily (e.g., Urban and Keitt 2001). That is, although interaction between two patches may be asymmetrical due to dominant environmental factors (e.g., wind direction, downhill direction, hydrologic flow) or different numbers of emigrants, these cannot be generated easily with standard GIS algorithms.

Measures of connectivity on landscape networks

A variety of graph-theoretic metrics have been developed to quantify networks in ecology (Table 17.2). Topological metrics characterize the basic structure of a graph using characteristics of nodes and edges, such as the average number of edges per node (degree or valence) or the ratio of the actual to possible number of edges (the gamma index γ: Forman 1995). Edge-weighted and node-weighted metrics recognize ecological differences between edges and nodes, respectively, by using attributes associated with edges and nodes. For example, in their topoecological index, Acosta *et al.* (2003) modified the gamma index γ to weight edges by a subjectively determined conservation status rating W ($0 \leq W \leq 1$). Node- and edge-weighted metrics use combined attributes of both edges and nodes, typically through identifying the node–edge–node "dumb-bell" formed by pairs of nodes and their connecting edge, such as the dispersal flux F (Urban and Keitt 2001). A "spatial interaction" metric that predicts the amount of movement between a pair of patches also can be computed as the sum of the products of the node weights normalized by an edge weight (see Table 17.2). Existing statistical methods such as Mantel's test (Mantel 1967) that measure the association between distance and an environmental variable can also be readily conducted using graphs (e.g., Arnaud 2003; Cottenie *et al.* 2003; Manel *et al.* 2003; Coulon *et al.* 2004). Dyer and Nason (2004) compared graph realizations to a binomial expectation to assess the significance of landscape structure.

An example landscape network

Here I briefly illustrate a landscape network developed for Canadian lynx (*Lynx canadensis*) in Colorado, USA, using the FunConn vi tools for

Table 17.2. *A listing of graph-theoretic metrics useful for quantifying landscape metrics in conservation science applications*

Type/category	Name	Formula	Source
Topological	Number of components or sub-graphs, G		Urban and Keitt (2001)
	Number of nodes, V		Urban and Keitt (2001)
	Number of links, E		Urban and Keitt (2001)
	Average node degree or valence		Forman (1995)
	Alpha index: ratio of actual to maximum number or loops		Bueno et al. (1995); Forman and Godron (1986); Lowe and Moryadas (1975)
	Gamma index γ (aka connectance): ratio of actual links to maximum possible links	Non-planar: $G = 2E/V(V-1)$ Planar: $G = E/3(V-2)$	Bueno et al. (1995); Forman and Godron (1986); Jordán et al. (2003); Lowe and Moryadas (1975)
	Topoecological index γ_{te}: the summation of the conservation status weights normalized by actual number of edges (from γ index)	$\gamma_{te} = E_W/E$	Acosta et al. (2003)
Edge-weighted	Accessibility: the "reachability" of a node to all other nodes for each of l edges or links	$A_i = \sum_{j=1}^{l} d_{ij}$	Lowe and Moryadas (1975)
Node-weighted	Sum of node weights s (e.g., population size, patch area) for v nodes	$S = \sum_{i=1}^{v} s_i$	Jordán (2003); Jordán et al. (2003)
	Sum of the products of node weights for pairs s_i, s_j, for l links	$S^* = \sum_{j=1}^{l} s_i s_j$	

(continued)

Table 17.2 *(cont.)*

Type/category	Name	Formula	Source
	Recruitment: importance of patch to recruitment potential flux—source strength (e.g., patch size) s_i for m connected nodes and a function related to habitat quality k_i	$R = \sum\limits_{i=1}^{m} s_i k_i$	Urban and Keitt (2001)
Node- and edge-weighted	Sum of the product of linked node weights (e.g., population size, patch size), normalized by link weight, L	$S^{**} = \sum\limits_{j=1}^{l} s_i s_j / L$	
	Dispersal flux: source strength (e.g., patch size) s_i and a function related to habitat quality k_i and probability p_{ij} of moving from node i to j	$F = \sum\limits_{i=2}^{m} \sum\limits_{j=1}^{i-1} p_{ij} s_i k_i$	Urban and Keitt (2001)
	Correlation length C_d: the average distance of a sub-graph given movement of distance d and radius of gyration r_i	$C_d = \dfrac{\sum\limits_{i=1}^{m} n_i r_i}{\sum\limits_{i=1}^{m} n_i}$	Keitt *et al.* (1997); Rothley and Rae (2005)

ArcGIS (Theobald *et al.* 2006). Four steps were required to generate an LN for lynx. First, a map of habitat quality was generated that specifies forage resources in terms of quality; the map is a function of the in situ vegetation or landcover type, proximity to patch edge (both to identify core areas within a patch and nearby resources), and disturbance from nearby land uses and activities (e.g., roads, noise, etc.). High-quality resources for lynx include high-elevation mesic conifer and aspen forests.

Second, habitat quality was functionally integrated or grouped based on the ability of lynx to move among different resources to create *habitat patches*. Given two locations (cells) with high habitat quality, it is more likely that locations nearby other high-quality locations will be used.

For a preliminary model, functionally defined patches were defined as "big enough" if at least 264 ha in size, and "close enough" using a moving window defined by a circle with radius 917 m. These parameters were derived by developing allometric relationships between body mass and home range size (Jetz *et al.* 2004).

Third, a friction surface was generated by inverting permeability values that reflected the relative ease or difficulty of a lynx moving through: forested land (easy), shrubland and grassland (moderately easy), and human-dominated land uses (difficult). A surface was computed that represented the cost-weighted distance likely needed to move among functionally defined patches. Finally, a landscape network was constructed from the raster representation of patches and cost-weights (Fig. 17.8). The resulting LN can then be analyzed to identify important, large patches (nodes), and edges that are important in maintaining overall landscape connectivity. An important advantage of an LN is that spatial queries can be conducted to select particular nodes or edges that might be modified for a given scenario, allowing realistic situations to be examined. For example, one could examine the potential effects on connectivity if a stretch of highway is to be widened. Edges could be intersected with a GIS data set of highways to identify which edges would be affected.

CONCLUSION

Here I have reviewed the basics of estimating effective distance in GIS and their use in graph-theoretic-based models, and introduced a refinement to standard methods called landscape networks. Landscape networks: (1) represent the landscape context in which the network is embedded so that the geometry (shape, size, and location) of patches and edges are explicit; (2) are constructed around functionally defined patches; (3) use estimates of effective distance that can include multiple pathways (edges) between a pair of patches; and (4) recognize "stepping stone" movement from patch to patch and employ planar graph algorithms for computational efficiency in GIS.

Landscape networks potentially offer conservation scientists two important conceptual advances when examining landscape connectivity. First, an LN provides a means of quantifying connectivity that goes beyond traditional metrics that measure isolation (e.g., nearest neighbor, proximity index), connectance, or contagion (Fagan 2002; Tischendorf *et al.* 2003). These traditional approaches examine the relationship of a patch i within the context of its adjacent or first-order neighbors N^I or within some

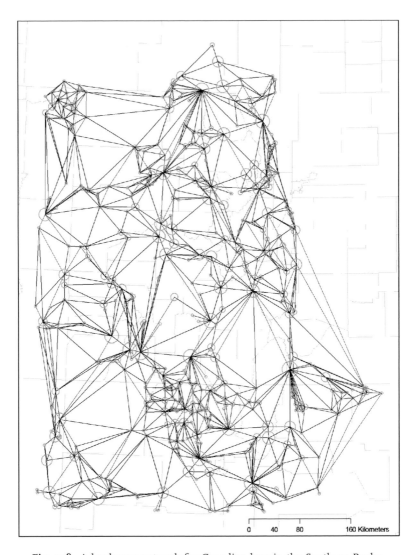

Fig. 17.8. A landscape network for Canadian lynx in the Southern Rocky Mountains in Colorado, New Mexico, and Wyoming, USA. The LN contains 1317 edges and 432 patches (mean 16 357 ha; SD 44 600). Nodes (circles) are located at patch centroids, and their size is proportional to patch area. Edges are represented as a line from node to node, but each edge is attributed with estimates of effective distance (e.g., Q_{min}, Q_{10}, Q_{25}, etc.). County boundaries are shown in gray for reference.

predefined neighboring distance N^d from the edge of patch i (Gustafson and Parker 1994; McGarigal et al. 2002; Bender et al. 2003). To generate a landscape-level metric, values are summarized over all N^I patches.

Graphs provide the ability to examine broader configurational aspects, to investigate $> N^I$. This ability is especially important for landscapes that are represented by a more detailed array of patches, such as when patches and movement at all scales are identified (e.g., dispersal between viable populations, seasonal migration between winter and summer ranges, weekly foraging, and daily movements among small fragments of resource patches). Although only initial suggestions were presented here about such methods, there is a great opportunity to make use of this advantage, and new techniques are likely to emerge such as based on circuit theory (McRae 2004).

The second conceptual advance offered by landscape networks is that they provide a means to examine directly the degree to which changes in a population are due to habitat loss or habitat fragmentation (Fahrig 2003). Because habitat is represented by a series of nodes, the effects of habitat loss may be modeled through node removal analysis, while possible fragmentation can be quantified directly by changes in edge topology, such as through edge removal analysis.

In addition, landscape networks offer a number of practical advantages to conservation scientists. Applied conservation planning often requires estimating models for multiple conservation targets (e.g., multiple species or processes), often with limited data and resources, and for a variety of possible alternative planning scenarios (e.g., forecasted growth, transportation master plans, mitigation measures for roadway construction, doubling of carbon dioxide climate change, etc.). Although improvements in parameter estimation for these models are needed, especially of landscape resistance, networks offer a robust and powerful model to examine possible conservation strategies (Fagan and Calabrese Chapter 12). Landscape networks and the ArcGIS v9-based tools used to create and analyze them (Theobald et al. 2006) will help conservation scientists respond to the critical need to examine effects of potential land-use change and possible mitigation actions on the movement of organisms and species of conservation concern.

ACKNOWLEDGEMENTS

I appreciate helpful and stimulating discussions about landscape connectivity with A. Bunn, K. Crooks, M. DiGiorgio, J. Grant, J. Kintsch,

B. McRae, M. Neel, J. Norman, C. Pague, C. Pyke, C. Ray, K. Rothley, B. Schwab, E. Treml, and R. Wostl. Research described here was supported by The Nature Conservancy's David H. Smith Fellowship, the Colorado Department of Transportation, the Southern Rockies Ecosystem Project, and US EPA STAR research grant No. CR-829095.

REFERENCES

Acosta, A., C. Blasi, M. L. Carranza, C. Ricotta, and A. Stanisci. 2003. Quantifying ecological mosaic connectivity and hemeroby with a new topoecological index. *Phytocoenologia* **33**:623–631.

Adriaensen, F., J. P. Chardon, G. De Blust, *et al.* 2003. The application of "least-cost" modeling as a functional landscape model. *Landscape and Urban Planning* **64**:233–247.

Ahuja, R. K., T. L. Magnanti, and J. B. Orlin. 1993. *Network Flows: Theory, Algorithms, and Applications*. Upper Saddle River, NJ: Prentice-Hall.

Arnaud, J. F. 2003. Metapopulation genetic structure and migration pathways in the land snail *Helix aspersa*: influence of landscape heterogeneity. *Landscape Ecology* **18**:333–346.

Baudry, J., and H. G. Merriam. 1988. Connectivity and connectedness: functional versus structural patterns in landscapes. *Proceedings of the 2nd International Seminar of the International Association for Landscape Ecology*, Munster, Germany, Pp. 23–28.

Bender, D. J., L. Tischendorf, and L. Fahrig. 2003. Using patch isolation metrics to predict animal movement in binary landscapes. *Landscape Ecology* **18**:17–39.

Bennett, A. F. 1999. *Linkages in the Landscape: The Role of Corridors and Connectivity in Wildlife Conservation*. Gland, Switzerland: International Union for the Conservation of Nature.

Berry, J. 1993. *Beyond Mapping: Concepts, Algorithms and Issues in GIS*. New York: John Wiley.

Boone, R. B., and M. L. Hunter. 1996. Using diffusion models to simulate the effects of land use on grizzly bear dispersal in the Rocky Mountains. *Landscape Ecology* **11**:51–64.

Boots, B. N. 1979. Weighting Thiessen polygons. *Economic Geography* **56**:248–259.

Bueno, J. A., V. A. Tsihrintzis, and L. Alvarez. 1995. South Florida greenways: a conceptual framework for the ecological reconnectivity of the region. *Landscape and Urban Planning* **33**:247–266.

Bunn, A. G., D. L. Urban, and T. H. Keitt. 2000. Landscape connectivity: a conservation application of graph theory. *Journal of Environmental Management* **59**:265–278.

Cantwell, M. D., and R. T. T. Forman. 1993. Landscape graphs: ecological modeling with graph theory to detect configurations common to diverse landscapes. *Landscape Ecology* **8**:239–255.

Chardon, J. P., F. Adriaensen, and E. Matthysen. 2003. Incorporating landscape elements into a connectivity measure: a case study for the speckled wood butterfly (*Pararge aegeria* L). *Landscape Ecology* **18**:561–573.

Coulon, A., J. F. Cosson, J. M. Angibault, et al. 2004. Landscape connectivity influences gene flow in a roe deer population inhabiting a fragmented landscape: an individual-based approach. *Molecular Ecology* 13: 2841–2850.

Cottenie, K., E. Michels, N. Nuytten, and L. DeMeester. 2003. Zooplankton metacommunity structure: regional vs. local processes in highly interconnected ponds. *Ecology* 84:991–1000.

Cramer, P. C., and K. M. Portier. 2001. Modeling Florida panther movements in response to human attributes of the landscape and ecological settings. *Ecological Modelling* 140:51–80.

Crooks, K. R. 2002. Relative sensitivities of mammalian carnivores to habitat fragmentation. *Conservation Biology* 16:488–502.

Dennis, R. L. H., T. G. Shreeve, and H. van Dyck. 2003. Towards a functional resource-based concept for habitat: a butterfly biology viewpoint. *Oikos* 102:417–426.

Douglas D. H. 1994. Least-cost path in GIS using an accumulated cost surface and slopelines. *Cartographica* 31:37–51.

Dyer, R. J., and J. D. Nason. 2004. Population graphs: the graph theoretic shape of genetic structure. *Molecular Ecology* 13:1713–1727.

Eastman, J. R. 1989. Pushbroom algorithms for calculating distances in raster grids. *Proceedings AUTOCARTO* 9:288–297.

Environmental Systems Research Institute. 2003. *ArcGIS v8 help*. Redlands, CA: ESRI.

Fagan, W. F. 2002. Connectivity, fragmentation, and extinction risk in dendritic metapopulations. *Ecology* 83:3243–3249.

Fahrig, L. 2003. Effects of habitat fragmentation on biodiversity. *Annual Review of Ecology, Evolution, and Systematics* 34:487–515.

Ferreras, P. 2001. Landscape structure and asymmetrical inter-patch connectivity in a metapopulation of the endangered Iberian lynx. *Biological Conservation* 100:125–136.

Forman, R. T. T. 1995. *Land Mosaics*. Cambridge, UK: Cambridge University Press.

Forman, R. T. T., and M. Godron. 1986. *Landscape Ecology*, New York: John Wiley.

Gardner, R. H., and E. J. Gustafson. 2004. Simulating dispersal of reintroduced species within heterogeneous landscapes. *Ecological Modelling* 171:339–358.

Gustafson, E. J., and R. Gardner. 1996. The effect of landscape heterogeneity on the probability of patch colonization. *Ecology* 77:94–107.

Gustafson, E. J., and G. R. Parker. 1994. Relationships between landcover proportion and indices of landscape spatial pattern. *Landscape Ecology* 7:101–110.

Haggett, P., and R. J. Chorley. 1969. *Network Analysis in Geography*. London: Edward Arnold.

Halpin, P. N., and A. G. Bunn. 2000. Using GIS to compute a least-cost distance matrix: a comparison of terrestrial and marine ecological applications. *Proceedings of the 2000 ESRI User Conference*, San Diego, CA.

Hanski, I. 1998. Metapopulation dynamics. *Nature* 396:41–49.

Hilty, J., and A. Merenlender. 2004. Use of riparian corridors and vineyards by mammalian predators in Northern California. *Conservation Biology* 18:126–135.

Hoctor, T. S., M. H. Carr, and P. D. Zwick. 2000. Identifying a linked reserve system using a regional landscape approach: the Florida ecological network. *Conservation Biology* 14:984–1000.

Jetz, W., C. Carbone, J. Fulford, and J. H. Brown. 2004. The scaling of animal space use. *Science* 306:266–268.

Johnson, A. R., J. A. Wiens, B. T. Milne, and T. O. Crist. 1992. Animal movements and population dynamics in heterogeneous landscapes. *Landscape Ecology* 7:63–75.

Joly, P., C. Morand, and A. Cohas. 2003. Habitat fragmentation and amphibian conservation: building a tool for assessing landscape matrix connectivity. *Comtes Rendues Biologies* 326:S132–S139.

Jordan, F. 2000. A reliability-theory approach to corridor design. *Ecological Modelling* 128:211–220.

Jordán F. 2003. Quantifying landscape connectivity: key patches and key corridors. Pp. 883–892 in E. Tiezzi, C. A. Brebbia, and J. L. Uso (eds.) *Ecosystems and Sustainable Development IV*. Southampton, UK: WIT Press.

Jordán F., A. Baldi, K. M. Orci, I. Racz, and Z. Varga. 2003. Characterizing the importance of habitat patches and corridors in maintaining the landscape connectivity of a *Pholidoptera transsylvanica* (Orthoptera) metapopulation. *Landscape Ecology* 18:83–92.

Keitt, T. H., D. L. Urban, and B. T. Milne. 1997. Detecting critical scales in fragmented landscapes. *Conservation Ecology* 1:4. Available online at http://www.consecol.org/vol1/iss1/art4

Knaapen, J. P., M. Scheffer, and B. Harms. 1992. Estimating habitat isolation in landscape planning. *Landscape and Urban Planning* 23:1–16.

Kramer-Schadt S., E. Revilla, T. Wiegand, and U. Breitenmoser. 2004. Fragmented landscapes, road mortality and patch connectivity: modelling influences on the dispersal of Eurasian lynx. *Journal of Applied Ecology* 41:711–723.

Krist F. J., and D. G. Brown. 1994. GIS modeling of Paleo-Indian period caribou migrations and viewsheds in Northeastern Lower Michigan. *Photogrammetric Engineering and Remote Sensing* 60:1129–1137.

van Langevelde, F. 2000. Scale of habitat connectivity and colonization in fragmented nuthatch populations. *Ecography* 23:614–622.

Larkin, J. L., D. S. Maehr, T. S. Hoctor, M. A. Orlando, and K. Whitney. 2004. Landscape linkages and conservation planning for the black bear in west-central Florida. *Animal Conservation* 7:23–34.

Lowe, J., and S. Moryadas. 1975. *The Geography of Movement*. Boston, MA: Houghton Mifflin.

MacArthur, R. H., and E. O. Wilson. 1967. *The Theory of Island Biogeography*. Princeton, NJ: Princeton University Press.

McGarigal, K., S. A. Cushman, M. C. Neel, and E. Ene. 2002. *FRAGSTATS: Spatial Pattern Analysis Program for Categorical Maps v3*. Amherst, MA: University of Massachusetts. Available online at: http://www.umass.edu/landeco/research/fragstats/fragstats.html

McRae, B. 2004. Integrating landscape ecology and population genetics: conventional methods and a new model. Ph.D dissertation, Northern Arizona University, Flagstaff, AZ.

Manel, S., M. K. Schwartz, G. Luikart, and P. Taberlet. 2003. Landscape genetics: combining landscape ecology and population genetics. *Trends in Ecology and Evolution* 18:189–197.

Mantel N. 1967. The detection of disease clustering and a generalized regression approach. *Cancer Research* 27:209–220.

Meegan, R. P., and D. S. Maehr. 2002. Landscape conservation and regional planning for the Florida panther. *Southeastern Naturalist* 1:217–232.

Michels, E., K. Cottenie, L. Neys, *et al.* 2001. Geographical and genetic distances among zooplankton populations in a set of interconnected ponds: a plea for using GIS modeling of the effective geographical distance. *Molecular Ecology* 10:1929–1938.

Miller, B., D. Foreman, M. Fink, *et al.* 2003. *Southern Rockies Wildlands Network Vision*. Golden, CO: Colorado Mountain Club Press.

Miller, H. J., and E. A. Wentz. 2003. Representation and spatial analysis in geographic information systems. *Annals of the Association of American Geographers* 93:574–594.

Olden, J. D., D. A. Jackson, and P. R. Peres-Neto. 2001. Spatial isolation and fish communities in drainage lakes. *Oecologia* 127:572–585.

Olden, J. D., R. L. Schooley, J. B. Monroe, and N. L. Poff. 2004. Context-dependent perceptual ranges and their relevance to animal movements in landscapes. *Journal of Animal Ecology* 73:1190–1194.

Paetkau, D., L. P. Waits, P. L. Clarkson, L. Craighead, and C. Strobeck. 1997. An empirical evaluation of genetic distance statistics using microsatellite data from bear (Ursidae) populations. *Genetics* 147:1943–1957.

Puth, L. M., and K. A. Wilson. 2001. Boundaries and corridors as a continuum of ecological flow control: lessons from rivers and streams. *Conservation Biology* 15:21–30.

Quinby, P., S. Trombulak, T. Lee, *et al.* 2000. Opportunities for wildlife habitat connectivity between Algonquin Provincial Park and the Adirondack Park. *Wild Earth* 10:75–80.

Ray, N., A. Lehmann, and P. Joly. 2002. Modeling spatial distribution of amphibian populations: a GIS approach based on habitat matrix permeability. *Biodiversity and Conservation* 11:2143–2165.

Ricketts, T. 2001. The matrix matters: effective isolation in fragmented landscapes. *American Naturalist* 158:87–99.

Roland, J., N. Keyghobadi, and S. Fownes. 2000. Alpine *Parnassius* butterfly dispersal: effects of landscape and population size. *Ecology* 81:1642–1653.

Rothley, K., and C. Rae. 2005. Working backwards to move forwards: graph-based connectivity metrics for reserve network selection. *Environmental Modeling and Assessment*, 10:107–113.

Rouget M., R. M. Cowling, R. L. Pressey, and D. M. Richardson. 2003. Identifying spatial components of ecological and evolutionary processes for regional conservation planning in the Cape Floristic Region, South Africa. *Diversity and Distributions* 9:191–210.

Russell, R. E., R. K. Swihart, and Z. Feng. 2003. Population consequences of movement decisions in a patchy landscape. *Oikos* 103:142–152.

Schadt, S., F. Knauer, P. Kaczensky, *et al.* 2002. Rule-based assessment of suitable habitat and patch connectivity for the Eurasian lynx. *Ecological Applications* 12:1469–1483.

Schippers, P., J. Verboom, J. P. Knaapen, and R. C. Apeldoorn. 1996. Dispersal and habitat connectivity in complex heterogeneous landscapes: an analysis with a GIS-based random walk model. *Ecography* 19:97–106.

Schooley, R. L., and J. W. Wiens. 2003. Finding habitat patches and directional connectivity. *Oikos* 102:559–570.

Servheen, C., J. S. Waller, and P. Sandstrom. 2001. Identification and management of linkage zones for grizzly bears between the large blocks of public land in the northern Rocky Mountains. *Proceedings of International Conference on Ecology and Transportation*, Pp.161–179.

Shurin, J. B., and J. E. Havel. 2002. Hydrologic connections and overland dispersal in an exotic freshwater crustacean. *Biological Invasions* 4:431–439.

Singleton, P. H., W. L. Gaines, and J. F. Lehmkuhl. 2002. *Landscape Permeability for Large Carnivores in Washington: A Geographic Information System Weighted-Distance and Least-Cost Corridor Assessment*, Research Paper PNW-RP-549 Portland, OR: US Department of Agriculture Forest Service.

Sutcliffe, O. L., V. Bakkestuen, G. Fry, and O. E. Stabbetorp. 2003. Modelling the benefits of farmland restoration: methodology and application to butterfly movement. *Landscape and Urban Planning* 63:15–31.

Taylor, P. D., L. Fahrig, K. Henein, and G. Merriam. 1993. Connectivity is a vital element of landscape structure. *Oikos* 68:571–573.

Theobald, D. M. 2001. Topology revisited: representing spatial relations. *International Journal of Geographical Information Science* 15:689–705.

Theobald D. M., and N. T. Hobbs. 2001. Functional definition of landscape structure using a gradient-based approach. Pp. 667–672 in J. M. Scott, P. J. Heglund, M. Morrison, *et al.* (eds.) *Predicting Plant and Animal Occurrences: Issues of Scale and Accuracy*. Covello, CA: Island Press.

Theobald, D. M., J. B. Norman, and M. R. Sherbune. 2006. *FunConn v1: Functional Connectivity Tools for ArcGIS v9*. Fort Collins, CO: Natural Resource Ecology Laboratory, Colorado State University.

Tischendorf, L., D. J. Bender, and L. Fahrig. 2003. Evaluation of patch isolation metrics in mosaic landscapes for specialist vs. generalist dispersers. *Landscape Ecology* 18:41–50.

Turchin, P. 1998. *Quantitative Analysis of Movement*. Sunderland, MA: Sinauer Associates.

Urban, D. L., and T. H. Keitt. 2001. Landscape connectedness: a graph theoretic perspective. *Ecology* 82:1205–1218.

Verbeylen, G., L. DeBruyn, F. Andriaensen, and E. Matthysen. 2003. Does matrix resistance influence Red squirrel (*Sciurus vulgaris* L. 1758) distribution in an urban landscape? *Landscape Ecology* 18:791–805.

Vos, C. C., and A. H. P. Stumpel. 1995. Comparison of habitat-isolation parameters in relation to fragmented distribution patterns in the tree frog (*Hyla arborea*). *Landscape Ecology* 11:203–214.

Vuilleumier, S., and R. Prelaz-Droux. 2002. Map of ecological networks for landscape planning. *Landscape and Urban Planning* 58:157–170.

Walker R., and L. Craighead. 1997. Analyzing wildlife movement corridors in Montana using GIS. *Proceedings of the ESRI User Conference 1997*. Available online at http://gis.esri.com/library/userconf/proc97/proc97/to150/pap116/p116.htm

Werner, C. 1968. The law of refraction in transportation geography: its multivariate extension. *Canadian Geographer* **7**:28–40.

Wiens, J. A. 1994. Habitat fragmentation: island v. landscape perspectives on bird conservation. *Ibis* **137**:S97–S104.

Wiens, J. A., N. C. Stenseth, B. Van Horne, and R. A. Ims. 1993. Ecological mechanisms and landscape ecology. *Oikos* **66**:369–380.

With, K. A., and T. O. Crist. 1995. Critical thresholds in species responses to landscape structure. *Ecology* **76**:2446–2459.

Zeiler, M. 1999. *Modeling our World*. Redlands, CA: ESRI Press.

Part III

Challenges and implementation
of connectivity conservation

Introduction: Don't fence me in

THOMAS LOVEJOY

Restoring connectivity on an adequate scale quite literally requires turning the concept of nature conservation inside out: from the conventional one of nature surviving in patches (and often in the least desirable parts) of human-dominated landscapes, to one in which human activity is imbedded in a natural matrix. Nothing less can ensure the future of the glorious diversity of life with which we share this planet. It is central to its future on land, in freshwater, and in the seas.

It is clear from this volume that there is much still to be learned about the science of connectivity. It is not an unalloyed "good" thing so that for example it can promote the dispersal of invasive species (Crooks and Suarez Chapter 18) and disease agents (McCallum and Dobson Chapter 19). Yet we do not have the luxury to take a linear approach and study the science of connectivity to a faretheewell before taking the actions to enhance connectivity. Rather science and action must go on in parallel with flexibility to modify in the light of new knowledge.

Consequently, connectivity and making it happen in a heavily modified world – where we neither fully understand the science, the socio-economics, nor the political challenges – becomes central to the conservation and sustainability agenda. Even more than in the past we must be prepared to be bold and try untested approaches, e.g., payment to farmers in lieu of commodity price supports to promote land conservation and connectivity. That means learning by doing on a scale never attempted and requiring budgets for monitoring and science as integral parts of connectivity projects. Without such elements, efforts to conserve connectivity could be fatally flawed. We must never be afraid to admit that we are experimenting and what we do not know.

Part of boldness means a willingness to think and act in ways we never have before. For example, most conservation action, including connectivity, tends to take place as individual projects, or at best at the landscape level. Far too little attention is paid to the condition of the overarching system until it begins to fail. Instead we need to track system condition so we can curb change and stress while there is still a safe buffer from failure. We need to think about conservation and connectivity beyond even the regional scale to the continental scale (Linden *et al.* 2004). We need to think about conserving enormous systems such as the entire Amazon, not just certain areas within it or even large corridors. Indeed we need to conserve and manage systems — like South Florida and its sheet flow of water — without which conservation units within would fail. Conservation must also explore ways to make the case for such bold moves, e.g., ecosystem services that go beyond but never belittle the glories of biodiversity itself.

Connectivity of human activities is affecting nature negatively. For example, the human/agricultural/industrial metabolism of the American Midwest has created the dead zone of the Gulf of Mexico (but one of 50-plus such examples worldwide and growing). Such anthropogenic connections provide every justification to think of connectivity in the reverse, namely the proactive conservation that is the subject of this volume. Indeed this could be a powerful way to help people recognize their own connectivity with nature.

Even without all the compelling reasons to restore connectivity, climate change alone would make it an imperative for conservation (Lovejoy and Hannah 2005; Noss and Daly Chapter 23; Soulé *et al.* Chapter 25). Paleoecology informs us that biological communities do not move as intact entities but rather the individual species move at different rates and in different directions so we can anticipate the communities we know will disassemble and novel ones will come into existence. Many protected areas will no longer retain the communities they were set aside to preserve. Rather we now understand that protected areas are the safe havens from which future biogeographic pattern will emerge — provided lack of connectivity is not an impediment.

Connectivity along with climate change will need the engagement of the scientific and conservation communities with the rest of society to an unprecedented degree. They both imply serious constraints in human behavior, but constraints and problems can, with creativity, become exciting opportunities. They are unlikely to do so, however, unless we really engage with other segments of society including residents, business, and government. We will need to be much better at learning their languages, and understanding their framework and concerns. Above all this must be undertaken with humility.

Lest all this seem hopelessly naive and optimistic, let's remember that connectivity projects have been appearing with increasing frequency in the last 15 years. There are trans frontier parks, corridors in various places in the Americas and elsewhere. There is certainly hope to be taken from the successes and much to be learned about what works and what doesn't as we plan the ambitious connectivity agenda ahead.

REFERENCES

Linden, E., T. Lovejoy, and J. Phillips. 2004. Seeing the forest: conservation on a continental scale. *Foreign Affairs* **83**(4):8–13.

Lovejoy, T. and L. Hannah (eds.) 2005. *Climate Change and Biodiversity*. New Haven, CT: Yale University Press.

18

Hyperconnectivity, invasive species, and the breakdown of barriers to dispersal

JEFFREY A. CROOKS AND ANDREW V. SUAREZ

INTRODUCTION

The conservation implications of connectivity arise at many spatial scales. At the regional or landscape level, decreasing connections between natural areas inhibit the movement of species dependent on those habitats. Therefore, current conservation efforts often focus on connecting systems and facilitating the exchange of organisms between otherwise isolated patches. This dispersal of individuals can benefit populations by promoting gene flow and decreasing local extinction risk. Although increasing connectivity at the regional level may have negative consequences, such as altering source—sink dynamics, preventing local adaptation, accelerating the transport of pathogens, and facilitating the localized spread of invaders (Simberloff *et al.* 1992), it is typically considered that the benefits of restoring connectivity outweigh the risks (Crooks and Sanjayan Chapter 1). Much of the current volume addresses the topic of maintaining or increasing connectivity of this type.

At a larger spatial scale, such as between continents, a different connectivity-related conservation concern arises. Because of the long distances involved, natural movement at these scales should be relatively rare. Species can naturally traverse long distances both passively (e.g., via ocean currents or air masses: Scheltema 1986; Censky *et al.* 1998; Ritchie and Rochester 2001; DiBacco *et al.* Chapter 8) and actively (e.g., through

Connectivity Conservation eds. Kevin R. Crooks and M. Sanjayan. Published by Cambridge University Press. © Cambridge University Press 2006.

migration: Møller *et al.* 2003), and those few species that do make long-distance treks are often the target of conservation efforts (Harrison and Bjorndal Chapter 9; Marra *et al.* Chapter 7). Currently, however, the majority of global species movement does not occur naturally. Rather, a vast array of anthropogenic transport mechanisms has arisen, allowing for rampant species invasions. This has resulted in systems with greatly inflated connectivity relative to natural, background levels.

This rapid expansion of invasion vectors has promoted an explosion of problems associated with the introduction of species. It is estimated that invasions in six countries (the USA, Britain, Australia, South Africa, India, and Brazil) cost over $300 billion per year in control efforts and damages (Pimentel 2002). Alien predators, parasites, competitors, and habitat modifiers have penetrated and wreaked havoc in ecosystems throughout the world (Bright 1998; Cox 1999; Baskin 2002; J. A. Crooks 2002). Africanized bees (*Apis mellifera*), gypsy moths (*Lymantria dispar*), brown tree snakes (*Boiga irregularis*), fire ants (*Solenopsis invicta*), zebra mussels (*Dreissena polymorpha*), and kudzu (*Pueraria montana*) are a familiar, but albeit small, sampling of the invaders that now abound throughout the world's ecosystems.

The global swapping of a few successful species is also leading to biotic homogenization of the world's ecosystems (Lockwood and McKinney 2001; Olden *et al.* 2004), resulting in an anthropogenic "New Pangaea" (Rosenzweig 2001). Contemporary times even have been referred to as the "Homogocene," an age of homogenization. Because invasions erode the uniqueness of systems, diversity at large spatial scales has been diminished. For example, invasions and urbanization in California promotes the success of a few human-associated bird species (~20%), at the expense of many sensitive species of high conservation value (>50%) (Blair 1996). More striking is the homogenization of fish faunas across the USA due to introductions. On average, US states now have 15 more fish species in common than during pre-European settlement (Rahel 2000). Interestingly, however, invasions can actually have neutral or positive effects on diversity at smaller spatial scales (Sax *et al.* 2002). Across the Pacific region as a whole, for example, avian species richness and endemism have been greatly reduced, but the average number of bird species on a per-island basis remains similar due to the successful establishment of a few widely dispersed introduced species (Steadman 1995; Case 1996). In places like San Francisco Bay, species extinctions appear not to have kept up with the remarkable pace of invasions, and it is very likely that more species are in that system now than were there 200 years ago (Cohen and Carlton 1998; Carlton *et al.* 1999).

Connectivity and invasions also intersect at the landscape level. In general, both habitat fragmentation and invasions represent major threats to the integrity and diversity of natural ecosystems, are increasing due to human activities, and are difficult to remedy once they have occurred (Wilcove *et al.* 1998). A tighter link also exists between invasions and habitat fragmentation. Relatively large numbers of invaders are typically found in smaller, more fragmented patches of natural habitats when compared to larger, more connected fragments. Similarly, highly modified habitats (such as the urban matrix) tend to support more invaders than less-developed areas (Hobbs and Huenneke 1993; Suarez *et al.* 1998; K. R. Crooks *et al.* 2004). This is commonly attributed to factors such as disturbance, habitat requirements of natives, and the relative increase in edge in small patches.

In this chapter, we examine increased connections caused by anthropogenic activities, and call this phenomenon hyperconnectivity. In terms of biological invasions, hyperconnectivity can result from both the provision of artificial vectors of species transport and the creation of expansive human-modified habitats that facilitate invaders. We first discuss the transport of invaders, including characterizations of invasion vectors, rates of invasion over time, and the effect of continued immigration on already established populations. We next consider how connectivity of habitats within recipient ecosystems influences invasion success, and conclude with a discussion of management implications related to hyperconnectivity and invasion.

HYPERCONNECTIVITY AND THE TRANSPORT OF INVADERS

The proliferation of invasion vectors

We live in an age of globalization. Changing economies and politics have meant new trading partners and the opening of world markets (Mack 2003). People and goods now move around the world with unprecedented ease. These connections by land, air, and sea move vast numbers of organisms in a global "ecological roulette" (Carlton and Geller 1993). Although the long-distance dispersal of species into a new locale can be a natural biological event, the manner and rate at which species now move around the globe is wholly unprecedented. Mosquitoes have been transported in used tires, brown tree snakes (*Boiga irregularis*) have hitched rides in airplane wheel wells, "killer algae" (*Caulerpa taxifolia*) have been dumped with aquarium water into the sea, lampreys (*Petromyzon marinus*) swam through canals to enter the Great Lakes, and kudzu (*Pueraria montana* var. *lobata*) was intentionally planted for ornamental purposes (Cox 1999; Ruiz and Carlton 2003). Here we consider the processes involved in the actual transport of invasive species.

In broad terms, organisms can be intentionally or unintentionally transported across natural barriers and released into the environment (Table 18.1). Intentional introductions occur for many different reasons, including food, sport, biological control, landscaping, and esthetics. An infamous example of intentional introduction is the millions of starlings (*Sturnus vulgaris*) in North America that owe their existence in the New World to the 60 birds released in 1890 by the first President of the North American Acclimatization Society (Lever 1992). Motivated by such factors as introducing all the birds mentioned by Shakespeare (in the case of the aforementioned starlings) and making foreign landscapes more familiar, acclimatization societies are responsible for the establishment of many invasive species in North America, Australia, and New Zealand (Lever 1992).

Biological control of pest organisms, achieved through the introduction of natural enemies, is another example of purposeful introductions, and it represents an entire subdiscipline of biology (DeBach 1974; Van Den Bosch 1982; Van Dreische and Bellows 1996). There are many examples of successful biological control (Flint and Dreisadt 1999), and standard practice dictates that the release of these agents receives intense scrutiny. There are, however, many examples of negative, non-target effects of biological control (Howarth 1991; Follett and Duan 1999). For example, the European weevil *Rhinocyllus conicus*, introduced to North America for control of exotic thistles, switched to native thistles where they reduce seed production and indirectly impact native insects through competition for food resources (Louda *et al.* 1997).

Biological control efforts on islands can be even more problematic. Host switching by parasitoids introduced to control non-native moths in Hawaii has lead to parasitism rates of about 20% across 54 endemic moth species (Henneman and Memmott 2001). Introductions of generalist predators provide even more stark examples of intentional releases gone awry, including a series of misguided attempts to control exotic rats (*Rattus* spp.) on Pacific islands (Laycock 1966). To deal with the rat problem, top predators such as monitor lizards (*Varanus indicus*) and mongoose (*Herpestes javanicus*) were intentionally introduced. However, rats are primarily nocturnal while the introduced predators are diurnal.

Table 18.1. *Vectors of invasion into ecosystems*

....

Intentional introductions of target species

Escape of species from containment
- Zoos and botanical gardens
- Landscaping and ornamentals
- Pets
- Farmed species
- Agriculture
- Aquaculture and mariculture
- Research

Release directly into the environment
- Forestry plants
- Plants for soil improvements (e.g., stabilization)
- Ornamental plants
- Animals for hunting and fishing
- Biological control
- "Freed" pets
- "Enrichment" of native biota (e.g., through acclimatization societies)
- Disposal of living packing material (e.g., seaweed for bait)
- Bait
- Research
- Reintroductions of natives

Release of non-target species ("accidental" releases)

Contaminants or hitchhikers associated with goods
- Produce
- Nursery plants
- Cut flowers
- Seed stock
- Soil
- Timber
- Aquaculture and mariculture species
- Packing material
- Mail and cargo

Contaminants or hitchhikers associated with transportation
- Cars, trucks, airplanes, etc.
- Machinery, equipment
- Dry ballast in ships
- Ballast water
- Ballast sediments
- Ship hull fouling (e.g., barnacles and mussels)
- Tourists, luggage
- Canals
- Roads

Hitchhikers associated with artificial structures
- Movement of maritime superstructures (e.g., oil rigs)
- Floating debris

....

Source: Adapted from Wittenburg and Cock (2001).

Subsequently, native species often became primary prey items, leading to the decline of many local birds and lizards (Case and Bolger 1991). Problems were compounded even further when other species were introduced to divert the predators' attention from natives. For example, the giant toad (*Bufo marinus*) was introduced to act as alternate prey for monitor lizards, but as the toads are poisonous, they have led to declines in species that eat them, including pet dogs and cats (Laycock 1966; Atkinson and Atkinson 2000).

Despite many examples of intentional introductions, the bulk of invasions result from "unintentional" introductions. A cautionary note about terminology is warranted here, however. Although many types of introductions are often called "accidental" or "unintentional," we know enough about their vectors to assess the relative likelihood of release of at least some potential invaders. There is little accidental about unloading a cargo of timber that is sure to be laden with foreign insects. This has management implications that rise above semantics, and it should be assumed that the operation of known vectors can release organisms. Therefore, the "unintentional" release of organisms through such activities should no longer be considered "accidents" (Moyle 1999).

Terminology aside, for an organism to successfully arrive in a recipient ecosystem by means other than targeted releases, a number of steps must occur. First, an organism must associate itself with a means of transport — an invasion vector. This often involves a species being moved in association with a physical mode of transportation (e.g., a boat or airplane), but man-made changes in the landscape, such as roads or canals, also can facilitate biological invasions. The likelihood of utilization of a vector by a species is affected by the abundance of an organism in its native range, its habitat preferences, and the characteristics of the vector. In the case of ballast water (used to maintain ship stability during ocean crossings), for example, those species most likely to be entrained are small organisms floating in the water column near ports. Second, the invader must survive during transit, where it will be subjected to a variety of biotic and abiotic pressures. In the ballast tank, a species is likely to encounter competitors and predators, and will be subjected to environmental conditions that may include the presence of toxins and lack of light. Finally, the transported organism must survive upon release into the new environment. This survival involves matching of conditions in source and recipient areas. A planktonic organism in ballast water from Guam will likely find disagreeable environmental conditions upon release in an Alaskan harbor.

The vectors that move organisms across biogeographic boundaries vary widely in size and efficacy. Among all vectors, ballast tanks appear to be the champion species movers (Carlton 1985, 1987; Carlton and Geller 1993). A single tank can hold many millions of liters of water, with astonishing densities of associated organisms. Densities per single liter of ballast water have been estimated at tens to hundreds for zooplankton, thousands to millions for phytoplankton, and billions for bacteria and viruses (Ruiz and Carlton 2003). This tremendous density translates into high species richness. Carlton and Geller (1993), studying ships entering Coos Bay, Oregon, conservatively identified 367 planktonic species in the ballast water. It has been estimated that at any one time, over 7000 species may be on the move in ballast tanks of ships plying the world's waters (Carlton 1999).

Patterns of vector operation often will vary markedly with time (Ruiz and Carlton 2003). These changes are primarily associated with techno-logical and socio-economic advances and can strongly influence the com-position of species being transported. Kiritani and Yamamura (2003) have examined temporal patterns of invasion in Japanese insects. Japan first opened to external influences in 1868, and from then until World War II, invasions were characterized by scale insects and mealybugs associated with a large-scale effort to import fruit trees. The 20 years after World War II saw an invasion of beetles associated with grains moved into the country to prevent food shortages. The third phase of invasion, from 1966 to 1985, was characterized by weevils that infest crops, turfs, vegetables, and ornamental trees, and the most recent wave by greenhouse pests such as thrips, aphids, and whiteflies.

Similar shifts have been seen in west coast marine systems of the USA (Carlton 1979; Ruiz and Crooks 2001; Wonham and Carlton 2005). The first primary vector of invasion was fouling, the transport of organisms living in or on the hulls of ships. The current practice of using metal hulls and antifouling paints has limited spread by this means, although it certainly continues and its impact may be underestimated (Fofonoff et al. 2003). The second wave of invasion came from the movement of com-mercially important animals, and more importantly, species associated with these intentionally transported organisms. Starting in the late nineteenth century, there were large-scale movements of both Japanese and Virginia oysters (Crassostrea spp.) onto the west coast, and many invasions can be traced to the mud, packing materials, and shells associa-ted with the movement of live oysters (Miller 2000). The most recent wave of invasion is due to transport with ballast water, and although this may be decreasing due to current regulations requiring ballast exchange, data for tracking invasion rates in response to management actions is limited (but see Drake and Lodge 2004). It also should be noted that there are many other vectors that have brought invasive species to this region (Carlton 1979; Chapman *et al.* 2003), such as the aquarium trade (Semmens *et al.* 2004) and algal packing material used for live seafood and bait (Carlton and Cohen 2003).

Despite our ability to detect broad patterns in species arrivals, predicting the precise timing of invasion via a particular vector is difficult. Many species likely utilize a vector soon after it becomes operational, but it is also possible that there are long lags in between the commencement of new means of invasion and successful use by invaders (J. A. Crooks 2005). In the case of species invading the Mediterranean Sea via the Suez Canal (Boudouresque 1999), the rate of appearance of new invaders has risen steadily since the time of opening in 1869 (Fig. 18.1). This may be due in part to changing conditions in the canal and Mediterranean, but stochastic factors are also likely at work. Another example is the zebra mussel in the Great Lakes. Despite ballast water being dumped into these lakes for decades, it was not until the 1980s that the mussel was first detected (Nalepa and Schloesser 1993). A potential problem with such cases, however, is that it is difficult to distinguish whether a lag was actually in the arrival phase (i.e., the species never utilized the vector) or in the establishment phase (i.e., the species arrived well before it got noticed) (J.A. Crooks and Soulé 1999; Costello and Solow 2003). Determining

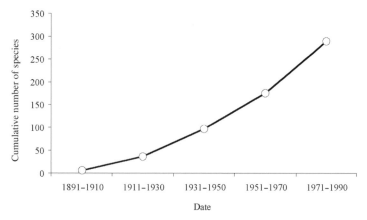

Fig. 18.1. Cumulative number of species invading the Mediterranean Sea via the Suez Canal, which opened in 1869. (Adapted from Boudouresque 1999.)

where the lag actually occurred is typically difficult, but it is important to recognize that the failure of a species to utilize a given vector does not indicate that it is incapable of doing so.

Anthropogenic influences on invasion rates

General principles suggest that probability of invader establishment should increase with (a) increased abundances of released organisms, (b) increased frequency of release, (c) increased numbers of released species, and (d) faster transfers (Ruiz and Carlton 2003). However, quantifying any of these relationships remains a great challenge in invasion biology because most of what we know about invasions comes from examination of already established invaders − ones that have successfully arrived as well as formed reproductive populations. These species have been subjected to the rigors associated with both the vector and the receiving environment, thus confounding the arrival and establishment phases of invasion. Despite this challenge, some patterns of invasion related to vector activity have emerged.

The first key question is the relationship between arrival and establishment, and determining what proportion of the inoculant pool will become successful invaders (e.g., Wonham et al. 2000, 2001). This number will be highly variable depending on the species transported, the vector, and the receiving environment (e.g., Cowie and Robinson 2003; Kraus 2003), but Williamson (1996) has suggested that approximately an order of magnitude fewer species will establish than arrive. There have been some attempts to quantify this. For intentionally introduced biological control agents, estimated success rates range between 13% and 34% (Hall and Ehler 1979; Ehler and Hall 1982). However, the success rate of unplanned, non-target invasions may be lower. In Japan, over 250 longhorned beetle species have been discovered by agricultural inspection, but only seven (3%) of those are considered established (Kiritani and Yamamura 2003). In the Yeayama Islands of Japan, 71 species of naturally migrating butterfly species have been observed, but only 10 (14%) are considered established (Kiritani and Yamamura 2003).

We can also ask how the rate of (successful) invasion has changed over time. There is no doubt that the total rate of invasion over the last few centuries far surpasses any rate seen previously, and estimated magnitudes of this difference are remarkable. For example, in Hawaii, it is estimated that one successful invertebrate colonization occurred every 50 000−100 000 years under natural conditions (Holt 1999). Currently, this anthropogenically inflated rate is estimated at one invasion every

18 days (Holt 1999). For ants specifically, despite winged dispersal of reproductives, no species naturally colonized the Hawaiian Islands. However, over 40 exotic ants are now established, most since World War II (Krushelnycky *et al.* 2005). For cladoceran crustaceans ("water fleas") in the Great Lakes, genetic evidence suggests that the anthropogenic rate of invasion is 50 000 times greater than the natural rate (Hebert and Cristescu 2002).

On more recent timescales, a wide variety of empirical data support constant or accelerating rates of both introduction of exotic species to the wild (Fig. 18.2) and establishment of successful invasive populations (Fig. 18.3). For example, recent syntheses of global reptile, amphibian, bird, and mammal invasion data demonstrate steadily increasing rates of release of these taxa (Kraus 2003) (Fig. 18.2). Similar patterns have also been seen for the introduction of terrestrial mollusks in Hawaii, with successful establishment rates that track those of introduction rates (Cowie and Robinson 2003). In aquatic systems, dramatic increases in successful invasions have been seen in many systems, including San Francisco Bay (Cohen and Carlton 1998), San Diego (J.A. Crooks 1998), the USA in general (Ruiz *et al.* 2000; Fofonoff *et al.* 2003; Fuller 2003), European coastlines (Ribera Siguan 2003), Port Phillip Bay in Australia (Thresher *et al.* 2000), and the Great Lakes (Ricciardi 2001). Although such patterns are undoubtedly affected by increased search effort and better taxonomy in recent years (Cohen and Carlton 1998), and perhaps by

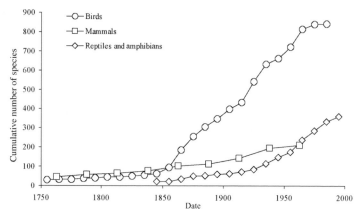

Fig. 18.2. Cumulative number of species introduced to the wild (globally). Data include both successful and unsuccessful invasions. (Adapted from Kraus 2003.)

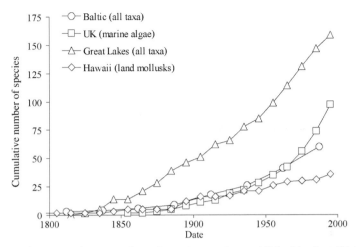

Fig. 18.3. Cumulative number of invasive species established in the wild for different geographic regions. Data from Ricciardi (2001), Leppäkoski *et al.* (2002), Cowie and Robinson (2003), and Ribera Siguan (2003).

the intrinsic dynamics of invasive populations (Costello and Solow 2003), there is little doubt that the rate of invader appearance in systems worldwide is on the rise (J. A. Crooks 2005).

There are several factors driving this increased invasion rate. Paradoxically, even though invaders are often associated with disturbance (see below), it has been suggested that improving environmental conditions in some highly degraded systems (such as urbanized bays) may have contributed to the increasing invasion rate (Great Lakes Environmental Research Laboratory 2002). For example, improving water quality in polluted and largely azoic parts of the Los Angeles/Long Beach Harbor allowed the invasion of crustacean bioeroders (gribbles) that destroyed wooden docks (Reish *et al.* 1980). The main force underlying the trend of increasing invasion rate, however, is clearly the frantic pace at which trade now occurs (Office of Technology Assessment 1993; Ruiz and Carlton 2003; Drake and Lodge 2004). In the USA, trade has been increasing exponentially, and over the next two decades it is expected to grow at about 6% per year (Levine and D'Antonio 2003).

Levine and D'Antonio (2003) discuss the relationship between invasions and trade, relating past imports to numbers of biological invaders and forecasting future invasion rates in the USA. The accumulation of successful invaders resulting from the operation of a specific vector turns out to be a problem similar to that of encountering new species when

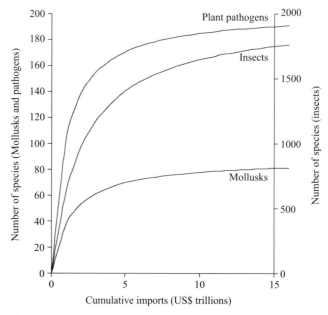

Fig. 18.4. Predicted relationship between cumulative imports and the accumulation of invaders. (Adapted from Levine and D'Antonio 2003.)

examining a series of samples from a community. In the case of species arriving on a ship, the per-ship probability of moving a new species declines as the number of ships increases. There is a limited pool of available species in the source region, and each ship is repeatedly sampling from the same pool. Thus, as more and more ships move, it is decreasingly likely that a new species will be sampled. The authors also modeled number of established exotic mollusks, plant pathogens, and insects as functions of cumulative imports since 1920, and used these models to forecast future invasions with projected increases in foreign imports (Fig. 18.4). The models conservatively predicted the establishment of three new terrestrial mollusks, five new plant pathogens, and 115 new insects in the USA over the next 20 years.

The dynamics of invasion also suggest that the rate of local appearance of new species can increase even without a concomitant increase in trade and vector activity. The "hub-and-spoke model" (Carlton 1996) demonstrates that as species get transported to new trade centers (the hubs), there will be an increased number of source regions from which invasions can radiate (the spokes). Unfortunately, this positive feedback suggests that new invasions will continue to be a problem even if vector activity levels off.

Vector activity and the population dynamics of established invaders

Hyperconnectivity is at the core of species initially arriving in a new location, but hyperconnectivity and the movement of individuals remains important as invasive populations grow and spread around the inoculation site. The process of utilizing an invasion vector, whether passively (e.g., on a boat) or actively (e.g., swimming through a canal), is likely to put checks on invasive populations. The decline in number of individuals caused by mortality incurred during transit and upon release are particularly important, and can result in relatively few individuals forming the new populations. Theory suggests that this likely will have genetic consequences related to population bottlenecks and founder effects, and the resulting restrictions in genetic diversity tend to negatively impact incipient populations (Mayr 1963; Soulé 1980; Mooney and Cleland 2001; Frankham Chapter 4).

The genetic factors associated with founding populations of invaders have important implications that highlight the need to consider invaders and their management on levels other than that of simply species presence and absence (Petit 2004). For example, just because an invader has already arrived in a system does not suggest that additional introductions of the same species are of no concern. Repeated inoculations will increase genetic flow and work against the inherent checks of small population size and low genetic diversity, thus increasing invasion success (J.A. Crooks and Soulé 1999; Mooney and Cleland 2001). Theory suggests that just one migrant per generation is sufficient to guard against losses in genetic diversity (Mills and Allendorf 1996), and thus there is a direct link between continued connectivity between systems and the potential for expansion of invasive populations. This should provide strong motivation for vector control above and beyond that related to the appearance of new species.

Despite this positive relationship between size of the founding colony and probability of successful establishment, very small numbers of individuals can eventually form large populations. Elephant seals, for example, were nearly exterminated and it is believed that a population of less than 20 individuals eventually recovered to over 100 000 seals (Hoelzel 1999). For invasive ants, evidence suggests that new populations can be founded with a single queen and as few as 10 workers (Tsutsui and Suarez 2003). There are also exceptions to the typical negative effects associated with small founding populations. For Argentine ants in North America, it has been suggested that a loss of genetic diversity resulting from the invasion process led to a decrease in the expression of intraspecific aggression in introduced populations (Holway *et al.* 1998; Tsutsui *et al.* 2000, 2003). This promotes the formation of expansive, competitively dominant super-colonies that have severe ecological impacts on natural communities (Holway *et al.* 2002a).

Connectivity also plays a role in the post-introduction range expansion of invaders. Species often spread via expanding fronts, and simple passive diffusion models have improved our understanding of this aspect of spread (Hengeveld 1989; Hastings *et al.* 2004). Invaders also can traverse long distances and over natural barriers by a process known as jump dispersal. The importance of this mode of movement has recently been highlighted by both theoretical and empirical work. Higgins and Richardson (1999) used simulations to demonstrate that long-distance dispersal by as little as 0.001% of propagules (seeds in this example) could increase the overall rate of spread by an order of magnitude. Suarez *et al.* (2001) reconstructed the spread of Argentine ants in the continental USA and determined that human-mediated jump-dispersal events accounted for a majority of the spread of this species. Natural rates of dispersal through a diffusion-like process averaged only around 200 m per year, yet this ant spread to over 275 counties in the USA within 50 years of its establishment.

In zebra mussels, human-mediated dispersal can consist of both advective (i.e., within-watershed), and jump dispersal (i.e., across-watershed) events (Johnson and Carlton 1996; Johnson and Padilla 1996). In an attempt to quantify the potential for spread through a primary vector for this species, Buchan and Padilla (1999) measured the rates and distances that recreational boaters traveled in Wisconsin. While a great majority of boaters traveled only short distances, primarily within watersheds, a small number of boaters traveled large distances among watersheds, providing ample opportunity for zebra mussels to establish new, distant foci from which to spread. Also, the extent to which species spread by human-mediated jump dispersal may influence the success of control strategies. For example, preventing the establishment of new infestations can greatly enhance control efforts relative to preventing spread from currently established populations (Moody and Mack 1988). In this light, human-derived hyperconnectivity among formerly isolated areas can greatly influence invasion rates and dynamics at wide-ranging spatial scales.

HABITAT CONNECTIVITY WITHIN INVADED LANDSCAPES

Ecosystem quality and invasion success

One of the most discussed topics in invasion biology has been the eco-system characteristics that confer ecological resistance to invasion (Levine and D'Antonio 1999; Lonsdale 1999). The general idea, espoused by Elton (1958), is that species-poor, disturbed systems should be particularly vulnerable to invasion. The recent literature shows substantial debate on the relationship between diversity and invasibility (and ecosystem proper-ties in general: Kaiser 2000; Loraeu *et al.* 2001; Naeem and Wright 2003). An emerging picture is that there is often a negative relationship between diversity and the success of invaders, but only when considered within a similar set of environmental conditions at relatively small spatial scales (Kennedy *et al.* 2002; Shea and Chesson 2002). Integrating over different habitat types at larger spatial scales, it appears that areas that are good for natives are also good for exotics, leading to positive correlations when a range of ecological conditions are considered (Stohlgren *et al.* 1999; Shea and Chesson 2002).

There have been similar discussions about the relationship between disturbance and invasion (Hobbs and Huenneke 1992; D'Antonio *et al.* 1999; Alpert *et al.* 2000). There has been much less experimental focus on this topic (but see Hobbs and Atkins 1988; Burke and Grime 1996; White *et al.* 1997), but a large body of empirical evidence supports the idea that disturbed habitats have more invaders (Elton 1958; D'Antonio *et al.* 1999). It is a common observation that degraded areas have a relative preponderance of invaders. For example, studies along roadsides show that the number of alien plants decreases with increasing distance from the road (Tyser and Worley 1992; Gelbard and Belnap 2003). In California, the invasive Argentine ant increases its abundance with elevated moisture, such as in urban areas and areas with runoff from development (Holway *et al.* 2002b). Also, the densities of non-native and human-commensal birds and mammals often rapidly decline away from urban or agricul-tural development (Blair 1996; Bolger *et al.* 1997; K.R. Crooks 2002; K.R. Crooks *et al.* 2004). Marine bays and estuaries, which tend to be characterized by anthropogenically reduced habitat quality relative to the open ocean, typically have more invaders than the open coast (e.g., Carlton 1979; Wasson *et al.* 2005).

It is difficult to attribute such patterns to environmental conditions alone, however. It is tempting to say that these areas are more invaded because they are more disturbed, but invasion patterns are confounded by the spatial distribution of vector operation (Cohen and Carlton 1998; Lonsdale 1999; Ruiz et al. 2000). It is expected that roadsides would have more invaders because more propagules would be transported there, such as on passing vehicles or through intentional plantings of ornamentals. Similarly, in the oceans, most of the transport mechanisms operate between one bay and another. Ships with their ballast and fouling move from port to port, and many intentional introductions and unintentional introductions of tag-along species occurred in protected coastal water bodies (Carlton 1979; Fofonoff et al. 2003). This would greatly favor the transport of estuarine rather than open-coast species.

Fragmentation is in the eye of the beholder

If we broadly accept that vectors and environmental conditions interact to make disturbed areas more vulnerable to invasion, then exotics and natives may tend to "view" an ecosystem differently. In essence, the establishment and spread of exotic organisms may be facilitated by the hyperconnected, degraded habitat that has resulted from expansion of urban and agricultural areas. What represents a very loosely connected, fragmented system to a native species depending on natural habitats may seem a highly connected, extensive system to an exotic associated with disturbed, urban, or agricultural settings. Smaller, more fragmented patches of remnant natural habitat appear particularly vulnerable to invasion. Moreover, the actual urban/agricultural matrix is suitable and perhaps even preferred habitat for invaders, while in many cases these heavily modified areas do not support large numbers of natives (Suarez et al. 1998; K.R. Crooks et al. 2004). Therefore the remnant natural patches that remain are truly isolated for these native species.

The well-studied scrub habitat remnants in urban San Diego, California, USA, provide an excellent example of how a system can be highly connected for some species but isolated for others. Many studies show that these habitat fragments are islands for many species of area-sensitive animals including scrub-specialist birds, mammals, and insects (Soulé et al. 1988; Bolger et al. 1997; Suarez et al. 1998; K.R. Crooks and Soulé 1999). For taxa as diverse as birds, mammalian carnivores, and ants, introduced species are much more common in the urban matrix than they are within the remnant patches of native vegetation (K.R. Crooks 2002; Holway et al. 2002b; K.R. Crooks et al. 2004). Their abundance therefore appears as an edge effect with densities peaking in urban sites and along urban edges and declining with distance into natural vegetation. In effect, the urban matrix, normally thought to decrease connectivity among isolated patches of habitat, is acting as continuous habitats and corridors for invasive species.

There is also some evidence to suggest that once invaded, ecosystems can be modified by established exotics to pave the way for further invasions. This process has been termed "invasional meltdown" (Simberloff and Von Holle 1999), and it suggests that positive feedbacks may develop that progressively decrease habitat quality for natives. For example, within an Argentinean national park, exotic plants more readily invade areas heavily browsed by exotic livestock than they do more intact areas (Veblen *et al.* 1992).

IMPLICATIONS OF HYPERCONNECTIVITY FOR CONSERVATION AND MANAGEMENT

Hyperconnectivity is homogenizing the world's biota and facilitating the invasion of species that are dramatically altering the structure and function of ecosystems. The best opportunities to develop sound management strategies that address this crisis will arise from understanding the process of invasion, characteristics of invaders, vectors of transport, and factors that influence vulnerability of ecosystems. Some important management principles that arise from a consideration of hyperconnectivity have been highlighted above. For example, invasions at the level of the gene emphasize that continued reintroductions of establishing invaders will serve to counter the natural factors working to limit invasion success. Also, the lag effect suggests that there may be long periods of time before a vector gets utilized or populations explode after successful invasion. These delayed responses hamper predictions regarding invader identity and invasion timing (J. A. Crooks 2005).

Above all, the overriding principle of invasion management should be that prevention is the best medicine, as it is notoriously difficult to deal with invaders after they have become established. A recent volume published as a result of a Global Invasive Species Programme conference (Ruiz and Carlton 2003) represents a comprehensive resource detailing specific characteristics and management of many different terrestrial and aquatic vectors. Therein, Ruiz and Carlton (2003) offer a useful framework for vector management. The first step in their management scheme is vector analysis, representing an assessment of how the actual transfer mechanisms work. This analysis should be done regionally, and will provide an indication of propagule supply on a location-by-location basis. The next phase is a vector strength assessment, which gauges the relative importance of a vector in causing invasions. This represents a key measure because in practice the number of actual invaders is more critical than the number of potential invaders (although the potential for lagged appearance of invaders should not be underappreciated). The next phase of management is vector interruption, representing the imposition of actual action. This can take a variety of different forms, including decreasing the probability of uptake, decreasing survival during transit, and/ or preventing release, and can be accomplished through measures such as screening, cleansing and quarantine. Also, in order to ensure effective interruption, stringent guidelines should be established and enforced to prevent the introduction of new species unless it can be demonstrated that the relocation will have a net benefit when economic, societal, and ecological factors are weighed. The last phase in this iterative management process is an assessment of the efficacy of the vector interruption. This can be evaluated in terms of both effects on propagule supply (through vector analysis) and ultimately successful invasions (using vector strength assessment). Again, the latter represents the more important proximate measure for management.

Monitoring programs are an essential component of vector management, as they provide the information needed to assess efficacy of the vector interruption and guide future management actions. Detecting decreases in invasion rates in the field, achieved through thorough monitoring and sound taxonomy, is the only way to gauge success of invasion prevention efforts (Ruiz and Carlton 2003). In addition, a good monitoring program may act as an early-warning system by detecting species before they become established or when they are at low enough densities that eradication remains a possibility (Wittenburg and Cock 2001).

A variety of constraints present themselves when undertaking vector management. The international nature of the invasion problem means that solutions should cut across national borders (Reaser et al. 2003). The potential conflict between environmental protection and economics also represents a major challenge to invasion management. Although approaches to some types of vector control will require technological and logistical advances (e.g., physical treatment of tremendous volumes of ballast water), it is often economic considerations that represent a major obstacle for even conceptually simple approaches. For example, considering ballast water as wastewater that would have to undergo treatment similar to that of sewage is a seemingly straightforward approach (Cohen and Foster 2000). However, the costs associated with treating large volumes of water at existing or new facilities make it an unpopular choice in the business sector, and there has been administrative reluctance to adopt this approach. Overall, truly effective invasion management will only be achieved when we adopt the precautionary principle and shift the burden of proof to those responsible for moving species in the first place (Mack *et al.* 2000). Such leaps, however, are typically very slow in coming (Dayton 1998).

Once an exotic has successfully invaded, any necessary regulatory action needed to initiate management action should be subject to a "rapid review and approval" process. The actual control measures to be employed will vary depending on the invader and the invaded ecosystem, and because control is often very costly and labor-intensive, only the most problematic of species are typically targeted (thus the emphasis on preventing invasions). Control often requires a brute-force approach and the "nasty necessity" of eradication (Temple 1990). The introduction of natural enemies represents another possible control measure, although this rarely achieves full elimination of an invader. Also, as biological control typically involves the intentional introduction of yet another invader, any such efforts should proceed with an abundance of caution.

Another potential means of controlling invader impact is managing the environment to promote natives at the expense of exotics. Despite the ample debate about the mechanisms of ecological resistance, we believe that observed negative relationships between habitat quality and invader success provide (yet another) reason to conserve and restore natural areas. Increasing connectivity of fragmented natural habitats may increase invasion resistance by improving habitat quality and increasing the size of natural areas. This will, however, present risk of spread of invasive species between otherwise separated patches (Simberloff and Cox 1987). This is apparent in riparian areas where connectivity is essential for upstream movement of fish, but aqueducts also may facilitate the spread of exotics to otherwise isolated watersheds (Pringle Chapter 10). Restoring lost connections also will minimize the influence of the surrounding urban/agricultural matrix on remnant natural areas by decreasing the overall contact of urban associates with the natural habitats via the edge.

Increasing public awareness of the problem of invasive species remains absolutely necessary in minimizing future introductions. Specifically, more effective education measures are necessary to teach the public of the economic costs associated with invasions and how exotics can undermine the structure and function of natural areas. Additional research on the ecology of biological invasions is also needed, although advances are being made. We are still a long way from determining what makes a species a successful invader – a necessary step in making invasion biology a more predictive science (Holway and Suarez 1999; Kolar and Lodge 2002). In addition, basic research assessing the current distribution and identity of exotic (as well as native) species, which will require high-resolution monitoring and careful taxonomy, is also urgently needed. Continued investigations on topics such as these will offer a more complete picture of mechanisms underlying invasion success and impacts, as well as maximize the efficacy of management efforts.

Despite the need to develop better tools for detecting and controlling exotics, it is clear that at its broadest level managing invasions will require managing connectivity. Human activities have artificially inflated connectivity and facilitated the current onslaught of invasions at different spatial scales. Globally, the abundance and efficiency of invasion vectors has begun to erode natural biogeographic boundaries, and to remedy this we must address the difficult issue of decreasing these connections in the face of ever-increasing globalization. At the landscape level, destruction and fragmentation of natural habitats have created hyperconnected urban and agricultural areas that promote invasion. Only by reversing this trend and increasing the size and connectivity of natural habitats will we be able to reconstitute systems that no longer favor exotics at the expense of natives.

ACKNOWLEDGEMENTS

We thank Kevin Crooks and M.A. Sanjayan for organizing this volume. The chapter benefited from thoughtful reviews provided by Marjorie Wonham, the Ecology Graduate Seminar at Colorado State University, and the IB 199 class at University of Illinois at Urbana–Champaign. We also thank the National Science Foundation (INT-0305660) for support provided to AVS. This work was also supported in part by the National Sea Grant College Program of the US Department of Commerce's National Oceanic and Atmospheric Administration under NOAA Grant NA04OAR4170038, project R/CZ-190A, through the California Sea Grant College Program; and in part by the California State Resources Agency. The views expressed herein do not necessarily reflect the views of any of these organizations.

REFERENCES

Alpert, P., E. Bone, and C. Holzapfel. 2000. Invasiveness, invasibility and the role of environmental stress in the spread of non-native plants. *Perspectives in Plant Ecology, Evolution and Systematics* 3:52–66.

Atkinson, I. A. E., and T. J. Atkinson. 2000. Land vertebrates as invasive species in the islands of the South Pacific Regional Environment Programme. Pp. 9–84 in G. Sherley (ed.) *Invasive Species in the Pacific: A Technical Review and Draft Regional Strategy.* Samoa: South Pacific Regional Environment Programme Publication Unit.

Baskin, Y. 2002. *A Plague of Rats and Rubbervines.* Washington, DC: Island Press.

Blair, R. B. 1996. Land use and avian species diversity along an urban gradient. *Ecological Applications* 6:506–519.

Bolger, D. T., A. C. Alberts, R. M. Sauvajot, *et al.* 1997. Response of rodents to habitat fragmentation in coastal southern California. *Ecological Applications* 7:552–563.

Boudouresque, C. F. 1999. The Red Sea–Mediterranean link: unwanted effects of canals. Pp. 213–235 in O. T. Sandlund, P. J. Schei, and A. Viken (eds.) *Invasive Species and Biodiversity Management.* Dordrecht, The Netherlands: Kluwer Academic Press.

Bright, C. 1998. *Life out of Bounds.* New York: W.W. Norton.

Buchan, L. A. J., and D. K. Padilla. 1999. Estimating the probability of long-distance overland dispersal of invading aquatic species. *Ecological Applications* 9:254–265.

Burke, M. J. W., and J. P. Grime. 1996. An experimental study of plant community invasibility. *Ecology* 77:776–790.

Carlton, J. T. 1979. History, biogeography, and ecology of the introduced marine and estuarine invertebrates of the Pacific coast of North America. Ph.D. dissertation, University of California, Davis, CA.

Carlton, J. T. 1985. Transoceanic and interoceanic dispersal of coastal marine organisms: the biology of ballast water. *Oceanography and Marine Biology: An Annual Review* 23:313–371.

Carlton, J. T. 1987. Patterns of transoceanic marine biological invasions in the Pacific Ocean. *Bulletin of Marine Science* 41:452–465.

Carlton, J. T. 1996. Pattern, process, and prediction in marine invasion ecology. *Biological Conservation* 78:97–106.

Carlton, J. T. 1999. The scale and ecological consequences of biological invasions in the World's oceans. Pp. 195–212 in O. T. Sandlund, P. J. Schei, and A. Viken (eds.) *Invasive Species and Biodiversity Management.* Dordrecht, The Netherlands: Kluwer Academic Press.

Carlton, J. T., and A. N. Cohen. 2003. Episodic global dispersal in shallow water marine organisms: the case history of the European shore crabs *Carcinus maenas* and *C. aestuarii. Journal of Biogeography* 30:1809–1820.

Carlton, J. T., and J. B. Geller. 1993. Ecological roulette: the global transport of non-indigenous marine organisms. *Science* 261:78–82.

Carlton, J. T., J. B. Geller, M. L. Reaka-Kudla, and E. A. Norse. 1999. Historical extinctions in the sea. *Annual Reviews of Ecology and Systematics* 30:515–538.

Case, T. J. 1996. Global patterns in the establishment and distribution of exotic birds. *Biological Conservation* 78:69–96.

Case, T. J., and D. T. Bolger. 1991. The role of introduced species in shaping the distribution and abundance of island reptiles. *Evolutionary Ecology* 5:272–290.

Censky, E. J., K. Hodge, and J. Dudley. 1998. Over-water dispersal of lizards due to hurricanes. *Nature* 395:556.

Chapman, J. W., T. W. Miller, and E. V. Coan. 2003. Live seafood species as recipes for invasion. *Conservation Biology* **17**:1386–1395.

Cohen, A. N., and J. T. Carlton. 1998. Accelerating invasion rate in a highly invaded estuary. *Science* **279**:555–558.

Cohen, A. N., and B. Foster. 2000. The regulation of biological pollution: preventing exotic species invasions from ballast water discharged into California coastal waters. *Golden Gate University Law Review* **30**:787–883.

Costello, C. J., and A. R. Solow. 2003. On the pattern of discovery of introduced species. *Proceedings of the National Academy of Sciences of the USA* **100**:3321–3323.

Cowie, R. H., and D. G. Robinson. 2003. Pathways of introduction of non-indigenous land and freshwater snails and slugs. Pp. 93–122 in G. M. Ruiz and J. T. Carlton (eds.) *Invasive Species: Vectors and Management Strategies*. Washington, DC: Island Press.

Cox, G. W. 1999. *Alien Species in North America and Hawaii*. Washington, DC: Island Press.

Crooks, J. A. 1998. The effects of the introduced mussel, *Musculista senhousia*, and other anthropogenic agents on benthic ecosystems of Mission Bay, San Diego. Ph.D. dissertation, Scripps Institution of Oceanography, University of California, San Diego, CA.

Crooks, J. A. 2002. Characterizing the consequences of invasions: the role of introduced ecosystem engineers. *Oikos* **97**:153–166.

Crooks, J. A. 2005. Lag times and exotic species: the ecology and management of biological invasions in slow-motion. *Ecoscience* **12**:316–329.

Crooks, J. A., and M. E. Soulé. 1999. Lag times in population explosions of invasive species: causes and implications. Pp. 103–125 in O. T. Sandlund, P. J. Schei, and A. Viken (eds.) *Invasive Species and Biodiversity Management*. Dordrecht, The Netherlands: Kluwer Academic Press.

Crooks, K. R. 2002. Relative sensitivities of mammalian carnivores to habitat fragmentation. *Conservation Biology* **16**:488–502.

Crooks, K. R., and M. E. Soulé. 1999. Mesopredator release and avifaunal extinctions in a fragmented system. *Nature* **400**:563–566.

Crooks, K. R., A. V. Suarez, and D. T. Bolger. 2004. Avian assemblages along a gradient of urbanization in a highly fragmented landscape. *Biological Conservation* **115**:451–462.

D'Antonio, C. M., T. L. Dudley, and M. Mack. 1999. Disturbance and biological invasions: direct effects and feedbacks. Pp. 429–468 in L. R. Walker (ed.) *Ecosystems of Disturbed Ground*. Amsterdam, The Netherlands: Elsevier.

Dayton, P. K. 1998. Reversal of the burden of proof in fisheries management. *Science* **279**:821.

DeBach, P. 1974. *Biological Control of Natural Enemies*. London: Cambridge University Press.

Drake, J. M., and D. M. Lodge. 2004. Global hot spots of biological invasions: evaluating options for ballast-water management. *Proceedings of the Royal Society of London B* **271**:575–580.

Ehler, L. E., and R. W. Hall. 1982. Evidence for competitive exclusion of introduced natural enemies in biological control. *Environmental Entomology* **11**:1–3.

Elton, C. S. 1958. *The Ecology of Invasions by Animals and Plants*. New York: John Wiley.

Flint, M. L., and S. H. Dreisadt. 1999. *Natural Enemies Handbook*. Berkeley, CA: University of California Press.

Fofonoff, P. W., G. M. Ruiz, B. Steves, and J. T. Carlton. 2003. In ships or on ships? Mechanisms of transfer and invasion for nonnative species to the coasts of North America. Pp. 152–182 in G. M. Ruiz, and J. T. Carlton (eds.) *Invasive Species: Vectors and Management Strategies*. Washington, DC: Island Press.

Follett, P. A., and J. J. Duan. 1999. *Nontarget Effects of Biological Control*. Dordrecht, The Netherlands: Kluwer Academic Publishers.

Fuller, P. L. 2003. Freshwater aquatic vertebrate introductions in the United States: patterns and pathways. Pp. 123–151 in G. M. Ruiz, and J. T. Carlton (eds.) *Invasive Species. Vectors and Management Strategies*. Washington, DC: Island Press.

Gelbard, J. L., and J. Belnap. 2003. Roads as conduits for exotic plant invasions in a semiarid landscape. *Conservation Biology* 17:420–432.

Great Lakes Environmental Research Laboratory. 2002. Exotic, invasive, alien, nonindigenous, or nuisance species: no matter what you call them, they're a growing problem. Available online at http://www.glerl.noaa.gov/pubs/ brochures/invasive/ansprimer.html

Hall, R. W., and R. W. Ehler. 1979. Rate of establishment of natural enemies in classical biological control. *Bulletin of the Ecological Society of America* 26:280–282.

Hastings, A., K. Cuddington, K. F. Davies, *et al.* 2004. The spatial spread of invasions: new developments in theory and evidence. *Ecology Letters* 8:91–101.

Hebert, P. D. N., and M. E. A. Cristecu. 2002. Genetic perspectives on invasions: the case of the Cladocera. *Canadian Journal of Fisheries and Aquatic Sciences* 59:1229–1234.

Hengeveld, R. 1989. *Dynamics of Biological Invasions*. New York: Chapman and Hall.

Henneman, M. L., and J. Memmott. 2001. Infiltration of a Hawaiian community by introduced biological control agents. *Science* 293:1314–1316.

Higgins, S. L., and D. M. Richardson. 1999. Predicting plant migration rates in a changing world: the role of long-distance dispersal. *American Naturalist* 153:464–475.

Hobbs, R. J., and L. Atkins. 1988. Effect of disturbance and nutrient addition on native and introduced annuals in plant communities in the western Australia wheat-belt. *Australian Journal of Ecology* 13:171–179.

Hobbs, R. J., and L. F. Huenneke. 1992. Disturbance, diversity, and invasion: implications for conservation. *Conservation Biology* 6:324–337.

Hoelzel, A. R. 1999. Impact of population bottlenecks on genetic variation and the importance of life-history: a case study of the northern elephant seal. *Biological Journal of the Linnean Society* 68:23–39.

Holt, A. 1999. An alliance of biodiversity, agriculture, health, and business interests for improved alien species management in Hawaii. Pp. 65–75 in O. T. Sandlund, P. J. Schei, and A. Viken (eds.) *Invasive Species and Biodiversity Management*. Dordrecht, The Netherlands: Kluwer Academic Press.

Holway, D. A., and A. V. Suarez. 1999. Animal behavior: an essential component of invasion biology. *Trends in Ecology and Evolution* **14**:328–330.

Holway, D. A., A. V. Suarez, and T. J. Case. 1998. Loss of intraspecific aggression in the success of a widespread invasive social insect. *Science* **282**:949–952.

Holway, D. A., L. Lach, A. V. Suarez, N. D. Tsutsui, and T. J. Case. 2002a. The ecological causes and consequences of ant invasions. *Annual Reviews of Ecology and Systematics* **33**:181–233.

Holway, D. A., A. V. Suarez, and T. J. Case. 2002b. Role of abiotic factors in governing susceptibility to invasion: a test with Argentine ants. *Ecology* **83**:1610–1619.

Howarth, F. G. 1991. Environmental impacts of classical biological control. *Annual Review of Entomology* **36**:485–509.

Johnson, L. E., and J. T. Carlton. 1996. Post-establishment spread in large-scale invasions: the relative roles of leading natural and human-mediated dispersal mechanisms of the zebra mussel, *Dreissena polymorpha*. *Ecology* **77**:1686–1690.

Johnson, L. E., and D. K. Padilla. 1996. Geographic spread of exotic species: ecological lessons and opportunities from the invasion of the zebra mussel, *Dreissena polymorpha*. *Biological Conservation* **78**:23–33.

Kaiser, J. 2000. Rift over biodiversity divides ecologists. *Science* **289**:1282–1283.

Kennedy, T. A., S. Naeem, K. M. Howe, et al. 2002. Biodiversity as a barrier to ecological invasion. *Nature* **417**:636–638.

Kiritani, K., and K. Yamamura. 2003. Exotic insects and their pathways for invasion. Pp. 44–67 in G. M. Ruiz and J. T. Carlton (eds.) *Invasive Species: Vectors and Management Strategies*. Washington, DC: Island Press.

Kolar, C. S., and D. M. Lodge. 2002. Ecological predictions and risk assessment for alien species. *Science* **298**:1233–1236.

Kraus, F. 2003. Invasion pathways for terrestrial vertebrates. Pp. 68–92 in G. M. Ruiz, and J. T. Carlton (eds.) *Invasive Species: Vectors and Management Strategies*. Washington, DC: Island Press.

Krushelnycky, P. D., L. L. Loope, and N. J. Reimer. 2005. The ecology, policy, and management of ants in Hawaii. *Proceedings of the Hawaiian Entomological Society* **37**:1–25.

Laycock, G. 1966. *The Alien Animals: The Story of Imported Wildlife*. New York: Ballantine Books.

Leppäkoski E., S. Gollasch, P. Gruska, et al. 2002. The Baltic: a sea of invaders. *Canadian Journal of Fisheries and Aquatic Science* **59**:1175–1188.

Lever, C. 1992. *They Dined on Eland: The Story of Acclimatization Societies*. London: Quiller Press.

Levine, J. M., and C. M. D'Antonio. 1999. Elton revisited: a review of evidence linking diversity and invasibility. *Oikos* **87**:15–26.

Levine, J. M., and C. M. D'Antonio. 2003. Forecasting biological invasions with increasing international trade. *Conservation Biology* **17**:322–326.

Lockwood, J. L., and M. L. McKinney. 2001. *Biotic Homogenization*. New York: Kluwer Academic/Plenum Press.

Lonsdale, W. M. 1999. Global patterns of plant invasions and the concept of invasibility. *Ecology* **80**:1522–1536.

Loreau, M., S. Naeem, P. Inchausti, *et al.* 2001. Biodiversity and ecosystem functioning: current knowledge and future challenges. *Science* **294**:804–808.

Louda, S. M., D. Kendall, J. Conner, and D. Simberloff. 1997. Ecological effects of an insect introduced for the biological control of weeds. *Science* 277:1088–1090.

Mack, R. N. 2003. Global plant dispersal, naturalization, and invasion: pathways, modes, and circumstances. Pp. 3–30 in G. M. Ruiz and J. T. Carlton (eds.) *Invasive Species. Vectors and Management Strategies.* Washington, DC: Island Press.

Mack, R. N., D. Simberloff, W. M. Lonsdale, *et al.* 2000. Biotic invasions: causes, epidemiology, global consequences, and control. *Ecological Applications* 10:689–710.

Mayr, E. 1963. *Animal Species and Evolution.* Cambridge, MA: Harvard University Press.

Miller, A. W. 2000. Assessing the importance of biological attributes for invasion success: eastern oyster (*Crassostrea virginica*) introductions and associated molluscan invasions of Pacific and Atlantic coastal systems. D.Env. dissertation, University of California, Los Angeles, CA.

Mills, L. S., and F. W. Allendorf. 1996. The one-migrant-per-generation rule in conservation management. *Conservation Biology* 10:1509–1518.

Møller P. R., J. G. Nielsen, and I. Fossen. 2003. Patagonian toothfish found off Greenland. *Nature* 421:599.

Moody, M. E., and R. N. Mack. 1988. Controlling the spread of plant invasions: the importance of nascent foci. *Journal of Applied Ecology* 25:1009–1021.

Mooney, H. A., and E. E. Cleland. 2001. The evolutionary impact of invasive species. *Proceedings of the National Academy of Sciences of the USA* 98:5446–5451.

Moyle P. B. 1999. Effects of invading species on freshwater and estuarine ecosystems. Pp. 177–194 in O. T. Sandlund, P. J. Schei and A. Viken (eds.) *Invasive Species and Biodiversity Management.* Dordrecht, The Netherlands: Kluwer Academic Press.

Naeem, S., and J. Wright. 2003. Disentangling biodiversity effects on ecosystem functioning: deriving solutions to a seemingly insurmountable problem. *Ecology Letters* 6:567–579.

Nalepa, T. F., and D. W. Schloesser. 1993. *Zebra Mussels: Biology, Impacts, and Control.* Boca Raton, FL: Lewis Publishers.

Office of Technology Assessment. 1993. *Harmful Non-Indigenous Species in the United States,* OTA Publication OTA-F-565. Washington, DC: US Government Printing Office.

Olden, J. D., N. L. Poff, M. R. Douglas, M. E. Douglas, and K. D. Fausch. 2004. Ecological and evolutionary consequences of biotic homogenization. *Trends in Ecology and Evolution* 19:18–24.

Petit, R. 2004. Biological invasions at the gene level. *Diversity and Distributions* 10:159–165.

Pimentel, D. 2002. *Biological Invasions: Economic and Environmental Costs of Alien Plant, Animal, and Microbe Species.* Boca Raton, FL: CRC Press.

Rahel, F. J. 2000. Homogenization of fish faunas across the United States. *Science* **288**:854–856.

Reaser, J. K., B. B. Yeager, P. R. Phifer, A. K. Hancock, and A. T. Gutierrez. 2003. Pp. 362–381 in G. M. Ruiz, and J. T. Carlton (eds.) *Invasive Species: Vectors and Management Strategies*. Washington, DC: Island Press.

Reish, D. J., D. F. Soule, and J. D. Soule. 1980. The benthic biological conditions of Los Angeles–Long Beach Harbors: results of 28 years of investigations and monitoring. *Helgolander Meeresuntersuchungen* **34**:193–205.

Ribera Siguan, M. A. 2003. Pathways of biological invasion of marine plants. Pp. 183–226 in G. M. Ruiz, and J. T. Carlton (eds.) *Invasive Species: Vectors and Management Strategies*. Washington, DC: Island Press.

Ricciardi, A. 2001. Facilitative interactions among aquatic invaders: is an invasional meltdown occurring in the Great Lakes? *Canadian Journal of Fisheries and Aquatic Sciences* **58**:2513–2525.

Ritchie, S. A., and W. Rochester. 2001. Wind-blown mosquitoes and the introduction of Japanese encephalitis into Australia. *Emerging Infectious Diseases* **7**:900–903.

Rosenzweig, M. L. 2001. The four questions: what does the introduction of exotic species do to diversity? *Evolutionary Ecology Research* **3**:361–367.

Ruiz, G. M., and J. T. Carlton. 2003. Invasion vectors: a conceptual framework for management. Pp. 459–504 in G. M. Ruiz, and J. T. Carlton (eds.) *Invasive Species: Vectors and Management Strategies*. Washington, DC: Island Press.

Ruiz, G. M., and J. A. Crooks. 2001. Marine invaders: patterns, effects, and management of non-indigenous species. Pp. 3–17 in P. Gallagher, and L. Bendell-Young (eds.) *Waters in Peril*. Dordrecht, The Netherlands: Kluwer Academic Publishers.

Ruiz, G. M, P. W. Fofonoff, J. T. Carlton, M. J. Wonham, and A. H. Hines. 2000. Invasion of coastal marine communities in North America: apparent patterns, processes, and biases. *Annual Reviews of Ecology and Systematics* **31**:481–531.

Sax, D. F., S. D. Gaines, and J. H. Brown. 2002. Species invasions exceed extinctions on islands worldwide: a comparative study of plants and birds. *American Naturalist* **160**:766–783.

Scheltema, R. S. 1986. On dispersal and planktonic larvae of benthic invertebrates: an eclectic overview and summary of problems. *Bulletin of Marine Science* **39**:290–322.

Semmens, B. X., E. R. Buhle, A. K. Salomon, and C. V. Pattengill-Semmens. 2004. A hotspot of non-native marine fishes: evidence for the aquarium as an invasion pathway. *Marine Ecology – Progress Series* **266**:239–244.

Shea, K., and P. Chesson. 2002. Community ecology as a framework for biological invasions. *Trends in Ecology and Evolution* **17**:170–176.

Simberloff, D., and J. Cox. 1987. Consequences and costs of conservation corridors. *Conservation Biology* **1**:63–71.

Simberloff, D., and B. Von Holle. 1999. Positive interactions of nonindigenous species: invasional meltdown? *Biological Invasions* **1**:21–32.

Simberloff, D., J. A. Farr, J. Cox, and D. W. Mehlman. 1992. Movement corridors: conservation bargains or poor investments? *Conservation Biology* **6**:493–504.

Soulé, M. E. 1980. Thresholds for survival: maintaining fitness and evolutionary potential. Pp. 151–169 in M. E. Soulé and B. A. Wilcox (eds.) *Conservation Biology: An Evolutionary-Ecological Perspective.* Sunderland, MA: Sinauer Associates.

Soulé, M. E., D. T. Bolger, A. C. Alberts, *et al.* 1988. Reconstructed dynamics of rapid extinction of chaparral-requiring birds in urban habitat islands. *Conservation Biology* 2:75–92.

Steadman, D. W. 1995. Prehistoric extinctions of Pacific island birds: biodiversity meets zooarchaeology. *Science* 267:1123–1131.

Stohlgren, T. J., D. Binkley, G. W. Chong, *et al.* 1999. Exotic plant species invade hot spots of native plant diversity. *Ecological Monographs* 69:25–46.

Suarez, A. V., D. T. Bolger, and T. J. Case. 1998. The effects of habitat fragmentation and invasion on the native ant community in coastal southern California. *Ecology* 79:2041–2056.

Suarez, A. V., D. A. Holway, and T. J. Case. 2001. Patterns of spread in biological invasions dominated by long-distance jump dispersal: insights from Argentine ants. *Proceedings of the National Academy of Sciences of the USA* 98:1095–1100.

Temple, S. 1990. The nasty necessity: eradicating exotics. *Conservation Biology* 4:113–115.

Thresher, R. E., C. L. Hewitt, and M. L. Campbell. 2000. Synthesis: introduced and cryptogenic species in Port Phillip Bay. Pp. 283–295 in C. L. Hewitt, M. L. Campbell, R. E. Thresher, and R. B. Martin (eds.) *Marine Biological Invasions of Port Phillip Bay, Victoria*, Centre for Research on Introduced Marine Pests, Technical Report 20. Hobart, Tasmania, Australia: CSIRO Marine Research.

Tsutsui, N. D., and A. V. Suarez. 2003. The colony structure and population biology of invasive ants. *Conservation Biology* 17:48–58.

Tsutsui, N. D., A. V. Suarez, D. A. Holway, and T. J. Case. 2000. Reduced genetic variation and the success of an invasive species. *Proceedings of the National Academy of Sciences of the USA* 97:5948–5953.

Tsutsui, N. D., A. V. Suarez, and R. K. Grosberg. 2003. Genetic diversity, asymmetrical aggression, and recognition in a widespread invasive species. *Proceedings of the National Academy of Sciences of the USA* 100:1078–1083.

Tyser, R. W., and C. A. Worley. 1992. Alien flora in grasslands along road and trail corridors in Glacier National Park, USA. *Conservation Biology* 6:253–262.

Van Den Bosch, R. 1982. *An Introduction to Biological Control.* New York: Plenum Press.

Van Driesche, R. D., and T. S. Bellows. 1996. *Biological Control.* New York: Chapman and Hall.

Veblen, T. T., M. Mermoz, C. Martin, and T. Kitzberger, 1992. Ecological impacts of introduced animals in Nahuel Huapi National Park, Argentina. *Conservation Biology* 6:71–83.

Wasson, K., T. Fenn, and J. S. Pearse. 2005. Habitat bias in marine invasions of central California. *Biological Invasions* 7:935–948.

White, T. A., B. D. Campbell, and P. D. Kemp. 1997. Invasion of temperate grassland by a subtropical annual grass across an experimental matrix of water stress and disturbance. *Journal of Vegetation Science* 8:847–854.

Wilcove, D. S., D. Rothstein, J. Dubow, A. Phillips, and E. Losos. 1998. Quantifying threats to imperiled species in the United States. *BioScience* **48**:607–615.

Williamson, M. B. 1996. *Biological Invasions*. London: Chapman and Hall.

Wittenburg, R., and M. J. W. Cock. 2001. *Invasive Alien Species: A Toolkit of Best Prevention and Management Practices*. Wallingford, UK: CAB International.

Wonham, M. J., and J. T. Carlton. 2005. Cool-temperate marine invasions at local and regional scales: the Northeast Pacific Ocean as a model system. *Biological Invasions* 7:369–392.

Wonham, M. J., J. T. Carlton, G. M. Ruiz, and L. D. Smith. 2000. Fish and ships: relating dispersal frequency to success in biological invasions. *Marine Biology* **136**:1111–1121.

Wonham, M. J., W. C. Walton, G. M. Ruiz, A. M. Frese, and B. S. Galil. 2001. Going to the source: role of invasion pathway in determining potential invaders. *Marine Ecology – Progress Series* **215**:1–12.

(19)

Disease and connectivity

HAMISH McCALLUM AND ANDY DOBSON

INTRODUCTION

Fragmentation of natural habitats has important effects on the viability and persistence of most free-living animal and plant species; the other chapters in this volume outline many of these effects in eloquent detail. In this chapter we focus our attention on the parasitic half of biodiversity and examine how the viability and persistence of pathogens and parasitic species are modified by fragmentation and reconnection of the patchy habitats in which their host species live. The problem can be addressed at a hierarchy of different scales, as almost by definition, parasites and pathogens are canonically "adapted" to live in the patchy environment defined by the individual hosts they live in (Dobson 2003). Life-history evolution in parasites is sharply defined by the twin processes of exploiting the patch of habitat in which you live (your host) and producing infective stages (your offspring), which have to then find new patches (hosts) to exploit. Movement between hosts for pathogens is similar in many ways to dispersal between patches for free-living organisms. The key difference is that all of the dispersal in pathogens is undertaken by transmission stages that are the effective offspring of the parasites that currently infect the host. So transmission between host patches is for parasites both birth and dispersal. Fragmentation of the host's habitat increases the average distances the parasites have to move between birth and successful colonization. However, in situations where host populations aggregate into remaining patches of a fragmented landscape, the transmission success of pathogens may increase and this may lead to increases in the prevalence of some parasites and pathogens.

Connectivity Conservation eds. Kevin R. Crooks and M. Sanjayan. Published by Cambridge University Press. © Cambridge University Press 2006.

Transmission of any infectious disease involves a connection between an infected and a susceptible host. It is therefore inevitable that processes leading to changes in connectivity between individuals or populations will have consequences for transmission and persistence of infectious disease. A very useful way of thinking about a host−pathogen system is as a network, where connections between the nodes represent contacts (Newman 2002). There is a large body of network theory (Strogatz 2001; Newman 2003) that can be drawn upon to predict how the invasion potential of an infectious agent may depend upon the connectivity properties of the network (Keeling 1999). In most of the existing theory, the nodes in the network represent individual organisms, which may be of different types (for example, susceptible, infected, or resistant). However, there is no reason why the same basic theory should not be applied at a metapopulation level, where the nodes represent patches of habitat, social groups, or subpopulations: connections represent movement of individuals between patches, and patches may be in a variety of different states (empty, occupied without disease, infected, or occupied but refractory to disease). For example, a recent network model investigated the transmission of mycoplasmal pneumonia between wards occupied by patients, and connected via shared caregivers (Meyers *et al.* 2003). Such a model is exactly analogous to a metapopulation connected via a mobile reservoir species.

The simplest conventional host−pathogen models divide the host population into susceptible, infected, and resistant individuals (SIR models: Anderson and May 1991) and use differential equations to track the changes through time in the number of individuals in each category. This approach implicitly assumes that each individual has an equal probability of encountering each other individual: connectivity is equal between all individuals. Network theory allows this obviously unrealistic assumption to be relaxed, so that the consequences of differing patterns of connectivity can be explored. The cost is that current methods allow only the final state of the epidemic to be investigated, whereas SIR models can provide information on the transient dynamics of the interaction (Newman 2003).

Both SIR models and network models share the important property of having a threshold transmission rate: if the probability of transmission is sufficiently high compared with the recovery rate, a large fraction of individuals (but not all individuals in either model) become infected following introduction of the disease. For such an epidemic to occur, the average number of secondary infections arising from each infected individual entering an uninfected host population (R_o in most infectious disease literature) must exceed 1. Otherwise the infection dies out after a relatively small number of transmission events. This threshold is of critical importance, because management actions that bring the system below the threshold will prevent an epidemic occurring, and conversely, if the system is brought above the threshold, the disease will emerge. An advantage of network models is that they deal explicitly with patterns of connectivity between individuals, and therefore can be used to explore the consequences of changes in connectivity.

One intuitively obvious result that emerges from almost all such models is the crucial role played in epidemic spread by the small proportion of highly connected nodes in such networks (May and Lloyd 2001; Newman 2002; Olinky and Stone 2004). Thus, targeting control interventions on such nodes is the optimal way to prevent or control an epidemic. However, without determining the structure of the entire network, identifying such nodes may not be an easy task. An ingenious suggestion, which has direct potential to assist in controlling disease spread either between individuals, or between subpopulations in a metapopulation, is to select an individual node at random, and then to treat random nodes connected to it (Cohen et al. 2003; Newman 2003). The structure of the network itself means that this strategy will have the effect of linking to the most highly connected nodes.

In the following sections, we first review the limited empirical evidence concerning the effect of habitat fragmentation on pathogen dynamics, before reviewing the phocine distemper outbreaks in the North Sea, which provide one of the best examples of the influence of population subdivision and connectivity on the dynamics of infectious disease in a wild animal. Next, we consider some of the practical implications of these results for the design of nature reserves and understanding disease emergence. Finally, we discuss the implications that connectivity may have on selection for pathogen resistance in hosts and the consequences this has in turn for pathogen dynamics.

EMPIRICAL EVIDENCE: DISEASE AND HABITAT FRAGMENTATION

There is little doubt that habitat fragmentation has increased the incidence of disease in some wildlife populations. Perhaps the most clearly established case is Lyme disease, a serious zoonotic bacterial disease in the northeastern USA (Allan et al. 2003). The disease is vectored by ticks that feed on a variety of mammals, including humans. Both the density of tick nymphs and the prevalence of infection in the nymphs are inversely related to forest patch area, leading to increased disease risk in highly fragmented landscapes. This pattern appears to be mediated by the changes that habitat fragmentation causes to mammalian communities in remnant forest patches. The white-footed mouse *Peromyscus leucopus* is the most competent reservoir for the bacterium. Whilst the diversity of the overall mammal community declines in small patches, the white-footed mouse appears to increase in density in small patches, probably because of reduced predation. This provides a dual benefit to the transmission of the bacterium. Not only does its most competent reservoir increase in abundance, but with a decreasing diversity of alternative hosts, a higher proportion of tick bites will be on that vector. This "dilution" effect for vector-transmitted pathogens with multiple hosts with differing susceptibilities is likely to be a general response of pathogens to host diversity (Dobson 2004).

Whether it is true in general that fragmentation increases disease incidence or disease threats is an important question that forms a major focus of this chapter. One of the few examples of disease-induced extinction in a metapopulation looked at plague outbreaks in prairie dog populations (Stapp *et al.* 2004). Black-tailed prairie dogs (*Cynomys ludovicianus*) are colonial rodents that inhabit the Great Plains of North America. When infected by plague *Yersinia pestis*, they suffer very high mortality that frequently leads to extinction of an entire colony. Colonies are naturally isolated by soil type, vegetation, and topography. Extinct colonies can become recolonized and infection can spread between colonies by movement of infected rodents (either prairie dogs themselves, or other rodent species) or by flea vectors. Together, these conditions provide an ideal situation in which to investigate the role of connectivity in disease-induced extinction within a metapopulation.

Using logistic modelling of data from 21 years of monitoring, Stapp *et al.* (2004) found that the best predictors of colony extinction were El Niño–Southern Oscillation (ENSO) events and colony area. Interestingly, both small and large colonies suffered higher rates of extinction relative to medium-sized colonies. High rates of extinction in large colonies may seem counterintuitive, but recent models suggest that epidemic-induced extinctions may indeed be more likely in populations that are far above the threshold density for disease persistence than in populations close to the threshold (Gerber *et al.* 2005).

Surprisingly, there was no evidence that colony isolation was related to the probability of extinction. However, there was evidence that connectivity played a role in disease-induced extinction. Colonies were more likely to go extinct if their neighbor went extinct, and if the neighboring colony was large. Furthermore, an interaction term indicated that these effects were synergistic. Thus, proximity to another colony suffering an epidemic, particularly a large one from which many emigrants could be expected, increased the risk of disease-induced extinction.

There is some evidence, though less convincing, that habitat fragmentation may increase disease impacts in some single-host pathogens. For example, the bacteria *Chlamydia pecorum* and *C. pneumoniae* are thought by some authors to be a significant threat to koala populations and the particular strains involved are thought to be specific to koalas (Phillips 2000). The suggestion that epidemic disease, exacerbated by habitat destruction, was threatening koala populations with extinction was made as early as 1937 (Pratt 1937). The prevalence of *Chlamydia* spp. in many koala populations is high, but overt disease appears to be triggered by the stress and overcrowding that are synergistic to habitat destruction and fragmentation (Melzer *et al.* 2000).

Habitat change and destruction can cause major changes in metapopulation structure. For example, in Australia, loss of native forests and increases in urbanization have caused flying fox (*Pteropus* spp.) to live in a relatively small number of very large roosts, many in urban areas, whereas formerly they occupied more, smaller roosts dispersed throughout the natural forest (Markus and Hall 2004). Flying foxes, in both Australia and Southeast Asia, have recently been found to be reservoirs of several emerging zoonotic viral diseases, including Hendra virus and Lyssavirus in Australia and Nipah virus in Southeast Asia (Mackenzie *et al.* 2001). There is little doubt that these viruses have recently increased in prevalence in human populations (Hyatt *et al.* 2004). Whether this is solely a result of habitat destruction and increased urbanization causing more bat—human contacts than previously or whether the viruses have increased in prevalence within the bat populations is unclear. What is clear is that combining a large number of small colonies together into a small number of large colonies will have major implications for the patterns of connectivity and network structure of the bat colonies and this will have profound consequences for pathogen transmission within the population.

The preceding examples are derived from terrestrial environments. Marine ecologists have long argued that marine systems are more highly connected and "open" than their terrestrial counterparts (Kinlan and Gaines 2003). Although this dogma has been challenged recently (Cowen *et al.* 2000; DiBacco *et al.* Chapter 8; Harrison and Bjorndal Chapter 9), there are certainly differences in the patterns of connectivity between terrestrial and marine systems that have major implications for the dynamics of host–pathogen systems (Harvell *et al.* 2004; McCallum *et al.* 2004). In particular rates of dispersal and of epidemic spread seem to be around two orders of magnitude faster in aquatic systems when compared to terrestrial systems (McCallum *et al.* 2003). This means that control of infectious disease in the ocean, or in riverine systems, will be considerably harder than in terrestrial systems. In most situations, the epidemiologists will be faced with "counting bodies on the beach" rather than trying to estimate whom to vaccinate, or whether to administer a cull that might create a firebreak. Nevertheless, studies of phocine distemper outbreaks in seal populations in the North and Baltic Seas have provided general and important empirical and theoretical insights into the influence of habitat structure and connectivity on pathogen dynamics.

PHOCINE DISTEMPER AND SEALS

In the late 1980s harp seals were spotted off the coast off Britain for the first time. They normally live in the Labrador channel between Greenland and Eastern Canada. The huge populations that live here support endemic infections of phocine distemper, a morbillivirus that is related to measles, rinderpest, and canine distemper (Heide-jorgensen *et al.* 1992). All of these viruses require large host populations to sustain their persistence. Bartlett (1960) and Black (1966) estimated that measles only persists in human populations in excess of half a million people. The higher birth rates of seals and canids probably replace the pool of susceptibles required to sustain these infections at a faster rate than occurs for humans, so the critical community sizes for phocine and canine distempers may be less than the half a million required for human measles. Nevertheless, models suggest that the critical community size for phocine distemper is considerably greater than the entire North Sea seal population (Swinton *et al.* 1998)

The appearance of harp seals off the coast of Britain was followed by reports of wide-scale deaths of grey and harbour seals around first the British coastline and then around the Baltic Sea; subsequent diagnosis showed this to be phocine distemper. The disease seemed to be mainly transmitted when the seals "haul out" on land to give birth to their pups and then to mate. The haul-outs occur in sheltered, isolated bays and on islands around the British coast; the limited number of these sites divides the seal population into large groups that are connected by immature males dispersing between haul-outs and looking for mating opportunities. The pattern of spread of phocine distemper around the coast of Britain and the Baltic Sea provided major insights into how subdivision of host populations affects the spread of epidemic diseases that cause sudden widespread mortality. Two key processes determined the spread of these epidemics: (1) the persistence time of the epidemic in each patch of habitat, which was largely determined by the size of the host population occupying the patch; and (2) the rate of spread of the pathogen between patches, which was largely determined by the dispersal rates of infected individuals who do not yet show symptoms of the disease. In the case of phocine distemper, no haul-out (habitat patch) was sufficiently large to sustain the infection, though the disease persisted for longer times at haul-outs with larger populations. It then spread almost linearly along the British coastline and across the North Sea into the Baltic before finally dying out. The final extinction occurred as the net seal population size was insufficient to produce enough new susceptible hosts to continuously maintain the pathogen.

A number of models were developed to examine the dynamics of phocine distemper (Grenfell et al. 1992; de Koeijer et al. 1998; Swinton et al. 1998). Many of these can be generalized and parameterized for other pathogens that cause either high mortality or lifetime immunity in the survivors. Many of the insights provided by the models also apply to the dynamics of other pathogens in fragmented populations, whether this fragmentation is natural or has been increased by human intervention. The first insight concerns the role of population fragmentation on pathogen persistence; simulations of the persistence time of the phocine distemper epidemic suggested it was persisting in the North Sea seal population for longer than would be predicted by a stochastic population model for a large well-mixed population of similar size to the North Sea population. However, the persistence time of the model increased considerably when the host population was divided into a number of subpopulations each weakly coupled together by occasional seal dispersal (Grenfell et al. 1992; Swinton et al. 1998). This is an important result, as it suggests that the persistence time of a pathogen may be considerably enhanced by fragmentation of the host population into a number of smaller subpopulations. This occurs because, in a well-mixed system, the post-epidemic trough in susceptibles that leads to fade-out of infection occurs in all places simultaneously. As fragmentation increases, epidemics in each patch become desynchronized, resulting in some infected patches always being available to propagate the infection. At sufficiently high levels of fragmentation, however, epidemics are likely to fade out on a single patch before they can be transmitted further.

The models developed by Swinton (1998) are particularly valuable here; they provide a means of calculating the persistence time (time to extinction) of a pathogen whose vital rates can be characterized by its incubation time, duration of infectivity, and transmission infectivity. The approach can be applied to any SI or SIR pathogen; Swinton's original derivation was developed from the work described on phocine distemper described above and assumed the pathogen moved between social groups of seals distributed in a linear fashion along a coastline. This original work was extended to two dimensions by Park *et al.* (2002); more detailed analyses by Barbour and Pugliese (2004) suggest the results are in close accordance with a considerably more complex model.

Swinton divides the problem of pathogen persistence in a fragmented population into two components. The first requires estimation of the minimal patch size, N_C, in which the pathogen will persist for sufficient time to allow transmission to another patch (or social group). If the patch (or group) size is larger than this then it is possible to calculate the second component, which is the modal time to transit between patches, T_M. The net persistence time of the pathogen, T_T, will then be given by a linear expression that simply adds the persistence time in an average patch, T_E, to the product of the modal transit time and the net number of patches minus 1, $(n-1)$.

$$T_T = T_E + (n-1)T_M. \tag{19.1}$$

The critical patch size (P_C) that will just allow the pathogen to persist until it manages to contaminate another patch is given by:

$$P_C = \frac{1}{(R_0 - 1)\rho} \log\left(\frac{1}{(1-\alpha)}\right). \tag{19.2}$$

Here $R_0 = \beta/\gamma$; where β is transmission rate of the pathogen, $1/\gamma$ is the average duration of infectivity, ρ is the rate of inter-patch mixing (relative to within-patch contact), and α is the probability a transition occurs between patches (this comes from the probability a transition will not occur $(1 - \alpha) = e^{-N_1/N_c}$). The time to extinction in an individual patch or social group that contains h hosts is given by

$$T_E = G \log(h) \left[\frac{\Gamma(m)}{(R_o - 1)} + \max(m, 1 - m) \right].$$ (19.3)

Here G is the "generation time" of the infection, where $G = 1/\delta + 1/\gamma$, where $1/\delta$ is the incubation period and m is the incubation time expressed as a proportion of the total time a host is infectious (thus $mG = 1/\delta$ and $(1 - m)G = 1/\gamma$). When there is a significant latent period ($m > 0.75$), then the correction term $\Gamma(m)$ is given by

$$\Gamma(m) = 1 + \tfrac{1}{2} + \left\{ [1 + 4m(1 - m)(R_o - 1)]^{\frac{1}{2}} - 1 \right\}.$$ (19.4)

The final expression needed is that for the modal transit time between patches, T_M. This is given by

$$T_M = \frac{G}{R_o - 1} \left[\ln \left(\frac{2}{\rho} \right) \right].$$ (19.5)

Here ρ is the ratio of between-patch to within-patch transition.

All of the above can be combined to give an expanded version of Eq. (19.1). Park et al. (2002) point out that Swinton's initial derivative assumes one-dimensional spread, as occurred with the phocine distemper along the coast of Britain. This can be readily modified to give an expression for persistence in two dimensions, as is more likely to occur as natural habitats are fragmented:

$$T_T = T_E + (n - 1)T_M^{\frac{1}{2}}.$$ (19.6)

The subtle insight to emerge here is that the transit time term is raised to a power of ½ due to the two-dimensional spatial spread of the pathogen. Two important insights arise from these equations. First, if the pathogen can establish in any patch ($R_o > 1$), then net persistence time varies inversely with R_o (see Eqs. (19.3) and (19.5)). This is fairly intuitive, as with a high R_o the disease spreads more rapidly and burns out the pool of susceptible individuals. Second, fragmentation of a habitat, or host population, into a metapopulation can significantly increase the persistence time of a pathogen. The time to extinction for a range of host populations fragmented into small groups is illustrated in Fig. 19.1 for both the one-dimensional and two-dimensional cases. The figure shows that persistence is increased by lower ratios of between- to within-patch transition, which reflects the slower dynamics of a pathogen that moves

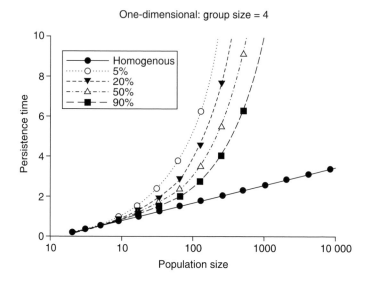

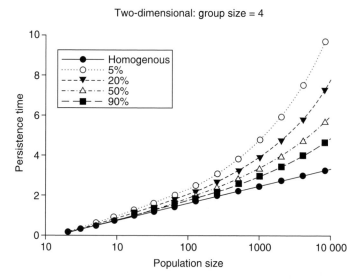

Fig. 19.1. The relationship between persistence time, host population size, and average social group size. The figure illustrates the case for two different social group sizes (4 and 10) and in each case the results are illustrated for populations living in a one- and two-dimensional habitat. The one-dimensional case corresponds to Swinton's (1998) study of seals living in social groups around a coastline. The two-dimensional case would correspond to social groups living in a savanna where each social group is bordered by several neighboring groups.

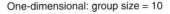

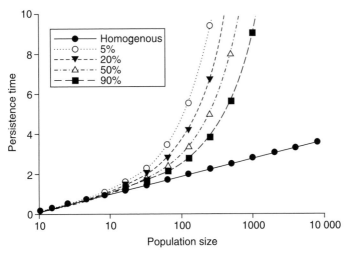

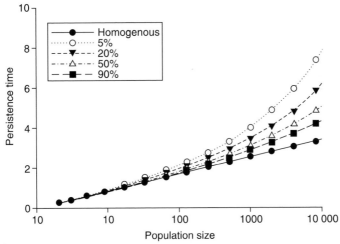

Fig. 19.1. *(cont.)*
In all cases, the lowest solid line (shortest persistence time) corresponds to
the persistence time of a single epidemic outbreak of the pathogen in a
homogeneous well-mixed population of the size indicated on the horizontal
axis. The other lines show the time to extinction when the population is divided
into groups of 4 or 10 individuals, with the level of transmission between
groups (relative to the average rate of within-group transmission) varying
between 5% and 90%. In general persistence time is shorter when there are
fewer larger social groups, and persistence is always less in two-dimensional
than in one-dimensional habitats.

slowly across the landscape. Interestingly, this effect is more marked in the one-dimensional case than in the two-dimensional case.

DOES CONNECTIVITY INCREASE THE RISK OF PATHOGEN-INDUCED EXTINCTION? THEORY

A potential negative consequence of establishing connectivity and corridors between fragmented populations is that they may permit the spread of epidemic disease (Hess 1994, Hess 1996). Hess (1996) developed a simple deterministic model closely based on the classic Levins (1969) metapopulation model. His model showed that increasing levels of connectivity lead to increasing numbers of patches occupied by the pathogen being modeled. At intermediate levels of connectivity, the total number of patches occupied by the host species decreased. It was this result that caused Hess to conclude that "movement of individuals among populations is a conservation strategy that may carry a greater risk than is recognised commonly." However, this result has a number of limitations. First, the model did not include any form of recovery at all: once a patch had become infected, it remained infected until local host extinction occurred. This assumption will not apply in many spatially subdivided populations; after an epidemic has passed through, the expected proportion of infected individuals remaining is so small that the disease becomes extinct through the stochastic process of fade-out, as occurs in the seal example described above, and the pathogen cannot again invade the population until the proportion of susceptible individuals has sufficiently increased. Second, Hess used a single-host/single-pathogen model, but in almost all cases where a pathogen has lead to an extinction threat, there have been one or more reservoir species present in the system (McCallum and Dobson 1995; Cleaveland et al. 2002; Gog et al. 2002). Reservoirs harbor the pathogen with little or no influence on their mortality and the pathogen can maintain a high force of infection even as the endangered species declines toward extinction.

When both reservoir hosts and refractory patches are included in a metapopulation model, increasing connectivity can have a number of different effects, depending on the properties of the hosts and pathogens. McCallum and Dobson (2002) developed a deterministic model with a pathogen and two host species, a reservoir and an "endangered" species that was so highly susceptible to the pathogen that it could not persist on a patch together with infected reservoirs. An important feature of our model was that patches occupied by reservoirs could become "resistant" following pathogen fade-out. Fade-out occurs when there are too few susceptible hosts remaining to maintain infection, so once it has occurred on a patch, the patch is "resistant," because the pathogen cannot invade that patch again until the proportion of susceptible hosts builds up sufficiently.

We found that high connectivity is not always detrimental to persistence of the endangered species. Most fundamentally, if a species is maintained according to the metapopulation paradigm with a dynamic equilibrium between extinction and colonization, too little connectivity will always lead to extinction, whatever the nature of any host–pathogen interaction. Aside from this, the possible outcomes depended on the relative colonization and extinction rates of the two host species, the infection rate of the pathogen, and the rate at which fade-out occurred. If the highly susceptible species had a higher ratio of colonization to extinction rate in the absence of disease than the reservoir, it could persist with the disease even at very high levels of connectivity by operating as a "fugitive" species, occupying patches before the reservoir and its pathogen could do so. A second means of coexistence at high levels of connectivity was that the susceptible species might use the sites in which fade-out had occurred as refuges. This was possible if the rate of fade-out was sufficiently high or the rate at which patches again became susceptible sufficiently low (Figure 19.2D below).

These results may apply in cases where the reservoir and the susceptible species both occupy the same patch type. The other possibility, investigated by Gog et al. (2002), is that the reservoirs may occupy the matrix between the patches occupied by the susceptible species. In their model, infection was transferred between patches by migration of infected hosts, but also occurred as an external pathogen "spillover" from the reservoir species in the matrix. The key result was that, unless the rate of spillover was extremely low, the benefit of increased connectivity between patches on patch occupancy by the endangered species was entirely positive. Where disease transmission occurs across patch boundaries, it is also clear that increased fragmentation will lead to higher rates of disease transmission.

Figure 19.2 contrasts the results of these three metapopulation models of disease and endangered species. A conclusion common to all is that too little connectivity will always lead to extinction of an endangered species, whether or not infectious disease is present. Too much connectivity is a major problem only in the special situation of the endangered host occupying the same patch type as the reservoir hosts. Given that

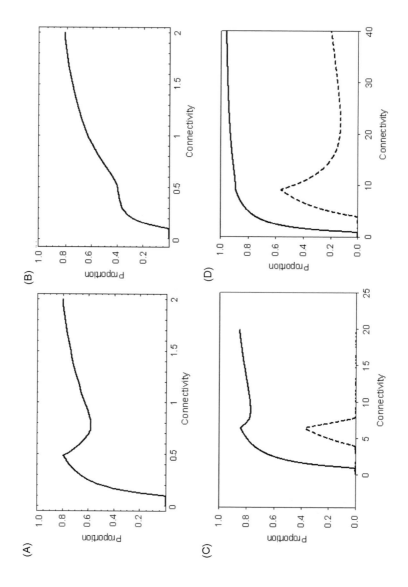

Fig. 19.2. The proportion of patches occupied by the hosts in host–pathogen metapopulation models, as a function of the connectivity between patches. (A) Hess (1996) model of a single-host/single pathogen system. Note the decline in occupied patches at intermediate levels of connectivity. (B) Gog et al. (2002) model with reservoirs occupying matrix around patches, but otherwise the same parameter values as in (A). In this example, the proportion of occupied patches increases monotonically with connectivity. For very low levels of pathogen influx from the matrix, there may be a decline in occupancy at intermediate connectivity, but this will be less pronounced than in (A). (C) McCallum and Dobson (2002) model with reservoirs and "target" species both occupying the same habitat patches; "targets" cannot coexist with infected reservoirs. Solid line, proportion of patches occupied by reservoirs; dashed line, proportion of patches occupied by "target" endangered species. In this case, where the ratio of colonization rate to extinction rate is higher for the reservoir than the target and the rate of loss of resistance is relatively high, too much connectivity can lead to extinction of the target. (D) As in (C), but with a lower rate of loss of resistance. The endangered species can now coexist with the reservoir and pathogen at high levels of connectivity.

anthropogenic influence has reduced connectivity in almost all terrestrial ecosystems through habitat fragmentation, these results suggest that concern about endemic pathogens should not preclude the establishment of corridor systems, as the host species must have coexisted with the pathogen at connectivity levels higher than those currently in existence. The situation with introduced pathogens (or introduced reservoirs) is less straightforward, because the host–pathogen community has not coexisted for an extended period at these higher levels of connectivity. To understand the effect of corridors in such systems, it is necessary to consider the colonization and extinction rates of both reservoir and endangered species, together with the persistence properties of the pathogen in the reservoir population. In aquatic systems, anthropogenic influence may have increased connectivity above natural levels, through transfer of ballast water, boat movements, and artificial waterways (DiBacco *et al.* Chapter 8; Crooks and Suarez Chapter 18), so the consequences of too much connectivity may require also more detailed consideration.

Similar problems of fragmentation arise when we examine the persistence of pathogens in host populations that are naturally fragmented into social groups that persist for periods of time that roughly equal life expectancy (Altizer *et al.* 2003). This is classically the case for many primate and ungulate species, but also occurs for the social carnivores such as lions and hyenas in East Africa. Within all these systems there is a trade-off between patch (or social group) size and pathogen persistence (Fig. 19.1). For any ratio of between- to within-patch transmission and duration of infectivity, pathogen persistence is maximized by an intermediate group size.

HABITAT FRAGMENTATION AND GENETIC RESISTANCE TO PATHOGENS

So far we have assumed that habitat fragmentation has its main impact on the transmission dynamics and persistence of pathogens in subdivided host populations. An important consequence of changes in the intensity of the interaction between host and parasite will be changes in the intensity of selection for resistance of the host to the presence of the pathogen. Some key details of this have been explored by Carlsson-Graner and Thrall (2002) in a plant–pathogen system. The work builds on a whole canon of work on interactions between plants and their pathogens that provides general insights into how host fragmentation affects both the genetics and dynamics of the interaction between hosts and parasites

(Burdon and Jarosz 1992; Thrall and Burdon 1997, 1999; Burdon and Thrall 1999). Field studies of a short-lived perennial plant, *Lychnis alpina* (L.), throughout its natural range in northern Sweden have observed different patterns of infestation with the anther smut fungus, (*Microbotryum violaceum*), a sterilizing pathogen that is transmitted during pollination. In the mountainous habitats where the hosts exhibit large continuous populations, 67% of discrete host populations are infected with the fungus, although disease prevalences within each population are low (7.8%). In contrast, when the hosts are distributed as patchy or small isolated populations on high coasts, or inland mountains, respectively, then the fraction of populations infected declines to 60% and 9%; while the proportion of individuals infected in each population increases to 15.5% and 26%, respectively.

Carlsson-Graner and Thrall (2002) suggest this pattern reflects an important trade-off between pathogen transmission and selection for resistance to infection. They then develop a spatial host–pathogen model that assumes a simple one-locus, two-allele Mendelian model for host resistance to a pathogen that induces lifetime sterility on infection. They assume that resistance carries a cost in the absence of the pathogen and that heterozygotes illustrate intermediate levels of resistance to either homozygote; resistance is simply expressed as reduced levels of transmission to an uninfected host, and the cost is expressed as a reduced level of fecundity in the absence of infection. The model can be readily parameterized from the field data for the *Lychnis–Microbotryum* system. The dynamics of the model can be examined for different levels of unsuitable habitat that effectively recreate different levels of fragmentation that range from one large contiguous population, to a few isolated small populations. The proportion of populations infected is always low when the host population is fragmented into small, isolated populations; this increases as the host population becomes larger and less fragmented (Fig. 19.3A). However, if genetic variability to pathogen susceptibility occurs in the hosts, then the proportion of host populations infected is always lower for any given level of fragmentation. Furthermore, with genetic variability in pathogen resistance, the prevalence of infection in infected populations significantly declines as populations become larger and more connected, whereas prevalence increases in the absence of such genetic variability (Fig. 19.3B). This latter effect is driven by more continuous selection for resistant alleles in the more connected populations that interact more frequently with the pathogen (Fig. 19.4).

(A) Fraction of sites with disease

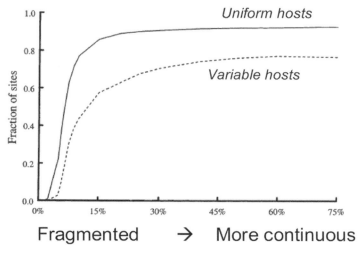

Fragmented → More continuous

(B) Disease prevalence in infected patches

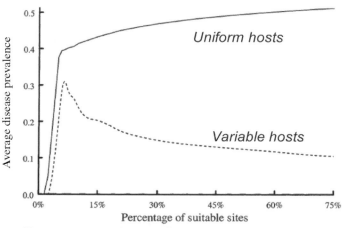

Fragmented → More continuous

Fig. 19.3. Disease patterns observed in the stochastic simulation developed by Carlsson-Graner and Thrall (2002). Each figure illustrates equilibrium conditions after 1000 generations. The *x*-axis illustrates the level of habitat fragmentation, which is highest when the percentage of suitable sites for colonization are low. The *y*-axis in (A) illustrates the fraction of sites in which the pathogen is recorded; in (B) the *y*-axis illustrates the proportion of individuals infected in patches that contain the pathogen. In both cases the solid line illustrates the case for a genetically uniform host and the dashed line illustrates the case for a host that is genetically variable in its resistance to the pathogen.

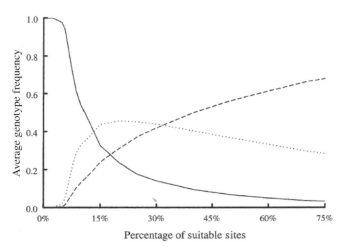

Fig. 19.4. The impact of habitat fragmentation on the frequency of homozygous susceptible (SS, solid line), homozygous resistant (RR, dashed line) and heterozygous (RS, dotted line) individuals in the simulations illustrated in Fig. 19.3.

The work on plant pathogens in a fragmented population provides a key insight that is missing from discussions of host–pathogen interactions that ignore co-evolutionary interactions. In a fragmented population, the reduction in the proportion of patches that are co-occupied by both hosts and pathogens will lead to a reduction in selection for resistance to the pathogen. As the habitat becomes more connected, the pathogen can certainly spread more easily, but its more continuous presence will in turn select for resistance in the host population that will help it reduce the impact of the pathogen. This would again argue that connectivity is not necessarily a problem when we consider the restoring of natural habitats by connecting currently isolated patches of habitat. While on the one hand this may allow pathogens to spread more easily, it will also allow genes that provide resistance to infection to spread and be maintained by more continuous selection.

CONCLUSIONS

The empirical and theoretical work we have discussed here clearly shows that patterns of connectivity have profound effects on the dynamics of infectious diseases; this has major implications for the way in which disease threats should be managed. Given that transmission is a function of connectivity, this is not an unexpected conclusion.

However, both theoretical and empirical studies show that the effect of changing connectivity on pathogen dynamics is often not monotonic: theory we discuss above suggests that the persistence time of pathogens in a metapopulation is greatest at intermediate levels of connectivity, and empirical evidence suggests the highest rates of plague-induced extinction occur in intermediate sized prairie dog colonies (Stapp *et al.* 2004). Similarly, Langlois *et al.* (2001) found that seroprevalence of Sin Nombre virus was highest in deer mice *Peromyscus maniculatus* populations with intermediate amounts of preferred habitat in their immediate vicinity.

How frequently human-induced habitat fragmentation leads to disease emergence or increased threats from pathogens remains an open question that requires further research. As we have discussed in this chapter, there is both theoretical and empirical evidence for the "dilution effect": the idea that loss of biodiversity through fragmentation often leads to increased disease transmission, particularly in vector-borne diseases. It is also clear that fragmentation tends to bring potential hosts into contact with novel pathogens (Daszak *et al.* 2000; Patz *et al.* 2004). However, empirical evidence on the role of habitat destruction or fragmentation in the emergence of even such a notorious pathogen as Ebola remains unclear (Leroy *et al.* 2004).

We think that the implications of parasites and pathogens for the design of nature reserve systems are important, but have been rather neglected. We have described some of the models that have addressed this question, but there are remarkably few data available. In one of the few empirical studies of the effects of reserves on parasites, the most abundant and least host-specific monogenean flatworm parasite in the system (*Lamellodiscus elegans*) was found to be about twice as abundant on the gills of white sea bream *Diplodus sargus* inside a marine reserve in the Mediterranean Sea offshore of France than it was outside the reserve (Sasal *et al.* 2004). More detailed studies on parasite communities inside and outside reserves and in reserves of differing sizes and degrees of isolation is urgently required.

Finally, a major challenge remaining is to better integrate evolutionary processes and multiple strains of both hosts and pathogens into our models of connectivity and epidemiology. Considerable progress has been made in plant epidemiology, as exemplified by the work of Burdon, Thrall, and co-workers discussed above. Many of the same principles should apply to animal diseases, but there is a clear need for more empirical work on wildlife diseases, backed up by deeper theoretical understanding.

REFERENCES

Allan, B. F., F. Keesing, and R. S. Ostfeld. 2003. Effect of forest fragmentation on Lyme disease risk. *Conservation Biology* **17**:267–272.

Altizer, S., C. L. Nunn, P. H. Thrall, *et al.* 2003. Social organization and parasite risk in mammals: integrating theory and empirical studies. *Annual Review of Ecology, Evolution, and Systematics* **34**:517–547.

Anderson, R. M., and R. M. May. 1991. *Infectious Diseases of Humans*. Oxford, UK: Oxford University Press.

Barbour, A. D., and A. Pugliese. 2004. Convergence of a structured metapopulation model to Levins's model. *Journal of Mathematical Biology* **49**:468–500.

Bartlett, M. S. 1960. The critical community size for measles in the US. *Journal of the Royal Statistical Society A* **123**:37–44.

Black, F. L. 1966. Measles endemicity in insular populations: critical community size and its evolutionary implication. *Journal of Theoretical Biology* **11**:207–211.

Burdon, J. J., and A. M. Jarosz. 1992. Temporal variation in the racial structure of flax rust (*Melampsora lini*) populations growing on natural stands of wild flax (*Linum marginale*): local versus metapopulation dynamics. *Plant Pathology* **41**:165–179.

Burdon, J. J., and P. H. Thrall. 1999. Spatial and temporal patterns in coevolving plant and pathogen associations. *American Naturalist* **153**:S15–S33.

Carlsson-Graner U., and P. H. Thrall. 2002. The spatial distribution of plant populations, disease dynamics and evolution of resistance. *Oikos* **97**:97–110.

Cleaveland S., G. R. Hess, A. P. Dobson, *et al.* 2002. The role of pathogens in biological conservation. Pp. 139–150 in P. J. Hudson, A. Rizzoli, B. Grenfell, and H. Heesterbeck (eds.) *The Ecology of Wildlife Diseases*. Oxford, UK: Oxford University Press.

Cohen, R., S. Havlin, and D. ben-Avraham. 2003. Efficient immunization strategies for computer networks and populations. *Physical Review Letters* **91**(247901):1–4.

Cowen, R. K., K. M. M. Lwiza, S. Sponaugle, C. B. Paris, and D. B. Olson. 2000. Connectivity of marine populations: open or closed? *Science* **287**:857–859.

Daszak, P., A. A. Cunningham, and A. D. Hyatt. 2000. Emerging infectious diseases of wildlife: threats to biodiversity and human health. *Science* **287**:443–449.

de Koeijer A., O. Diekmann, and P. Reijnders. 1998. Modelling the spread of phocine distemper virus among harbour seals. *Bulletin of Mathematical Biology* **60**:585–596.

Dobson, A. P. 2003. Metalife! *Science* **301**:1488–1490.

Dobson, A. P. 2004. Population dynamics of pathogens with multiple hosts. *American Naturalist* **164**:S64–S78.

Gerber, L., H. McCallum, K. Lafferty, J. Sabo, and A. Dobson. 2005. Exposing extinction risk analysis to pathogens: is disease just another form of density dependence? *Ecological Applications* **15**:1402–1414.

Gog, J., R. Woodroffe, and J. Swinton. 2002. Disease in endangered metapopulations: the importance of alternative hosts. *Proceedings of the Royal Society of London B* **269**:671–676.

Grenfell, B. T., M. E. Lonergan, and J. Harwood. 1992. Quantitative investigations of the epidemiology of phocine distemper virus (PDV) in European common seal populations. *Science of the Total Environment* 115:15–29.

Harvell, D., R. Aronson, N. Baron, *et al.* 2004. The rising tide of ocean diseases: unsolved problems and research priorities. *Frontiers in Ecology and the Environment* 2:375–382.

Heide-jorgensen M. P., T. Harkonen, R. Dietz, and P. M. Thompson. 1992. Retrospective of the 1988 European seal epizootic. *Diseases of Aquatic Organisms* 13:37–62.

Hess, G. R. 1994. Conservation corridors and contagious disease: a cautionary note. *Conservation Biology* 8:256–262.

Hess, G. R. 1996. Disease in metapopulation models: implications for conservation. *Ecology* 77:1617–1632.

Hyatt, A. D., P. Daszak, A. A. Cunningham, H. Field, and A. R. Gould. 2004. Henipaviruses: gaps in the knowledge of emergence. *EcoHealth* 1:25–38.

Keeling, M. J. 1999. The effects of local spatial structure on epidemiological invasions. *Proceedings of the Royal Society of London B* 266:859–867.

Kinlan, B. P., and S. D. Gaines. 2003. Propagule dispersal in marine and terrestrial environments: a community perspective. *Ecology* 84:2007–2020.

Langlois, J. P., L. Fahrig, G. Merriam, and H. Artsob. 2001. Landscape structure influences continental distribution of hantavirus in deer mice. *Landscape Ecology* 16:255–266.

Leroy, E. M., P. Rouquet, P. Formenty, *et al.* 2004. Multiple Ebola virus transmission events and rapid decline of central African wildlife. *Science* 303:387–390.

Levins, R. 1969. Some demographic and genetic consequences of environmental heterogeneity for biological control. *Bulletin of the Entomological Society of America* 15:237–240.

Mackenzie, J. S., K. B. Chua, P. W. Daniels, *et al.* 2001. Emerging viral diseases of Southeast Asia and the Western Pacific. *Emerging Infectious Diseases* 7:497–504.

Markus, N., and L. Hall. 2004. Foraging behaviour of the black flying-fox (*Pteropus alecto*) in the urban landscape of Brisbane, Queensland. *Wildlife Research* 31:345–355.

May, R. M., and A. L. Lloyd. 2001. Infection dynamics on scale-free networks. *Physical Review E* 64(066112):1–4.

McCallum, H. I., and A. P. Dobson. 1995. Detecting disease and parasite threats to endangered species and ecosystems. *Trends in Ecology and Evolution* 10:190–194.

McCallum, H. I, and A. Dobson. 2002. Disease, habitat fragmentation and conservation. *Proceedings of the Royal Society of London B* 269:2041–2049.

McCallum, H., C. D. Harvell, and A. Dobson. 2003. Rates of spread of marine pathogens. *Ecology Letters* 6:1062–1067.

McCallum, H. I., A. Kuris, C. D. Harvell, *et al.* 2004. Does terrestrial epidemiology apply to marine systems? *Trends in Ecology and Evolution* 19:585–591.

Melzer, A., F. Carrick, P. Menkhorst, D. Lunney, and B. St John. 2000. Overview, critical assessment, and conservation implications of koala distribution and abundance. *Conservation Biology* 14:619–628.

Meyers, L. A., M. E. J. Newman, M. Martin, and S. Schrag. 2003. Applying network theory to epidemics: control measures for *Mycoplasma pneumoniae* outbreaks. *Emerging Infectious Diseases* 9:204–210.

Newman, M. E. J. 2002. Spread of epidemic disease on networks. *Physical Review* E 66(016128):1–11.

Newman, M. E. J. 2003. The structure and function of complex networks. *Siam Review* 45:167–256.

Olinky, R., and L. Stone. 2004. Unexpected epidemic thresholds in heterogeneous networks: the role of disease transmission. *Physical Review* E 70:030902(R).

Park, A. W., S. Gubbins, and C. A. Gilligan. 2002. Extinction times for closed epidemics: the effects of host spatial structure. *Ecology Letters* 5:747–755.

Patz, J. A., P. Daszak, G. M. Tabor, *et al.* 2004. Unhealthy landscapes: policy recommendations on land use change and infectious disease emergence. *Environmental Health Perspectives* 112:1092–1098.

Phillips, S. S. 2000. Population trends and the koala conservation debate. *Conservation Biology* 14:650–659.

Pratt, A. 1937. *The Call of the Koala*. Melbourne, Australia: Robertson and Mullens.

Sasal, P., Y. Desdevises, E. Durieux, P. Lenfant, and P. Romans. 2004. Parasites in marine protected areas: success and specificity of monogeneans. *Journal of Fish Biology* 64:370–379.

Stapp, P., M. F. Antolin, and M. Ball. 2004. Patterns of extinction in prairie dog metapopulations: plague outbreaks follow El Niño events. *Frontiers in Ecology and the Environment* 2:235–240.

Strogatz, S. H. 2001. Exploring complex networks. *Nature* 410:268–276.

Swinton, J. 1998. Extinction times and phase transitions for spatially structured closed epidemics. *Bulletin of Mathematical Biology* 60:215–230.

Swinton, J., J. Harwood, B. T. Grenfell, and C. A. Gilligan. 1998. Persistence thresholds for phocine distemper virus infection in harbour seal *Phoca vitulina* metapopulations. *Journal of Animal Ecology* 67:54–68.

Thrall, P. H., and J. J. Burdon. 1997. Host–pathogen dynamics in a metapopulation context: the ecological and evolutionary consequences of being spatial. *Journal of Ecology* 85:743–753.

Thrall, P. H., and J. J. Burdon. 1999. The spatial scale of pathogen dispersal: consequences for disease dynamics and persistence. *Evolutionary Ecology Research* 1:681–701.

Maintaining and restoring connectivity in landscapes fragmented by roads

ANTHONY P. CLEVENGER AND JACK WIERZCHOWSKI

INTRODUCTION

Transportation networks and systems are vital to today's economy and society (Button and Hensher 2001). Not only do roads provide for safe and efficient movement of goods and people across cities and continents, throughout the world they have become a permanent part of our physical, cultural, and social environment (Robinson 1971; Lay 1992). Roads and their networks are one of the most prominent human-made features on the landscape today (Sanderson *et al.* 2002). Compared to polygonal blocks of built areas, road systems are linear and etched into the landscape to form a woven network of arteries that maintain the pulse of societies. However, as road networks extend across the landscape and their weave intensifies, natural areas become increasingly fragmented and impoverished biologically (Forman *et al.* 2003).

Although less studied compared to other agents of fragmentation, roads cause changes to wildlife habitat that are more extreme and permanent than other anthropogenic sources of fragmentation (Forman and Alexander 1998; Spellerberg 2002). Road networks and systems not only cause conspicuous changes to physical landscapes, but also alter the patterns of wildlife and the general function of ecosystems within these landscapes (Swanson *et al.* 1988; Transportation Research Board 1997;

Connectivity Conservation eds. Kevin R. Crooks and M. Sanjayan. Published by Cambridge University Press. © Cambridge University Press 2006.

Olander *et al.* 1998). Busy roads can be barriers or filters to animal movement (Hels and Buchwald 2001; Rondinini and Doncaster 2002; Chruszcz *et al.* 2003) and in some cases the leading cause of animal mortality (Maehr *et al.* 1991; Jones 2000; Kaczensky *et al.* 2003). Sustainable transportation systems must provide effectively for natural processes and biodiversity, and safe and efficient human mobility.

Over the last decade, federal land management and transportation agencies have become increasingly aware of the effects of roads on wildlife. Significant advances in our understanding of these impacts have been made; however, the means to adequately mitigate these impacts have been slower in coming. Effective wildlife fencing and crossing structures can significantly reduce many harmful impacts of roads on wildlife populations (Kistler 1998; Clevenger *et al.* 2001a; Cain *et al.* 2003). Yet currently there is limited knowledge on the design of wildlife crossing systems that promote sustainable wildlife populations and functioning ecosystems (Transportation Research Board 2002a). Knowledge of the locations of primary wildlife crossings and/or problem areas is the first step in planning mitigation on highways; however, few methodological approaches to identify and prioritize these key areas have been explored. Departments of Transportation, resource agencies, universities, and non-governmental organizations have attempted to fill this gap by conducting workshops, often with Department of Transport sponsorship; biologists, researchers, and regulatory specialists come together in a workshop setting to make decisions on conservation and connectivity needs based on analysis of best available environmental data (see Beier *et al.* Chapter 22). Anticipated population growth and ongoing highway investments in most regions, coupled with the resounding concern for maintaining large-scale, landscape connectivity, have generated increasing interest in wildlife crossings or mitigation passages as conservation tools.

The impact of transportation systems on wildlife ecology and remedial actions to counter these effects is an emerging science. Research remains scarce on the influence of road systems on habitat fragmentation and the conservation value of crossings in restoring connectivity (Spellerberg 2002; Transportation Research Board 2002b; Forman *et al.* 2003). Moreover, research has primarily focused at the level of individuals and single species (Guyot and Clobert 1997; Ortega and Capen 1999; Gibbs and Shriver 2002). Key questions remain regarding population- and community/ecosystem-level impacts of roads and the benefits of wildlife crossings to reduce those impacts (Clevenger and Waltho 2000, 2005; Underhill and Angold 2000).

In this chapter, we discuss roads as agents of habitat fragmentation and means of restoring connectivity across roads with wildlife crossings. Developing strategies to mitigate road impacts begins with a step-wise approach that contemplates project goals and context, followed by decisions regarding specific placement and design of the measures implemented. We describe some general guidelines used in the planning process, information needs, and practical applications. As part of a project evaluating mitigation measures along the Trans-Canada Highway in Banff National Park, Alberta, Canada, we present several geographic information system (GIS)-based approaches to model animal movements for planning sustainable transportation projects.

ROADS AS AGENTS OF HABITAT FRAGMENTATION

The impact of roads on wildlife has been the focus of many studies (Canters 1997; Evink et al. 1999; Forman et al. 2003). These studies show that roads affect wildlife in numerous ways (Reijnen and Foppen 1994; Vos and Chardon 1998; Sweanor et al. 2000). Apart from the most obvious direct impact of wildlife mortality, roads affect habitat by changing habitat condition and levels of connectivity.

Habitat change

Road construction and improvements result in habitat loss for wildlife by transforming natural habitats to pavement and cleared roadsides or verges. Some species are more vulnerable to habitat loss than others. For example, species such as wide-ranging carnivores, with large area requirements, relatively low densities, and low reproductive rates, tend to be the most sensitive to road-induced habitat loss (Trombulak and Frissell 2000; Carroll et al. 2001; Carroll Chapter 15), although road networks also affect other taxa, such as herpetofauna (Vos and Chardon 1998; Gibbs and Shriver 2002). Road construction can increase the amount of edge habitat in a landscape due to the long thin shape of roads, resulting in a decrease in the amount of habitat for interior, edge-sensitive species. Metapopulation theory suggests that more mobile species are better able to tolerate habitat loss (Hanski 1999), yet mortality of individuals in the matrix habitat (e.g., road corridors) does not typically figure into metapopulation theory (Taylor et al. Chapter 2; Moilanen and Hanski Chapter 3). Studies have shown that when mortality is high in the matrix habitat, highly mobile species are actually more vulnerable to habitat loss (Carr and Fahrig 2001; Gibbs and Shriver 2002).

Roads and disturbance from traffic also can reduce the quality of nearby habitat. Disturbance from roads can affect wildlife behaviorally and numerically. Behavioral responses can be of two types: (1) an avoidance response (avoidance zone) associated with regular or constant disturbance from high-volume highways, and (2) avoidance due to irregular, individual disturbances generally found on roads with less traffic. Numerical responses from roads consist of a decrease in abundance or density of breeding individuals. There are many examples of numerical responses by wildlife, primarily birds, to an array of road types and traffic disturbance (Van der Zande et al. 1980; Reijnen and Foppen 1994).

Alternatively, wildlife can be attracted to road corridors or roads themselves for a variety of reasons, often as a result of conditions related to habitat (nesting, living space) or food resources. For example, higher use by raptors and ravens of roadsides compared to adjacent habitat was due to the greater availability of perch sites and wide verges (Knight and Kawashima 1993; Meunier et al. 2000). Road construction also can create high-quality habitat where food resources are more abundant compared to adjacent areas. Lush forage created by fencing along medians and verges attracts herbivores, from microtine rodents to large mammals such as deer and elk. Locally abundant small-mammal populations living in these fenced-off areas become targets for avian and terrestrial predators such as owls, hawks, coyotes, and foxes. With prey and their predators foraging close to traffic in the road corridor, collisions with vehicles are inevitable, thus resulting in roadside carrion and attracting aerial and terrestrial scavengers if not promptly removed.

Connectivity

Landscape connectivity is the degree to which the landscape facilitates animal movement and other ecological flows (Forman 1995; Bennett 1999; Taylor et al. Chapter 2). High levels of landscape connectivity occur when the matrix areas of the landscape comprise relatively benign types of habitats without barriers, thus allowing organisms to move freely (Tischendorf and Fahrig 2000). Reduced landscape connectivity and impeded movements due to roads may result in higher mortality, lower reproduction, and ultimately smaller populations and lower population viability (Gerlach and Musolf 2000; Keller and Lagiardèr 2003). These deleterious effects have underscored the need to maintain and restore essential movements of wildlife across roads, particularly those with high traffic volumes.

Fragmentation effects caused by roads begin as individual animals become reluctant to move across roads to access mates or otherwise preferred habitats for food and cover. This aversion to roads is generally attributed to road features (traffic volume, road width) or habitat changes caused by the road. High-volume roads have the greatest impact in blocking animal movements (Brody and Pelton 1989; Rondinini and Doncaster 2002; Chruszcz et al. 2003). Yet secondary highways and unpaved roads can impede animal movements as well (DeMaynadier and Hunter 2000; Develey and Stouffer 2001; Laurance et al. 2004). Ultimately, the barrier effect of a road (i.e., the "hardness" of the edge) will impact populations differently depending on species behavior, dispersal ability, habitat needs, and population densities (Lima and Zollner 1996; Cassady St Clair 2003). For example, an open road corridor with grass-covered verges can be a formidable barrier to a forest-specialist small mammal regardless of mortality risks due to vehicles on the roadway. In the last decade, studies have begun assessing barrier permeability and dispersal success of animals in a patch-matrix landscape (Lima and Zollner 1996; Tewksbury et al. 2002).

Although roads can limit movement of some taxa, they can potentially facilitate dispersal and range extensions of others, including both native and non-native species, although such evidence is limited (Spellerberg and Gaywood 1993). Studies in Holland and Australia have reported increased movement along roads by small mammals and invertebrates, suggesting they are able to connect core habitats using road verge habitat (Vermeulen 1994; Straker 1998). The US Interstate Highway System, bordered by dense grass, has created an extensive array of potential avenues for dispersal by grassland fauna throughout the country (Huey 1941; Getz et al. 1978). Although large carnivores often occur at lower densities in highly roaded areas, predators can utilize roads for movement because they provide ease of travel and greater access to prey (Koehler and Brittell 1990; Thurber et al. 1994; James and Stuart-Smith 2000). The extent to which roads influence the distribution and abundance of non-native fauna is poorly known (May and Norton 1996; Forman et al. 2003). In Australia, non-native cane toads were more abundant on road surfaces, using them as dispersal corridors, than in many surrounding habitats (Seabrook and Dettman 1996). In general, despite roads being a dominant and permanent landscape feature, relatively few studies have investigated road systems and their role in habitat fragmentation (see Gibbs 1998).

MAINTAINING AND RESTORING CONNECTIVITY ACROSS ROADS

Mitigation of road impacts on connectivity

One of the earliest recommendations to arise from studies of habitat fragmentation was that habitat patches linked by a corridor of similar habitat are likely to have greater conservation value than isolated fragments of similar size (Diamond 1975). This early recommendation was based entirely on theory of island biogeography (MacArthur and Wilson 1967). Since then, there has been widespread interest in corridors as conservation measures (Saunders and Hobbs 1991; Beier and Noss 1998; Bennett 1999; this volume).

Wildlife crossings are designed to link critical habitats and provide safe movement of animals across busy roads. Typically they are combined with high fencing and together are proven measures to reduce road-related mortality of wildlife and restore movements (Foster and Humphrey 1995; Clevenger *et al.* 2001a; Cain *et al.* 2003). In recent years there has been an increase in construction of crossings in North America and worldwide (McGuire and Morrall 2000; Goosem *et al.* 2001; Bank *et al.* 2002). The US Transportation Equity Acts of the last decade have enabled mitigation passages to be part of the early stages of highway project planning and thus more are being built today (US Department of Transportation 1999; Marshik *et al.* 2001). With the reauthorization of the Transportation Equity Act in August 2005, we will continue to see more wildlife crossings in future highway construction and improvement projects in the USA.

Function and performance of wildlife crossings

The principles of corridor theory can be applied to wildlife crossings to help assess how well they function. Until now, the general idea of how well a crossing ultimately performs has not gone far beyond the simplest level of scrutiny — if animals use it, then it is working and must be functional. Answering this question is difficult, although it appears to be simple and straightforward. There are many interpretations of what functional and well-performing wildlife crossings should do. Beier and Noss (1998) reported that generalizations about the conservation value of habitat corridors remain elusive because of the species-specific nature of the problem. The same is true for wildlife crossings, as there is no common answer to the questions "Do wildlife crossings provide connectivity?" or "Are wildlife crossings performing"?

These questions only make sense when referring to a particular focal or target species (Crooks and Sanjayan Chapter 1; Taylor *et al.* Chapter 2; Noss and Daly Chapter 23). However, we know that species do not function in isolation but are components of ecological systems that inherently fall into the category of organized complexity (Allen and Starr 1982; O'Neill *et al.* 1986). Consequently, any single-species mitigation structure may have cascading effects (some positive, some negative) on other non-target species. If mitigation measures for habitat connectivity are to succeed, then it is paramount that a multi-taxonomic approach be adopted to evaluate the efficacy of such mitigation on non-target species as well. If the goal of wildlife crossings is to maintain biological diversity at multiple levels of organization (Noss 1990; Redford and Richter 1999), then means of evaluating measures of crossing structure efficacy can become quite complex.

Measuring wildlife crossing performance and conservation value

Most studies of wildlife crossings have simply described the number of species using crossings and their frequency of use (Foster and Humphrey 1995; Taylor and Goldingay 2003; Ng *et al.* 2004). Others have used passage data as a dependent variable to identify factors that facilitate passage by wildlife (Yanes *et al.* 1995; Rodríguez *et al.* 1996; Clevenger and Waltho 2000, 2005). Few studies have actually measured performance of mitigation in meeting design goals (Woods 1990; Clevenger *et al.* 2001a; Cain *et al.* 2003). Until now, virtually all studies have been focused at the level of individuals and suggested benefits at the population level. We are not aware of any studies that have empirically addressed whether wildlife crossings enhance or diminish the population viability of species impacted by roads.

After several decades of increased activity building wildlife crossings, engineers and land managers still lack guiding principles as a large void exists in devising functional designs based on criteria that are relevant to real conservation decisions. This is largely because few studies have rigorously evaluated the efficacy of wildlife crossings (Romin and Bissonette 1996; Forman *et al.* 2003). There are approximately 100 highway passages built specifically for wildlife in North America (Evink 2002). Yet seldom are monitoring programs part of mitigation projects. When monitoring is conducted, rarely is it designed to evaluate performance based on pre- and post-construction tests (Hardy *et al.* 2003). Thus, results from most studies are based on anecdotal information. Furthermore, monitoring is generally for short periods and fails to address the need for wildlife to respond to such large-scale landscape change. Such adaptation periods can take several years depending on the species as they experience, learn, and adjust their own behaviors to the wildlife crossings (Opdam 1997; Clevenger *et al.* 2002a). Small sampling windows, typical of 1- or 2-year monitoring programs, are too brief, can provide spurious results, and do not adequately sample the range of demographic and behavioral variability in most wildlife populations.

Wildlife crossings are in essence site-specific movement corridors strategically placed over a deadly matrix habitat of pavement and high-speed vehicles. Consequently, crossings that function as habitat or landscape connectors should allow for the following: (1) movement within populations and genetic interchange; (2) biological requirements of finding food, cover, and mates; (3) dispersal from maternal ranges and recolonization after long absences; (4) redistribution of populations in response to environmental changes and natural disasters; and (5) long-term maintenance of metapopulations, community stability, and ecosystem processes. These functions encompass the three levels of biological organization, i.e., genes, species—population, community—ecosystem (see Noss 1990).

Measuring the conservation value of wildlife crossings is a complex and time-consuming task. Nonetheless, it is important to have clear goals and objectives for proper assessment. Up until now, few monitoring efforts were designed to test specific hypotheses (Forman *et al.* 2003). Hypothesis testing will aid in better understanding whether crossings enhance the population viability of species. Some questions may be broad or general and require answers from several scales and perspectives. General principles have to be well founded, and they are often based on intensive studies of the life histories of animals in local environments. The hierarchy concept also recognizes that effects of environmental stresses from roads and traffic can reverberate through other levels, often in unpredictable ways, as secondary and cumulative effects.

We provide an example of how a monitoring and assessment project might be implemented using the following eight steps. These guidelines provide a framework that can be used to design monitoring schemes to evaluate the conservation value of wildlife crossings:

1. *Establish goals and objectives.* What are the mitigation goals? In many cases, the goals are to reduce wildlife—vehicle collisions and/or reduce barrier effects to movement and maintain genetic interchange.
2. *Establish baseline conditions.* To develop a mitigation scheme it will be important to determine the extent, distribution, and intensity of road

impacts to wildlife in the area of concern. The impacts may consist of mortality, habitat loss, habitat fragmentation (reduced movements), or some combination thereof. In most cases, the conditions occurring pre-construction/mitigation will comprise the baseline.

3. *Identify specific questions to be answered by monitoring.* These questions will be formulated from the goals and objectives identified in Step 1 and conditions identified in Step 2. Some questions might include: Is road-related mortality increasing or decreasing? Is animal movement across the road increasing or decreasing? Are animals able to disperse and populations carry out migratory movements? Are populations residing in the transportation corridor stable and reproducing? Before implementing a monitoring program it will help if transportation and land managers can agree on specific benchmarks and thresholds at which management actions may or may not intervene. For example, ≥50% reduction in road-kill would be acceptable, but <50% reduction would initiate management actions to improve mitigation performance, including highway-related mitigation (e.g., reinforcing fencing) or driver-related actions (e.g., reducing traffic speed, animal detection systems, motorist awareness).

4. *Select indicators.* Identify indicators at multiple levels of biological organization that correspond to goals and objectives identified in Step 1 and questions in Step 3. For example: *genetic*, gene flow and genetic structure; *population–species*, demographic processes such as dispersal, survivorship, mortality; *community–ecosystem*, herbivory and predation rates.

5. *Identify control and treatment areas.* If pre-construction/mitigation data are not available, then control areas (unmitigated road sections) may be used to compare indicators with treatments (mitigated roads).

6. *Design and implement a monitoring plan.* Applying principles of experimental design, select sites for monitoring the identified goals and objectives from Step 1 and questions in Step 3. Although treatments and controls should be replicated, that may not always be possible.

7. *Validate relationships between indicators and benchmarks.* Detailed research carried out over the long-term will be needed to determine how well the selected indicators correspond to the mitigation goals and objectives.

8. *Analyze trends and patterns, and recommend management actions.* One analysis strategy is to construct a time series, a sequence of measurements typically taken at successive points in time. Time series

analysis includes a broad spectrum of exploratory and hypothesis testing methods that have two main goals: to identify the nature of the phenomenon represented by the sequence of observations (e.g., mortalities, successful crossings), and forecasting future trends and patterns. Analyses of this type will provide a more accurate assessment of the biological value of the measures over the long term and whether changes in mitigation are warranted.

WILDLIFE CROSSING PLACEMENT

The previous section has shown that decision-making in the design of effective wildlife crossings has been hampered by lack of study. Determining placement of wildlife crossings is even more of a challenge given the few methodological approaches. Transportation planning has generally considered a one-dimensional linear zone along the highway. Thus the engineering and design dimensions have been the main concern for planners. However, we know the ecological effects of roads are far greater than the road itself and can be immense and pervasive (Forman and Alexander 1998; Trombulak and Frissell 2000). Due to the broad landscape context of road systems, it is essential to incorporate landscape patterns and processes in the planning and construction process (Forman 1987). When used in a GIS environment, regional or landscape level connectivity models can facilitate the identification and delineation of barriers and corridors for animal movement (van Bohemen et al. 1994; Bekker et al. 1995). This provides for the development of a more integrated land-management strategy.

In this section, we look at questions regarding spatially explicit mitigation planning and their methodologies, focusing primarily on an unmitigated section of the Trans-Canada Highway in Banff National Park, Alberta (phase 3B) (Fig. 20.1). Our case study area represents an exceptionally problematic area of the Greater Rocky Mountain ecological network (Noss et al. 1996; Page et al. 1996). This work was part of a larger research project aimed at evaluating highway mitigation measures and road impacts to wildlife where a major transportation corridor bisects a critically important protected area. We developed several GIS-based approaches to model animal movements across the Trans-Canada Highway. The modeling exercise varied depending on model objectives and research questions.

Situated approximately 120 km west of Calgary, Banff is the most heavily visited national park in North America with over 5 million visitors

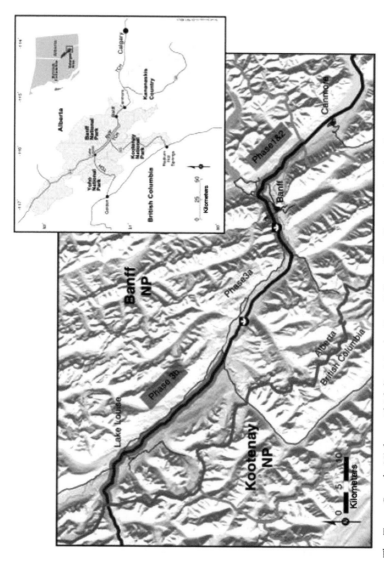

Fig. 20.1. The Trans-Canada Highway (thick line) in the Bow Valley of Banff National Park, Alberta, Canada. The phases represent the sequence of highway mitigation; phase 1, 2, and 3a is 45 km of fenced highway with wildlife crossings; phase 3b is 30 km of unfenced highway without mitigation at present. Inset map shows Banff National Park and the Trans-Canada Highway with respect to other mountain national parks, provincial parks (Kananaskis Country), and roads in the Central Canadian Rocky Mountains.

per year (Parks Canada, unpublished data). The Trans-Canada Highway is a major transportation corridor that bisects Banff and neighboring Yoho National Park (Fig. 20.1). Annual average daily traffic volume at the park east entrance was over 16 000 vehicles per day in 2003 and volumes have increased 40% in the last 10 years (Parks Canada, unpublished data). Since the 1980s, fencing and wildlife crossings (overpasses and underpasses) have been installed along 45 of the 70 km of the Trans-Canada Highway in Banff (Woods 1990; McGuire and Morrall 2000). These mitigated sections of highway are referred to as phase 1, 2, and 3A (see Fig. 20.1).

In 2005, expansion to four lanes with construction of fencing and wildlife crossings began on the remaining 30 km section between Castle Mountain junction and the border with British Columbia (phase 3B). Our modeling efforts focused on this unmitigated section of highway, specifically to identify and prioritize locations for mitigating highway impacts on wildlife habitat connectivity and road-related mortality.

The first modeling approach, although not specifically applied to guide mitigation on the Trans-Canada Highway, explores a real-life transportation dilemma, where wildlife crossings need to be built on a stretch of highway, but baseline information is lacking and time constraints do not allow for pre-construction data collection. To do this, we developed two expert-based models to identify cross-highway habitat linkages and potential locations for mitigation measures, and compared these predictions to an empirically based habitat model.

The second and third approaches are real Banff applications. They are based on the use of empirical data and reflect a situation where pre-construction data were available to develop linkage models from species' habitat use patterns and spatial analysis using a GIS. We created models at two different spatial scales that simulate animal movements across the Trans-Canada Highway. In the first, we created regional-scale, spatially explicit species movement models for identifying potential highway mitigation areas along the unmitigated section of the Highway, and to help assess whether placement of the existing wildlife crossings on the mitigated stretch of the Highway was appropriate (see Fig. 20.1). These models were generalized, and predicted crossing locations accordingly had a relatively wide margin of error (= 1000 m).

In the third and last approach, movement modeling was taken one step further. We modeled movements at the local level, which is the appropriate scale and resolution for accurately predicting the locations of key highway crossing areas and future mitigation placement. Recommendations for identifying and prioritizing future placement of wildlife crossings on the unmitigated Trans-Canada Highway were based on this work (Clevenger *et al.* 2002a).

Expert-based linkage models

Planning the most suitable location for wildlife crossings is typically derived from road-kill information (Evink 1996); however, where animals unsuccessfully cross roads are not necessarily the same locations where they are able to successfully cross (Clevenger *et al.* 2002a). Other means of locating crossings might use data obtained from monitoring animal movements along roads (Evink 1996; Kobler and Adamic 1999; Thompson 2003). Rarely are good pre-construction data available before construction or sufficient time given to acquire these data.

Expert information can be used to develop simple, predictive, habitat linkage models in a relatively short period of time (Marcot 1986; Clevenger *et al.* 2002b; Yamada *et al.* 2003). The objective of this exercise was to determine the accuracy of expert-based models and whether they might be a useful tool for mitigation planning under data and time constraints. For a single species, black bears, we developed three different but spatially explicit habitat models to identify linkage areas across the Trans-Canada Highway. One model was based on empirical data, while the other two were based on expert opinion and expert literature. We used the empirical model as a yardstick to measure the accuracy of the two expert-based models (Clevenger *et al.* 2002b). We selected black bears to model habitat use and identify linkage areas because we had sufficient empirical data to build a habitat model and enough data from crossings and mortality locations to validate it.

Methods

Empirical model We developed the empirical habitat model by first determining the habitat requirements of black bears in the study area using radiolocation data and a suite of biophysical variables, such as elevation, aspect, and slope. A habitat suitability model was then developed using a resource selection function and probability of occurrence (PO) classes. We defined four PO categories: low (<25%), moderate (25–50%), high (50–75%), and very high (>75%) probabilities that bears would occupy an area. We generated a stratified random sample of points ($n = 580$) to compare with the biophysical variables within each of the PO categories (see Clevenger *et al.* 2002b for details).

Expert models   Expert habitat models were developed as weighted linear combinations of each model's layers (biophysical variables) obtained by (a) expert opinion or (b) review of the literature on black bear habitat requirements. We used the pair-wise comparison method developed by Saaty (1977) in the context of a decision-making process known as the analytical hierarchy process (Rao *et al.* 1991; Eastman *et al.* 1995). The procedures for carrying out the expert modeling process are described in detail in Clevenger *et al.* (2002b). Seasons were defined based on the biological needs of bears: pre-berry (den exit to July 15) and berry (July 16 to den entry). Five habitat variables were selected by experts to be used in the analysis: elevation, slope, aspect, greenness, and distance to nearest drainage.

Linkage zone identification and data analysis   We identified highway crossing/habitat linkage zones for the empirical and expert models based on the assumption that: (1) the probability of a bear crossing a highway increases in areas where the highway bisects high-quality bear habitat; and (2) the highest probability of crossings occur in areas where topographic and landscape features are conducive to lateral, cross-valley movements. For each model we generated four classes of linkage zones based on juxtaposition of habitat and human development adjacent to the highway (see Clevenger *et al.* 2002b). We then tested our linkage zone predictions for each of the three models using an independent data set of 37 black bear crossing and mortality locations. These were acquired by intensive radiotracking of movements and by mortality locations obtained by a spatially accurate (<3 m error), global positioning system (GPS) unit. We generated random points of highway crossings, equal in sample size to the actual crossing data, and calculated the distances from both sets of points to the linkage zones predicted for each model.

Results and discussion

Our tests showed that the empirical habitat model was statistically sound. The overall cross-validated classification accuracy was 87% and the model correctly classified 79% of the radiolocations into prime black bear habitat. Through the statistical analysis of crossing data and random points we found that the Class III linkages (sections of the Highway that crossed prime bear habitat and were ≥250 m away from any permanent human development) were most accurate for mapping cross-highway movement for all three models; crossing and mortality locations were significantly closer to Class III linkages predicted by the models than random points.

Our findings confirmed that the expert-literature-based model was consistently more similar and conformed to the empirical model better than the pre-berry and berry expert-opinion-based models. These results were based on the test of distribution of the actual crossing and mortality locations in relation to the modeled linkages, the descriptive characteristics of the Class III linkages, the measure of agreement between models, and the measure of agreement between model linkage zones (see Clevenger *et al.* 2002b).

We believe the poor predictive power of the expert-opinion-based model can be attributed to an overestimation of the importance of riparian habitat, as compared to the opinions expressed in the literature. Another possible explanation is that the expert-literature model is based on an analytical process (data collected, statistically analyzed, and summarized), whereas the expert-opinion model is based on information taken from how experts perceive attributes from memory and experience.

Regional-scale movement models

Broad-scale GIS-based linkage models have been developed to evaluate habitat fragmentation resulting from human activities and to identify predominant landscape permeability patterns for wildlife (Servheen and Sandstrom 1993; Carroll Chapter 15). GIS weighted-distance and least-cost corridor analysis also has been used to evaluate landscape permeability, primarily for large carnivores (Walker and Craighead 1997; Kobler and Adamic 1999; Singleton *et al.* 2002; Theobald Chapter 17). The spatial resolution of nearly all of the above models was relatively large (≥ 1000 m) as some linkage and corridor maps covered large regions, entire states, or several contiguous states (e.g., northern US Rocky Mountains).

Simple individual-based movement models have been used successfully to simulate responses of animals to their habitat and terrain features (Boone and Hunter 1996; Turchin 1998; Tracey Chapter 14). We modeled movement patterns of large mammals at a regional scale in the Central Canadian Rocky Mountains using rules for simulated movements based on habitat quality and permeability of landscape elements. We were interested in using a GIS to determine whether easily available spatial data can successfully describe key linkages and crossing areas for large mammals across busy transportation corridors.

Specifically the aims of this work were: (1) to develop regional habitat suitability models for four wide-ranging large mammal species (black bear, grizzly bear, moose, and elk); and (2) create regional-scale movement models for the four species, indicating the location of potential mitigation based on the intersection of predicted pathways with transportation corridors (Clevenger *et al.* 2002b). We also identified potential locations for future mitigation on phase 3B of the Highway, and compared the placement of existing wildlife crossing structures on phases 1, 2, and 3A to the predicted regional movement pathways. The results are being used to provide land managers with an empirical assessment of the impediments transportation corridors pose to the regional movement patterns of wildlife in an exceptionally problematic area of the Rocky Mountain cordillera.

Methods

Habitat model   We modeled regional scale movements of black bear, grizzly bear, moose, and elk and identified their potential linkage areas across the Trans-Canada Highway. These species were selected because: (1) they exhibit long-ranging movement patterns and potential for interactions with transportation corridors in the study area (Noss *et al.* 1996; Carroll *et al.* 2001), (2) sufficient empirical location data were available to construct predictive spatial models of habitat suitability, and (3) empirical crossing and mortality data were available to independently test the models. Habitat suitability models were developed using a resource selection function, as described for the expert models earlier. We stratified the radiolocation data for bears into pre-berry and berry seasons, and for ungulates into summer (moose: May–October; elk: April–October) and winter (moose: November–April; elk: November–March) seasons.

Movement model   We based the movement component of the model on the least-cost movement principle (Theobald Chapter 17) and quantified the effects of slope angle and orientation (with respect to movement direction) on movement pathway. We used the habitat probability surfaces for the habitat component of the movement model. We simulated the movement pathways for each species by identifying 11 potential entry and exit points located outside the Bow Valley and the Trans-Canada Highway transportation corridor. Entry and exit points were situated in high-quality, valley bottom habitat, the most likely population source areas from which animals would be expected to disperse. For any given pair of entry–exit points there were three iterations resulting in three different pathways. The first iteration simulated the least-cost movement pathway with no obstructions imposed. In the second iteration, the first pathway was blocked, forcing the creation of a new pathway distinct from the original. In the third iteration, the first two pathways were blocked and an alternative route taken. These three distinct model runs produced primary, secondary, and tertiary movement pathways.

Highway crossing zone analysis We mapped the potential wildlife crossing zones along the Trans-Canada Highway by calculating the number of simulated pathway intersections along the highway. Our method often identified long sections of highway, which may be too generalized when recommending the placement of wildlife crossings. We addressed this problem by modifying the model to analyze 1-km-long segments of highway. Given the 120-m pixel size of habitat and topography layers and the obtained fit of the habitat models, we considered 1 km to be the minimum segment length we could safely use.

Model testing We tested the accuracy of highway crossing zones predicted by black bear and elk movement models with an independent set of empirical crossing and mortality points. Data on grizzly bear and moose crossing and mortality were insufficient to test their models. Empirical crossing and mortality locations were defined as described earlier for expert models, but included snowtracking data for elk and moose. The method of testing also was the same as the previous model, i.e., whether empirical crossing points were randomly distributed with respect to the distance to the predicted crossing zones created by the models.

Within the fenced part of the Trans-Canada Highway (phases 1, 2, 3a: Fig. 20.1), we evaluated correlations of the predicted regional movement patterns with frequency of use of the existing wildlife crossings by the four target species (Clevenger *et al.* 2002a). We also identified the potential locations for highway mitigation on the unfenced phase 3B section of the Highway by plotting the pathway crossing frequencies by 1-km segments. We defined high crossing frequency segments as those that registered the number of intersections greater than the given distribution's mean value.

Results and discussion

The overall cross-validated classification accuracies and habitat model validation tests suggested that all of the models showed a reasonably good fit with the empirical data. In the black bear model there was strong statistical evidence that the empirical bear crossing and mortality locations were closer to predicted high- and moderate–high-frequency crossing zones than expected by chance. Similarly, empirical elk winter crossing and mortality locations were closer to the predicted high-frequency crossing zones than random points, but there was no difference between empirical points and the moderate–high-frequency crossing zone locations. We concluded that the model and empirical data correlated well (Clevenger *et al.* 2002a).

We plotted the number of cumulative primary pathways and total pathways (primary, secondary, and tertiary) in relation to the existing wildlife crossings along the mitigated, fenced part of the Trans-Canada Highway. The predicted primary pathway crossing frequencies on km 0–24 of the Highway showed a close association with the empirical data for wildlife crossing use by the four large mammal species. We found a close association between total pathway crossing frequencies and observed wildlife crossing use on the same section of highway; however, this pattern was not as strong as the primary pathway crossings.

Primary pathway crossing frequencies between km 25 and 50 also showed a strong association with the empirical data for wildlife use of crossing structures. There were no highway segments with greater predicted than empirical crossings, nor were there any high predicted crossings in areas without crossing structures. Total pathway crossing frequencies compared to crossing structure use were nearly identical to the primary pathway crossing frequencies.

The models identified several areas along the 0–50 km mitigated section of highway that were noteworthy in terms of their importance for wildlife movement. These predicted crossing locations were in agreement with the rank-ordered importance of wildlife crossings as indicated by usage by all wildlife species (Clevenger *et al.* 2002a).

At the species level, the pattern of movement across the entire length of the Trans-Canada Highway (km 0–86) as predicted by the species' movement models was consistent, varied slightly, and overall was similar to that described above at the group level. To roughly assess potential locations for wildlife crossings along the unmitigated section of the Highway (km 50–86), we weighted the four species equally, utilized the cumulative movement patterns generated by the models, and examined the intersection of primary and total pathways with the highway. Eight locations were indicated by high frequencies of predicted primary crossings across the highway (Fig. 20.2).

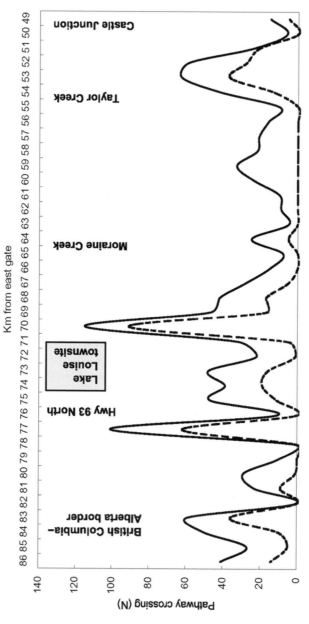

Fig. 20.2. Predicted distribution and frequency of cumulative movement pathways by four large mammal species along the unmitigated section (km 50–86) of the Trans-Canada Highway as indicated by primary (intermittent line) and total pathways (solid line). Geographic landmarks on the Highway are shown on top.

Planning of wildlife crossing placement

There are few methodological approaches to identify the placement of wildlife crossings along road corridors and even fewer ways to determine how to space them. The placement of crossings has generally been related to location, i.e., riparian corridors, road-kill hotspots, or wildlife travel or migration routes (Bekker et al. 1995; Evink 1996; Iuell 2003). Generally, wildlife crossings are spaced at 1.5 to 2.0 km intervals (Evink 1996; Marshik et al. 2001; Clevenger et al. 2002a).

We suggest a mitigation planning scheme that consists of (1) locating crossing structures in the area of key crossing zones as predicted by the models, and (2) locating additional crossings so that there is at least 1.5 km between all crossings. The proposed spacing interval is not empirically based, but knowledge of species' movement patterns, variability in movement patterns over time, and predictability of the impacts of landscape change will help guide planning.

Our results also suggest that by providing additional crossing opportunities in areas not identified by the model, the structures will be used if positioned and designed properly. To maximize connectivity across roads for multiple large mammal species, road construction schemes should include a diversity of crossing structures of mixed size classes. This strategy will likely provide greater permeability of roads by accommodating a variety of species and behavioral profiles (Clevenger and Waltho 2005). Lastly, to improve the permeability of roads for small- and medium-sized mammals, we recommend that small culverts (smaller than wildlife crossing structures) be placed at intervals of 150–300 m to provide sufficient opportunities for smaller animals to avoid crossing busy roads. We also recommend a mixed size class of culverts to accommodate the greatest variety of species possible (Clevenger et al. 2001b).

Local-scale movement models

Broad-scale linkage models have been important tools to identify critical habitats for conservation and to integrate land use planning with large-scale conservation priorities. These mapping efforts have evaluated how road networks may undermine the integrity of large ecological networks, like the Yellowstone-to-Yukon ecoregion (L. Craighead, unpublished data). However, a key improvement to the linkage model concept for highway mitigation planning would require fine-scale resolution models that incorporate local habitat and highway-specific parameters necessary for placement of mitigation passages.

After creating regional-scale models to predict animal movements in our larger study area, we then focused specifically on the section of highway soon to be upgraded with crossings and fencing. Animal movement at the local level is likely to be influenced by the location and intensity of sound sources and viewshed variables such as vegetation density and distribution and hiding cover (Servheen and Sandstrom 1993; Reijnen and Foppen 1994; Mace *et al.* 1996). Spatial models of sound propagation and terrain visibility that take into account local topography, vegetation type, and vegetation density would improve accuracy. In the final exercise, we modeled animal movements across our focal, unfenced section of the Trans-Canada Highway. These models differed from the regional models by inclusion of high-resolution, digital data layers for the upper Bow River Valley transportation corridor and the incorporation of sound and viewshed components.

Methods

Habitat model development   We developed habitat probability surfaces for five large mammal species (elk, moose, black bear, grizzly bear, and wolf) in the unmitigated phase of the Trans-Canada Highway, phase 3B (Fig. 20.1). Seasons were defined as in the regional model; for wolves they were the same as for elk.

The habitat selection analysis was carried out as described for the previous models. Most of our databases were at 1:50 000 scale or less. To facilitate local-level habitat and movement analyses, we created data sets comparable in resolution to the empirical data used to test the models, i.e., GPS-derived crossing and mortality data. All data sets were created at 10-m pixel resolution for the entire study area. We developed 21 biophysical variables in a GIS format grouped into six categories (Clevenger *et al.* 2002a).

Movement model development   As in the regional model earlier, we used individual-based models with rules for simulated movements based on habitat quality and permeability of landscape elements. We simulated movement patterns using 12 entry and exit points located on the periphery of the area (Fig. 20.3). Nine entry–exit points coincided with the movement corridors identified in the previous regional models; three secondary points were located at the entry to three prominent side valleys within the Bow Valley. Movements were simulated from all possible combinations of entry and exit points, as in the regional models.

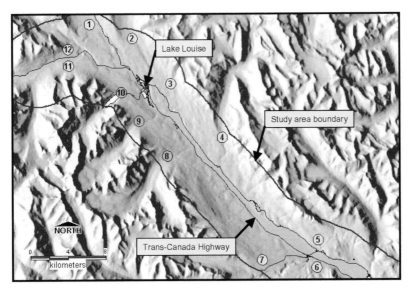

Fig. 20.3. Location of 12 entry—exit points used for local-scale modeling of large mammal movements in the upper Bow Valley transportation corridor, Banff National Park, Alberta, Canada. The study area is bordered by thick black line.

Highway crossing zone analysis We mapped the potential wildlife crossing zones on the Trans-Canada Highway by calculating the number of simulated pathway intersections with segments of highway 200 m in length. We believed the 200-m segment to be small enough to match the resolution of the model and long enough to provide the flexibility in selecting locations for wildlife crossings. We ran the model for each species and season separately, as well as for the cumulative, all-species, all-season model. We considered the all-species iteration to be the most useful for planning highway mitigation measures, such as wildlife crossings, as they should reflect the multi-species, year-round habitat, and movement conditions in the valley.

Model testing We tested the accuracy of highway crossing zones predicted by black bear, elk, and combined-species models as in the previous models, using empirical data and comparing their distributions with random points. We defined high-frequency crossing segments as in the regional model.

Results and discussion

We generated nine habitat suitability models from the five species data sets. The overall cross-validated classification accuracies and validation tests suggested that all models showed a reasonably good fit with the empirical data. We found that the empirical black bear crossing/mortality locations were significantly closer to the modeled high-frequency crossing zones than expected by chance. There was no significant difference between the distances of the random and empirical elk winter crossing locations to the modeled high-frequency crossing zone; however, crossing locations on average were closer to the modeled high-frequency crossing zones than the random points. The empirical locations were significantly closer to the modeled high-frequency crossing zones for the cumulative model pathway intersections than expected by chance.

The increased resolution of spatial databases and the introduction of sound and viewshed models led to a marked improvement in the spatial resolution and reliability of the Trans-Canada Highway linkage model. It is worth stressing the wide applicability of the local-scale models to other planning issues in mountainous environments. The models could be applied to other human infrastructure, such as railways, trails, or other road systems. It is our long-term goal to adapt the models to allow the incorporation of spatial databases (e.g., forest cover data), thus extending its applicability to provincial lands in British Columbia and Alberta.

Management implications

Expert-based models

There are several advantages to the expert-based techniques presented from our work. There are an assortment of GIS tools designed for model building purposes that are readily available today. GIS applications such as Idrisi (Clark University, Worcester, MA, USA), and ArcView (Environmental Systems Research Institute, Redlands, CA, USA) are relatively inexpensive and easy to use. Idrisi has decision support procedures as a program module built into the geographic analysis system. Remotely sensed data, digital landcover data, and habitat suitability maps are readily accessible, frequently updated, and refined for individual users or government agencies (Smith 1999; Serrano et al. 2002). Further, empirical data from field studies of many wildlife species, particularly game species, are obtainable in most countries where road mitigation practices are implemented. The use of the Saaty's pair-wise comparison matrix requires little training and ensures consistency in developing relative weights in the development of the expert-based models (Saaty 1977). This procedure is available in the Idrisi software package.

Both expert model types we presented can provide a useful tool for resource and transportation planners charged with determining the location of mitigation passages for wildlife when baseline information is lacking and when time constraints do not allow for pre-construction data collection. Regarding the latter, we spent approximately 2 months developing the empirical and expert models. More than half of that time was dedicated to developing the more complex, data-intensive, empirical black bear habitat model. We do not advocate modeling linkage zones using exclusively expert information if empirical data are available. However, we do encourage others with empirical data for model building and testing to develop expert models concurrently so that their findings may be contrasted with ours.

Multi-scale, regional, and local models

Simulation studies have greatest value when computer models can be coupled with field studies, both to calibrate model parameters and to test or confirm model predictions (Bennett 1999; Tracey Chapter 14). Statistical tests of our models showed that animal movement simulations consistently conformed to the empirical data on failed and successful highway crossings. Results from a mortality model we developed suggested an association between road crossing/mortality locations and areas of high-quality habitat (Clevenger et al. 2002a). This corroborates the primary assumption of the movement model, i.e., that movement follows a least-cost path defined by the juxtaposition of high-quality habitat patches. The above results validate the use of the GIS-based linkage model in mapping wildlife crossing zones on a major highway.

We recognize there are shortcomings of the work presented. Due to the large spatial scale (pixel size = 120 m) our regional models were generalized, and predicted crossing locations accordingly have a wide margin of error. Nevertheless, we feel they can be valuable tools for identifying locations of important bottlenecks or fracture zones at a regional scale. As suggested above, once these are identified, smaller, local-scale features of the landscape, including possible wildlife concerns and engineering constraints, can be evaluated to select the most appropriate site for wildlife crossings (Iuell 2003). Modeling of local-scale movements using high-resolution data that include fine-scale elements of landscape conditions will provide greater precision and weight in planning mitigation in transportation projects.

In these two exercises, we have equally weighted all four species. However, some management strategies may give higher precedence to key species of conservation concern (Mills *et al.* 1993; Lambeck 1997). Adjustments can be made to the models by weighting individual species according to management priorities. There are an assortment of techniques for the development of weights; however, one of the most promising appears to be that of pair-wise comparisons developed by Saaty (1977) as used in the expert-based model. It is particularly appealing because it serves as an excellent vehicle for discussion of the criteria and objectives involved and their relative strengths (Starfield and Herr 1991; Llewellyn *et al.* 1996).

CONCLUSION

During the past 30 years, the environmental impacts of transportation have been addressed through policy initiatives, planning and analysis, new programs, and new technologies (Transportation Research Board 1997; Button and Hensher 2001). Wildlife conservation and habitat connectivity concerns have received little attention by transportation agencies until now, because the primary concern of most transportation agencies is regulatory compliance of federal and state laws, e.g., the US National Environmental Policy Act and Endangered Species Act (Evink 2002). Traffic and roads are strongly implicated in many of the major environmental problems we face today: air and water pollution, fragmented natural habitats, wildlife and biodiversity losses, and urban sprawl. During the next 25 years, significant growth and changes in North America's population and economy are expected to occur. The impacts of roads on natural environments and the means of mitigating such damage are undoubtedly one of the most important land and wildlife conservation challenges of this new century.

Landscape-level thinking and ecosystem-level initiatives are becoming more common at federal, state, and provincial departments of transportation (US Department of Transportation 2006). These concerns have also percolated up to the US Congress. For the first time, the new transportation bill requires all state transportation departments to consult with resource agencies at the beginning of the planning process, if roads are built with federal money. Also, for the first time, the bill considers wildlife—vehicle collisions to be a major safety issue and allocates federal money for fencing, wildlife crossings, and other measures to reduce wildlife-related accidents.

Healthy and well-functioning ecosystems are vital to the protection of our diverse biological resources and to sustaining the economies and communities that rely on their products and benefits (Luck *et al.* 2003). Federal and state transportation agencies have recognized that early stakeholder involvement and identification of issues and areas of concern is essential if their projects are to be environmentally sustainable. Recent developments in state-wide, GIS-based information for transportation planning and mapping priority habitat conservation needs provides an unprecedented opportunity to coordinate ecological and transportation networks at a state-wide scale.

State transportation plans such as Statewide Transportation Improvement Program (STIP) identify critical areas for infrastructure investments in the short and long term. Similarly, state natural resource agencies are developing comprehensive wildlife conservation strategies that address wildlife and habitat conservation issues (Anonymous 2004; The Biodiversity Partnership 2006). The marrying of transportation and ecological networks would significantly advance environmental streamlining. Integrating these plans would help ensure that habitat conservation and connectivity concerns appear at the beginning of the planning process and guide transportation and land management actions. Looking at the broader picture instead of reacting to a specific project is certainly a novel approach for transportation practitioners. Mapping ecological and transportation corridors will help better understand stakeholder concerns, prioritize agency objectives, and incorporate landscape patterns and processes in the planning and construction process (Forman 1987). An effort of this type would greatly enhance interagency collaboration while working toward a common goal — sustainable surface transportation.

We have described several modeling approaches in this chapter that can be used to identify fracture or conflict zones between transportation and habitat corridors. GIS-based connectivity models are becoming more popular with transportation and natural resource agencies charged with identifying and delineating barriers and corridors for animal movement (Singleton *et al.* 2002). Nonetheless, surprisingly few states have actually integrated transportation and ecological networks using the wealth of data available in a GIS format (Smith 1999). These models can be particularly valuable tools for transportation planners and land managers as they are proactive, provide for the development of a more integrated land-use strategy by taking into account different land management practices, and help prioritize habitat conservation concerns.

It is our hope that the site-specific, one-dimension (linear), sectional road planning approach traditionally used by transportation practitioners quickly succumbs to more integrated, larger-scale methodological schemes that contemplate landscape patterns and ecological connectivity (Bennett 1999; Iuell 2003). However, this cannot be realized without political and agency support. The emerging principles of road ecology are providing useful guidelines and best practices for mitigating road impacts on ecological connectivity, but they ultimately need to become embedded in federal and state administrative policies and legal frameworks. High-quality targeted research precedes effective applications. The need for more science-based knowledge for decision-making is urgent and unprecedented, as an aggressive transportation program is being carried out across the land. This will provide a sound scientific basis for effective planning, policy, and implementation. Perhaps more important, it will inspire confidence in individuals, agencies, and society as a whole that transportation impacts on wildlife, connectivity, and biodiversity loss is worthy of substantial and continuing investment.

ACKNOWLEDGEMENTS

The research was funded by research contracts with Parks Canada, Public Works and Government Services Canada, and grants from the Wilburforce Foundation's Yellowstone-to-Yukon Science program. We thank Bryan Chruszcz and Kari Gunson for their assistance. We sincerely thank Kevin Crooks for his many helpful comments on the chapter. Jeff Tracey and Jerald Powell helped streamline the chapter. The Woodcock Foundation, Wilburforce Foundation, and US Humane Society provided support for the lead author during the preparation of this chapter.

REFERENCES

Allen, T. F. H., and T. B. Starr (eds.) 1982. *Hierarchy: Perspectives for Ecological Complexity.* Chicago, IL: University of Chicago Press.

Anonymous. 2004. *Comprehensive Wildlife Conservation Strategies State Progress Report* Washington, DC: No. 2, March 2004. International Association of Fish and Wildlife Agencies. Available online at http://www.teamingwithwildlife. org/

Bank, F. G., C. L. Irwin, G. L. Evink, *et al.* 2002. *Wildlife Habitat Connectivity across European Highways,* Office of International Programs, Publication No. FHWA-PL-02-011. Washington, DC: Federal Highway Administration.

Beier, P., and R. Noss. 1998. Do habitat corridors provide connectivity? *Conservation Biology* 12:1241–1252.

Bekker, H., V. D. B. Hengel, and H. van der Sluijs. 1995. *Natuur over Wegen* [Nature over motorways]. Delft, The Netherlands: Ministry of Transport, Public Works and Water Management.

Bennett, A. F. 1999. *Linkages in the Landscape: The Role of Corridors and Connectivity in Wildlife Conservation*. Gland, Switzerland: International Union for the Conservation of Nature.

Boone, R. B., and M. L. Hunter. 1996. Using diffusion models to simulate the effects of land use on grizzly bear dispersal in the Rocky Mountains. *Landscape Ecology* 11:51–64.

Brody, A. J., and M. R. Pelton. 1989. Effects of roads on black bear movements in western North Carolina. *Wildlife Society Bulletin* 17:5–10.

Button, K., and D. A. Hensher. 2001. *Handbook of Transportation Systems*. New York: Pergamon Press.

Cain, A. T., V. R. Tuovila, D. G. Hewitt, and M. E. Tewes. 2003. Effects of a highway and mitigation projects on bobcats in Southern Texas. *Biological Conservation* 114:189–197.

Canters, K. (ed.) 1997. *Habitat Fragmentation, and Infrastructure*. Delft, The Netherlands: Ministry of Transport, Public Works and Water Management.

Carr, L. W., and L. Fahrig. 2001. Effect of road traffic on two amphibian species of differing vagility. *Conservation Biology* 15:1071–1078.

Carroll, C., R. F. Noss, and P. Paquet. 2001. Carnivores as focal species for conservation planning in the Rocky Mountain region. *Ecological Applications* 11:961–980.

Cassady St. Clair, C. 2003. Comparative permeability of roads, rivers, and meadows to songbirds in Banff National Park. *Conservation Biology* 17:1151–1160.

Chruszcz, B., A. P. Clevenger, K. Gunson, and M. Gibeau. 2003. Relationships among grizzly bears, highways, and habitat in the Banff–Bow Valley, Alberta, Canada. *Canadian Journal of Zoology* 81:1378–1391.

Clevenger, A. P., and N. Waltho. 2000. Factors influencing the effectiveness of wildlife underpasses in Banff National Park, Alberta, Canada. *Conservation Biology* 14:47–56.

Clevenger, A. P., and N. Waltho. 2005. Performance indices to identify attributes of highway crossing structures facilitating movement of large mammals. *Biological Conservation* 121:453–464.

Clevenger, A. P., B. Chruszcz, and K. Gunson. 2001a. Highway mitigation fencing reduces wildlife–vehicle collisions. *Wildlife Society Bulletin* 29:646–653.

Clevenger, A. P., B. Chruszcz, and K. Gunson. 2001b. Drainage culverts as habitat linkages and factors affecting passage by mammals. *Journal of Applied Ecology* 38:1340–1349.

Clevenger, A. P., B. Chruszcz, K. Gunson, and J. Wierzchowski. 2002a. *Roads and Wildlife in the Canadian Rocky Mountain Parks: Movements, Mortality and Mitigation*, Final Report. Banff, Alberta, Canada: Parks Canada.

Clevenger, A. P., J. Wierzchowski, B. Chruszcz, and K. Gunson. 2002b. GIS-generated expert based models for identifying wildlife habitat linkages and mitigation passage planning. *Conservation Biology* 16:503–514.

DeMaynadier, P. G., and M. L. Hunter. 2000. Road effects of amphibian movements in a forested landscape. *Natural Areas Journal* 20:56–65.

Develey, P. F., and P. C. Stouffer. 2001. Effects of roads on movements by understory birds in mixed-species flocks in central Amazonian Brazil. *Conservation Biology* 15:1416–1422.

Diamond, J. M. 1975. The island dilemma: lessons on modern biogeographic studies for the design of natural reserves. *Biological Conservation* 7:129–146.

Eastman, J. R., W. Jin, P. A. K. Kyem, and J. Toledano. 1995. Raster procedures for multi-criteria/multi-objective decisions. *Photogrammetric Engineeering and Remote Sensing* 61:539–547.

Evink, G. L., 1996. Florida Department of Transportation initiatives related to wildlife mortality. Pp. 278–286 in G. L. Evink, D. Zeigler, P. Garrett, and J. Berry (eds.) *Highways and Movement of Wildlife: Improving Habitat Connections and Wildlife Passageways across Highway Corridors.* Tallahassee, FL: Florida Department of Transportation.

Evink, G. 2002. *Interaction between Roadways and Wildlife Ecology: a Synthesis of Highway Practice,* National Cooperative Highway Research Program Synthesis No. 305. Washington, DC: Transportation Research Board.

Evink, G., P. Garrett, and D. Zeigler (eds.) 1999. *Proceedings of the 3rd International Conference on Wildlife Ecology, and Transportation.* Tallahassee, FL: Florida Department of Transportation.

Forman, R. T. T. 1987. The ethics of isolation, the spread of disturbance, and landscape ecology. Pp. 213–229 in M. G. Turner (ed.) *Landscape Heterogeneity, and Disturbance.* New York: Springer-Verlag.

Forman, R. T. T. 1995. *Land Mosaics: The Ecology of Landscapes and Regions.* Cambridge, UK: Cambridge University Press.

Forman, R. T. T., and L. E. Alexander. 1998. Roads and their major ecological effects. *Annual Reviews of Ecology and Systematics* 29:207–231.

Forman, R. T. T., D. Sperling, J. Bissonette, *et al.* 2003. *Road Ecology: Science and Solutions.* Washington, DC: Island Press.

Foster, M. L., and S. R. Humphrey. 1995. Use of highway underpasses by Florida panthers and other wildlife. *Wildlife Society Bulletin* 23:95–100.

Gerlach, G., and K. Musolf. 2000. Fragmentation of landscape as a cause for genetic subdivision in bank voles. *Conservation Biology* 14:1066–1074.

Getz, L. L., F. R. Cole, and D. L. Gates. 1978. Interstate roadsides as dispersal routes for *Microtus pennsylvanicus. Journal of Mammalogy* 59:208–213.

Gibbs, J. P. 1998. Amphibian movements in response to forest edges, roads, and streambeds in southern New England. *Journal of Wildlife Management* 62:584–589.

Gibbs, J. P., and G. Shriver. 2002. Estimating the effects of road mortality on turtle populations. *Conservation Biology* 16:1647–1652.

Goosem, M., Y. Izumi, and S. Turton. 2001. Efforts to restore habitat connectivity for an upland tropical rainforest fauna: a trial of underpasses below roads. *Ecological Management and Restoration* 2:196–202.

Guyot, G., and J. Clobert. 1997. Conservation measures for a population of Hermann's tortoise, *Testudo hermanni,* in southern France bisected by a major highway. *Biological Conservation* 79:251–256.

Hanski, I. 1999. *Metapopulation Ecology.* Oxford, UK: Oxford University Press.

Hardy, A., A. P. Clevenger, M. Huijser, and G. Neale., 2003. An overview of methods and approaches for evaluating the effectiveness of wildlife crossing structures: emphasizing the science in applied science. Pp. 319–330 in C. L. Irwin, P. Garrett, and K. McDermott (eds.) *Proceedings of the International Conference on Ecology and Transportation.* Raleigh, NC: Center for Transportation and the Environment, North Carolina State University.

Hels, T., and E. Buchwald. 2001. The effect of road kills on amphibian populations. *Biological Conservation* **99**:331–340.

Huey, L. M. 1941. Mammalian invasion via the highway. *Journal of Mammalogy* **22**:383–385.

Iuell, B. (ed.) 2003. *Wildlife, and Traffic: A European Handbook for Identifying Conflicts and Designing Solutions.* Utrecht, The Netherlands: KNNV Publishers.

James, A. R. C., and A. K. Stuart-Smith. 2000. Distribution of caribou and wolves in relation to linear corridors. *Journal of Wildlife Management* **64**:154–159.

Jones, M. E. 2000. Road upgrade, road mortality and remedial measures: impacts on a population of eastern quolls and Tasmanian devils. *Wildlife Research* **27**:289–296.

Kaczensky, P., F. Knauer, B. Krze, *et al.* 2003. The impact of high speed, high volume traffic axes on brown bears in Slovenia. *Biological Conservation* **111**:191–204.

Keller, I., and C. R. Lagiardèr. 2003. Recent habitat fragmentation caused by major roads leads to reduction of gene flow and loss of genetic variability in ground beetles. *Proceedings of the Royal Society of London B* **270**:417–423.

Kistler, R. 1998. Wissenschaftliche Begleitung der Wildwarnanlagen Calstrom WWA-12-S July 1995–November 1997: Schlussbericht. Zurich, Switzerland: Infodienst Wildbiologie and Ökologie.

Knight, R. L., and J. Y. Kawashima. 1993. Responses of raven and red-tailed hawk populations to linear right-of-ways. *Journal of Wildlife Management* **57**:266–271.

Kobler, A., and M. Adamic., 1999. Brown bears in Slovenia: identifying locations for construction of wildlife bridges across highways. Pp. 29–38 in G. L. Evink, P. Garrett, and D. Zeigler (eds.) *Proceedings of the 3rd International Conference on Wildlife Ecology and Transportation.* Tallahassee, FL: Florida Department of Transportation.

Koehler, G. M., and J. D. Brittell. 1990. Managing spruce-fir habitat for lynx and snowshoe hares. *Journal of Forestry* **88**:10–14.

Lambeck, R. J. 1997. Focal species: a multi-species umbrella for nature conservation. *Conservation Biology* **11**:849–856.

Laurance, S. G. W., P. C. Stouffer, and W. F. Laurance. 2004. Effects of road clearings on movement patterns of understory rainforest birds in Central Amazonia. *Conservation Biology* **18**:1099–1109.

Lay, M. G. 1992. *Ways of the World.* New Brunswick, NJ: Rutgers University Press.

Llewellyn, D. W., G. P. Shaffer, N. J. Craig, *et al.* 1996. A decision-support system for prioritizing restoration sites on the Mississippi River alluvial plain. *Conservation Biology* **10**:1446–1455.

Lima, S. L., and P. A. Zollner. 1996. Towards a behavioural ecology of ecological landscapes. *Trends in Ecology and Evolution* **11**:131–135.

Luck, G. W., G. C. Daily, and P. R. Erlich. 2003. Population diversity and ecosystem services. *Trends in Ecology and Evolution* **18**:331–336.

MacArthur, R. H., and E. O. Wilson. 1967. *The Theory of Island Biogeography.* Princeton, NJ: Princeton University Press.

Mace, R. D., J. S. Waller, T. L. Manley, L. J. Lyon, and H. Zuuring. 1996. Relationships among grizzly bears, roads and habitat in the Swan Mountains, Montana. *Journal of Applied Ecology* 33:1395–1404.

Maehr, D. S., E. D. Land, and M. E. Roelke. 1991. Mortality patterns of panthers in southwest Florida. *Proceedings of the Annual Conference of Southeast Game and Fish and Wildlife Agencies* 45:201–207.

Marcot, B. G. 1986. Use of expert systems in wildlife-habitat modeling. Pp. 145–150 in J. Verner, M. L. Morrison, and C. J. Ralph. (eds.) *Wildlife 2000: Modeling Habitat Relationships of Terrestrial Vertebrates.* Madison, WI: University of Wisconsin Press.

Marshik, J., L. Renz, J. L. Sipes, D. Becker, and D. Paulson. 2001. Preserving spirit of place: U.S. Highway 93 on the Flathead Indian Reservation. *Proceedings of the International Conference on Ecology and Transportation* 2001:244–256.

May, S. A., and T. W. Norton. 1996. Influence of fragmentation and disturbance on the potential impact of feral predators on native fauna in Australian forest ecosystems. *Wildlife Research* 23:387–400.

McGuire, T. M., and J. F. Morrall. 2000. Strategic highway improvements to minimize environmental impacts within the Canadian Rocky Mountain national parks. *Canadian Journal of Civil Engineering* 27:523–532.

Meunier, F. D., C. Verheyden, and P. Jouventin. 2000. Use of roadsides by diurnal raptors in agricultural landscapes. *Biological Conservation* 92:291–298.

Mills, L. S., M. E. Soulé, and D. F. Doak. 1993. The keystone-species concept in ecology and conservation. *BioScience* 43:219–224.

Ng, S. J., J. W. Dole, R. M. Sauvajot, S. P. Riley, and T. J. Valone. 2004. Use of highway undercrossings by wildlife in southern California. *Biological Conservation* 115:499–507.

Noss, R. F. 1990. Indicators for monitoring biodiversity: a hierarchical approach. *Conservation Biology* 4:355–364.

Noss, R. F., H. B. Quigley, M. G. Hornocker, T. Merrill, and P. Paquet. 1996. Conservation biology and carnivore conservation in the Rocky Mountains. *Conservation Biology* 10:949–963.

Olander, L. P., F. N. Scatena, and W. L. Silver. 1998. Impacts of disturbance initiated by road construction in a subtropical cloud forest in the Luquillo Experimental Forest, Puerto Rico. *Forest Ecology and Management* 109:33–49.

O'Neill, V., D. L. DeAngelis, J. B. Waide, and T. F. H. Allen. (eds.) 1986. *A Hierarchical Concept of Ecosystems.* Princeton, NJ: Princeton University Press.

Opdam, P. F. M. 1997. How to choose the right solution for the right fragmentation problem? Pp. 55–60 in K. Canters (ed.) *Habitat Fragmentation, and Infrastructure.* Delft, The Netherlands: Ministry of Transportation, Public Works and Water Management.

Ortega, Y. K., and D. E. Capen. 1999. Effects of forest roads on habitat quality for ovenbirds in a forested landscape. *Auk* 116:937–946.

Page, R., S. Bayley, J. D. Cook, *et al.* 1996. *Banff–Bow Valley: At the Crossroads. Summary Report for the Banff–Bow Valley Task Force.* Ottawa, Ontario, Canada: Canadian Heritage.

Rao, M., S. V. C. Sastry, P. D. Yadar, *et al.* 1991. *A Weighted Index Model for Urban Suitability Assessment — GIS Approach.* Mumbai, India: Bombay Metropolitan Regional Development Authority.

Redford, K. H., and B. D. Richter. 1999. Conservation of biodiversity in a world of use. *Conservation Biology* 13:1246–1256.

Reijnen, R., and R. Foppen. 1994. The effects of car traffic on breeding bird populations in woodland. I. Evidence of reduced habitat quality for willow warblers (*Phylloscopus trochilus*) breeding close to a highway. *Journal of Applied Ecology* 31:85–94.

Robinson, J. 1971. *Highways and our Environment.* San Francisco, CA: McGraw-Hill.

Rodríguez, A., G. Crema, and M. Delibes. 1996. Use of non-wildlife passages across a high-speed railway by terrestrial vertebrates. *Journal of Applied Ecology* 33:1527–1540.

Romin, L. A., and J. A. Bissonette. 1996. Deer–vehicle collisions: status of state monitoring activities and mitigation efforts. *Wildlife Society Bulletin* 24:276–283.

Rondinini, C., and C. P. Doncaster. 2002. Roads as barriers to movement for hedgehogs. *Functional Ecology* 16:504–509.

Saaty, T. L. 1977. A scaling method for priorities in hierarchical structures. *Journal of Mathematical Psychology* 15:234–281.

Sanderson, E. W., M. Jaiteh, M. A. Levy, *et al.* 2002. The human footprint and the last of the wild. *BioScience* 52:891–904.

Saunders, D. A., and R. J. Hobbs. 1991. *Nature Conservation,* vol. 2, *The Role of Corridors.* Chipping Norton, NSW, Australia: Surrey Beatty and Sons.

Seabrook, W. A., and E. B. Dettman. 1996. Roads as activity corridors for cane toads in Australia. *Journal of Wildlife Management* 60:363–368.

Serrano, M., L. Sanz, J. Puig, and J. Pons. 2002. Landscape fragmentation caused by the transportation network in Navarra (Spain): two-scale analysis and landscape integration assessment. *Landscape and Urban Planning* 58:113–123.

Serveheen, C., and P. Sandstrom. 1993. Ecosystem management and linkage zones for grizzly bears and other large carnivores in the northern Rocky Mountains in Montana and Idaho. *Endangered Species Bulletin* 18:1–23.

Singleton, P. H., W. L. Gaines, and J. F. Lehmkuhl. 2002. *Landscape Permeability for Large Carnivores in Washington: A Geographic Information System Weighted-Distance and Least-Cost Corridor Assessment.* Research Paper PNW-RP-549. Portland, OR: US Department of Agriculture Forest Service.

Smith, D. 1999. Identification and prioritization of ecological interface zones on state highways in Florida. Pp 209–230 in G. L. Evink, P. Garrett, and D. Zeigler (eds.) *Proceedings of the 3rd International Conference on Wildlife Ecology and Transportation.* Tallahassee, FL: Florida Department of Transportation.

Spellerberg, I. F. 2002. *Ecological Effects of Roads.* Plymouth, UK: Science Publisher Inc.

Spellerberg, I. F., and M. J. Gaywood. 1993. *Linear Features: Linear Habitats and Wildlife Corridors,* English Nature Research Report No. 63. Peterborough, UK: English Nature.

Starfield, A. M., and A. M. Herr. 1991. A response to Maguire. *Conservation Biology* 5:435.

Straker, A. 1998. Management of roads as biolinks and habitat zones in Australia. Pp. 181–188 in G. Evink, D. Zeigler, and J. Berry (eds.) *Proceeding of the International Conference on Wildlife Ecology and Transportation*. Tallahassee, FL: Florida Department of Transportation.

Swanson, F. J., T. K. Kratz, N. Caine, and R. G. Woodmansee. 1988. Landform effects on ecosystem patterns and processes. *BioScience* **38**:92–98.

Sweanor, L. L., K. A. Logan, and M. G. Hornocker. 2000. Cougar dispersal patterns, metapopulation dynamics, and conservation. *Conservation Biology* **14**:798–808.

Taylor, B. D., and R. L. Goldingay. 2003. Cutting the carnage: wildlife usage of road culverts in north-eastern New South Wales. *Wildlife Research* **30**:529–537.

Tewksbury, J. J., D. J. Levey, N. M. Haddad, *et al.* 2002. Corridors affect plants, animals, and their interactions in fragmented landscapes. *Proceedings of the National Academy of Sciences of the USA* **99**:12923–12926.

The Biodiversity Partnership. 2006. *Home page*. Available online at http://www.biodiversitypartners.org/bioplanning/elements.shtml

Thompson, L. M. 2003. Abundance and genetic structure of two black bear populations prior to highway construction in eastern North Carolina. M.Sc. thesis, Knoxville, TN: University of Tennessee.

Thurber, J. M., R. O. Peterson, T. D. Drummer, and S. A. Thomasma. 1994. Gray wolf response to refuge boundaries and roads in Alaska. *Wildlife Society Bulletin* **22**:61–68.

Tischendorf, L., and L. Fahrig. 2000. On the usage and measurement of landscape connectivity. *Oikos* **90**:7–19.

Transportation Research Board. 1997. *Toward a Sustainable Future: Addressing the Long-Term Effects of Motor Vehicle Transportation on Climate and Ecology*. Washington, DC: National Academy Press.

Transportation Research Board. 2002a. *Environmental Research Needs in Transportation*. Conference Proceedings No. 28. Washington, DC: National Academy Press.

Transportation Research Board. 2002b. *Surface Transportation Environmental Research: A Long-Term Strategy*. Special Report No. 268. Washington DC: National Academy Press.

Trombulak, S. C., and C. A. Frissell. 2000. Review of ecological effects of roads on terrestrial and aquatic communities. *Conservation Biology* **14**:18–30.

Turchin, P., 1998. *Quantitative Analysis of Movement*. Sunderland, MA: Sinauer Associates.

Underhill, J. E., and P. G. Angold. 2000. Effects of roads on wildlife in an intensively modified landscape. *Environmental Review* **8**:21–39.

US Department of Transportation. 1999. *Transportation Equity Act for the 21st Century*. Washington, DC: Federal Highway Administration.

US Department of Transportation. 2006. *Federal Highway Administration Report*. Available online at http://www.fhwa.dot.gov/environment/ecosystems/ecoinitc.htm

Van Bohemen, H., C. Padmos, and H. de Vries. 1994. Versnippering-ontsnippering: Beleid en onderzoek bij verkeer en waterstaat. *Landschap* **1994**:15–25.

Van der Zande, A. N., W. J. Ter Keurs, and W. J. van der Weijden. 1980. The impact of roads on the densities of four birds species in an open field habitat: evidence of a long-distance effect. *Biological Conservation* **18**:299–321.

Vermeulen, H. J. W. 1994. Corridor function of a road verge for dispersal of stenotopic heathland ground beetles (Carabidae). *Biological Conservation* **69**:339–349.

Vos, C. C., and J. P. Chardon. 1998. Effects of habitat fragmentation and road density on the distribution pattern of the moor frog *Rana arvalis*. *Journal of Applied Ecology* **35**:44–56.

Walker, R., and F. L. Craighead. 1997. Analyzing wildlife movement corridors in Montana using GIS. *Proceedings of the 1997 ESRI User Conference*. Redlands, CA: Environmental Sciences Research Institute. Available online at http://www.esri.com/library/userconf/proc97/proc97/to150/pap116/p116.htm

Woods, J. G. 1990. *Effectiveness of Fences and Underpasses on the Trans-Canada Highway and Their Impact on Ungulate Populations*. Banff, Alberta, Canada: Banff National Park Warden Service.

Yamada, K., J. Elith, M. McCarthy, and A. Zerger. 2003. Eliciting and integrating expert knowledge for wildlife habitat modelling. *Ecological Modelling* **165**:251–264.

Yanes, M., J. M. Velasco, and F. Suárez. 1995. Permeability of roads and railways to vertebrates: the importance of culverts. *Biological Conservation* **71**:217–222.

21

Where to draw the line: integrating feasibility into connectivity planning

SCOTT A. MORRISON AND MARK D. REYNOLDS

INTRODUCTION

The long-term persistence of populations in fragmented landscapes depends on connectivity among disjunct habitat patches. Retaining or restoring habitat corridors has become a dominant strategy for maintaining connectivity in fragmented landscapes (e.g., Bennett 1999; Groves 2003; this volume), especially when the surrounding matrix is hostile to dispersing wildlife. The biological merits of any particular corridor, however, will depend upon a variety of factors, including the ecology of the targets (species, communities, natural processes) it is intended to serve, as well as the specific attributes of the corridor itself, the habitat matrix in which it is embedded, and the core areas it is connecting. Often, the effects of corridors on target biota are unknown. In some cases, corridors may be ineffective or even counter-productive (e.g., Simberloff and Cox 1987; Hess 1994; Dobson *et al.* 1999; Crooks and Suarez Chapter 18). Biological advantages and disadvantages of corridors have been discussed elsewhere (e.g., Beier and Noss 1998; Groves 2003; Crooks and Sanjayan Chapter 1). Here, we focus on how feasibility considerations factor into the decision-making process of conservation practitioners deciding where best to invest resources in connectivity conservation.

In an ideal world, wildlife corridor planning would occur with detailed knowledge of biological resource needs, multiple options for corridor

Connectivity Conservation eds. Kevin R. Crooks and M. Sanjayan. Published by Cambridge University Press. © Cambridge University Press 2006.

design, unlimited resources for implementation, and cooperative land-owners. More often, virtually nothing is known about plant and animal movement needs within a landscape, human land uses have already greatly constrained corridor options, conservation funding is inadequate, and landowners and political entities are recalcitrant (Swenson and Franklin 2000). It is in that reality that corridor planning and implementation must proceed.

In this chapter, we focus on landscapes undergoing intensive and rapid fragmentation by irrevocable human land-use impacts — situations where the opportunity to preserve connectivity is essentially now or never, and decision-making is fraught with scant information and high uncertainty (e.g., urbanizing southern California). For simplicity, we assume that in most instances habitat corridors are desirable, and that any adverse effects of corridors can be mitigated through design and appropriate management. In such landscapes, there is a great urgency to abate the threats to whatever natural connectivity remains, but the financial resources to do so are far less than what is required. It is here that we face the question: where specifically should we invest in corridors? Given scarce funding and limited capacity, which of the potentially many connections represent the best conservation investment? Unfortunately, today's conservation practitioners — who are often not biologists — are left to address that question without the support of analyses necessary to make informed prioritization decisions.

A TYPICAL CONNECTIVITY CONUNDRUM

To illustrate, we consider a hypothetical landscape that has undergone extensive habitat conversion to human land uses (Fig. 21.1). Natural habitat has been fragmented into five blocks of habitat (core reserve areas A–E). Because the plants, animals, and ecosystems that we seek to conserve in this landscape evolved with fewer barriers to gene flow, movement, dispersal, and ecosystem processes than those now present in this human-modified landscape, we assume that connectivity for the full suite of species native to this area is desirable. Often in a case like this, representative focal species for connectivity are identified (e.g., wide-ranging species that may not be accommodated by core reserve areas alone), their habitat requirements for movement are estimated, and the biological corridors assumed to offer the best connectivity for the most species are delineated (corridors S–Z).

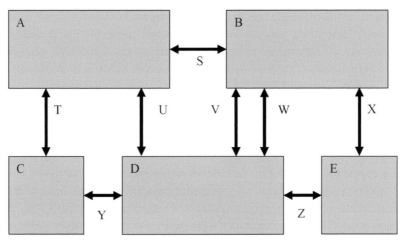

Fig. 21.1. A hypothetical fragmented landscape. Gray boxes (A–E) represent existing or potential core reserve areas; arrows (S–Z) represent potential corridor routes; white represents matrix.

At some point, the biologically derived corridor plan must be shared with decision-makers and conservation practitioners (e.g., public agencies, government officials, non-governmental conservation organizations). The practitioners, in turn, may analyze the land ownership, regulatory, and jurisdictional overlays traversed by the "biologically preferred" corridor(s) in order to develop strategies for maintaining connectivity (e.g., land acquisition, acquisition of development and other rights through conservation easements, habitat restoration, infrastructure modification, land-use regulation and zoning, financial incentives for compatible land uses). The arduous work of "implementation" – of building partnerships, raising funds, influencing policy, protecting and restoring lands – begins (see Beier *et al.* Chapter 22).

Yet, it is not unusual for the conservation process to break down before any corridors are actually secured. Why? Because, typically, the funding and capacity required for implementing the biologically preferred solution far surpasses the resources that are actually available or foreseeable. The strain of resource limitation is an underlying source of several difficulties:

- *Corridor implementation (i.e., protection, restoration, management) bears high real cost.* Land values in human-dominated landscapes, for example, are often high in matrix areas where corridors traverse, relative to less-impacted areas where core habitat might be secured. Indirect costs of conservation also may be high; implementation entails effective engagement with multiple stakeholders, specialized expertise, transactional expenses, and so on, and that capacity generally must be supported or provided by the conservation practitioner. Once the corridor is protected, maintaining corridor function requires substantial management costs above the initial investment. We suspect that the extraordinary costs of implementing corridors are often underestimated by connectivity biologists; lines on maps, even when drawn by experts, do not necessary represent a solution that is actionable.

- *Corridor implementation bears significant opportunity cost.* Resources invested in one place are not available to spend elsewhere. Although such trade-offs are inherent in most if not all aspects of conservation (e.g., decisions related to the conservation of core habitat), the opportunity cost of conserving linkages is arguably higher because the intrinsic habitat value of the land protected for corridors may be considerably lower than that which could be conserved for core areas. Where the rate of habitat conversion is rapid or accelerating, investing in one place may require sacrificing the opportunity to preserve connectivity elsewhere.
- *Corridor implementation bears risks.* Rarely is there adequate information — biological or otherwise — to design corridors with reasonable assurance of sustained functionality. Especially where fragmentation is recent or still under way, risks associated with functionality cannot be fully evaluated until long after the decision to invest in the corridor has been made. Moreover, in areas undergoing habitat conversion there is the risk that the corridor will be severed before it is completely protected. For the most part, linkages are of value only if core reserve habitat has been secured and if the connection can be completed; partial corridors are of uncertain and limited value, especially in landscapes with an inhospitable matrix.

The initial enthusiasm behind an ambitious connectivity plan may cool as the practitioner confronts these realities: implementation of multiple corridors is often not possible due to financial or capacity constraints; threats to connectivity are not evenly distributed in space, time, or severity and that variability is difficult to parameterize; opportunities are also not synchronous, in that high-priority conservation lands may not be available for protection (e.g., if landowners are unwilling sellers). As the practitioner realizes the need for further prioritization within and among corridors, she circles back to a fundamental biological question: How much value do we get for any particular corridor, anyway? How critical is corridor X? Is corridor X more important than corridor Y, or than augmenting core reserve area A? Would investment in one, the other, or *neither* be the more responsible allocation of scarce conservation resources? If she is trying to maintain the possibility of movement of individuals of a focal species across the landscape, one corridor option may cost $1 million and take an estimated 5 years to implement, whereas another option would cost $5 million and take 1 year to implement. What if a third option would cost $5 million, would take 5 years, but would work for two focal species? Where should she invest? These are the types of questions that routinely confront connectivity practitioners, and with the continued fragmentation of habitat, will come to dominate the conservation dialogue of this century. Most existing approaches to connectivity planning, however, provide practitioners with little guidance to address these questions.

FRAMING THE TRADE-OFFS

All conservation decisions are trade-offs of some sort. Consequently, the question of where, how, and when to invest conservation resources represents a significant strategic decision of conservation organizations and public agencies responsible for stewardship of scarce funds. With regard to connectivity, our growing recognition of the adverse effects of habitat fragmentation coupled with our still fundamental lack of understanding of those effects would argue for a precautionary, "better safe than sorry" approach to conservation, erring on the side that more connectivity is better (e.g., Noss 2003; Crooks and Sanjayan Chapter 1). Realistically, however, not all connectivity that is important can be protected. Because resources for securing connectivity are limited, connectivity biologists must become very specific about what connectivity is required and where. The conservation of connectivity demands that precaution be balanced by parsimony.

Better tools are needed to evaluate and communicate the biological trade-offs among different corridor options. It may take acknowledging that most connectivity will likely be lost to help focus research questions on trade-offs. What specifically are the consequences of lack of connectivity for a population or an ecosystem process? Is connectivity for a focal species in this place *desired*, or is it *imperative*? If biologists do not assist with such assessments and prioritizations in real-time, decisions concerning where (or whether) to act will be made in a manner resembling a triage determination, most likely by non-biologists likely daunted by the enormity of the implementation challenge and using entirely different criteria (like cost alone). A characterization of alternative potential corridors in a common metric (e.g., contribution to population viability of a focal species) would be helpful, to allow a comparison of relative returns of different conservation investment options. For the evaluation of biological trade-offs among corridor alternatives to be meaningful, however, it must be conducted within the context of feasibility considerations; otherwise, implementation costs would be unrealistically unbounded.

Although connectivity planning must focus on biological outcomes, the delivery of those outcomes depends on the implementation of corridors, which in turn ultimately rests on social, political, economic, and other feasibility factors. Typically, such factors enter into the conservation planning process, if at all, *after* the biological assessment has determined the preferred route through the landscape. Yet, corridor plans with the most cost-efficient and feasible estimations will be the most likely to be implemented. In fact, they may be the *only* corridors with a good chance of implementation. Just as there are general biological principles that guide the design of linkages in the absence of comprehensive field data, so too are there general principles concerning the relative likelihood of successful implementation. Feasibility considerations include development potential, land values, parcel size, number of parcels, land-use patterns and trends, access to development infrastructure, and so on. In areas where numerous options for routing of a corridor exist, consideration of implementation factors concurrently with the biological factors may allow for the identification of corridors that optimize the biological benefit as well as the feasibility of implementation.

In some cases, feasibility considerations themselves might effectively guide the planning process. For example, mountain lions (*Felis concolor*) may move through a landscape using a variety of pathways. If that is so, why not first determine what corridor is most feasible to implement and then ask whether that pathway would function adequately for mountain lion? An examination of ownership patterns may reveal swaths of lands between core areas that are divided by only a few large landholdings instead of many small parcels. All else being equal, planners may wish to steer a linkage through the larger landholdings in order to minimize the total number of landowners and potential transactions involved in the implementation process. Given the difficulties of funding and staffing conservation, it may behoove conservation practitioners to first conduct a rapid but serious evaluation of the connectivity options that maximize feasibility; the benefits of a corridor's relative feasibility may outweigh its less-than-optimal biological benefits if the biologically preferred option is considerably more costly. In other words, rather than investing a great deal of planning and analytical effort into the ideal placement of a corridor from a biological perspective, it could be more fruitful to first identify a range of feasible corridors and then simply ask if any of the practical options will meet the minimum biological needs. It is important to note that basic "seat of the pants" planning approaches (e.g., review of aerial photographs, rapid field reconnaissance: Noss and Daly Chapter 23) may not necessarily provide an adequate overview of the status and trends of relevant feasibility factors. For example, remaining habitat that appears promising for corridors may already have been foreclosed upon by earlier land-use planning decisions or approved development projects not yet implemented.

An emphasis on feasibility in conservation planning should not be construed as a subjugation of biology. On the contrary, we suspect that the integration of non-biological data into connectivity planning analyses will have the effect of honing biological questions and assumptions. Obviously, a corridor plan that is feasible to implement but that has unacceptably low biological value should not be pursued. But, if there is an option that is far more likely to be implemented than others, analysis and hypothesis testing should be focused on evaluating its effectiveness at attaining ecological goals. Indeed, that option may provide a welcome means of narrowing an otherwise bewildering and abstract array of ecological questions into specific hypotheses grounded by well-defined comparison. If that opportunity for connectivity were to be seized, for which species would we expect the resulting corridor to function? For which species will it not? For those not served by the corridor, how problematic is that specific lack of connectivity for population viability? If it is determined that other species need to be serviced by a corridor, then other corridor scenarios might be evaluated to address their needs, and the hypothesis generation could continue. Or, should corridor implementation costs be deemed too high, then perhaps other management strategies need to be developed to ensure target viability. Now constrained by the same reality, connectivity biologists might come to the planning table prepared to assist planners decide — in the context of all the conservation needs — which corridors are and which are not good conservation investments. By characterizing connectivity priorities as trade-offs between biological value, cost, and risk, connectivity conservationists would come to provide invaluable assistance to conservation practitioners.

An eye toward implementation early in the planning phase may make the difference between a corridor that remains conceptual and a corridor that becomes protected. The following three case studies illustrate (1) the challenges of corridor implementation in compromised landscapes and the importance of multidisciplinary and integrated planning, (2) an approach for evaluating conservation trade-offs to inform decision-making, and (3) an approach to infuse connectivity and implementation considerations into the design of regional conservation reserve networks.

IMPLEMENTATION IN PRACTICE: THE TENAJA CORRIDOR

In the early 1990s, The Nature Conservancy and partners initiated an effort to connect the Santa Rosa Plateau Ecological Reserve in Riverside County, CA, with US Forest Service lands approximately 6 km away (Fig. 21.2). The intervening matrix was small (~15 ha) parcels, undergoing conversion from a landscape dominated by chaparral to one of estate homes, ranchettes, and agriculture. A team of biologists designed a corridor based on the hypothesized habitat requirements of myriad focal species. A land tenure map overlain on the corridor design indicated which parcels required conservation attention. Implementation strategies ranged from fee-acquisition to conservation easements to planned development. Because of the high land costs in the area, a central implementation strategy was intended to be a "Conservation Buyer Program," in which The Nature Conservancy would acquire a priority parcel; if a portion of the parcel was relatively less critical for the functionality of the corridor, The Nature Conservancy would allow a private party to purchase the property and build a home on that less critical portion, so long as a conservation easement protected the conservation value of the remainder (hence, "conservation buyer"). In addition to helping finance protection of the corridor, the conservation buyer approach was intended to transform landowners into stakeholders interested the corridor's success. Thus began the implementation of the Tenaja Corridor, with a strong biological foundation and a promising primary implementation mechanism.

What The Nature Conservancy hoped to achieve in the Tenaja Corridor differs from what has happened. The parcels to be protected using the conservation buyer approach were encumbered by utility assessments (fees), that were apportioned based on the development potential of the lands prior to the conservation acquisition or easement. In other words, owners of each parcel in the utility district were required to pay an annual fee for those utilities, regardless of whether those utilities were being used

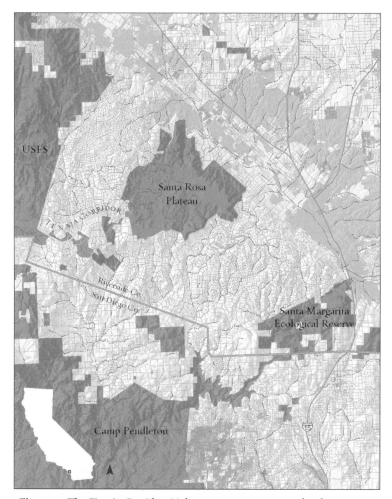

Fig. 21.2. The Tenaja Corridor. Light gray areas represent development and present-day parcelization; dark gray areas represent local, state, federal, and private conservation lands. Inset indicates location within the State of California.

or ever would be used. As lands were acquired, The Nature Conservancy, or the subsequent conservation buyer, was required to pay the utility assessments. The assessments rendered the conservation buyer tracts very difficult to market. The Nature Conservancy could have lobbied to have the assessments waived on the conservation parcels, but to do so would effectively transfer the costs of those public bond obligations onto other neighbors in the community service district. Transferring these costs might have engendered a local backlash to corridor conservation efforts.

The risk of backlash was considered to be unacceptable because community support of the corridor, and tolerance of the wildlife using it, is arguably as essential to the corridor's long-term functionality as the protected lands themselves.

Could some of these obstacles have been avoided? Perhaps. The corridor was designed as the shortest distance to neighboring public conservation management lands to the west. But, there are also significant public lands to the south, toward San Diego County where parcels do not have the utility assessments that have hindered implementation. A southern linkage, however, would have been longer. How would the different corridor routes compare, in terms of ecological value versus economic or feasibility cost? Looking back, it is difficult to predict how the planning team would have designed the corridor if these feasibility constraints were factored in with the biological considerations. Then as now, we lack the clear framework for evaluating trade-offs inherent in connectivity conservation investments.

Ten years later, The Nature Conservancy is still implementing the Tenaja Corridor. Over 526 ha of corridor lands have been protected, but the connection between the reserves remains incomplete and there are still many more priority parcels than can be afforded. Larger properties have been parcelized into 5-, 10-, or 20-acre (2-, 4- or 8-ha) lots; 20-acre parcels sell for $1 million; home prices currently increase at a rate of ~30% per year. With hindsight it is clear that planners underestimated the rate and severity of the looming habitat conversion. It had been assumed, for example, that steep slopes would be "conserved by default" because they were "undevelopable"; it has since been made clear that steep habitat is indeed developable. A simple extrapolation of the trajectory of development currently under way in the area suggests that unless parcels have explicit conservation protection, they will unlikely provide permeable habitat in the future for most native species. This is an important precautionary note for planners facing similar corridor implementation scenarios: assume a worse case for the matrix and plan accordingly. As the race between conservation and conversion of habitat continues, some parcels prioritized and acquired early in implementation have become dead-ends for connectivity due to development; others that were once lower priority have increased in ecological valuable as the surrounding matrix quality erodes. One way or another, feasibility factors will influence corridor design.

While habitat within the corridor parcels does appear to be utilized by some focal species (mountain lion, for example, are tracked within those tracts), the matrix is still in the early stages of build-out. Increases in houses, roads, people, pets, and livestock will undoubtedly make the corridor increasingly difficult for wildlife to navigate. Until the matrix parcels have reached their maximum build-out potential (which will be effectively hardscape), and the fragmented ecosystem has equilibrated to that condition, the functionality of the corridor cannot be known.

The Tenaja Corridor is a relatively small portion of one of the 60 linkages/linkage complexes identified in the "Missing Linkages" analysis for the northern (i.e., US) half of the California South Coast Ecoregion (Penrod et al. 2001; Beier et al. Chapter 22). The Missing Linkages map, depicting corridors necessary for the protection of regional ecological connectivity, represents the collective recommendations of biologists and others from a broad coalition of local, state, and federal agencies, conservation interests, and academia. To some degree, the Tenaja Corridor experience calibrates the challenge of implementing connectivity in highly fragmented, urbanizing regions. Strategic, efficient, and dynamic planning is imperative if connectivity is to be salvaged or maintained in such landscapes.

INTEGRATING FEASIBILITY AND BIOLOGY IN THE SANTA CLARA RIVER VALLEY

One approach of characterizing trade-offs among corridor options is illustrated by a recent corridor planning effort for the Santa Clara River valley north of Los Angeles, CA. The analysis, conducted as a group Master's thesis project at the University of California—Santa Barbara Donald Bren School for Environmental Science and Management (Casterline et al. 2003), expanded upon work from a "Missing Linkages" experts' workshop (Penrod et al. 2001; Beier et al. Chapter 22) that identified the need to protect corridors between the Los Padres National Forest (henceforth, the northern node) and South Mountain and the Santa Susana Mountains (the southern node). It was assumed that the southern node would provide a stepping stone for biota moving from the northern node to the Simi Hills and onward to the public protected areas of the Santa Monica Mountains (Fig. 21.3). The Missing Linkages analysis identified the need for corridors across the Santa Clara River valley, which has undergone extensive agricultural conversion between the expanding urban centers of Santa Paula and Fillmore. The linkages would ideally function for the full suite of native species and ecological processes of the region.

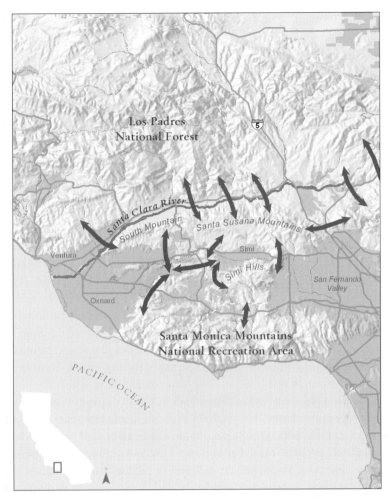

Fig. 21.3. The "Missing Linkages" of the Santa Clara River valley. Gray areas represent development; arrows represent the connectivity priorities identified by the Missing Linkages Initiative (see text). Inset indicates location within the State of California.

The initial goal in the Casterline *et al.* (2003) analysis was a corridor plan that would provide for the movement needs of multiple taxa. A number of planning and implementation obstacles were identified. First, although there is a great diversity of species to be served by corridors in this region, there are very few data on the regional movements or habitat requirements of any of those potential focal species. So, rather than plan for all biodiversity at the onset, the authors sought to identify movement corridors for wide-ranging, habitat generalist carnivores. Second, the possibility of identifying corridors by means of field reconnaissance was limited, because the study area is extensive and largely inaccessible (most of the land is privately owned). A GIS-modeling approach was adopted to overcome this limitation. Finally, it was expected that the economic cost of implementing any corridor in the study area was going to be high, because it is impossible to travel between the northern and southern node without traversing lands of intensive human use. Land uses, land values, developmental pressures, and regulatory restrictions vary across the study area, however; this study aimed to incorporate that variability into the planning process.

The authors sought to identify a number of potential corridors and compare them based on biological value and feasibility of implementation. To do so, they conducted a least-cost path analysis (e.g., Walker and Craighead 1997; Theobald Chapter 17) for a "composite" generalist species modeled from habitat preferences of the mountain lion, bobcat (*Lynx rufus*) and gray fox (*Urocyon cinereoargenteus*). The biological cost surface was parameterized using a scoring system that ascribed rank values to different landcover and vegetation attributes. Paths could then be identified between the northern and southern nodes that would minimize the traversal of less suitable habitat. Because of the uncertainty inherent in the model inputs, a sensitivity analysis was conducted, in which the model parameters were allowed to change randomly, but still within biologically realistic bounds. From this analysis multiple candidate corridors were generated, each with a biological value score associated with it. Each "biologically derived" candidate corridor traversed a complex landscape of human land uses. To provide an indicator of relative implementation cost for each candidate corridor, a parcel map with a cost-per-area attribute was overlain over the candidate corridors. The biological benefits and a crude approximation of implementation costs could then be summed and compared for each candidate corridor (Fig. 21.4). Ecological benefit and implementation cost form two axes, where points are individual corridor scores. A conservation practitioner might strive to implement corridor options that have higher ecological benefits (i.e., lower ecological "cost") and lower implementation cost, which, incidentally, might not be the candidate corridor with the highest ecological benefit.

Although the analysis of Casterline *et al.* (2003) is not without shortcomings (e.g., some of the parameters used in the biological model are confounded, some of the assumptions used in estimating economic costs were overly simplified), the integrated approach they describe is

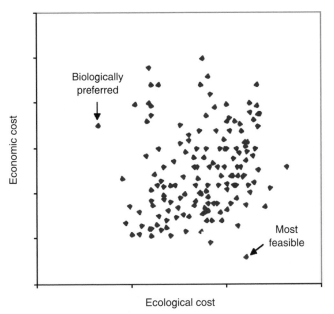

Fig. 21.4. A biology versus feasibility trade-off evaluation model. Each point represents a computer-modeled wildlife corridor through a landscape; each corridor is depicted in terms of its ecological cost (estimated by total habitat suitability) and the economic (implementation) cost estimated for that pathway (see text). Candidate corridors nearest the origin may best optimize biological as well as feasibility considerations. (Adapted from Casterline *et al.* 2003.)

a straightforward means of characterizing and visualizing trade-offs in corridor design, so that conservation decision-making can be more informed. The use of GIS tools is now commonplace in connectivity planning as a cost-effective way of developing linkage plans for complex planning areas, where multiple connectivity options exist and field data are sparse. It is imperative for connectivity planners to advance such analyses beyond purely biological modeling to a frontier where informative implementation data are integrated in the design phase.

PLANNING AND IMPLEMENTING NETWORKS: NATURAL COMMUNITIES CONSERVATION PLANNING

Planning for connectivity is arguably most efficient and effective when it is a central element of regional conservation planning and analyses, rather than an afterthought. Natural Communities Conservation Planning in southern California illustrates one such approach to that integration. In 1991, the State of California passed the Natural Communities Conservation Planning Act, in part to overcome the administrative and biological shortcomings of piecemeal enforcement of the US Endangered Species Act (i.e., one species, one project, and one permit at a time). Under Natural Communities Conservation Planning, wildlife agencies issue endangered species "take" permits in return for a regional conservation plan that protects important habitats from development. Through this approach, local, state, and federal agencies collaborate to produce a "conservation blueprint" for an area's natural communities, targeting common as well as sensitive species. The blueprint, a result of extensive scientific analysis and review, represents a network of core habitat areas and corridors that is hypothesized to support viable occurrences of focal species and communities in the region. Thus, the interdependence of functional cores and connectors are addressed from the onset.

A key innovation of the Natural Communities Conservation Planning is that the blueprint is coupled with an implementation mechanism administered through the local land-use planning process. For example, within the planning areas of San Diego city and county (Fig. 21.5), land-use regulatory agencies evaluate proposed land-use changes based on the blueprint. Development projects that will impact natural habitat must mitigate their impact by preserving habitat areas within the biologically higher-valued swaths depicted on the blueprint. Development projects that are proposed on properties within the biologically higher-valued areas incur a greater mitigation obligation than development projects within the less biologically valuable areas. Local, state, and federal governments share in the implementation, as they each have to fund additional conservation within the reserve plan. Natural Communities Conservation Planning is no panacea; its effectiveness is debated regarding the protections it provides for current or future imperiled species, the adequacy of its monitoring and management, and so on. However, this public/private, market-based mechanism does spread the burden of implementation more broadly among stakeholders, creating conservation actors and funds and focusing them on a common conservation vision. That focus, combined with the supporting regulatory incentives and disincentives, helps overcome the considerable capacity and financial limitations typically faced by conservation practitioners implementing connectivity plans. Moreover, it helps reduce risks associated with connectivity implementation; when the implementation responsibility is shared and the diversity of conservation mechanisms is broadened,

Fig. 21.5. Design of the coastal San Diego County Natural Communities Conservation Planning reserve. Gray areas indicate development; black areas indicate the existing and proposed reserve network design. Natural Communities Conservation Planning programs are currently under way for the eastern portion of the county. Inset indicates location within the State of California.

the likelihood of completing a connection is improved, as is the probability that it will function once established.

CONNECTIVITY CONSERVATION IN PRACTICE

If we are to maintain wide-ranging species and large-scale ecological phenomena in highly fragmented landscapes we need focused efforts aimed at stitching together some sort of ecological connectivity. In rapidly fragmenting landscapes like southern California, opportunities for providing connectivity are dwindling, and agencies and organizations need to act quickly and effectively despite huge gaps in data and funding. In most cases, the fate of connectivity rests on the actions or inactions of conservation practitioners. Focused information is required to impel action. California State Parks' decision to implement the Coal Canyon corridor (discussed in Beier *et al.* Chapter 22) provides an illustrative example: the decision was based in large part on the planners' ability to certify that (1) *that* particular corridor was the most feasible option, (2) the corridor would function once implemented, and (3) there would be deterioration in the quality of existing public conservation lands if the corridor were *not* implemented (R. Rayburn, pers. comm.). That characterization of potential adverse effects on existing regional investments in conservation (and those investments are substantial) allowed the implementation of the corridor to be cast as a "maintenance cost" for those lands, and that helped place the cost of this expensive corridor in the proper context.

Maintaining connectivity is high-stakes conservation that requires innovation and risk-taking informed by science-based modeling and monitoring. Risks inherent in resource allocation decisions are magnified in highly fragmented landscapes where connectivity is precarious. Land with high potential corridor value may be land with relatively low intrinsic habitat quality. Questions about the wisdom of investing limited resources in such lands are destined to become increasingly difficult, so we must become better equipped to answer them. New approaches for evaluating and communicating trade-offs between biological priorities and feasibility are needed. Developing those approaches will require not only a better understanding of dispersal, habitat suitability and landscape-scale ecological processes but also a fuller appreciation of how economic, social, and political constraints factor into the practice of conservation (e.g., Kremen *et al.* 1999). Connectivity biologists can support the evaluation of trade-offs in corridor conservation by setting explicit goals for target species; evaluating the scale at which connectivity is threatened (e.g., local, regional, continental); assessing the degree to which a corridor helps meet the goals or abate the threats; and parameterizing the feasibility of protecting that corridor. Enhancing connectivity planning by including feasibility information, however coarse or subjective, will go a long way toward bridging the current disconnect between planning and implementation, thereby improving the chances that a plan will actually be implemented. In a rapidly urbanizing world, delaying connectivity decisions — whether because of inaction or inefficient planning — may foreclose the opportunity for connectivity among reserves fated to exist within an impermeable matrix. It is that urgency which begs for a more integrated planning approach, focused on the relationship between feasible options and biological outcomes. The relevance of connectivity conservation biology to the practice of corridor conservation rests on its ability to help elucidate the essential from the ideal, and from that, what is actually possible.

ACKNOWLEDGEMENTS

Order of authorship was determined by coin toss. We thank Brian Cohen for the map figures. We thank M. Casterline, C. Costello, E. Fegraus, E. Fujioka, L. Hagan, C. Mangiardi, M. McGinnis, M. Riley, and H. Tiwari for helpful discussions, as well as D. Cameron, K. Crooks, P. Kareiva, G. LeBuhn, R. Shaw, P. V. Smith, J.A. Stallcup, and C. Tredennick for review of earlier versions of the manuscript.

REFERENCES

Beier, P., and R. F. Noss. 1998. Do habitat corridors provide connectivity? *Conservation Biology* 12:1241–1252.
Bennett, A. F. 1999. *Linkages in the Landscape: The Role of Corridors and Connectivity in Wildlife Conservation.* IUCN, Gland, Switzerland: International Union for the Conservation of Nature.
Casterline, M., E. Fegraus, E. Fujioka, *et al.* 2003. Wildlife corridor design and implementation in southern Ventura County. Unpublished group M.Sc. thesis project, Bren School for Environmental Sciences, University of California–Santa Barbara, Santa Barbara, CA.
Dobson, A., K. Ralls, M. Foster, *et al.* 1999. Corridors: reconnecting fragmented landscapes. Pp. 129–170 in M. E. Soulé and J. Terborgh (eds.) *Continental Conservation: Scientific Foundations of Regional Reserve Networks*, Washington, DC: Island Press.
Groves, C. R. 2003. *Drafting a Conservation Blueprint: A Practitioner's Guide for Planning for Biodiversity.* Covelo CA: Island Press.
Hess, G. R. 1994. Conservation corridors and contagious disease: a cautionary note. *Conservation Biology* 8:256–262.
Kremen, C., V. Razafimahatratra, R. P. Guillery, *et al.* 1999. Designing a new national park in Madagascar based on biological and socio-economic data. *Conservation Biology* 13:1055–1068.
Noss, R. 2003. A checklist for wildlands network designs. *Conservation Biology* 17:1270–1275.

Penrod, K., R. Hunter, and M. Merrifield. 2001. Missing linkages: restoring connectivity to the California landscape. Proceedings of a workshop held November 2, 2000, San Diego, CA.

Simberloff, D. S., and J. Cox. 1987. Consequences and costs of conservation corridors. *Conservation Biology* 1:63–71.

Swenson, J. J., and J. Franklin. 2000. The effects of future urban development on habitat fragmentation in the Santa Monica Mountains. *Landscape Ecology* 15:713–730.

Walker, R., and L. Craighead. 1997. Least-cost-path corridor analysis: analyzing wildlife movement corridors in Montana using GIS. *1997 ESRI International User's Conference Proceedings.*

22

South Coast Missing Linkages: restoring connectivity to wildlands in the largest metropolitan area in the USA

PAUL BEIER, KRISTEEN L. PENROD, CLAUDIA LUKE,
WAYNE D. SPENCER, AND CLINT CABAÑERO

INTRODUCTION

The South Coast Ecoregion encompasses 3.4 million ha or roughly 8% of California. Lying west of the Sonoran and Mohave Deserts and south of the Santa Ynez and Transverse Ranges, the ecoregion extends about 320 km south into Baja California, Mexico (Fig. 22.1). California's most populated ecoregion, it has the dubious distinction of being the most threatened hotspot of biodiversity in the USA, with over 400 species of plants and animals considered at risk by government agencies and conservation groups (Hunter 1999). Despite a human population of over 19 million (2000 census), the South Coast Ecoregion has many large wildland areas, mostly in more rugged and higher-elevation habitats within the Los Padres, Angeles, San Bernardino, and Cleveland National Forests, Santa Monica Mountains National Recreation Area, Marine Corps Base Camp Pendleton, and several State Parks. Although each wildland core area would benefit from expansion, increased protection, and restoration, each enjoys some degree of protection from urban expansion, and few if any major new wildland areas are likely to be designated. Therefore we focus on the previously neglected portion of a wildland network, namely the linkages between core areas.

Connectivity Conservation eds. Kevin R. Crooks and M. Sanjayan. Published by Cambridge University Press. © Cambridge University Press 2006.

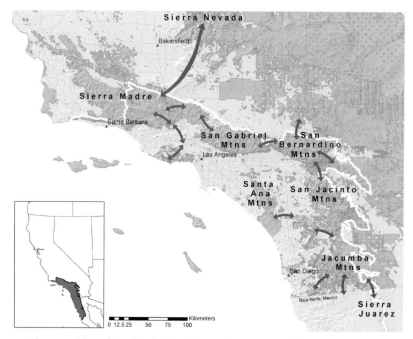

Fig. 22.1. Map of the South Coast Ecoregion (inset) and the 15 priority linkages for South Coast Missing Linkages.

Although numerous conservation efforts are underway, such as California's Natural Communities Conservation Plans (Polak 2001), these efforts do not span the ecoregion, and do not conserve ecosystem processes and functions that operate over grand scales – such as top–down regulation by large predators or gene flow between core areas lying in different planning jurisdictions. To address these gaps, each of us has worked on independent projects to conserve and connect large wildland areas where large-scale processes can operate in a semblance of their natural rhythms. Since 2000, we have worked together on an ecosystem-wide effort – the South Coast Missing Linkages project.

The South Coast Missing Linkages project began with a state-wide workshop in November 2000, sponsored by The Nature Conservancy, US Geological Survey, California State Parks, California Wilderness Coalition, and San Diego Zoo. Over 200 land managers and biologists from throughout California identified 232 actual or potential linkages needed to sustain ecosystem processes in protected wildlands. South Coast Wildlands was formed in early 2001 with an Executive Director, a Board, a team of Science Advisors, and the goal of conserving essential linkages throughout the South Coast Ecoregion. South Coast Wildlands brought together under the umbrella of the South Coast Missing Linkages project a variety of agencies and organizations already engaged in various linkage conservation efforts. We worked with these partners to develop a standardized set of methods for conserving a network of protected wildlands for the region.

Widespread and increasing urbanization in most linkage planning areas constrained conservation options and added urgency to the planning process. We experienced an understandable urge to use expert opinion to quickly map conservation targets − a "seat-of-the-pants" approach (Noss and Daly Chapter 23). However, three ideas compelled us to develop a set of scientific rule-based procedures for delineating what we call our Linkage Design. First, despite our confidence that maps based on expert opinion would lead to sound conservation decisions, when we experimented with more formal methods, we discovered some options that we had overlooked (see also Cowling *et al.* 2003). Second, a model is transparent. Landowners, developers, conservation investors, and decision-makers demand strong support for recommendations to conserve particular areas. If they doubt one or more assumptions, parameter estimates, data layers, or decision rules, they can rerun the model and see if it makes a difference. Finally, rule-based procedures allow formal sensitivity analysis, a valuable tool for conservation planning.

In this chapter, we describe the South Coast Missing Linkages project's science-based, collaborative approach to linkage planning in the largest urban area in the USA. Our goal is to provide one promising recipe for designing plans that conserve and restore connectivity in real landscapes. These methods were developed predominantly by the authors and incorporate a variety of geographic information system (GIS) methods developed by others. Significant elements have been incorporated from conservation efforts by our partners, most notably, the workshop-based approach developed by the San Diego State University Field Station Programs.

This chapter is a broad overview to be supplemented by additional papers on the mechanics and results of prioritization, permeability, and habitat analyses. Because our focus is on a science-based approach, we ignore important considerations of history and organizational theory; future papers will describe false starts, historical lessons, and the interplay among biological foundations, conservation design, and conservation delivery. We have already achieved a number of successes with this approach, but acknowledge that it is a work in progress. We adamantly hope that others will improve on our efforts.

The following sections are numbered and titled as prescriptions, because we hope others will use them as an outline for future efforts. Steps 2 through 7 correspond to the six steps in linkage design suggested by Beier and Loe (1992), which have proven rather discrete and chronological in practice as well as in concept. We add a new Step 1 – coalition building – which is logically first and permeates all other steps.

STEP 1: BUILD A COALITION

Key elements in developing a coalition for South Coast Missing Linkages include serving as a catalyst, engaging partners, holding organizational meetings, forming a steering committee, and developing an inclusive workshop-based approach to conservation planning.

Serving as a catalyst

Conserving a wildland network on a regional scale requires strong collaboration among land management agencies, conservation groups, transportation and resources agencies, sovereign Native American tribes, and others. As the smallest of these entities, South Coast Wildlands serves as a catalyst – an agent that develops synergy among various larger partners. We believe that a small group like South Coast Wildlands can best fill this role because implementing the vision of a connected ecoregional wildland network is our sole focus and *raison d'être*, rather than one of many priorities vying for attention. Furthermore, most other agencies have internal priorities that would favor some linkages (e.g., linkages that serve lands owned or managed by the agency) that could make them an inappropriate lead agency for a regional effort.

Engaging partners

The statewide Missing Linkages workshop had five major sponsors (above). By organizing this successful conference, the nascent South Coast Wildlands earned the respect of these partners. More important, it became obvious at the workshop that all of the management and conservation agencies considered the workshop simply a first step in linkage conservation. The idea had become mainstream and could command enormous energy if an effective plan were in place. The workshop report (Penrod *et al.* 2001) – with the logos of these sponsors on the cover – was distributed to most agencies and consulting firms in California and received front-page coverage in most California daily newspapers during its August 2001 release. Ten days later, capitalizing on this publicity, South Coast Wildlands convened a meeting among the original sponsors, plus other organizations potentially interested in linkage conservation in the South Coast Ecoregion. At that meeting, we outlined the proposal, distributed a brief concept paper, asked for feedback, and solicited and received commitments of time and funding to the effort. From the outset, this was presented as a collaboration, not as a project of South Coast Wildlands with others as junior partners.

Partners now include scientific and educational agencies (Conservation Biology Institute, San Diego State University Field Station Programs, San Diego Zoo, US Geological Survey), federal land management agencies (National Park Service, US Forest Service), state agencies (California State Parks, Department of Fish and Game, Resources Agency, Santa Monica Mountains Conservancy), and conservation non-governmental organizations (NGO) (California State Parks Foundation, California Wilderness Coalition, The Nature Conservancy, The Wildlands Conservancy). Each partner allows use of the organization's name and logo on reports and in publicity events, and provides some form of support (not always funding). In addition, we have excellent working relationships with entities that are not yet partners, including Native American tribes, county planning departments, local land conservancies, Bureau of Land Management, the California Department of Transportation, Pronatura (a Mexican conservation NGO) and Conabio (Mexico's federal Comisión nacional para el conocimiento y uso de la biodiversidad).

Steering committee

In August 2001, we formed a steering committee with representatives from each major partner. The steering committee holds monthly conference calls to ensure that South Coast Missing Linkages is integrated with other efforts, most notably the Natural Communities Conservation Plans being developed by the Resources Agency. The steering committee has averted potentially serious misunderstandings and has kept the project on-track and visible to participating groups and agencies.

Workshops

As described below, we used workshops to engage partners in many aspects of linkage planning. By including developers and their consulting biologists, as well as our natural allies, we demonstrated that our process is transparent, inclusive, and honest. When participants saw that their input is genuinely sought and used, they tended to adopt the effort as their own. We have involved partners in every aspect of the process because the plan will not succeed if South Coast Wildlands simply asks partners to implement its plan. Only by collaborating from start to finish will all players fully engage in implementation.

As with any collaboration, our partnership has faced difficulties. Dwindling resources and staff time have prevented some partners from providing resources necessary to conduct some analyses. Perhaps the most problematic issue has been the rare plea to dispense with time-consuming science and get our products out faster. However, we have managed to keep our focus on the big picture and these distractions have not disrupted our working relationships nor changed our commitment to a scientific approach.

STEP 2: SELECT CORE AREAS AND PRIORITIZE LINKAGES

We initiated Beier and Loe's (1992) first step in science-based connectivity planning – identifying cores in need of linkages – at the November 2000 Missing Linkages workshop. A core area was defined as a large wildland with reasonable prospect for retaining its wild character for the foreseeable future, including large military installations, but excluding sovereign tribal lands. The process was minimally selective; all proposed linkages in California were accepted as long as core areas were identified. For Mexico, where no large protected areas occur within 100 km of the international border but many large wildlands still exist, the nearest areas of natural habitat $>2000 \, km^2$ were used as core areas. In Mexico, Pronatura and Conabio have enthusiastically greeted our initiative, shared their plans and data, and are working to ensure that cross-border linkages will connect to protected areas in Mexico. Conservation Biology Institute (linkage manager for the cross-border linkages) is our primary liaison with Mexico.

Realizing that resources were insufficient to take immediate, effective action on all 60 linkages in the ecoregion and nine additional linkages connecting to wildlands in other ecoregions, we proposed 12 linkages for conservation action. Almost immediately, advocates for particular wildlands not directly served by those 12 linkages lobbied to have the list changed or expanded. Obviously, a defensible prioritization process was needed, and only a transparent process open to all partners would suffice.

Following a process inspired by Pressey *et al.* (1994), Pressey and Taffs (2001), and Noss *et al.* (2002), South Coast Wildlands invited all partners to send representatives to a prioritization workshop, at which each linkage was scored in two dimensions – biological importance and vulnerability. Participants assigned highest priority to linkages that fell in the upper right quadrant (most important, highest threat). Seven criteria were used to assess biological importance: sizes of the two core areas (35% weighting), the degree to which the linkage facilitated connection to other ecoregions, or was essential to the utility of "downstream" linkages (20%), habitat quality in the smaller core area (20%), existing width and habitat quality in the linkage (10%), the degree to which the linkage connects the ocean to salmonid nursery habitat, or would reduce contaminants, sediment, and insolation of riverine habitat (8%), and the degree to which the linkage might allow for seasonal migration or facilitate range shifts in response to climate change (7%). The seventh criterion was a debit of 10 points for each riverine linkage that lacked upland habitat, was over 10 km long, and had an average width narrower than 200 m. This debit distinguished between true landscape-level linkages and those linkages that, while technically "connecting" large core areas, would not facilitate movement of wide-ranging carnivores or other upland vertebrates due to frequent road crossings, severe edge effects (noise and light pollution, garbage-dumping and other disturbance, conflicts with pets), and low diversity and integrity of natural habitats.

The weighting among scores reflects an emphasis on ecosystem processes and top carnivores, and thus area was more important than the particular habitats or habitat quality in core areas. The full 35 points for size of core areas was awarded to linkages that would connect two large ($>2000 \, km^2$ each) wildlands. We assigned lower scores for a linkage between large and medium-sized (60 to 2000 km^2) wildlands, or between large and small ($<60 \, km^2$) wildlands, down to a low of 0 points for a linkage between two small wildlands. The 2000 km^2 and 600 km^2 thresholds correspond to the minimum areas required to support puma (Beier 1993) or bobcats (*Felis rufus*: Crooks 2002), respectively, over the short term. In addition to being among the most area-sensitive species in the ecoregion, these high-level carnivores are important regulators of ecosystem function (Terborgh *et al.* 1999). In addition, one or both of these two species occurred in all core areas, and were thus more appropriate than species such as peninsular bighorn (*Ovis canadensis cremnobates*) that were present only in some core areas.

The relatively low weight given to current habitat conditions reflected our optimism that if we could avoid urbanization of large degraded wildlands, we would conserve at least the opportunity to confront the restoration challenges. The relatively low weight for response to climate change was hotly debated. All participants agreed that global change will have profound impact on biodiversity. However there was considerable scientific uncertainty about the direction, seasonality, and magnitude of changes in temperature and precipitation expected in our ecoregion, with corresponding uncertainty as to which linkages would best facilitate range shifts. To earn the 7 points for this criterion, a linkage had to span an elevation gradient $> 650\,$m or two major life zones.

At the November 2000 conference, persons describing each linkage had rated the severity of each of several types of threat to the linkage on a scale of 1 (low) to 5 (high). For our assessment of vulnerability, we used the higher of the threat scores for urbanization or roads. We ignored other threats, such as off-road vehicle use or agricultural conversion, on the grounds that these threats are relatively reversible compared to urbanization and roads.

Determining the criteria and scoring system was an iterative process during which participants gradually reached consensus on the conceptual underpinnings of the gestalt ratings that each person held at the start of the process. Scoring the 69 linkages went quite quickly once these issues were resolved. The biological importance scores were clustered in two groups, with 22 linkages scoring as most important. Twelve of these 22 priority linkages had high vulnerability ratings ($\geq$4), and thus emerged as conservation priorities. In addition, we added three linkages with moderate (3) vulnerability scores in areas where our partners had already begun conservation planning. We offer several related justifications for this departure from our prioritization scheme. First, the importance–vulnerability algorithm is not responsive to real-world opportunities, and should be used to inform, but not dictate, conservation decisions (Noss *et al.* 2002). These opportunities related not only to the particular linkages involved, but also to maintaining and strengthening the coalition needed to conserve these linkages. Finally, in one case, acquisition efforts were half complete, and we reasoned that a quick victory would help maintain partner enthusiasm for the full program.

These 15 linkages are the focus of our current efforts (Fig. 22.1). They include nine linkages within the South Coast Ecoregion and six linkages between ecoregions (including Baja California as an ecoregion). The core areas served by these linkages include all the obvious major wildlands in and adjacent to the ecoregion, such as the San Gabriel Mountains, San Bernardino Mountains, San Jacinto Mountains, Anza-Borrego desert lands, and Santa Monica Mountains. The smallest core area is the Otay Mountain area of southern San Diego County ($\sim$150 km^2). The longest linkage spans over 80 km of the privately owned Tehachapi Mountains to connect the large protected wildlands in the Sierra Madre to those in the

Sierra Nevada. The two shortest linkages serve core areas separated only by a freeway and a few small private parcels. For other linkages, the edges of the two protected cores are 6 to 24 km apart.

Each linkage was adopted by a partner organization to serve as its "linkage manager," or the entity most responsible for planning that linkage. South Coast Wildlands, San Diego State University Field Station Programs, National Park Service, The Nature Conservancy, California State Parks, US Forest Service, and Conservation Biology Institute serve as linkage managers or co-managers.

STEP 3: SELECT FOCAL SPECIES FOR EACH LINKAGE

Although our ultimate goal is to conserve ecosystem function, we designed linkages to serve the needs of particular focal species. We used a focal species approach for the practical reason that we do not know how to conduct permeability analysis or design a linkage (Step 4) in a way that directly conserves ecosystem processes in the core areas. We acknowledge that our approach could result in linkages that allow movement of focal species between core areas, but that might fail to conserve natural patterns and mechanisms of gene flow, pollination, seed dispersal, interspecific interactions, energy flow, and nutrient cycling. We do not take this risk lightly. However, given the pace of urbanization, we cannot wait for answers to these questions. We can immediately exploit the focal species approach, which has the further advantage that species-based management is accepted and supported by managers, decision-makers, and public opinion (Lambeck 1997; Miller *et al.* 1999; Carroll *et al.* 2001; Bani *et al.* 2002; Noss and Daly Chapter 23).

To minimize the disconnect between focal species and ecosystem processes, we sought a variety of focal species for each linkage, including species that are closely related to ecosystem function or sensitive to linkage loss, such as indicator species, keystone species, area-sensitive species, and umbrella species (Miller *et al.* 1999; Coppolillo *et al.* 2004). For instance, a linkage that serves focal species such as puma (*Puma concolor*) conserves one necessary condition for top-down trophic regulation. Similarly, we hope that a linkage designed to serve a plant species with limited seed dispersal will conserve that process for less dispersal-limited species.

Our suite of focal species also included a few "orthogonal" species, i.e., a species that occurs within the linkage but not necessarily in the core areas. Planning for such species can help ensure that linkages maintain ecological integrity and are not sterile gauntlets through which other species must pass. Thus, although most of our focal species were "species that need the linkage" (to pass between core areas), the orthogonal taxa represented "species the linkage needs" (to ensure its integrity). For example the little pocket mouse (*Perognathus longimembris*) occurs on fine sandy soils in arid valleys between major mountain ranges, but not in the mountains themselves. Its sensitivity to human barriers, such as roads and concrete ditches, made it a good focal species for ensuring linkage integrity between the mountainous uplands. We did not give rare or threatened species special priority as focal species, although some, such as San Joaquin kit fox (*Vulpes macrotis mutica*), were chosen because they met other criteria. Rarity in itself does not make a species a good keystone, umbrella, or indicator species.

Focal species were selected by participants in five workshops, each organized around one to four linkages in geographic proximity. Participants included land managers, planners, consulting biologists, California Department of Fish and Game staff, and experts on species, habitats, and conservation plans in the linkage area. Selected taxonomic experts gave presentations on what was known about various species habitat connectivity requirements and suggested some initial candidate focal species. Participants then sorted into taxonomic workgroups to select focal species. South Coast Wildlands and the collaborating linkage manager provided detailed instructions on how to select focal species and emphasized how these species would be used to design and justify the linkage, and to serve as indicators of linkage function over time.

Participants were asked to select species that (a) require inter-core dispersal at the scale of *this* landscape for metapopulation persistence, (b) have a localized distribution at the spatial scale of this landscape, (c) have short or habitat-restricted dispersal movements, (d) represent a surrogate for an important ecological process (e.g., predation, pollination, fire regime), (e) need connectivity to avoid genetic divergence of a now-continuous population, (f) might change from being ecologically dominant to ecologically trivial if connectivity were lost, (g) is an important pollinator or seed-disperser, or would suffer reproductive failure if it lost the service of a fragmentation-sensitive pollinator or seed-disperser, or (h) is reluctant to traverse barriers (e.g., culverts under roads) and would be a useful umbrella for other species sharing this trait. Workgroups tried to include focal species that varied with respect to habitat specialization and dispersal distances, but were asked to limit the number of species chosen to fewer than six per taxonomic group. Workgroups reviewed the lists of other workgroups to eliminate redundant species, i.e., species that seemed unlikely to add to the linkage design in light of other included species. In deciding which of two species to consider, we retained the species whose habitat needs and local distribution were better known.

A total of 109 species were identified in all 15 linkages, including 26 plants, 25 invertebrates, 18 amphibians and reptiles, 4 fish, 20 birds, and 16 mammals. Although some species (usually plants or invertebrates) were selected for a single linkage, the average focal species appeared on lists of 2.7 linkages (range 1 to 15 linkages). Puma, mule deer (*Odocoileus hemionus*), and badger (*Taxidea taxus*) each appeared on lists for 14-15 linkages. Steelhead (*Oncorhynchus mykiss*), western pond turtle (*Clemmys marmorata*), and western toad (*Bufo boreas*) each appeared on nine or more lists. On average 19 focal species were identified per linkage (range 14 to 32).

STEP 4: CREATE A DETAILED LINKAGE DESIGN

We developed a multi-stage procedure for identifying priority lands for conservation in each linkage. The first three stages (A−C below) reflect the different types of focal species and the considerable variation in ecological knowledge available for each. For appropriate species (A), we used least-cost corridor analysis (B) to identify lands likely to facilitate movement. Patch size and configuration analysis (C) was then used to evaluate whether each focal species could persist and move through the union of least-cost paths, and to expand that union as needed. The final stage (D) added a buffer to accommodate edge effects, ecological uncertainty, metapopulation dynamics, and processes and species omitted from the analysis.

A: Determine whether least-cost corridor analysis is appropriate to identify lands that best facilitate movement of each focal species, or their genes, between the two core areas

Least-cost corridor analysis (LCCA) is a GIS-based method of estimating the optimal location of a landscape linkage between core protected areas based on estimates or assumptions about how a focal species responds to various landscape features that can be reflected in digital map layers (Singleton *et al.* 2002). Because least-cost *corridor* analysis identifies all pixels with low travel costs, it produces a swath that can include more than one alternative path, and is thus superior to least-cost *path* analysis, which yields a single path one pixel in width for its entire length (Theobald Chapter 17). Other alternative approaches to LCCA are presented by

Bani *et al.* (2002), Tracey (Chapter 14), Carroll (Chapter 15), and Noss and Daly (Chapter 23). We chose LCCA because we lacked detailed data needed for the more sophisticated alternatives (such as movement of radiotagged animals, or parameter estimates for spatially explicit population viability models).

Although the most quantitative and flashy tool in our toolbox, LCCA is the most data-demanding. It is also inappropriate for some focal species. To guard against inappropriate use of this tool, we used it only for species that met all three of the following criteria. First, we must know enough about the movement of the species, or the movement of its obligate pollinators and seed-dispersers, to estimate cost-weighted distance using the data layers available to us. For example, although steelhead and arroyo chub (*Gila orcutti*) are confined to streams, the GIS stream layer is not detailed enough to indicate which stream stretches have aboveground flow (most blue-line streams in the area do not), or what barriers might exist to movements. Second, the species must occur, or have historically occurred, in both core areas to be linked, such that restoration is feasible, and the species or its genes must be capable of moving between the cores (although not necessarily within a single generation). This excluded the orthogonal species from LCCA. Third, the timescale of the species' gene flow between core areas must be shorter than, or not much longer than, the timescale at which currently mapped vegetation layers are likely to be replaced by disturbance events and other environmental variation. This condition excluded focal species such as Engelmann oak (*Quercus engelmannii*), for which gene flow would only occur over many hundreds of years. This criterion would not be needed for a LCCA that included dynamic vegetation maps reflecting vegetation response to disturbance or climate change.

In each linkage, about half of the focal species (including reptiles, amphibians, birds, and mammals, but no fish, invertebrates, or plants) met our criteria for conducting LCCA. We considered the needs of the other species via habitat suitability analysis (Section C, below).

B: For appropriate focal species, conduct least-cost corridor analysis (LCCA)

We conducted LCCA using four GIS data layers that were readily available and likely to influence movement of many animals: vegetation/land use, topographic feature (ridge, canyon bottom, flat, or slope), elevation (classes defined by each species expert), and road density (km of paved road per km^2). Land use (urban, agriculture, disturbed) and paved road density are intended to encompass all the human activities that affect suitability of linkage habitat. Although other measures (densities of humans, livestock, pets, off-road vehicles) seem attractive, most of these are probably highly correlated with urban land uses or paved road density, and none is readily available in GIS format.

For each focal species subject to LCCA, we asked a biologist studying that species or a closely related species to estimate the relative importance of each factor for habitat use by the animal. Recognizing that it is impossible to disentangle the influence of vegetation from that of topography and elevation, we instructed the rater to think of vegetation as the factor that integrates the influence of topography and elevation in a way that is most important to the species. We also stressed the priority of vegetation because there is a much larger literature on selection of vegetation types than on responses to the other factors. Thus the weights for elevation, topography, and roads reflected only their *additional* influence on animal habitat preference; in some cases this resulted in 0% weights for these factors.

The biologist also scored the various vegetation/land-use classes, elevation classes, topographic classes, and road-density classes with respect to animal preference on a scale of 1 (highest preference) to 10 (strongest avoidance). Because Clevenger *et al.* (2002; see also Clevenger and Wierzchowski Chapter 20) found that expert-based models that did not include a literature review performed significantly worse than literature-based expert models, we asked raters to first assemble the literature on habitat selection by the focal species and closely related species, and we offered assistance in gathering those papers.

Although these scores (weights in the equations below) were used to parameterize a LCCA, we asked raters for habitat preference scores rather than *permeability* or *travel cost* scores. We made this decision because experts are much more consistent in rating habitat suitability than in rating ability to move through a habitat (B. McRae and P. Beier, unpublished data on ratings of habitat suitability and permeability provided by six puma experts). Furthermore, there is a large literature on habitat use and preference, but almost no literature on permeability or travel cost in various habitats.

We used California Fire and Resource Assessment Program (FRAP) landcover/land-use data as the source for our vegetation layer, US Geological Survey 30-m digital elevation models (DEM) for our elevation layer, and a topographic feature layer derived from elevation and slope models using Weiss' (2000) topographic position and landform algorithm. Because our fieldwork showed that the only widely available digital road layer (TIGER Line files - Census 2000) failed to differentiate between unimproved roads and paved roads, we used road data from Thomas Brothers, Inc. and 1-m aerial imagery to modify these files to create a paved road density layer. We did not distinguish among types of paved roads (e.g., freeway versus two-lane highway) nor among roads with differing traffic volumes.

Our LCCA was similar to that of Singleton *et al.* (2002) except that we used an additive model rather than a multiplicative one. (We do not claim superiority for our additive model; we are currently assessing whether the two approaches produce different maps.) Pixel size was 0.09 ha (30-m grid) in each linkage except one in which data availability forced us to use 1-ha cells. For each species, each pixel was assigned a travel cost,

$$C = \sum_{i=1}^{4} w_i \cdot s_j, \tag{22.1}$$

where w_i = the weight assigned to factor i (e.g., vegetation type or road density), and s_j = the score assigned to class j (e.g., to the particular vegetation type or road-density class in that pixel). To estimate the cost of movement from the edge of one core area, we assigned each pixel a cost-weighted distance,

$$D = \min \sum_{i=1}^{k} C_i, \tag{22.2}$$

where k = the number of pixels along a path from the focal cell to the largest block of suitable habitat (as defined by California Department of Fish and Game 2002) within one core area. Superimposing (adding) the cost-weighted distances from the two core areas produced a map depicting, for all pixels in the linkage area, the average cost-weighted distances from the two core areas (Fig. 22.2; see also Theobald Chapter 17). We tentatively accepted the lowest percentile of cost-weighted distances that formed a continuous swath of pixels between cores. This was typically 1% or 2% of the linkage area (the smallest rectangle enclosing both cores).

The least-cost corridor for each species was sent to one or more species experts and persons familiar with the landscape, who reviewed the model structure and outputs, and recommended a percentile (e.g., the most permeable 2% of pixels) that would allow movement of the focal species (Quinby *et al.* 1999). Although this recommended percentile sometimes was higher than the lowest percentile that produced a continuous swath,

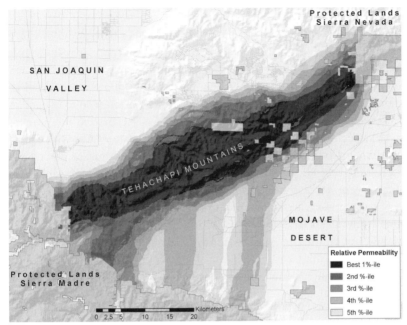

Fig. 22.2. Cost-weighted distance map for puma between protected lands in the southern Sierra Nevada core area and southeastern Sierra Madre core area, highlighting pixels with the lowest total cost, in 1-percentile increments. Percentiles are based on a rectangle encompassing both cores. Because our procedures will always produce a least-cost corridor, even if the "best" corridor does not facilitate animal movement, species experts reviewed each map and recommended the smallest fraction of pixels that would ensure animal movement. In this case, the best 0.7% (a subset of the 1% pixels) was considered a sufficient linkage for this species.

the Linkage Design (which reflected needs of additional species and a minimum width) always encompassed the expert-recommended minimum. If needed, we would have expanded the Linkage Design to accommodate an individual focal species, but our multiple-species approach made it unnecessary to engage in this subjective process (Quinby *et al.* 1999).

We combined the maps of all species to produce a union of least-cost corridors (ULCC) that encompassed the entire least-cost corridor of each species. We decided not to map the ULCC using different tones to indicate the number of species served by different parts of the ULCC, on the grounds that this would not promote our policy of "No species left behind." In most cases, the ULCC formed a single band between the core areas.

LCCA will always produce a least-cost corridor, even if the "best" corridor crosses a freeway, aqueduct, or other obvious barrier to movement of focal species. To address this, any competent practitioner will conduct fieldwork to identify such barriers and recommend appropriate restoration or mitigation. However, we caution practitioners about a more subtle pitfall. A transportation agency or developer may be tempted to use LCCA to simulate the impact of adding a road or a golf course to the heart of a linkage area. It is important to understand that LCCA will almost certainly produce the same map – complete with the road or golf course within the best 1% – because this area will still be more permeable than the adjacent housing tract or reservoir. Because someone would inevitably misinterpret such a result as indicating "no impact on connectivity," it is best to avoid such abuse of LCCA altogether. Put another way, LCCA should not be used to evaluate scenarios about landscape features (such as a particular highway) that occur at a finer scale than the inputs into the analysis (such as road density, which is only crudely related to any particular road).

C: Conduct habitat suitability analysis

A least-cost corridor does not necessarily encompass habitat patches large enough to support viable populations, nor are such patches necessarily within the dispersal distance of the focal species. To evaluate the effectiveness of each ULCC to provide connectivity for all focal species (including orthogonal species and other species for which LCCA was not conducted), we mapped the distribution and size of suitable habitat patches for each focal species. We used suitability scores provided by experts, or extracted from literature review or the California Wildlife-Habitat Relationships database (California Department of Fish and Game 2002) to identify suitable habitat in the planning area. We considered a cluster of pixels large enough to support 50 individuals as a *potential population center*, rounding up to the nearest order of magnitude in hectares (e.g., we rounded 2 ha to 10 ha, and 650 ha to 1000 ha). This rounding avoids belabored inferences from published estimates of home range size or density. Similarly we considered a cluster of suitable pixels large enough to support more than two individuals (again rounded to the nearest order of magnitude) as a *potential habitat patch* if it was within twice the species' mean dispersal distance from a potential population center. We chose twice the mean dispersal distance because estimates of dispersal distance are based on small samples (thus missing extreme events) and are biased low (because researchers lose track of individuals that move beyond the researcher's search radius: Barrowclough 1978). The rare dispersals longer than the known mean can be responsible for significant gene flow or demographic rescue (Brown and Kodric-Brown 1977). Thus using the mean would cause potentially important patches to be considered "useless." When data were lacking, we used the home range size and dispersal distances for other species in the same genus or family from studies in the most similar ecoregion.

Typically, most potential population centers and habitat patches fell within a core or the ULCC; the others were considered *candidates* for addition to the ULCC. We added a candidate population center or habitat patch to the ULCC if that addition (a) decreased the total amount of unsuitable habitat that an individual animal would have to traverse in a journey between core areas, or (b) provided a route with greater dominance of potential population centers (instead of potential habitat patches). If the focal species could fly across urban or agricultural areas, the center or patch was added as a disjunct stepping-stone. For other species, we added pixels of native vegetation (or agricultural land if insufficient native vegetation was present) to connect the area to the ULCC.

D: Impose minimum widths on each ULCC

Portions of some linkages were narrow due to the distribution of urbanized or agricultural lands. We expanded any constriction points along the ULCC to a width of 2 km by adding pixels of natural vegetation, or, when there was insufficient natural vegetation, agricultural land (on which natural vegetation should be restored). We did not add pixels of urbanized land, however, and this often precluded expansion to 2 km. When possible we used additions to increase the diversity of topographic elevation and aspect within the linkage, reasoning that this would increase the utility of the linkage during future climate changes.

There are many reasons why linkages should be wide. (1) Many smaller animals, such as salamanders and lizards, will take dozens of generations to cross between core areas, and thus need enough area to support resident metapopulations over time. (2) For species whose needs are not well represented by our focal species, a wide area will help ensure availability of appropriate habitat or habitat elements (e.g., host plants, pollinators, roosting sites). (3) Contaminants, sediments, and nutrients can reach streams from distances >1 km (Maret and MacCoy 2002; Scott 2002; Naicker *et al.* 2003), and fish, amphibians, and aquatic invertebrates often are more sensitive to land use at the watershed scale than at the scale of narrow riparian buffers (Goforth 2000; Fitzpatrick *et al.* 2001; Stewart *et al.* 2001; Wang *et al.* 2001; Scott 2002; Willson and Dorcas 2003; Pringle Chapter 10). (4) A wide linkage buffers against edge effects (pets, lighting, noise, nest predation, nest parasitism). (5) Fire is a natural disturbance factor in the South Coast Ecoregion, and a wide linkage allows for a semblance of a natural fire regime to operate with minimal constraints from adjacent urban areas. (6) A wide linkage enhances the ability of the biota to respond to climate change. (7) Harrison (1992) suggests that a linkage for a species that needs to live in (as opposed to move through) the corridor should be approximately the square root of half an individual home range area.

Harrison's (1992) reasoning provides an attractive argument for a width of 1 to 2.5 km to accommodate badgers, coyotes, or bobcats (home range sizes reported by Goodrich and Buskirk 1998; Riley *et al.* 2003). However, these species probably could use a narrower linkage 6-24 km long (i.e., the lengths of most of our linkages) that provided a combination of live-in and pass-through habitat. None of these arguments provide rigorous support for 2 km (or any other value) as a minimum width. We chose 2 km as a reasonable width that probably achieves 5 of these 7 goals, although it may be too narrow to allow a fire regime that simulates natural conditions (goal 5) or enable biotic response to climate change (goal 6).

The Linkage Design

For each linkage, we use the term *Linkage Design* for the map depicting the buffered ULCC. In most of our linkages, the Linkage Design was a relatively narrow swath 6-24 km long and 2-3 km wide along most of its length, with occasional constrictions to accommodate existing urban development. But several Linkage Designs encompassed broader areas for part of their length, including large patches that can function as stepping stones or even core areas for even the most area-demanding focal species.

A narrative accompanying the Linkage Design map described the extent to which the Linkage Design serves the needs of each focal species. Although the Linkage Design offers the best chance of facilitating movement of each species, we have to admit that our best may not be good enough for some focal species in some linkages. For example, grasslands have been almost entirely lost to development in several linkage areas, making it difficult to create a corridor for badgers. For the remaining focal species, we hypothesize that, even after urbanization of areas outside the Linkage Design, focal species or their genes would move between core areas in a way that ensures species viability. In non-scientific parlance, this hypothesis can be expressed as "If we build it, they will come." For orthogonal species, we hypothesize that the species would persist within the Linkage Design after urban build-out. We discuss testing these hypotheses in Step 7.

STEP 5:   SPECIFY RESTORATION OPPORTUNITIES AND MANAGEMENT NEEDS

Linkage managers used high-resolution aerial photos and fieldwork to identify restoration opportunities and management needs (e.g., road and aquatic barriers, land-use patterns) for each Linkage Design. The fieldwork was especially valuable. For instance, high-resolution air photos suggested that an oil refinery was blocking a potential linkage, but fieldwork showed the facility to be abandoned and posted for sale. In another case, lush riparian vegetation on the air photo proved to be thickets dominated by the invasive exotics tamarisk (*Tamarix ramosissima*) and giant reed (*Arundo donax*). Biologists walked each aquatic linkage and photographed and measured dams, siphons, and encroachments. Highway edges were photographed, and existing crossing structures measured. Sites where improved road crossings could be constructed were identified. In rural areas, biologists noted the local styles in fencing, outdoor recreation, lighting, livestock husbandry, and pet control. Locations of important features were recorded with global positioning systems (GPS). We provided a narrative and accompanying photos to document these existing conditions.

The narrative also included recommendations regarding land use, domestic livestock, pets, off-road vehicles, artificial night lighting, and recreational activities. As appropriate, we proposed restoration of native vegetation, removal of aquatic barriers, rehabilitation of mined areas, and, most especially, improvement of permeability across major roads. High traffic volumes on Southern California freeways for the last 30–50 years have made these roads into especially formidable barriers. For example, California highways 40–60 years old markedly diminish gene flow among bobcat and coyote populations (Riley *et al.* 2006), produce genetic divergence similar to that produced by 15 km of inappropriate habitat between populations of desert bighorn sheep *Ovis canadensis nelsoni* (Epps *et al.* 2005), and are associated with genetic discontinuities similar to that produced by the rock and ice of the Sierra Crest between puma populations (Ernest *et al.* 2003).

(A)

(B)

Fig. 22.3. (A) At the bottom of the fill slope, 0.6-m diameter pipes (not visible) accommodate the flow of Cherry Canyon, the largest non-urbanized drainage crossing Interstate 5 along the linkage between the eastern and western Sierra Madre. A bridge here would serve many focal species. (B) Several pumas have been killed in vehicle collisions on this portion of Interstate 15, where the freeway crosses the Santa Ana–Palomar linkage. Because the freeway is already cut into bedrock here, an underpass is not feasible, but a vegetated overpass would facilitate movement of most focal species.

Thirteen of the 15 linkages were crossed by freeways up to ten lanes wide. Only two of these 13 freeway segments had crossing structures that facilitate movement of terrestrial species. Our LCCA and habitat suitability analyses deliberately ignored the location and quality of existing freeway crossing structures, none of which had been located or built to facilitate wildlife movement. Because such structures are easier to create, relocate, and improve than native vegetation, topography, and urban areas, we viewed them as landscape elements that should respond to animal movement patterns, rather than vice versa. We caution conservationists undertaking similar efforts not to let locations of existing road-crossing structures channelize their thought processes and skew their recommendations away from biological optima.

Where more than one biological optimum was apparent, we considered existing culverts and crossing structures within the Linkage Design as places where improved structures could be constructed at lower cost (Fig. 22.3A). Anecdotal information (e.g., road-kills, game trails, animal sign) also helped suggest locations for crossing structures. In some locations, we recommended vegetated overpasses (Fig. 22.3B), or converting vehicle underpasses into wildlife underpasses (Fig. 22.4). Where a highway crosses a linkage for several kilometers, we recommended multiple crossing structures spaced as close as 2 km apart (see Clevenger and Wierzchowski (Chapter 20) for discussion of siting and monitoring crossing structures).

We made bold recommendations for maintenance, enhancement, and construction of wildlife crossing structures, but in discussions with transportation agencies, we did not ask for immediate construction of major improvements. Instead we emphasized the opportunity for the agency to implement meaningful mitigation measures when they next add lanes or otherwise upgrade these freeway segments. Although improvements may not occur for a decade or more, we hope that once connectivity is restored, genomes of all affected species will rapidly recover.

STEP 6:   PARCEL-LEVEL MAPS AND IMPLEMENTATION

Throughout our reports and meetings, we have emphasized the importance of connecting two core areas for the sake of biodiversity in all its dimensions. We have to remind even our most sympathetic friends that this is not just an effort "to get the puma across the road." Although roads emerged as the most important potential barrier in every linkage, the best-designed crossing structures only make sense if they are appropriately sited, and if the wildlands between the road and each core area are conserved. Although pumas are an important focal species, whatever linkages we conserve over the next decade will provide all the connectivity any species will enjoy for the next century or more.

Fig. 22.4. An interchange on the Riverside Freeway (SR 91) being converted into a wildlife crossing in February 2003, to facilitate movement along the linkage between the Santa Ana Mountains and the Chino Hills. Although this is not one of our 15 priority linkages, this illustrates the feasibility of the enhancements that we will recommend in some linkage areas. California State Parks is investing $1.5 million to restore natural vegetation and the Coal Canyon stream channel through the underpass.

To promote this broad view, our written reports described the likely biodiversity consequences of losing the linkage and the conservation investments in the core areas that would lose or gain ecological value due to success or failure of this project. The value of state and regional parks, National Forest land, and private reserves in these areas reaches in the billions of dollars, and a relatively modest investment in connective habitats can help ensure their continued value. We also described how linkage protection would advance other conservation efforts in the area.

Although conservation decisions, such as purchases of easements or land, or changes in zoning, will be made at the parcel level, our printed reports offered no recommendations more site-specific than the Linkage Design map and descriptions of improved highway crossing structures. We made a strategic decision to exclude from our published reports any parcel maps and any data on size, value, zoning, or ownership of parcels.

We believe that publishing such data could be counter-productive because media, developers, landowners, and others are likely to focus on the parcel map. Arguments about individual parcels would distract from the scientific and conservation message embodied in the Linkage Design.

Our partners are currently in the process of translating the Linkage Design into priority parcels for conservation action. Partners select priority parcels, and discuss appropriate conservation measures at small workshops at which politically sensitive discussions can take place. For instance, partners can discuss the biological and economic trade-offs of omitting specific parcels from the conservation plan, or of allowing trophy home development on a few key parcels in the Linkage Design, or whether easements, purchase, or zoning would be the most appropriate tool for conserving the linkage value of a particular parcel. These compromises are sometimes disconcerting, but we recognize that decisions to make conservation investments lie with the investors rather than the scientists, and that conservation delivery involves an expanded set of skills compared to conservation design (see Morrison and Reynolds Chapter 21).

The role of science, and of South Coast Wildlands, will not terminate with the release of the 15 Linkage Conservation Plans for the South Coast Ecoregion. We envision a series of implementation meetings at which partners will interactively build scenarios using South Coast Wildlands' biological expertise, photodocumented descriptions of potential barriers, and GIS layers (including 1-m resolution air photos, parcels, zoning, and administrative boundaries). Immediate feedback from scientists on the likely biological consequences of various decisions will help the conservation community make scientifically sound decisions.

Ongoing conservation activities with linkage managers have provided opportunities for enhancing and supporting linkage conservation. For example, the South Coast Conservation Forum, a coalition of county, state, and federal agencies, universities, and NGOs, was recently formed to advise the Department of Defense on reducing urban encroachment and conflicts with military training maneuvers on Marine Corps Base Camp Pendleton. On the basis of information we provided, Department of Defense recognized the linkage as an important mitigator of long-term impacts to sensitive species in this planning area. This effort may effectively protect the western third of the Santa Ana–Palomar Mountains linkage, one of the 15 priority linkages. Similarly, South Coast Wildlands collaborated with other conservation groups to suggest reconfigurations of the proposed reserve system for the Western Riverside County Multiple Species Habitat Conservation Plan. That plan offered better species protection at less cost. Riverside County has incorporated some of our recommendations, which may help secure two of our linkages.

Public outreach is also an important part of implementation. Our interim products are of interest and utility not only to partners, but also to citizens, media, and conservation educators. These materials include maps of conservation designs, biological attributes, and restoration opportunities. We make these available as rapidly and as widely as possible through our website and on CD-ROM. We have also prepared two types of visual journey through each linkage: (1) a flyover animation consisting of color aerial photographs draped over a digital elevation map, and (2) an interactive US Geographical Survey 1:24 000 topographic map of the Linkage Design hyperlinked to digital photographs taken from the ground to simulate a walk through the linkage.

STEP 7: DESIGN AND IMPLEMENT A MONITORING PROGRAM

As described in Step 4, each Linkage Design map, with accompanying recommendations for management and restoration, embodies one or more testable hypotheses regarding focal species. To advance the science of linkage planning, we intend to design monitoring programs that address these hypotheses. Design of a monitoring program must address several related challenges, including formulating testable predictions, securing long-term funding, implementing improvements (e.g., a new crossing structure, restoring vegetation), and collecting data.

Deriving testable predictions from the vague hypothesis that "the Linkage Design benefits focal species" first requires selecting an appropriate dependent variable, such as numbers of linkage passages by individual animals, or demographic or genetic traits of the populations in the core areas (Beier and Loe 1992; Beier and Noss 1998). Movement studies should attempt to confirm whether movements between core areas occur often enough to influence population viability, and that in a landscape without linkages such movements would occur too rarely to benefit the population. Beier and Noss (1998) recommended a Before–After–Control–Impact–Pairs design to maximize strength of inference from these minimally replicated landscape experiments. Two types of control sites are feasible. For example if we are restoring a linkage between two core areas that are apparently isolated from each other, the control site could be either two well-connected core areas or another pair of disjunct cores for which no restoration is planned. We believe the strongest inferences would flow from having both types of controls, but finding suitably matched sites in this rapidly changing landscape will not be easy.

Peculiar funding issues arise because pre- and post-treatment data may need to be collected over the course of many years. Any ecologist who has undertaken studies on vertebrate response to forest treatments can attest to the agony of collecting pre-treatment data and waiting years for well-intentioned management agencies to start and complete treatments. Research on linkage function will be a similar waiting game. Recruiting researchers to conduct independent research on plant and animal populations in linkages with an eye for repeating these studies in 10 to 20 years may be one solution. Finally, interpreting the results of a monitoring program will be complicated by inevitable differences (due to compromise and errors) between planned treatments and treatments as implemented. We do not view any of these problems as insurmountable, and we hope to design a monitoring framework that is rigorous, robust to these difficulties, and relevant to implementing biotic linkages in a real landscape.

Although the long-term (decades to centuries) effectiveness of each linkage is the most important response variable for adaptive management, we also recommend research to document indicators of short-term (months to years) success of each linkage. In most cases, this will involve documenting animal use via camera traps, tracks, scats, trapping, or other surveys. For instance, if adjacent habitat is suitable, a new highway-crossing structure should start to be used by focal species within 1–3 years after construction (N. Dodd, Arizona Game and Fish Department, unpublished data from SR-260 study). Failure to observe such use would indicate that either the design of the structure, or that some other element in the linkage, is defective. Such information should promptly inform improvements in other linkages.

CONCLUSIONS

The remaining large wildlands of the South Coast Ecoregion form an archipelago of natural open space within one of the world's largest metropolitan areas. Until the recent dramatic surge in human domination of this landscape, these wild areas formed one ecological system. We envision a future interconnected system of natural space, and we offer our approach as a biologically defensible and repeatable procedure to design conservation linkages.

Hallmarks of the South Coast Missing Linkages project have been the development of rigorous quantitative methods to prescribe linkage conservation needs and the highly collaborative nature of the planning effort. This approach (1) spans jurisdictional boundaries and promotes the partnerships needed to implement landscape connectivity at this scale, (2) garners greater visibility from agencies and focuses disparate conservation efforts on a coordinated regional plan that appeals to the public and to the agencies, (3) increases the effectiveness of partners working at local scales, (4) increases rigor and objectivity and provides products that are defensible in touchy political and social arenas, and (5) enhances communication by providing beautiful and easily comprehensible graphic outputs for agencies and the public.

We trust that our approach will be copied, tweaked, and improved by others. Arizona has initiated a similar effort with two promising innovations. First, the Federal Highway Administration and Arizona Department of Transportation were involved as lead agencies from the start. Because these agencies are such a critical part of implementing any solution, having them involved in a meaningful way (developing the agenda and providing web-hosting and GIS support for the initial workshop) augurs well for the Arizona effort. Second, at Arizona's initial state-wide workshop, participants were asked to provide the data needed for prioritization, as well as list focal species for each linkage. Obtaining data on biological importance has enabled Arizona to prioritize linkages more rapidly. Although the lists of focal species obtained at the initial workshop are less comprehensive than those developed in California, Arizona may be able to select additional focal species for the priority linkages more quickly, perhaps by an e-mail appeal to knowledgeable persons, followed by dialogs between experts (either one at a time or via conference call) and a highly skilled staff person. Thus Arizona consolidated Steps 2 and 3, and jump-started Step 4.

Arizona's effort is led by a coalition of agencies (chiefly Arizona Game and Fish Department, The Arizona Department Of Transportation, the Federal Highway Administration, and US Forest Service) rather than an NGO. This has advantages (prominent roles for and buy-in from the transportation agencies, more financial stability than a tiny NGO), but we do see a risk in having no analog to South Coast Wildlands. Not one person in Arizona goes to work each day with the sole goal of advancing connectivity in the state. Although the commitment of each transportation agency has been genuine and impressive, will it be sustained as political administrations change, or when key players must pay attention to other priorities? An excellent step toward minimizing this risk was taken in the 2005 reauthorization of the US Transportation Efficiency Act. The law requires the Federal Highway administration and state transportation agencies using federal dollars to consult with state wildlife agencies at the initial stages of project planning. It also permits use of federal dollars to pay the salary of a state liaison. This could ensure that each state would have a staff person in their conservation agency whose primary job is to be engaged in consulting with transportation agencies.

We cannot overemphasize the importance of investing in building and maintaining relationships. Development of technical plans to overcome barriers to animal movement must be matched by efforts to build and maintain linkages among all the players. We advise similar efforts to budget ample time to engage partners, especially including extra time and effort for relationships that span sovereign boundaries. It is not sufficient to e-mail invitations to Mexican and Native American tribal agencies. International travel can be difficult and relatively expensive. Tribal sovereignty and ways of doing business must be respected.

Our effort has received considerable publicity, virtually all of it positive. We believe media exposure has been helpful and urge other efforts to use public-relation specialists in partner agencies to generate and sustain positive publicity. Participants in a workshop get positive reinforcement when they see a news story on the event or on the release of the workshop report several months later. Agencies (such as a transportation agency) gain confidence about moving in a new, "greener" direction when they are publicly praised for their action; it is especially useful if high-ranking officials are featured in press releases. Reporters tend to be sympathetic filters, especially in the early stage of identifying pairs of core areas in need of connectivity (with connective areas only vaguely defined). Developing rapport with reporters, partners, and the public at this time can help set a positive tone for later stories about specific implementation measures.

Although insects, plants, and birds need connectivity, the large four-legged furred creatures will probably be the first to suffer when connectivity is lost, and they are often the best flagships to "sell" a linkage design. We urge practitioners to emphasize the needs of flagships (including reptiles and other non-mammals) to garner public support.

Large mammals also tend to lend themselves to least-cost corridor analysis (LCCA). Because LCCA produces crisp and persuasive GIS outputs, it is tempting to use LCCA for all focal species. However, we advise careful matching of analytical tools to the species' natural history and the data available for each species. For example, although LCCA is appropriate for some highly sedentary birds analyzed on a coarse landscape, for most birds a pixel-to-pixel permeability analysis would not pass the "laugh test" for either scientific or lay audiences, both of which know that most birds can fly over dozens of pixels of inappropriate habitat. We offer patch size and configuration analysis (Step 4C) as a way to meaningfully consider the needs of diverse species, including those for which LCCA is not appropriate.

Finally, in an ecoregion less urbanized than the South Coast of California, we advise that the Linkage Design (Step 4) should rarely be a narrow hard-line corridor. Simberloff *et al.* (1992) suggested that connectivity could best be obtained by managing "the entire landscape ... as a matrix supporting the entire biotic community." Although massive urbanization in our landscape precluded this option in many of our linkages, we did pursue this option in those portions of those linkages

Fig. 22.5. The confluence of four highways, a railroad line, high-voltage power lines, and microwave communication towers, as seen from the edge of the California Aqueduct, which moves water 440 km from the Sacramento River delta into the Los Angeles Basin. Our project intends to add one more layer of infrastructure to this scene by protecting and restoring the ridge in the background, which provides the only wildland link between the Santa Susana Mountains (off left edge of the photo) and the San Gabriel Mountains (off right edge of photo).

where it was feasible. We envy those who have the luxury of managing broad swaths for permeability throughout their ecoregion.

The USA's largest metropolitan area has a human infrastructure without equal on the planet. People, water, information, electric power, gas, automobiles, and trains move across this landscape with remarkable efficiency (Fig. 22.5). Our goal is to create a "green infrastructure" that is commensurate with these other types of infrastructure. We pray that the quality of our effort befits this global hotspot of biodiversity.

ACKNOWLEDGEMENTS

We acknowledge all of our partners in the South Coast Missing Linkages project for working to develop meaningful outputs that are useful products in conservation planning. We thank workshop participants and other experts for their time in providing valuable information on focal species. This effort would not have been possible without funding from The Wildlands Conservancy, The State Resources Agency, US Forest Service, California State Parks Foundation, Environment Now, and in-kind support from Zoological Society of San Diego, The Nature Conservancy, San Diego State University Field Station Programs, and Conservation Biology Institute. We thank Sedra Shapiro, Executive Director of the San Deigo State University Field Station Programs, for her work in developing the format for the workshop-based planning approach and application of planning results.

We give special thanks to the steering committee for the South Coast Missing Linkages Project: Madelyn Glickfeld (The Resources Agency California Legacy Project), Gail Presley (California Department of Fish and Game), Therese O'Rourke (US Fish and Wildlife Service), Rick Rayburn (California State Parks), Ray Sauvajot (National Park Service), and Tom White (US Forest Service). Finally, we recognize our partners on the South Coast Missing Linkages Project: The Wildlands Conservancy, The Resources Agency, US Forest Service, California State Parks, California State Parks Foundation, National Park Service, San Diego State University Field Stations Program, Environment Now, The Nature Conservancy, Conservation Biology Institute, Santa Monica Mountains Conservancy, California Wilderness Coalition, Wildlands Project, Zoological Society of San Diego Center for Reproduction of Endangered Species, Pronatura, Conabio, and Universidad Autonoma de Baja California.

Data sheets used in workshops and spreadsheets used in prioritizing linkages in California and Arizona are available from the authors on request.

REFERENCES

Bani, L., M. Baietto, L. Bottoni, and R. Massa. 2002. The use of focal species in designing a habitat network for a lowland area of Lombardy, Italy. *Conservation Biology* 16:826–831.

Barrowclough, G. F. 1978. Sampling bias in dispersal studies based on finite area. *Journal of Field Ornithology* 4:333–341.

Beier, P. 1993. Determining minimum habitat areas and corridors for cougars. *Conservation Biology* 7:94–108.

Beier, P., and S. Loe. 1992. A checklist for evaluating impacts to wildlife movement corridors. *Wildlife Society Bulletin* 20:434–440.

Beier, P., and R. F. Noss. 1998. Do habitat corridors provide connectivity? *Conservation Biology* 12:1241–1252.

Brown, J. H., and A. Kodric-Brown. 1977. Turnover rates in insular biogeography: effect of immigration on extinction. *Ecology* 58:445–449.

California Department of Fish and Game. 2002. *CWHR v 8.0 Personal Computer Program.* Sacramento, CA: California Interagency Wildlife Task Group.

Carroll, C., R. F. Noss, and P. C. Paquet. 2001. Carnivores as focal species for conservation planning in the Rocky Mountain region. *Ecological Applications* 11:961–980.

Clevenger, A. P., J. Wierzchowski, B. Chruszcz, and K. Gunson. 2002. GIS-directed, expert-based models for identifying wildlife habitat linkages and planning mitigation passages. *Conservation Biology* 16:503–514.

Coppolillo, P., H. Gomez, F. Maisels, and R. Wallace. 2004. Selection criteria for suites of landscape species as a basis for site-based conservation. *Biological Conservation* 115:419–430.

Cowling, R. M., R. L. Pressey, R. Sims-Castley, *et al.* 2003. The expert or the algorithm? – comparison of priority conservation areas in the Cape Floristic Region identified by park managers and reserve selection software. *Biological Conservation* 112:147–167.

Crooks, K. 2002. Relative sensitivities of mammalian carnivores to habitat fragmentation. *Conservation Biology* 16:488–502.

Epps, C. W., P. J. Palsbøll, J. D. Wehausen, *et al.* 2005. Highways block gene flow and cause a rapid decline in genetic diversity of desert bighorn sheep. *Ecology Letters* 8:1029–1038.

Ernest, H. B., W. M. Boyce, V. C. Bleich, *et al.* 2003. Genetic structure of mountain lion (*Puma concolor*) populations in California. *Conservation Genetics* 4:353–366.

Fitzpatrick, F. A., B. C. Scudder, B. N. Lenz, and D. J. Sullivan. 2001. Effects of multi-scale environmental characteristics on agricultural stream biota in eastern Wisconsin. *Journal of the American Water Resources Association* 37:1489–1508.

Goforth, R. R. 2000. Local and landscape-scale relations between stream communities, stream habitat and terrestrial land cover properties. *Dissertation Abstracts International Part B: Science and Engineering* **8**:3682.

Goodrich, J. M., and S. W. Buskirk. 1998. Spacing and ecology of North American badgers (*Taxidea taxus*) in a prairie-dog (*Cynomys leucurus*) complex. *Journal of Mammalogy* **79**:171–179.

Harrison, R. L. 1992. Toward a theory of inter-refuge corridor design. *Conservation Biology* **6**:292–295.

Hunter, R. 1999. *South Coast Regional Report: California Wildlands Project Vision for Wild California*. Davis, CA: California Wilderness Coalition.

Lambeck, R. J. 1997. Focal species: a multi-species umbrella for nature conservation. *Conservation Biology* **11**:849–856.

Maret, T. R., and D. E. MacCoy. 2002. Fish assemblages and environmental variables associated with hard-rock mining in the Coeur d'Alene River Basin, Idaho. *Transactions of the American Fisheries Society* **131**:865–884.

Miller, B., R. Reading, J. Strittholt, *et al.* 1999. Using focal species in the design of nature reserve networks. *Wild Earth* **8**:81–92.

Naicker, K., E. Cukrowska, and T. S. McCarthy. 2003. Acid mine drainage arising from gold mining activity in Johannesburg, South Africa and environs. *Environmental Pollution* **122**:29–40.

Noss, R. F., C. Carroll, K. Vance-Borland, and G. Wuerthner. 2002. A multicriteria assessment of the irreplaceability and vulnerability of sites in the Greater Yellowstone Ecosystem. *Conservation Biology* **16**:895–908.

Penrod, K. L., R. Hunter, and M. Merrifield. 2001. *Missing Linkages: Restoring Connectivity to the California Landscape*. California Wilderness Coalition, The Nature Conservancy, US Geological Survey, Center for Reproduction of Endangered Species, and California State Parks. San Diego, CA: San Diego Zoo.

Polak, D. 2001. *Natural Community Conservation Planning (NCCP): The Origins of an Ambitious Experiment to Protect Ecosystems*. Sacramento, CA: California Research Bureau, California State Library.

Pressey, R. L., and K. H. Taffs. 2001. Scheduling conservation action in production landscapes: priority areas in western New South Wales defined by irreplaceability and vulnerability to vegetation loss. *Biological Conservation* **100**:355–376.

Pressey, R. L., I. R. Johnson, and P. D. Wilson. 1994. Shades of irreplaceability: towards a measure of the contribution of sites to a reservation goal. *Biodiversity and Conservation* **3**:242–262.

Quinby, P., S. Trombulak, T. Lee, *et al.* 1999. *Opportunities for Wildlife Habitat Connectivity between Algonquin Provincial Park and the Adirondack Park*. Powassan, Ontario, Canada: Ancient Forest Exploration and Research.

Riley, S. P. D., R. M. Sauvajot, T. K. Fuller, *et al.* 2003. Effects of urbanization and habitat fragmentation on bobcats and coyotes in southern California. *Conservation Biology* **17**:566–576.

Riley, S. P. D., J. P. Pollinger, R. M. Sauvajot, *et al.* 2006. A southern California freeway is a physical and social barrier to gene flow in carnivores. *Molecular Ecology* **15**:1733–1741.

Scott, M. C. 2002. Integrating the stream and its valley: land use change, aquatic habitat, and fish assemblages (North Carolina). *Dissertation Abstracts International Part B: Science and Engineering* **63**:51.

Simberloff, D., J. A. Farr, J. Cox, and D. W. Mehlman. 1992. Movement corridors: conservation bargains or poor investments? *Conservation Biology* **6**:493–504.

Singleton, P. H., W. L. Gaines, and J. F. Lehmkuhl. 2002. Landscape permeability for large carnivores in Washington: a geographic information system weighted-distance and least-cost corridor assessment. Research Paper PNW-RP-549. Portland, OR: US Department of Agriculture Forest Service.

Stewart, J. S., L. Wang, J. Lyons, J. A. Horwatich, and R. Bannerman. 2001. Influences of watershed, riparian-corridor, and reach-scale characteristics on aquatic biota in agricultural watersheds. *Journal of the American Water Resources Association* **37**:1475–1488.

Terborgh, J., J. A. Estes, P. Paquet, *et al.* 1999. The role of top carnivores in regulating terrestrial ecosystems. Pp. 39–64 in M. E. Soulé, and J. Terborgh (eds.) *Continental Conservation*. Covelo, CA: Island Press.

Wang, L., J. Lyons, P. Kanehl, and R. Bannerman. 2001. Impacts of urbanization on stream habitat and fish across multiple spatial scales. *Environmental Management* **28**:255–266.

Weiss, A. D. 2000. A GIS algorithm for topographic position index. Poster presented at ESRI Users' Conference, San Diego, CA. Available online at aweiss@induscorp.com

Willson, J. D., and M. E. Dorcas. 2003. Effects of habitat disturbance on stream salamanders: implications for buffer zones and watershed management. *Conservation Biology* **17**:763–771.

23

Incorporating connectivity into broad-scale conservation planning

REED F. NOSS AND KATHLEEN M. DALY

INTRODUCTION

Conservation planning, as usually practiced, involves two fundamental tasks: (1) selection of sites that will collectively meet a set of (generally static) conservation goals, and (2) designing a network of sites that has a high probability of maintaining dynamic biodiversity and natural processes over time (Noss and Cooperrider 1994; Scott and Csuti 1997; Margules and Pressey 2000; Groves 2003). The second task addresses the issue of persistence, i.e., the long-term viability of populations and ecological functions. Connectivity is important in this context because, for example, although no single reserve might maintain a viable population of a particular species, a well-connected network of reserves might contain a viable population or metapopulation. Hence, a connected system of reserves can potentially be a whole greater than the sum of its parts (Noss and Harris 1986).

The computerized site-selection methods (algorithms) that have become the tools of the trade of conservation planning were not designed to address viability and, by themselves, poorly address connectivity and other issues of reserve design for persistence of features (Briers 2002; Cabeza 2003). In many cases the suite of sites selected by an algorithm to efficiently meet representation targets for species or ecosystems will fail to meet the needs of wide-ranging species, permit the continued operation

Connectivity Conservation eds. Kevin R. Crooks and M. Sanjayan. Published by Cambridge University Press. © Cambridge University Press 2006.

of natural processes, or allow faunas and floras to adjust to climate change. Nevertheless, some algorithms (e.g., Andelman *et al.* 1999) can be programmed to cluster rather than disperse sites across the landscape, and population viability analyses of focal species can be paired with selection algorithms to improve their ability to address persistence issues (Carroll *et al.* 2003a; Carroll Chapter 15). New methods to address habitat quality and spatial configuration directly during the site-selection process show promise (Cabeza 2003). In addition, the incorporation of spatial surrogates of ecological and evolutionary processes into the suite of features considered by selection algorithms may enhance their ability to meet long-term goals (Pressey *et al.* 2003).

Connectivity, despite its controversial history in academic circles (Noss 1987a; Simberloff and Cox 1987; Hobbs 1992; Simberloff *et al.* 1992; Rosenberg *et al.* 1997; Beier and Noss 1998; Haddad *et al.* 2000; Noss and Beier 2000; Crooks and Sanjayan Chapter 1), is well accepted among conservation planners as a key consideration in the design of reserve networks. Corridors (variably called landscape linkages, connectors, greenways, etc.) are the most popular means to achieve connectivity. The empirical literature on this topic, though still sparse, is growing rapidly and generally supports the notion that well-designed corridors function to provide demographic connectivity between populations (Beier and Noss 1998; Haddad and Tewksbury Chapter 16). There is little agreement, however, on what methods should be used to identify or design corridors. Methods have ranged from simply drawing a line or swath between two areas of conservation interest to relatively sophisticated modeling and empirical approaches.

Our goals for this chapter are to review and evaluate (1) ways of identifying and designing corridors, (2) ways of providing connectivity other than corridors, and (3) ways for improving the methods used for connectivity planning. Our focus is on terrestrial reserve network designs. Aquatic and hydrologic connectivity, and the methods used to incorporate connectivity in marine reserve design, are discussed in DiBacco *et al.* (Chapter 8), Harrison and Bjorndal (Chapter 9), Pringle (Chapter 10), and Neville *et al.* (Chapter 13). We use the term "corridor" interchangeably with linkages and related terms and in the broad sense of a landscape feature that may provide functional connectivity. We recognize that corridors can be identified at many spatial and temporal scales (Noss 1991; Dobson *et al.* 1999), but we concentrate on a broad spatial scale, ranging from regions millions of hectares in size to subcontinental areas spanning multiple regions. These are the scales on which metapopulation viability of wide-ranging species, adjustment of multiple taxa to climate change, and many ecological and evolutionary processes must be examined (see Soulé *et al.* Chapter 25). We conclude that multifaceted methods for corridor design are often more defensible than methods based on limited criteria and analyses, and that increasing use of quantitative habitat and population modeling, combined with extensive field research, will make corridor identification and design more reliable.

WAYS OF IDENTIFYING AND DESIGNING BROAD-SCALE CORRIDORS

Scientists, planners, and conservationists have applied a wide variety of methods to identify and design corridors. The variation in methodology can be traced both to technical issues (e.g., whether geographic information systems and associated modeling tools were available) and to the functions of connectivity of interest in particular cases. Of the many potential functions of connectivity (Noss 1993a), scientists and especially conservation planners have paid most attention to providing for daily and seasonal movements of animals and facilitating dispersal, gene flow, and rescue effects. Much less attention has been given to designing corridors that will allow for range shifts of species or maintaining flows of ecological processes.

The recent emphasis on ecosystem-level conservation (Franklin 1993; Noss *et al.* 1995) notwithstanding, when it comes to determining the necessary configuration of reserve networks, there is no substitute for considering the life-history requirements of particular species. Connectivity is determined by the intersection of an organism's life history and the structure of the landscape (Taylor *et al.* Chapter 2). Hence, it is a very species-specific and landscape-specific property. We have found that large carnivores and other wide-ranging, area-sensitive mammals are often appropriate focal species for addressing connectivity and other issues of landscape configuration on broad spatial scales (e.g., Tracey Chapter 14, Carroll Chapter 15). For wide-ranging mammals, connectivity is mainly an issue of circumventing barriers to movement (e.g., highways, developed areas) and minimizing human-caused mortality (e.g., hunting, trapping, vehicle collisions) (Bennett 1991, 1999; Beier 1993, 1995).

The approaches to corridor design reviewed in this chapter are not mutually exclusive. A linkage designed on the basis of several approaches and multiple criteria usually will be more defensible than one based on a single approach. Comprehensive planning for connectivity should involve several spatial scales, for example the three scales of region, landscape, and local project (Ruediger 2001). First, an examination of regional geography, vegetation, species distributions, and land-use patterns will help determine potential linkages between major wild areas at larger spatial scales. Habitat and population models can be applied to address issues of metapopulation persistence for wide-ranging species (Paquet and Hackman 1995; Noss *et al.* 1996; Carroll *et al.* 2001, 2002, 2003a, 2003b; Carroll Chapter 15). Corridors recognized on this scale can then be refined by intra-regional or landscape analyses that incorporate more detailed data on land use, habitat, and locations of road-kills (Smith 1999; Clevenger *et al.* 2003) to focus attention on population persistence, seasonal movements, habitat fragmentation, and ecological processes such as disturbance and hydrology. Then, project-level decisions can be informed by small-scale studies that examine local habitat use, movements by individual animals, and particular movement barriers or highway crossings (Clevenger and Waltho 2000; Craighead *et al.* 2002; Clevenger and Wierzchowski Chapter 20).

Looking at the literature on conservation planning, much of which is unpublished, we recognize three basic approaches to the design of broad-scale linkages: (1) intuitive or "seat-of-the-pants" approaches; (2) empirical approaches; and (3) modeling approaches, as well as many combined approaches.

Seat-of-the-pants approaches

By "seat-of-the-pants" we mean approaches that are intuitive, opportunistic, or otherwise based on subjective best-guesses, existing knowledge, or expert opinion. Such approaches are not necessarily inferior to more technical approaches to identifying corridors. Sometimes a simple but well-informed guess of an appropriate corridor may approximate the results of sophisticated, time-consuming (and expensive) analyses. The most obvious corridor to the human eye may be the route that is most apparent to wildlife. For example, radiotelemetry studies of mountain lions (*Puma concolor*) show that dispersing lions use travel routes that are situated along clearly recognizable, natural corridors in the landscape (Beier 1995). However, in other cases human perceptions of connectivity are arguably quite different from how animals view the landscape. While humans might focus on the visible connectivity of habitats, for example as evident from aerial photographs, animals might cue in on scent trails or other sensory inputs (Lidicker 1999). Considering that areas identified as linkages can become the focus of expensive land acquisitions or engineering solutions, it is crucial that planners identify linkages that serve the goal of functional connectivity, e.g., by enhancing population viability of target species in the habitat patches that are connected (Beier and Noss 1998; Carroll Chapter 15). If corridors are created or protected in the wrong places and prove ineffective, populations could become locally extinct, funds will have been wasted, and conservationists could lose credibility. Nevertheless, data are always limited and uncertainty is usually high in conservation decision-making. Hence, best-guesses by biologists and other knowledgeable individuals should not be discounted in the absence of more definitive information.

Shortest or most direct routes

In conceptual drafts of reserve networks, planners often identify the shortest and most direct or "logical" routes between core areas as corridors. For example, this simple approach was used in early reserve designs for Florida (Noss and Harris 1986; Noss 1987b; Harris and Gallagher 1989), the Oregon Coast Range (Noss 1993b), and the Rocky Mountains (Alliance for the Northern Rockies (undated), as cited in Noss and Cooperrider 1994), albeit taking in consideration landscape features such as riparian networks and undeveloped areas with low road density. Other examples of corridors identified by the apparently most direct routes are the MesoAmerican Biological Corridor (originally called Paseo Pantera: Marynowski 1992), early regional network designs associated with the Wildlands Project, and the conceptual design of the Conception Coast Project in California (E.Inlander, pers. comm.). With subsequent analyses, initial corridor proposals can be refined. We suspect that best-guesses are most likely to be accurate in landscapes with little natural habitat remaining and where options for connections between core areas are limited.

Only remaining routes

The proactive planning of corridors before regional fragmentation takes place is more effective than trying to restore connectivity to a fragmented landscape. In many areas throughout the world, however, corridors of submarginal habitat are all that remain (Kubes 1996). For example, in Southern California the connection between Chino Hills State Park and Cleveland National Forest is bisected by Highway 91; a narrow culvert underneath was used by some, but not all, individual mountain lions (Beier 1995) and other mammals. This "choke-point" was being additionally degraded by private development along the access roads by the freeway interchange. A coalition of conservationists was successful in persuading the state of California to acquire the private land, resulting in the protection of the Coal Canyon biological corridor, the closing of a freeway exit ramp, and the ongoing restoration of the underpass area to natural vegetation (see Beier *et al.* Chapter 22). In some regions, such as Florida, riparian areas (which regularly flood) represent virtually the last remaining natural habitats, as upland areas have been appropriated for agricultural and urban uses.

As another example, after a lengthy court battle conservationists in Ontario, Canada succeeded in persuading the province to create a network of "natural core areas," "natural linkage areas," and "countryside areas" along the Oak Ridges Moraine, just north of the Greater Toronto Metropolitan Area. Linkages were based on strips of remaining natural and semi-natural habitat and were mandated to be at least 2 km wide, except where precluded by existing development. Although linkages were intended to incorporate features that "support movement of plants and animals between Natural Core Areas and along river valleys and stream corridors" (Ontario Ministry of Municipal Affairs and Housing 2001), no particular species were named in the plan, nor were any data on animal movement along the moraine explicitly considered.

In such highly modified landscapes, corridors may contain disturbed vegetation with high densities of exotic species. Nevertheless, native plants and animals may still be able to use these corridors for dispersal or resident habitat. Although connectivity is not their intended or primary purpose, windbreaks, fencerows, drainage ditches, riparian strips, or road rights-of-way (verges) may serve as corridors for some species (Bennett 1990; Lorenz and Barrett 1990; Merriam and Lanoue 1990; Wegner and Merriam 1990; Noss 1993a; Johnson 1999). Nevertheless, in some cases these de facto corridors (especially roadsides) may be population sinks (Trombulak and Frissell 2000).

Routes incorporating sites of conservation interest

Conservation planners often link sites of intrinsic conservation interest into connected networks. Also, corridors sometimes result from regulations that require protection of certain features, such riparian vegetation (i.e., to maintain water quality) completely separate from their value as linkages (Kubes 1996). Rivers and streams often guide animal movement (Harris 1984; Noss 1993a; Walker and Craighead 1997). It is a reasonable assumption that intelligent animals will follow a path of least resistance through the landscape, and riparian areas or adjacent slopes often provide for energetically efficient movement. Sufficiently wide riparian corridors encompass a gradient of ecosystems that may facilitate movement of numerous terrestrial vertebrate species (Harris *et al.* 1996). Some species do not inhabit or travel along riparian corridors, thus upland corridors and cross-gradient corridors also should be protected to meet the needs of a wider range of taxa (Noss 1993a). Encompassing complete environmental gradients in corridors will often require substantial restoration.

The Lower Rio Grande Valley Wildlife Corridor is an ambitious effort by local landowners, US Fish and Wildlife Service, Texas Parks and Wildlife Department, National Audubon Society, and the Nature Conservancy of Texas to preserve and restore a continuous corridor of natural habitat within southern Texas, along the last 275 river miles (440 km) of the Rio Grande (Buckler *et al.* 2002). Located at the convergence of the Mississippi and the Central flyways, the refuge and surrounding areas are used by many migratory birds. The corridor now contains 33 000 ha acquired from willing sellers and protects more threatened and endangered species and species of concern than any other refuge or national park in North America (Chapman 2002). The refuge was created by linking a few remaining islands of natural habitat together into a near-continuous swath, but not specifically to provide for wildlife movement or other functions of connectivity.

Routes based on expert knowledge of focal species

In many cases linkage designs concentrate on the needs of one or more focal species (Lambeck 1997). The focal species approach requires reasonably comprehensive information on habitat use, behavior, dispersal ability, and other aspects of a species' natural history. As considered in this section, the focal species approach can be applied to connectivity planning on the basis of existing knowledge and without new research, albeit this method is usually inferior to the empirical and modeling approaches considered later in this chapter.

In the Rocky Mountains and other regions with relatively large remaining core areas of wild habitat, landscape linkages and proposed protected areas are often defined by the needs of large carnivores (Noss *et al.* 1996; Carroll Chapter 15), although it should not be assumed that these species will serve as effective umbrellas for all taxa or under all conditions. In much of the world, including most of the USA, large carnivores have been extirpated or are present in severely reduced populations. Planners may still choose to base conservation plans on the needs of these species, in order to assess their potential for recovery and to identify critical linkages before they are lost (Noss and Harris 1986; Noss 1987b; Merrill *et al.* 1999; Carroll *et al.* 2003b; Foreman *et al.* 2003). In other cases, network designs have been based on species that are still present in the landscape, though not as area-demanding as large carnivores (e.g., Noss *et al.* 1999).

Large ungulates are also good candidates for regional-scale focal species. Some linkages designed for large carnivores may also benefit ungulates, but in other cases the movement patterns of migratory ungulates differ from those of carnivores and must be considered directly (Berger 2004). For many ungulates, such as elk (*Cervus elaphus*), pronghorn antelope (*Antilocapra americana*), and bighorn sheep (*Ovis canadensis*), the greatest need for corridors occurs during annual migrations between summer and winter ranges. For example, elk often make seasonal movements, especially in mountainous regions (Adams 1982), and studies of elk migration in the Greater Yellowstone Ecosystem now span more than five decades (Smith and Robbins 1994; Berger 2004). A compelling example of connectivity planning based on the needs of a migratory ungulate is a recent analysis of long-distance migration of North America's only extant endemic ungulate, the pronghorn. This study (Berger 2004) was analytic rather than based solely on existing knowledge, however, so it is considered in the "Empirical and modeling approaches" section.

Many species require connectivity to several different habitat types or resources used in various years or times of year (Forman 1995). Species with complex social structures or those that rely on dispersed food resources are likely to fare poorly in linear strips of habitat (Lindenmayer and Nix 1992). Large mammals, amphibians, and many other species need access to water. For example, some species, such as the Lower Keys marsh rabbit (*Sylvilagus palustris hefneri*), require not only marsh habitat but upland-forest corridor habitats for dispersal (Forys and Humphrey 1996). In a study of arboreal marsupials in southeastern Australia, Lindenmayer and Nix (1992) observed that wildlife corridors that contained a variety of topographic positions supported more species and a greater abundance of animals than sites confined to a single topographic position, such as a mid-slope or ridge.

Combinations of expert-based approaches

Linkage designs often incorporate combinations of expert-based approaches, for example incorporating expert opinion regarding multiple focal species, conservation opportunities, landscape data, and consideration of the potential effects of climate change and other ecological processes. Early "map charette" workshops held by the Wildlands Project throughout North America involved experts mapping out cores and linkages with a set of focal species (generally large carnivores) in mind. More recently, 40 experts identified connectivity zones between core reserves and along freshwater systems and coastal areas in Nova Scotia, using maps on land use, roads, topography, vegetation, and some species' distributions (Beazley et al. 2000). Linkages were selected primarily on the basis of conservation opportunism, strategic location, or special features (e.g., rare landforms). This approach is quick and dirty, but can be useful if one recognizes that results are limited by available knowledge. Participants generally select areas that they know or perceive to have conservation value, but may neglect lands for which ecological information or knowledge is lacking, or lands that are currently ecologically compromised but hold potential for restoration (Beazley et al. 2002).

In 2000 the California Wilderness Coalition, California State Parks, US Geological Survey, the San Diego Zoo, and The Nature Conservancy held the "Missing Linkages" Conference in California, which gathered 160 experts from public agencies, advocacy groups, consulting firms, and academia to examine potential wildlife linkage zones through California (see Beier et al. Chapter 22). The experts identified about 300 wildlife corridors thought to be vital to California's wildlife populations (Penrod et al. 2000). Linkage priorities were based on the combined knowledge of the experts present and incorporated subjective information on presence of species, threats, opportunities for acquisition and support, and existence of supporting data. These linkages are now being refined on a regional scale in the South Coast ecoregion of California with detailed focal species habitat mapping, least-cost path analyses, and field work to identify specific movement barriers (Luke et al. 2004; Beier et al. Chapter 22).

Satellite imagery and digital orthophotos can be used to identify corridors that were not identified by experts (Gulinck et al. 1991), such as a study that identified potential corridors for elephant (Elephas maximus) migrations (Kachhwaha 1993). Similarly, even after five separate linkage analyses for grizzly bear (Ursus arctos) in the northern Rocky Mountains of the USA, a color infrared satellite image illustrated several potential linkages that had not been identified by other efforts (Ruediger 2001).

Consideration of how linkages might be designed to facilitate adjustment of species to climate change is another expert-based approach that holds merit. Climate is the primary factor determining the geographic distributions of species and major vegetation types (Holdridge 1947; Brown and Lomolino 1998). The potential exists for significant reductions in the geographic extent of some ecosystems and changes in the distributions and abundances of individual species with global warming (Malcolm and Pitelka 2000). Global climate change may make even large reserves unsuitable for current resident species.

Most taxa are limited in their ability to respond through natural selection to climate change (e.g., Coope 1979); rather, most species have shifted their distributions in response to past changes in climate (Noss 2001). The current rate of warming is rapid, however, perhaps ten times faster than the warming at the end of the recent glacial maximum (Brown and Lomolino 1998). Fragmentation of habitat intensifies the impact of climate change by making it more difficult for species to move toward the poles or to higher elevations in response to warming. A species' ability to shift distribution in response to environmental change is determined by a combination of its inherent dispersal capacity and both natural and human barriers to movement. Because of the uncertainty about how climate change will affect natural communities and how species will respond, in addition to computational challenges, climate change models have not been incorporated explicitly in reserve network designs. Nevertheless, general principles for responding to climate change have been proposed. For example, in addition to protecting as many unfragmented landscapes as possible, it will generally be helpful to maintain habitat linkages parallel to latitudinal, elevational, and coastal–inland gradients, minimize artificial barriers to dispersal, and maintain continuity of species' populations across their present geographical ranges (Hobbs and Hopkins 1991; Noss 1993a, 2001; Bennett 1999).

Climatic change is one of several ecological processes that require consideration of landscape connectivity (Soulé et al. Chapter 25). One interesting example of where inferences about the connectivity of ecological processes helped inform conservation planners comes from southern California. Most of the many rare, endemic species of the Coachella Valley inhabit unstable, shifting sand dunes. This habitat condition is generated and maintained by strong winds flowing eastward through the San Gorgonio Pass between the San Bernardino and San Jacinto Mountains, then down the Coachella Valley through several pathways. Conservationists have identified these critical "wind corridors" and fear that development in the corridors could disrupt their function, and in some areas, it apparently already has (Barrows 1996; Noss et al. 2001). This information is being used to help guide conservation planning in the valley.

Empirical and modeling approaches

Although the approaches to corridor planning reviewed above have proven useful in many cases, their relative subjectivity and lack of rigorous scientific methods and documentation create considerable uncertainty over their ultimate value to conservation. In general, assessments that rely on expert opinion, as opposed to observation, experiment, and analysis, are vulnerable to criticism from scientists as well as from members of the public who oppose conservation. In an evaluation of alternative habitat models for black bear (*Ursus americana*), Clevenger *et al.* (2002) found that models based on species–habitat associations documented in the technical literature perform nearly as well as empirical models and much better than models based solely on expert opinion (see also Clevenger and Wierzchowski Chapter 20). In an evaluation of site selection based on expert opinion versus a systematic site-selection algorithm in South Africa, Cowling *et al.* (2003) found that a "wish list" developed by park managers was not very effective or efficient in achieving conservation goals, excluded large areas most in need of conservation, and failed to provide a basis for scheduling implementation or exploring alternatives. On the other hand, experts often have "in their heads" relevant information that is not contained in digital databases. Wherever possible, expert-based approaches should be complemented by more rigorous empirical and modeling approaches. This is not a matter of one replacing the other – rather, expert opinion and scientific analyses should inform each other in a step-wise, iterative fashion.

Although empirical versus modeling studies are somewhat distinct, we consider them together in this section because many studies combine the two. We suggest that a combination of modeling and empirical (e.g., field validation) approaches is an ideal way to develop a basis for accurate predictions concerning the effects of alternative landscape configurations or management activities on populations. Nevertheless, different types of empirical data are appropriate for different modeling applications and for different scales and measures of connectivity (Fagan and Calabrese Chapter 12).

Routes based on observations of animal presence, movements, or signs

Direct observation or evidence of individuals using a particular corridor regularly is a valid reason for including that corridor in a network design. Many researchers have used animal survey techniques to demonstrate the value of particular corridors to wildlife. For example, Sieving *et al.* (2000) surveyed 24 forested corridors in Chile using passive and song-playback censuses in order to distinguish vegetated corridors functioning as dwelling habitat versus those used only for short-distance movements. Similarly, Mönkkönen and Mutanen (2003) evaluated the utility of riparian corridors in boreal forest landscapes in Finland as habitats and dispersal routes for forest-associated moths. Corridors were strips of forest 30–70 m wide bordered by clearcuts or regeneration stands. The abundance of moths in corridors was generally equal to that observed in forest interiors and edges, suggesting that the corridors serve as breeding habitats or dispersal routes for the moths.

Twenty years of sighting data were used to characterize corridors used by Nubian ibex (*Capra ibex nubiana*) in Israel (Shkedy and Saltz 2000). By categorizing each sighting record as belonging to a corridor or a core zone based on the total number of sightings in its vicinity, the authors identified three main core populations, a natural corridor connecting two of them, and a less obvious corridor connecting the third population.

A pronghorn antelope migration route has been documented (by researchers directly following the herds over the past ten years, and by examination of archeological records) that stitches together the ancient ecological connections between the Grand Teton and Greater Yellowstone ecosystems and the Wyoming Basin (Red Desert) (Fig. 23.1). Researchers identified a small set of natural and anthropogenic corridor constrictions (bottlenecks); land managers, non-profit organizations, and land trusts are attempting to keep these bottlenecks from becoming impermeable barriers. The Wyoming Wildlife Federation, Wildlife Conservation Society, and Wyoming Outdoor Council mapped elk, mule deer (*Odocoileus hemionus*), and pronghorn winter ranges, elk feeding grounds, current and historical migration routes, and current barriers (fences, major roads, energy development) in the region. They subsequently proposed Congressional designation of a National Migration Corridor from the Yellowstone and Grand Teton National Parks down to the Red Desert. Berger (2004) demonstrated that pronghorn still migrate up to 550 km (round trip) annually from Grand Teton National Park to the Upper Green River Basin. This migration requires the use of historic, but narrow corridors (0.1–0.8 km wide) that have existed for at least 5800 years and exceed travel distances of many African ungulates, including elephants (*Loxodonta africana*). Unfortunately, the accelerated leasing of public lands for energy development threatens to truncate these migrations.

Snow and sand track monitoring are frequently used to document animal movements (Scheick and Jones 1999; Singleton and Lehmkuhl 1999). The results of multi-year monitoring of tracks and other signs

Fig. 23.1. After mapping existing and historical wildlife migration corridors for pronghorn, mule deer, and elk, a coalition of groups in Wyoming is attempting to protect these corridors from future development and restore historic routes from Yellowstone to the Red Desert. (Reprinted with permission from Wyoming Outdoor Council and Erik Lindquist, Terra Nova.)

of mountain lions and black bears in southeastern Arizona were used, in part, to design corridors in the Sky Islands Wildlands Network Design (Foreman *et al.* 2000). Others have recorded wildlife use of underpasses and culverts using movement-triggered cameras, snowtracking, and other techniques that document individual animal movement, but no studies have empirically addressed whether specific wildlife crossings have enhanced or diminished the population viability of species impacted by roads (Foster and Humphrey 1995; Clevenger and Waltho 2000; Smith 2003; Clevenger and Wierzchowski Chapter 20).

Use of trained scat-detection dogs is another promising method for identifying wildlife movement routes. Fecal analysis is a powerful technique for identifying species and individuals, estimating population size, and assessing physiological stress, reproductive status, sex ratio, home range, paternity, and kinship (Kohn 1999; Ernest et al. 2000). Trained scent dogs, which typically locate many times more scats than trained human observers, can greatly augment the utility of DNA techniques and might be the only way to obtain information on the presence or absence of some endangered species (Smith et al. 2001, 2003). This non-invasive technique is ideal for studying species that are elusive, wide-ranging, and relatively rare (MacKay 2003). Scat-detection dogs have been used to study a variety of terrestrial mammalian carnivores and are being tested for owls and right whales (Eubalaena spp.) (Meadows 2002; P. MacKay, pers. comm.). Although scat analysis is much less intrusive than radiotracking, mark–recapture, and other field techniques for studying wildlife movements, disadvantages have been noted, e.g., (1) the logistics of scat collection for a reliable population estimate throughout a large study area are highly challenging and require considerable field effort; (2) laboratory analyses of scat-derived DNA are problematic (Taberlet et al. 1999); and (3) scat analysis provides little of the complementary data that radiotracking yields such as behavioral observations, mortality rates and causes, and dispersal (Mech and Barber 2002).

Similar advantages and disadvantages have been noted for another non-invasive survey technique: the use of hair-snag stations. Typically, hair-snag sites are selected based on criteria such as proximity to game trails or abundance of food resources for the species of interest. After selecting a site, a strand of barbed wire is set at an appropriate distance above the ground and wrapped around trees to create a corral, within which scents, food, or other lures are placed. An animal in the vicinity will pick up the scents and, when investigating, will snag some of its hair. The hair sample is then subjected to DNA analysis (US Department of the Interior, US Geological Survey, undated). Hair-snag stations can be used in the same way as scat-detection dogs to identify the species and individuals using a particular corridor. For example, Wills and

Vaughan (2002) used hair-snag stations to monitor travel corridors used by black bears in the Great Dismal Swamp National Wildlife Refuge in Virginia and determine spatial and seasonal patterns of highway crossings.

Routes based on radiotelemetry and other marking of animals

Radiotelemetry has been used to determine wildlife movement routes in many regions, for example Florida (Land and Lotz 1996; Roof and Wooding 1996; Eason and McCown 2002), Montana (Waller and Servheen 1999), and California (Beier 1993, 1995). Telemetry studies are useful not only to investigate habitat use by individual animals, but can also provide information on dispersal distances and behavior, which are essential considerations in linkage design (Maehr *et al.* 2002). Direct observation and radiotracking of individuals helped the Canadian Parks and Wilderness Society (CPAWS)—Yukon determine migration corridors for caribou (*Rangifer tarandus*) and Dall sheep (*Ovis dalli*) (Farnell and Russell 1984; CPAWS—Yukon 2000). The availability of lightweight global positioning system (GPS) units now allows tracking of individual animals by satellite at short intervals (every 3—5 h) (Merrill *et al.* 1998; Mech and Barber 2002; Nelson *et al.* 2004; Tracey Chapter 14). Although radiotelemetry is an extremely useful tool for determining movements of individuals, it is expensive (especially in the case of GPS collars) and, therefore, is rarely applied to multiple species in a study landscape.

Unlike many other methods, radiotelemetry is able to provide direct evidence that animals are using a linkage or are successfully (or unsuccessfully) crossing roads (Beier 1995; Eason and McCown 2002; Clevenger and Wierzchowski Chapter 20). Gibeau *et al.* (2002) used radiotelemetry to study the effects of highways on grizzly bear movement in the Bow River watershed in Alberta, Canada, and found that one highway was effectively a barrier. They also discovered that bears cross highways at specific locations, allowing recommendations for wildlife crossings. In the Tensas River Basin in northern Louisiana, 19 threatened Louisiana black bears (*Ursus americanus luteolus*) were radiotracked for 18 months to determine use of four major habitat patches and existing wooded corridors (Anderson 1997). The corridors were rivers, bayous, and ditches bordered by wooded strips 5—75 m wide. It was determined that 52% of the patch-to-patch movements of male bears and 100% of patch-to-patch movements of females were within these narrow wooded corridors.

The movement distances of some mammals illuminated by radio-telemetry are remarkable. For example, mountain lions dispersing from the San Andres Mountains crossed up to 100 km of unsuitable habitat, including two animals that crossed the Rio Grande. Several mountain lions translocated over 400 km in New Mexico returned to their original territories (Logan *et al.* 1996; Sweanor *et al.* 2000).

A relatively new technique for marking and recapture of animals is the injection of passive integrated transponder devices (PIT tags) into the bodies of animals. These tiny, coded markers, which are dormant until activated by a reader (much like supermarket bar codes), permit positive identification of individual animals (Gibbons and Andrews 2004). Therefore, they can be used to document that an animal has moved from the point of original capture to any points where the animal is recaptured. The use of automated readers makes PIT tags virtually as useful as radiotelemetry for determining movements in some situations. For example Boarman *et al.* (1998) used PIT tags and automated monitoring devices to document the use of drainage culverts under a major highway by desert tortoises (*Gopherus agassizii*). Monitoring devices were placed at both ends of the culvert, which recorded individuals as they entered and exited the culvert. We note that remote cameras also have been used successfully to document movement of animals through culverts and other crossings (e.g., Foster and Humphrey 1995).

Routes based on least-cost path analysis

Modeling studies, especially when coupled with field data, have demonstrated value for identifying and designing linkages, especially at broad spatial scales. Given the availability of user-friendly geographic information systems (GIS), least-cost path analysis has become a popular method for identifying corridors (Theobald Chapter 17). "Cost" in this sense is the estimated cost to the animal or population, not economic cost (but see Morrison and Reynolds (Chapter 21) for evaluation of economic or implementation costs associated with a traditional least-cost pathway for carnivores in southern California). Least-cost path analysis is used to identify potential travel routes, along which an animal would have the best chance of survival according to a determined "cost surface." The cost surface is based on a habitat suitability model – the higher the suitability, the lower the predicted cost of moving through the landscape. The model does not predict movements of animals; rather, it predicts the success of an animal, should it choose a certain route along which to travel (American Wildlands 2000). Hence, least-cost path analysis can provide the basis for recommendations about where to protect land for linkages and how and where to mitigate wildlife mortality within these areas (e.g., Theobald Chapter 17; Beier *et al.* Chapter 22). Nevertheless, results from least-cost path analysis are uninformative about demographic connectivity; the analysis will always find a pathway, even if it is impossible for an individual to get from one place to another, unless impenetrable barriers are programmed into the models (Carroll Chapter 15; Beier *et al.* Chapter 22; D. Smith, pers. comm.).

The cost surface has been considered an index of risk to the animal or its converse, security and food availability. For example, a least-cost path for grizzly bear movement was identified largely on the basis of habitat quality and human disturbance (Craighead *et al.* 2002). A similar approach has been used at a finer scale to model probable highway crossing points for grizzly bears in Slovenia (Kobler and Adamic 1999). In Washington and adjacent British Columbia, Singleton *et al.* (2002) estimated landscape permeability for wolves (*Canis lupus*), wolverine (*Gulo gulo*), lynx (*Lynx canadensis*), and grizzly bear based on habitat models. Landscape permeability was defined as "the quality of a heterogeneous land area to provide for passage of animals" (Singleton and Lehmkuhl 1999). By using suitability models corresponding to breeding habitat, the authors were able to identify areas that are likely to support source populations, as well as the least-cost paths between areas and weighted distances (i.e., estimates of the total movement cost for animals traveling outward from particular source areas). This broad evaluation provides an estimation of relative potential for animal passage across the entire landscape, including the identification of potential barriers to movement.

In cases where core or dispersal habitat for a particular species within a potential linkage is lacking, least-cost path analysis can still show the path of least resistance to movement. Expansive areas between the Adirondacks and Algonquin Provincial Park did not meet accepted criteria for either core or dispersal habitat for wolves (Harrison and Chapin 1997, 1998; Paquet *et al.* 1999). Yet least-cost path analysis was useful for identifying paths of least resistance for wolves that might attempt to move between the parks (Quinby *et al.* 2000) (Fig. 23.2). These paths would be logical locations for habitat restoration to enhance the probability of successful dispersal (Carroll Chapter 15).

Least-cost path analysis also was applied to moose (*Alces alces*), American marten (*Martes americana*), and Northern goshawk (*Accipiter gentilis*) for conservation planning in Nova Scotia (Beazley *et al.* 2000a, 2000b). As suggested by Harrison (1992), the minimum width of linkage zones was based on home-range diameters for each species, ensuring that linkages contain enough suitable habitat for the species to occupy them on

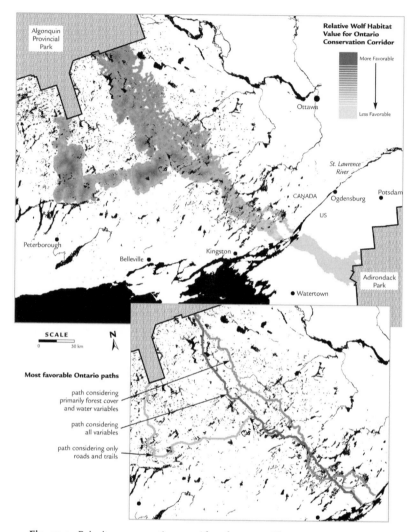

Fig. 23.2. Priority conservation corridors between Algonquin Provincial Park and Adirondack Park, based on the top 5% of wolf habitat along most favorable corridor paths. (Reprinted with permission from Quinby *et al.* 2000.)

a potentially permanent basis. For those species sensitive to edge effects, the home-range size should be buffered by additional habitat. Estimates of home-range diameters for moose (1250 m), lynx (1900 m), and marten (360 m), plus a buffer of 300–600 m, have been used to determine minimum widths of connectivity zones, also taking into consideration the availability of preferred habitat types within these zones (Spowart and Samson 1986; Beazley 1998).

A decision-support model was used to identify large areas of ecological significance and landscape linkages in a state-wide network for Florida (Hoctor *et al.* 2000). Building the decision-support model involved three steps: (1) identification of priority ecological areas, based on multiple ecological criteria; (2) selection of hubs, i.e., areas with the highest potential for ecological integrity; (3) identification of linkages. Five linkage types were identified: coastal to coastal, riverine to riverine, upland to upland, riverine to coastal, and cross-basin hub to hub. Least-cost path analysis was used to identify the optimal paths for a suite of species for each linkage type, then paths were widened to include all contiguous cells of native habitat or lower-intensity land uses up to a width of 25% of the linkage length. Public comment and expert review resulted in addition of a few corridors and widening some corridors to enhance their effectiveness, as well as the removal of some areas at the request of private landowners (University of Florida *et al.* 1999; D. Smith, pers. comm.). The authors recognized that the model was not sufficient for consideration of long-term viability of wide-ranging species (such as the Florida panther (*Puma concolor coryi*) and Florida black bear (*Ursus americanus floridanus*)) and recommended species-specific analyses to determine the potential for connectivity of particular populations and metapopulations.

Overall, least-cost path analysis provides a snapshot of current habitat conditions. Because they are static (albeit spatially explicit) representations of habitat quality, such models are poorly suited for consideration of population persistence and source—sink dynamics over time. They also provide little information on the potential demographic value of particular sites and linkages within a broader spatial context.

Routes based on spatially explicit population modeling

Spatially explicit population models (SEPMs) offer several advantages over static habitat models (Carroll Chapter 15). With SEPMs, the locations of habitat patches, individuals, and other features are incorporated explicitly in the model, and the effects of changing landscape structure on population dynamics can be predicted (Dunning *et al.* 1995). SEPMs allow extrapolation from the behavior of individuals to landscape-scale patterns and processes (South *et al.* 2002). A general advantage of SEPMs is that they provide qualitative insights into factors, such as variance in population size, that are difficult to explore using static models (Carroll *et al.* 2003b).

A combination of static habitat-suitability models (e.g., in the form of resource selection functions: Boyce and McDonald 1999) and SEPMs (e.g., PATCH: Schumaker 1998) has proven especially useful for evaluating source–sink and connectivity issues on a regional scale (Noss et al. 2002; Carroll et al. 2003a, 2003b; Carroll Chapter 15). For example, an evaluation of carnivore population viability in the Greater Yellowstone Ecosystem, as part of a more comprehensive conservation assessment, revealed potential demographic connectivity to central Idaho, which would be lost if recent rates of development (especially road-building) continue for the next 25 years (Noss et al. 2002).

Importantly, the spatially explicit output of PATCH and similar SEPMs allows planners to evaluate connectivity under alternative future scenarios and visualize the effects of landscape change, habitat restoration, or alternative management strategies (Turner et al. 1995; Carroll Chapter 15). For example, one can assess the demographic consequences to an entire metapopulation of preserving, destroying, or restoring a particular corridor (Carroll et al. 2002). On a regional scale, the ideal connectivity is continuous source habitat. As a potential linkage is degraded to largely sink habitat or non-habitat, functional connectivity declines.

SEPM predictions carry considerable uncertainty. The sensitivity of SEPM output to dispersal parameters, which are poorly known for many species (Bullock et al. 2002), makes predictions about such critical phenomena as natural colonization or recolonization from distant sources questionable (South et al. 2002; Carroll et al. 2003b). The various types of SEPMs differ in their representation of the spatial structure of populations and in their representation of the dispersal behavior of individuals; hence, their predictions concerning dispersal and functional connectivity often vary (South et al. 2002). Like static habitat-suitability models, SEPMs require validation, at some point or another, by independent field data. SEPMs developed on regional scales are unlikely to be helpful for management decisions at purely local scales, albeit they can help managers appreciate the value of local areas within a broader spatial context.

CONNECTIVITY BY MEANS OTHER THAN CORRIDORS

Much of the controversy over the value of corridors in conservation can be traced to the different spatial scales people have in mind when they use the term "corridor" (Noss 1991). The first use of the term "corridor" in biology was apparently by biogeographers, who had in mind huge areas (such as the Bering land bridge or the isthmus of Panama) that permit the spread of taxa from one region to another (Brown and Lomolino 1998). More recently, many biologists seem to have in mind narrow conduits such as hedgerows and riparian strips, whereas regional conservation planners usually think much bigger. In this chapter, although we focus on broad-scale planning, we have discussed corridors ranging from small wildlife-crossing structures to broad landscape linkages, as this entire spatial spectrum requires consideration in planning. At a regional scale, a suitable corridor for a wide-ranging species may be the landscape matrix, so long as road density and other measures of (in)security are low enough to provide safe passage. Studies of small animals (e.g., butterflies) have shown that a favorable landscape matrix can greatly reduce the effective isolation of habitat patches (Ricketts 2001). On the other hand, with the expanding human population and associated development, it would be dangerous to assume that the landscape matrix lacking formal protection will remain suitable for animal movement (Soulé et al. Chapter 25).

Every corridor has a matrix — the area of less suitable or unsuitable habitat it adjoins. Activities in the matrix can affect the overall integrity of the corridor (Taylor et al. Chapter 2). Pesticides and fertilizers can drift from fields into corridors, affecting plants and animals (Forman 1995). Adjacent roads can pollute the corridors and may be a major source of mortality for wildlife. Corridors close to human developments are more accessible to humans (including poachers), as well as their pets, which can negatively impact corridor utility (Arango-Velez and Kattan 1997; Crooks and Soulé 1999). A study in West Africa found that hunting pressure accounted for 87% of the variation in extinction rates among reserves (Brashares et al. 2001). Threat abatement, such as limiting hunting in zones between core areas, might be more effective for allowing dispersal in some cases than trying to channelize animal movement into corridors (Harrison and Bruna 1999).

In some cases "stepping stones" of habitat in the matrix between core areas are a viable alternative to continuous corridors, especially for species that evolved within naturally patchy landscapes (Forman 1995; Schultz 1998; Bennett 1999). An implicit assumption is that the matrix is permeable to movement. Stepping stones are most likely to be useful in cases where animals (1) can detect a stepping stone from a core area or other source; (2) are not constrained or directed by habitat boundaries; (3) are unwilling to enter corridors that are dominated by edge habitat; or (4) experience high levels of predation within corridors (Haddad 2000). For many migratory birds, waterfowl, raptors, and butterflies, a corridor may be composed of stepping stones of habitat connecting seasonal core habitats. Restoration of whooping cranes (*Grus americana*) has largely been based on stepping stones of habitat islands along their 4800-km migration path (Harris 1988). In the Rio Santa Cruz watershed in Arizona and Sonora, Mexico, Nabhan and Donovan (2000) characterized the migratory pollinator fauna and determined which species utilized various stepping stone habitats (see Ricketts *et al.* (Chapter 11) for a discussion of connectivity for pollinators).

Another potential alternative to corridors is to capture and translocate animals between isolated populations. An IUCN workshop on the population and habitat viability of African wild dogs (*Lycaon pictus*) in Southern Africa recommended a "constellation" of wild-dog refuges (Mills *et al.* 1998). The possibility of natural dispersal between these refuges is remote, and many of them would have to be fenced. Wild dog packs live at low densities and range over very large areas. Consequently, even those living within large protected areas may travel outside reserve borders where they encounter threats associated with human activity. Persecution is the most serious threat to wild dogs; outside protected areas mortality can be as high as 92% (Rasmussen 1997). Therefore, the experts concluded that a managed metapopulation with artificial immigration and emigration was the best option for the continued viability of wild dogs. In general, however, the benefit of moving animals among sites is questionable, especially for wide-ranging carnivores, which tend to move great distances from release sites, resulting in drastically reduced survival (Linnel *et al.* 1997). Additionally, translocation of individuals of one species does not address maintenance of ecological processes or provide opportunities for range shifts of multiple species in the face of climate change.

CONCLUSIONS

In this chapter we have focused on methods for maintaining or restoring connectivity among core areas in reserve networks, with an emphasis on broad (i.e., regional) scales of planning. Our chosen topic should not be construed as an argument that connectivity is the most critical consideration in conservation planning, although we would suggest that connectivity is fundamental when planning goals are comprehensive and include consideration of the population viability of species sensitive to habitat fragmentation. Importantly, connectivity between reserves should not be considered a substitute for the conservation of large core areas. Rather, corridors are an important complement, not an alternative, to the critical strategy of establishing large and multiple reserves (Noss 1992). Without secure core areas, most corridors have little value. When species are distributed as metapopulations, regional persistence may depend largely on the continued occupancy of extinction-resistant core areas (Wilson *et al.* 2002).

We conclude with some summary guidelines for improving the methods used for connectivity planning:

- Poorly designed corridors have costs. If corridors are placed in the wrong places and prove ineffective, populations could go extinct, funds will have been wasted, and conservationists may lose credibility.
- Multifaceted methods for corridor design are preferable to simpler approaches, within reason. A linkage designed on the basis of several approaches and multiple criteria usually will be more defensible than one based on a single approach.
- Corridor-design approaches that are purely intuitive, opportunistic, or otherwise based on subjective best-guesses, existing knowledge, or expert opinion are not always inferior to more technical approaches, but they are usually less reliable and will be easier for others to criticize. Increased use of quantitative habitat and population modeling, combined with extensive field studies, will make corridor identification and design more reliable and scientifically defensible.
- An integration of spatially explicit models with expert knowledge in a reserve-network design can assist in the prioritization of sites and their designation as cores, corridors, and compatible matrix lands.
- The life-history requirements of particular species must be considered in corridor planning. Connectivity is a species-specific and landscape-specific property, and for conservation purposes it must be functional connectivity, not just structural (i.e., apparent to the human observer). Nevertheless, attempting to consider, in detail, the needs of a large suite of species would be overwhelming. Attention should be given to the species most sensitive to fragmentation in the region and on the appropriate spatial scale.
- Most corridors have been designed to allow for movement of animals within home ranges or for dispersal among populations. More attention should be given to designing corridors that will allow for range shifts of species (as in response to climate change), maintain flows of ecological processes, and maintain an ecological theater within which the evolutionary play (*sensu* Hutchinson 1965) can continue as it has for millions of years.

- A comprehensive conservation plan will often have to consider several spatial scales, i.e., general planning on a regional scale but zooming in for higher-resolution analysis of high-priority corridors or particularly troublesome corridor bottlenecks or highway crossings.
- Conservationists should recognize that connectivity can be provided by means other than corridors. For example, a suitable landscape matrix and/or stepping-stone habitat patches may allow movement for many species. However, strict management and protection measures need to be enforced.

Once potential corridors or connectivity zones have been identified through one or more of the approaches described in this chapter, considerations of cost and ease of protection often come into play. It is useful for conservation planners to provide multiple options and explicitly recognize the biological and socio-economic trade-offs of selecting alternative corridors. For example, a less-than-optimal corridor that is composed of two or three large, easily acquired parcels may be preferable to the "best" corridor that goes through a checkerboard of hundreds of small parcels, because the latter would require much more money, time, and effort to protect (see Morrison and Reynolds Chapter 21). With limited dollars, conservation must be as cost-efficient as possible. Nevertheless, we suggest that biological requirements be given primacy, and that trade-offs be considered later and not be allowed to compromise population viability or other ecological values. Although it has become politically incorrect in some circles to admit, we are truly living in a time of biological crisis, such that maintaining life on Earth and its potential for further evolution trumps all other societal priorities.

REFERENCES

Adams, A. W. 1982. Migration. Pp. 301–321 in J. W. Thomas and D. E. Toweill (eds.) *Elk of North America: Ecology and Management.* Harrisburg, PA: Stackpole Books.

American Wildlands. 2000. *Corridors of Life: Weaving a Web of Wildlife Habitat in the Northern Rockies.* Bozeman, MT: American Wildlands.

Andelman, S., I. Ball, F. Davis, and D. Stoms. 1999. *SITES V 1.0: An Analytical Toolbox for Designing Ecoregional Conservation Portfolios.* The Nature Conservancy.

Anderson, D. R. 1997. Corridor use, feeding ecology, and habitat relationships of black bears in a fragmented landscape in Louisiana. M.Sc. thesis, University of Tennessee, Knoxville, TN.

Arango-Velez, N., and G. H. Kattan. 1997. Effects of forest fragmentation on experimental nest predation in Andean cloud forest. *Biological Conservation* 81:137–143.

Barrows, C. 1996. An ecological model for the protection of a dune ecosystem. *Conservation Biology* 10:888–891.

Beazley, K. F. 1998. A focal species approach to biodiversity management in Nova Scotia. Ph.D. thesis, Dalhousie University, Halifax, Nova Scotia, Canada.

Beazley, K., R. Long, and P. MacKay. 2000. *Nova Scotia Wild Lands and Wild Seas Mapping Workshop, 1999*. Burlington, VT: Greater Laurentian Wildlands Project.

Beazley, K., P. Austin-Smith, Jr., and M. Rader. 2002. Toward completing a protected areas system for Nova Scotia: Terrestrial and Marine. Pp. 516–530 in N. Munro and M. Willison (eds.) *Challenges to National Parks and Protected Areas: Learning from the Past, Looking to the Future*, Proceedings of the 4th International Conference on Science and the Management of Protected Areas. Wolfville, Canada: Science and the Management of Protected Areas Association.

Beier, P. 1993. Determining minimum habitat areas and habitat corridors for cougars. *Conservation Biology* 7:94–108.

Beier, P. 1995. Dispersal of juvenile cougars in fragmented habitat. *Journal of Wildlife Management* 59:228–237.

Beier, P., and R. F. Noss. 1998. Do habitat corridors provide connectivity? *Conservation Biology* 12:1241–1252.

Bennett, A. F. 1990. *Habitat Corridors: Their Role in Wildlife Management and Conservation*. Melbourne, Victoria, Australia: Arthur Rylah Institute for Environmental Research, Department of Conservation and Environment.

Bennett, A. F. 1991. Roads, roadsides and wildlife conservation: a review. Pp. 99–117 in D. A. Saunders and R. J. Hobbs (eds.) *Nature Conservation*, vol. 2, *The Role of Corridors*. Chipping Norton, NSW, Australia: Surrey Beatty and Sons.

Bennett, A. F. 1999. *Linkages in the Landscape: The Role of Corridors and Connectivity in Wildlife Conservation*. Gland, Switzerland: IUCN.

Berger, J. 2004. The last mile: how to sustain long-distance migration in mammals. *Conservation Biology* 18:320–331.

Boarman, W. I., M. L. Beigel, G. C. Goodlet, and M. Sazaki. 1998. A passive integrated transponder system for tracking animal movements. *Wildlife Society Bulletin* 26:886–891.

Boyce, M. S., and L. L. McDonald. 1999. Relating populations to habitats using resource selection functions. *Trends in Ecology and Evolution* 14:268–272.

Brashares, J. S., P. Arcese, and M. K. Sam. 2001. Human demography and reserve size predict wildlife extinction in West Africa. *Proceedings of the Royal Society of London B* 268:2473–2478.

Briers, R. A. 2002. Incorporating connectivity into reserve selection procedures. *Biological Conservation* 103:77–83.

Brown, J. H., and M. V. Lomolino. 1998. *Biogeography*, 2nd edn. Sunderland, MA: Sinauer Associates.

Buckler, D., D. Papoulias, G. Ozuna, D. Woodward, M. Flora, and L. Ditto. 2002. *Water-Resources Issues in the Lower Rio Grande Valley — Below Falcon Reservoir to the Gulf of Mexico Subarea,* Field Coordinating Committee Fact Sheet 4. Columbia, MO: US Department of Interior.

Bullock, J. M., R. E. Kenward, and R. S. Hails (eds.) 2002. *Dispersal Ecology.* Oxford, UK: Blackwell Science.

Cabeza, M. 2003. Habitat loss and connectivity of reserve networks in probability approaches to reserve design. *Ecology Letters* 6:665–672.

Carroll, C., R. F. Noss, and P. C. Paquet. 2001. Carnivores as focal species for conservation planning in the Rocky Mountain region. *Ecological Applications* 11:961–980.

Carroll, C., R. F. Noss, and P. C. Paquet. 2002. *Rocky Mountain Carnivore Project Final Report,* Final Report to World Wildlife Fund Canada. Corvallis, OR: Conservation Science, Inc.

Carroll, C., R. F. Noss, P. C. Paquet, and N. H. Schumaker. 2003a. Integrating population viability analysis and reserve selection algorithms into regional conservation plans. *Ecological Applications* 13:1773–1789.

Carroll, C., M. K. Phillips, N. H. Schumaker, and D. W. Smith. 2003b. Impacts of landscape change on wolf restoration success: planning a reintroduction program based on static and dynamic spatial models. *Conservation Biology* 17:536–548.

Chapman, J. 2002. Too little, but not too late. *Mesquite Review* **April/May.**

Clevenger, A. P., and N. Waltho. 2000. Factors influencing the effectiveness of wildlife underpasses in Banff National Park, Alberta, Canada. *Conservation Biology* 14:47–56.

Clevenger, A. P., J. Wierzchowski, B. Chruszcz, and K. Gunson. 2002. GIS-generated, expert-based models for identifying wildlife habitat linkages and planning mitigation passages. *Conservation Biology* 16:503–514.

Clevenger, A. P., B. Chruszcz, and K. E. Gunson. 2003. Spatial patterns and factors influencing small vertebrate fauna road-kill aggregations. *Biological Conservation* 109:15–26.

Coope, G. R. 1979. Late Cenozoic fossil Coleoptera: evolution, biogeography, and ecology. *Annual Reviews of Ecology and Systematics* 10:247–267.

Cowling, R. M., R. L. Pressey, R. Sims-Castley, *et al.* 2003. The expert or the algorithm? — comparison of priority conservation areas in the Cape Floristic Region identified by park managers and reserve selection software. *Biological Conservation* 112:147–167.

CPAWS—Yukon, 2000. *Peel River Watershed Study: The Wind, Snake and Bonnet Plume.* Whitehorse, Yukon: Canadian Parks and Wilderness Society—Yukon Chapter and World Wildlife Fund Canada.

Craighead, A. C., E. A. Roberts, and F. L. Craighead. 2002. Bozeman pass wildlife linkage and highway safety study. In *Proceedings of the International Conference on Ecology and Transportation.* Center for Transportation and the Environment, Raleigh, NC, pp. 405–422.

Crooks, K., and M. E. Soulé. 1999. Mesopredator release and avifaunal extinctions in a fragmented system. *Nature* **400**:563–566.

Dobson A. P., K. Ralls, M. Foster, *et al.* 1999. Corridors: reconnecting fragmented landscapes. Pp. 129–170 in M. E. Soulé, and J. Terborgh (eds.) *Continental Conservation: Scientific Foundations of Regional Reserve Networks.* Washington, DC: Island Press.

Dunning, J. B., Jr., D. J. Stewart, B. J. Danielson, *et al.* 1995. Spatially explicit population models: current forms and future uses. *Ecological Applications* 5:3–11.

Eason, T. H., and J. W. McCown. 2002. Black bear movements and habitat use relative to roads in Ocala National Forest, Florida. In *Proceedings of the International Conference on Ecology and Transportation,* Center for Transportation and the Environment. Raleigh, NC, Pp. 397–404.

Ernest, H. B., M. C. T. Penedo, B. P. May, M. Syvanen, and W. M. Boyce. 2000. Molecular tracking of mountain lions in the Yosemite Valley region in California: genetic analysis using microsatellites and faecal DNA. *Molecular Ecology* 9:433–441.

Farnell, R., and D. Russell. 1984. *Wernecke Mountain Caribou Studies 1980–1982,* condensed report. Yukon Territorial Government, Canada: Department of Renewable Resources.

Foreman, D., K. Daly, B. Dugelby, *et al.* 2000. *Sky Islands Wildlands Network Conservation Plan.* Tucson, AZ: The Wildlands Project.

Foreman, D., K. Daly, R. F. Noss, *et al.* 2003. *New Mexico Highlands Wildlands Network Vision.* Richmond, VT: The Wildlands Project.

Forman, R. T. T. 1995. *Land Mosaics: The Ecology of Landscapes and Regions.* Cambridge, UK: Cambridge University Press.

Forys, E. A., and S. R. Humphrey. 1996. Home range and movements of the lower keys marsh rabbits in a highly fragmented habitat. *Journal of Mammalogy* 77:1042–1048.

Foster, M. L., and S. R. Humphrey. 1995. Use of highway underpasses by Florida panthers and other wildlife. *Wildlife Society Bulletin* 23:95–100.

Franklin, J. F. 1993. Preserving biodiversity: species, ecosystems or landscapes. *Ecological Applications* 3:202–205.

Gibbons, J. W., and K. M. Andrews. 2004. PIT tagging: simple technology at its best. *BioScience* 54:447–454.

Gibeau, M. L., A. P. Clevenger, S. Herrero, and J. Wierzchowski. 2002. Effects of highways on grizzly bear movement in the Bow River Watershed, Alberta, Canada. In *Proceedings of the International Conference on Ecology and Transportation.* Center for Transportation and the Environment, Raleigh, NC, pp. 458–472.

Groves, C. R. 2003. *Drafting a Conservation Blueprint: A Practitioner's Guide to Planning for Biodiversity.* Washington, DC: Island Press.

Gulinck H., O. Walpot, P. Janssens, and I. Dries. 1991. The visualization of corridors in the landscape using SPOT data. Pp. 9–17 in D. A. Saunders and R. J. Hobbs (eds.) *Nature Conservation,* vol. 2, *The Role of Corridors.* Chipping Norton, NSW, Australia: Surrey Beatty and Sons.

Haddad, N. 2000. Corridor length and patch colonization by a butterfly, *Junonia coenia. Conservation Biology* 14:738–745.

Haddad, N. M., D. K. Rosenberg, and B. R. Noon. 2000. On experimentation and the study of corridors: response to Beier and Noss. *Conservation Biology* 14:1543–1545.

Harris, J. (ed.) 1988. *Proceedings of the International Crane Foundation Workshop*, 1–10 May 1987, Qiqihar, Heilongjiang Province, People's Republic of China.

Harris, L. D. 1984. *The Fragmented Forest: Island Biogeographic Theory and the Preservation of Biotic Diversity*. Chicago, IL: University of Chicago Press.

Harris L. D., and P. B. Gallagher. 1989. New initiatives for wildlife conservation: the need for movement corridors. Pp. 11–34 in G. MacKintosh (ed.) *Preserving Communities and Corridors*. Washington, DC: Defenders of Wildlife.

Harris, L. D., T. Hoctor, D. Maehr, and J. Sanderson. 1996. The role of networks, and corridors in enhancing the value and protection of parks and equivalent areas. Pp. 173–197 in R. G. Wright (ed.) *National Parks and Protected Areas: Their Role in Environmental Protection*. Cambridge, MA: Blackwell Scientific Publications.

Harrison, D. J., and T. G. Chapin. 1997. *An Assessment of Potential Habitat for Eastern Timber Wolves in the Northeastern United States and Connectivity with Habitat in Southeastern Canada*, Working Paper No. 7. New York: Wildlife Conservation Society.

Harrison, D. J., and T. G. Chapin. 1998. Extent and connectivity of habitat for wolves in eastern North America. *Wildlife Society Bulletin* 26:767–775.

Harrison, R. L. 1992. Toward a theory of inter-refuge corridor design. *Conservation Biology* 6:293–295.

Harrison, S., and E. Bruna. 1999. Habitat fragmentation and large scale conservation: what do we know for sure? *Ecography* 22:225–232.

Hobbs, R. J. 1992. The role of corridors in conservation: solution or bandwagon? *Trends in Ecology and Evolution* 7:389–392.

Hobbs, R. J., and A. J. M. Hopkins. 1991. The role of conservation corridors in a changing climate. Pp. 283–290 in D. A. Saunders and R. J. Hobbs (eds.) *Nature Conservation*, vol. 2, *The Role of Corridors*. Chipping Norton, NSW, Australia: Surrey Beatty and Sons.

Hoctor, T. S., M. H. Carr, and P. D. Zwick. 2000. Identifying a linked reserve system using a regional landscape approach: the Florida ecological network. *Conservation Biology* 14:984–1000.

Holdridge, L. R. 1947. Determination of world plant formations from simple climatic data. *Science* 105:367–368.

Hutchinson, G. E. 1965. *The Ecological Theater and the Evolutionary Play*. New Haven, CT: Yale University Press.

Johnson, C. W. 1999. *Conservation Corridor Planning at the Landscape Level: Managing for Wildlife Habitat*, National Biology Handbook, Part 614.4. Washington, DC: US Department of Agriculture.

Kachhwaha, T. S. 1993. Temporal and multisensor approach in forest-vegetation mapping and corridor identification for effective management of Rajaji National Park, Uttar Pradesh, India. *International Journal of Remote Sensing* 14:3105–3114.

Kobler, A., and G. Adamic. 1999. Brown bears in Slovenia: identifying locations for construction of highways in Slovenia. Pp. 29–38 in G. L. Evink, P. Garrett, and D. Zeigler (eds.) *Proceedings of the 3rd International Conference on Wildlife Ecology and Transportation*. Tallahassee, FL: Florida Department of Transportation.

Kohn, M. H., E. C. York, D. A. Kamradt, G. Haught, and R. M. Sauvajoy. 1999. Estimating population size by genotyping faeces. *Proceedings of the Royal Society of London B* 266:657–663.

Kubes, J. 1996. Biocentres and corridors in a cultural landscape: a critical assessment of the territorial system of ecological stability. *Landscape and Urban Planning* 35:231–240.

Lambeck, R. J. 1997. Focal species: a multi-species umbrella for nature conservation. *Conservation Biology* 11:849–856.

Land, D., and M. Lotz. 1996. Wildlife crossing designs and use by Florida panthers and other wildlife in southwest Florida. Pp. 379–386 in *Trends in Addressing Wildlife Mortality*, Proceedings of the transportation-related wildlife mortality seminar, FL-ER-58-96. Tallahassee, FL: Florida Department of Transportation.

Lidicker, W. Z. 1999. Responses of mammals to habitat edges: an overview. *Landscape Ecology* 14:333–343.

Lindenmayer, D. B., and H. A. Nix. 1992. The elements of connectivity where corridor quality is variable. *Landscape Ecology* 4:157–170.

Linnell, J. D. C., J. Odden, M. E. Smith, R. Aanes, and J. Swenson. 1997. Translocation of carnivores as a method for problem animal management: a review. *Biodiversity Conservation* 6:1245–1257.

Logan K. A., L. L. Sweanor, T. K. Ruth, and M. G. Hornocker. 1996. *Cougars of the San Andres Mountains, New Mexico*, Final Report, Federal Aid in Wildlife Restoration Project W-128-R. Santa Fe, NM: New Mexico Department of Game and Fish.

Lorenz, G. C., and G. W. Barrett. 1990. Influence of simulated landscape corridors on house mouse (*Mus musculus*) dispersal. *American Midland Naturalist* 123:348–356.

Luke, C., K. Penrod, C. Cabañero, *et al.* 2004. *South Coast Missing Linkages Project: A Linkage Design for the Santa Ana–Palomar Connection*. San Diego, CA: San Diego State University Field Station Programs, and Idyllwild, CA: South Coast Wildlands. Available online at http://www.scwildlands.org

MacKay, P. 2003. Dogs join wildlife researchers in Vermont. *Northern Woodlands* 10:14.

Maehr, D. S., E. D. Land, D. B. Shindle, O. L. Bass, and T. S. Hoctor. 2002. Florida panther dispersal and conservation. *Biological Conservation* 106:187–197.

Malcolm, J. R., and L. F. Pitelka. 2000. *Ecosystems and Global Climate Change: A Review of Potential Impacts on US Terrestrial Ecosystems and Biodiversity*. Washington, DC: Pew Center on Global Climate Change.

Margules, C. R., and R. L. Pressey. 2000. Systematic conservation planning. *Nature* 405:243–253.

Marynowski, S. 1992. Paseo Pantera. *Wild Earth Special Issue*: 71–74.

Meadows, R. 2002. Scat-sniffing dogs. *ZooGoer* Sept/Oct 31:22–27.

Mech, L. D., and S. M. Barber. 2002. *A Critique of Wildlife Radio-Tracking and its Use in National Parks: A Report to the US National Park Service.* Jamestown, ND: US Geological Survey, Northern Prairie Wildlife Research Center.

Merriam, G., and A. Lanoue. 1990. Corridor use by small mammals: field measurement for three experimental types of *Peromyscus leucopus*. *Landscape Ecology* 4:123–131.

Merrill, S. B., L. G. Adams, M. E. Nelson, and L. D. Mech. 1998. Testing releasable GPS collars on wolves and white-tailed deer. *Wildlife Society Bulletin* 26:830–835.

Merrill, T., D. J. Mattson, R. G. Wright, and H. B. Quigley. 1999. Defining landscapes suitable for restoration of grizzly bears (*Ursus arctos*). *Biological Conservation* 87:231–248.

Mills, M. G. L., S. Ellis, R. Woodroffe, et al. 1998. *Population and Habitat Viability Assessment for the African Wild Dog* (Lycaon pictus) *in Southern Africa.* Apple Valley, MN: IUCN/SSC Conservation Breeding Specialist Group.

Mönkkönen, M., and M. Mutanen. 2003. Occurrence of moths in boreal forest corridors. *Conservation Biology* 17:468–475.

Nabhan G. P., and J. Donovan. 2000. *Nectar Trails for Pollinators: Designing Corridors for Conservation.* Tucson, AZ: Arizona-Sonora Desert Museum.

Nelson, M. E., L. D. Mech, and P. F. Frame. 2004. Tracking of white-tailed deer migration by global positioning system. *Journal of Mammalogy* 85:505–510.

Noss, R. F. 1987a. Corridors in real landscapes: a reply to Simberloff and Cox. *Conservation Biology* 1:159–164.

Noss, R. F. 1987b. Protecting natural areas in fragmented landscapes. *Natural Areas Journal* 7:2–13.

Noss, R. F. 1991. Landscape connectivity: different functions at different scales. Pp. 27–39 in W. E. Hudson (ed.) *Landscape Linkages and Biodiversity.* Washington, DC: Island Press.

Noss, R. F. 1992. The Wildlands Project: land conservation strategy. *Wild Earth* Special Issue:10–25.

Noss R. F. 1993a. Wildlife corridors. Pp. 43–68 in D. S. Smith and P. C. Hellmund (eds.) *Ecology of Greenways.* Minneapolis, MN: University of Minnesota Press.

Noss, R. F. 1993b. A conservation plan for the Oregon Coast Range: some preliminary suggestions. *Natural Areas Journal* 13:276–290.

Noss, R. F. 2001. Beyond Kyoto: forest management in a time of rapid climate change. *Conservation Biology* 15:578–590.

Noss, R. F., and P. B. Beier. 2000. Arguing over little things: a reply to Haddad et al. *Conservation Biology* 14:1546–1548.

Noss, R. F., and A. Cooperrider. 1994. *Saving Nature's Legacy: Protecting and Restoring Biodiversity.* Washington, DC: Defenders of Wildlife and Island Press.

Noss, R. F., and L. D. Harris. 1986. Nodes, networks, and MUM's: preserving diversity at all scales. *Environmental Management* 10:299–309.

Noss, R. F., E. T. LaRoe, and J. M. Scott. 1995. *Endangered Ecosystems of the United States: A Preliminary Assessment of Loss and Degradation,* Biological Report No.28. Washington, DC: US Department of the Interior National Biological Service.

Noss, R. F., H. B. Quigley, M. G. Hornocker, T. Merrill, and P. C. Paquet. 1996. Conservation biology and carnivore conservation in the Rocky Mountains. *Conservation Biology* 10:949–963.

Noss, R. F., J. R. Strittholt, K. Vance-Borland, C. Carroll, and P. Frost. 1999. A conservation plan for the Klamath–Siskiyou ecoregion. *Natural Areas Journal* 19:392–411.

Noss, R. F., E. Allen, G. Ballmer, *et al.* 2001. *Independent Science Advisors' Review: Coachella Valley Multiple Species Habitat Conservation Plan (MSHCP/NCCP).* Corvallis, OR: Conservation Science, Inc.

Noss, R. F., C. Carroll, K. Vance-Borland, and G. Wuerthner. 2002. A multicriteria assessment of the irreplaceability and vulnerability of sites in the Greater Yellowstone Ecosystem. *Conservation Biology* 16:895–908.

Ontario Ministry of Municipal Affairs and Housing. 2001. *Oak Ridges Moraine Conservation Plan.* Toronto, Ontario, Canada: Ministry of Municipal Affairs and Housing.

Paquet, P. C., and A. Hackman. 1995. *Large Carnivore Conservation in the Rocky Mountains: A Long-Term Strategy for Maintaining Free-Ranging and Self-Sustaining Populations of Carnivores.* Toronto, Ontario, Canada: World Wildlife Fund-Canada.

Paquet, P. C., J. R. Strittholt, and N. L. Staus. 1999. *Wolf Reintroduction Feasibility in the Adirondack Park.* Corvallis, OR: Conservation Biology Institute.

Penrod, K., R. Hunter, and M. Merrifield. 2000. *Missing Linkages: Restoring Connectivity to the California Landscape.* San Diego, CA: California Wilderness Coalition, The Nature Conservancy, US Geological Survey, Center for Reproduction of Endangered Species, and California State Parks.

Pressey, R. L., R. M. Cowling, and M. Rouget. 2003. Formulating conservation targets for biodiversity pattern and process in the Cape Floristic Region, South Africa. *Biological Conservation* 112:99–127.

Quinby, P., S. Trombulak, T. Lee, *et al.* 2000. Opportunities for wildlife habitat connectivity between Algonquin Provincial Park and the Adirondack Park. *Wild Earth* 10:75–80.

Rasmussen, G. S. A. 1997. *Conservation Status of the Painted Hunting dog* Lycaon pictus *in Zimbabwe.* Harare, Zimbabwe: Ministry of Environment and Tourism, Department of National Parks and Wildlife Management.

Ricketts, T. H. 2001. The matrix matters: effective isolation in fragmented landscapes. *American Naturalist* 158:87–99.

Roof, J., and J. Wooding. 1996. *Evaluation of S.R. 46 Wildlife Crossing,* Florida Cooperative Wildlife Research Unit Technical Report No.54. Gainesville, FL: University of Florida.

Rosenberg, D. K., B. R. Noon, and E. R. Meslow. 1997. Biological corridors: form, function and efficacy. *BioScience* 47:677–687.

Ruediger, B. 2001. *Draft Report on Wildlife Linkage Habitat.* Missoula, MT: Interagency Grizzly Bear Working Group.

Scheick, B., and M. Jones. 1999. Locating wildlife underpasses prior to expansion of highway 64 in North Carolina. Pp. 247–250 in G. L. Evink, P. Garrett, and D. Zeigler (eds.) *Proceedings of the 3rd International Conference on Wildlife Ecology and Transportation.* Tallahassee, FL: Florida Department of Transportation.

Schultz, C. B. 1998. Dispersal behavior and its implications for reserve design in a rare Oregon butterfly. *Conservation Biology* 12:284–292.

Schumaker, N. H. 1998. *A User's Guide to the PATCH Model*, EPA/600/R-98/135. Corvallis, OR: US Environmental Protection Agency, Environmental Research Laboratory.

Scott, J. M., and B. Csuti. 1997. Gap analysis for biodiversity survey and maintenance. II. Pp. 321–340 in M. L. Reaka-Kudla, D. E. Wilson, and E. O. Wilson (eds.) *Biodiversity: Getting the Job Done*. Washington, DC: National Academy Press.

Shkedy, Y., and D. Saltz. 2000. Characterizing core and corridor use by Nubian ibex in the Negev Desert, Israel. *Conservation Biology* 14:200–206.

Sieving, K. E., M. F. Willson, and T. L. De Santo. 2000. Defining corridor functions for endemic birds in fragmented south-temperate rainforest. *Conservation Biology* 14:1120–1132.

Simberloff, D., and J. Cox. 1987. Consequences and costs of conservation corridors. *Conservation Biology* 1:63–71.

Simberloff, D., J. A. Farr, J. Cox, and D. W. Mehlman. 1992. Movement corridors: conservation bargains or poor investments? *Conservation Biology* 6:493–504.

Singleton, P. H., and J. F. Lehmkuhl. 1999. Assessing wildlife habitat connectivity in the Interstate-90 Snoqualmie Pass corridor, Washington. Pp. 75–83 in G. L. Evink, P. Garrett, and D. Zeigler (eds.) *Proceedings of the 3rd International Conference on Wildlife Ecology and Transportation*. Tallahassee, FL: Florida Department of Transportation.

Singleton, P., W. Gaines, and J. Lehmkuhl. 2002. Using a weighted distance and least-cost corridor analysis to evaluate regional scale large carnivore habitat connectivity in Washington. *Proceedings of the International Conference on Ecology and Transportation*. Raleigh, NC, 583–594.

Smith, B. L., and R. L. Robbins. 1994. *Migrations and Management of the Jackson Elk Herd*, Technical Report Series, Research Bulletin No.199. Washington, DC: US Department of the Interior.

Smith, D. A., K. Ralls, B. Davenport, B. Adams, and J. E. Maldonado. 2001. Canine assistants for conservationists. *Science* 291:435.

Smith, D. A., K. Ralls, A. Hurt, *et al.* 2003 Detection and accuracy rates of dogs trained to find scats of San Joaquin kit foxes (*Vulpes macrotis mutica*). *Animal Conservation* 6:339–346.

Smith, D. J. 1999. Identification and prioritization of ecological interface zones on state highways in Florida. Pp. 209–330 in Q. L. Evink, P. Garrett, and D. Zeigler (eds.) *Proceedings of the 3rd International Conference on Wildlife Ecology and Transportation*. Tallahassee, FL: Florida Department of Transportation.

Smith, D. J. 2003. The ecological effects of roads: theory, analysis, management, and planning considerations. Ph.D. dissertation. University of Florida, Gainesville, FL.

South, A. B., S. P. Rushton, R. E. Kenward, and D. W. Macdonald. 2002. Modelling vertebrate dispersal and demography in real landscapes: how does uncertainty regarding dispersal behaviour influence predictions of spatial population dynamics? Pp. 327–349 in J. M. Bullock, R. E. Kenward, and R. S. Hails (eds.) *Dispersal Ecology*. Oxford, UK: Blackwell Science.

Spowart, R., and F. Samson. 1986. Carnivores: inventory and monitoring of wildlife habitat. Pp. 475–496 in A. Cooperrider, R. Boyd, and H. R. Stuart (eds.) *Inventory and Monitoring of Wildlife Habitat*. Denver, CO: US Department of the Interior, Bureau of Land Management.

Sweanor, L. L., K. A. Logan, and M. C. Hornocker. 2000. Cougar dispersal patterns, metapopulation dynamics, and conservation. *Conservation Biology* 14:798–808.

Taberlet, P., L. P. Walts, and G. Luikart. 1999. Noninvasive genetic sampling: look before you leap. *Trends in Ecology and Evolution* 14:323–327.

Trombulak, S. C., and C. A. Frissell. 2000. Review of the ecological effects of roads on terrestrial and aquatic communities. *Conservation Biology* 14:18–30.

Turner, M. G., G. J. Arthaud, R. T. Engstrom, *et al.* 1995. Usefulness of spatially explicit population models in land management. *Ecological Applications* 5:12–16.

University of Florida's Department of Landscape Architecture, Department of Urban, and Regional Planning GeoPlan Center, Department of Wildlife Ecology and Conservation Program in Landscape Ecology. 1999. *The Florida Statewide Greenways System Planning Project: Recommendations for the Physical Design of a Statewide Greenways System*. Gainesville, FL: University of Florida.

US Department of the Interior, US Geological Survey. Undated. *Setting up a Hair Snag Station*. Available online at http://www.nrmsc.usgs.gov/research/NCDEhairsnag.htm

Walker, R., and L. Craighead. 1997. Analyzing wildlife movement corridors in Montana using GIS. In *Proceedings of the 1997 International Environmental Sciences Research Institute. Users' Conference*, Redwood, CA.

Waller, J., and C. Servheen. 1999. Documenting grizzly bear highway crossing patterns using GPS technology. Pp. 21–24 in G. L. Evink, P. Garrett, and D. Zeigler (eds.) *Proceedings of the 3rd International Conference on Wildlife Ecology and Transportation*. Tallahassee, FL: Florida Department of Transportation.

Wegner, J., and G. Merriam. 1990. Use of spatial elements in a farmland mosaic by a woodland rodent. *Biological Conservation* 54:263–276.

Wills, J., and M. Vaughan. 2002. A method to monitor travel corridor use by black bears along the Eastern boundary of the Great Dismal Swamp National Wildlife Refuge. In *Proceedings of the International Conference on Ecology and Transportation*. Center for Transportation and the Environment, Raleigh, NC, pp. 529–532.

Wilson, R. J., S. Ellis, J. S. Baker, *et al.* 2002. Large-scale patterns of distribution and persistence at the range margins of a butterfly. *Ecology* 83:3357–3368.

24

Escaping the minimalist trap: design and implementation of large-scale biodiversity corridors

JAMES SANDERSON, GUSTAVO A. B. DA FONSECA,
CARLOS GALINDO-LEAL, KEITH ALGER,
VICTOR HUGO INCHAUSTY, KARL MORRISON,
AND ANTHONY RYLANDS

INTRODUCTION

Our natural world is on the verge of a profound loss of biological diversity (Crooks and Sanjayan Chapter 1). Although the economic, cultural, and spiritual costs of this ecological impoverishment are enormous and irreversible, from a human point of view extinction's denouement appears to be "slow-motion." This slow-motion results in a limited recognition of its urgency and the very little time we have to prevent it from occurring. As evident in this volume, the threats cut across multiple scales of ecological organization, from genes and species all the way to ecological processes. To face this complex challenge, action plans to avoid extinction must become more comprehensive, including strategies to preserve both areas and ecological and evolutionary processes, as well as those targeted to avoid the foreseeable extinction of particular threatened species.

One comprehensive regional-scale approach with great promise for effective conservation is based on the concept of "biodiversity conservation corridors," a large-scale planning region where actions are taken to integrate representation and viability of species, ecosystems, and ecological and evolutionary processes in a scenario of explicitly defined human

Connectivity Conservation eds. Kevin R. Crooks and M. Sanjayan. Published by Cambridge University Press. © Cambridge University Press 2006.

needs. The biodiversity conservation corridor approach shifts focus from a local to a regional scale, and represents an ambitious attempt to make protected area networks that are sufficient for species survival besides promoting an optimum allocation of resources to conserve biodiversity at the least economic cost to society (Salwasser *et al.* 1987). The purpose of this chapter is to describe the general principles that define biodiversity conservation corridors and to present considerations for their design and implementation. Conservation International's regional conservation planners are implementing these concepts and guidelines to secure, manage, and, when necessary, restore naturally functioning landscapes that support ecological and evolutionary processes, while at the same time ensuring the preservation of target species.

Expanding the conservation approach

As our understanding of ecological scales, patterns, and processes has improved, both the "where" and "how" of biodiversity conservation have expanded accordingly. At first, areas were given protection for their spiritual, recreational, economic, or esthetic characteristics. Other early conservation efforts that concentrated on the protection of populations of charismatic or economically important species were essentially esthetic or utilitarian. Pioneering conservation strategies sought merely to preserve minimum representative samples of different habitats and ecosystems in protected areas such as parks and reserves. The strategy was largely opportunistic and the areas relatively small. Seeking to minimize conflicts and save on implementation costs, even the rather young discipline of conservation biology, founded in the early 1980s, was largely based on questions concerning the minimum critical size of populations and habitats. Overall, the first generation of conservation biologists was minimalist in its approach to preventing biodiversity loss.

As experience accumulated, however, data showed that these traditional approaches were insufficient to stem or resolve the plethora of problems affecting species and native habitats. Even some of the best-known and relatively large protected areas were failing to maintain viable populations over the long term. For example, ecological decay in many parks in North America has resulted in the loss of many vertebrate species, particularly those with low population densities or high temporal variation in their population numbers (Newmark 1985, 1995).

More importantly, managers and conservationists realized that many site-specific strategies, such as protected areas under a variety of management schemes, were "sitting ducks" in the landscape when faced with the scale of threats to their long-term survival. The tendency was to limit focus to existing protected areas and their buffer zones. As "set-asides," park management dealt only with the most immediate and proximate threats, treating only the symptoms of threats which often have their origins in circumstances prevailing in other areas of the region. Managers, for example, could not anticipate nor deal with a sudden flux of land-hungry migrants displaced by shifts in socio-economic dynamics operating hundreds of kilometers away. Likewise, human-induced climate change, a regional and global-scale phenomenon, threatened the very existence of protected areas, particularly those surrounded by highly modified land-scapes. By focusing only on sites, managers also missed opportunities to improve management capacity and funding for the areas under their control.

This shift in emphasis and strategy led to an expansion of focus from a few individuals of a local population of a critically endangered species to encompass thousands of species and hundreds of diverse habitats. The conservation of biodiversity requires that threats be addressed at their source, and that conservation strategies encompass major ecological processes that allow for the long-term viability not only of wildlife popula-tions, but for the protected areas themselves. Several lessons have con-tributed to this shift. Large population sizes, numbering in the thousands, are critical. Large populations can only be maintained in very large areas, and the insidious effects of habitat and population fragmentation are today a major field of research in conservation biology. Additionally, local ecosystem degradation has regional and global impacts. Landslides, flooding, and pollution are some examples of disrupted ecosystem func-tions with effects on species and economies over enormous areas. Lastly, strategies for conservation at the appropriate scale increase considerably the opportunities for landscape planning, which can address the sources of threats to biodiversity and ecological functions not just their effects (Noss and Daly Chapter 23).

There is now widespread agreement among conservation biologists that this "shift" to a regional approach toward conservation planning is vital to maintain all critical components of biodiversity (Galindo-Leal and Bunnell 1995; Ayres *et al.* 1997; Fonseca *et al.* 1998; Sanderson and Harris 1998; Soulé and Terborgh 1999; Zavaleta *et al.* 2001; Groves *et al.* 2002; Groves 2003; this volume). The biodiversity conservation corridor concept addresses regional biodiversity concerns comprehensively, helping managers to better understand and prioritize conservation policies, research, and actions. Indeed, the conceptualization and realization of biodiversity conservation corridors has become a hallmark program in countries where Conservation International works.

Conservation realities mandate a broader focus

Conservationists have often been reluctant to incorporate lands occupied by humans into species conservation plans because of the potential for social disputes and associated costs of managing conflicts. However, in many circumstances, there is no other option than to broaden the focus of conservation planning to the larger landscape. Conservation actions can then expand or shift entirely to realms physically distant from the core of a protected area. While the level of strategic complexity increases with the move to get out of the trenches, the expansion of the scope of the conservation target enables planners and managers to include long-term habitat requirements in their thinking, an opportunity not usually available when the focus is on a single site such as a protected area. If the political and social environment is conducive for compromises, this may also enable better integration of conservation and development objectives. For instance, by recognizing that the economic potential of land varies greatly over landscapes and by using public policy to match ecological requirements with incentives for low-impact land uses, positive conservation outcomes can be achieved with much lower social costs than have been previously thought possible, while at the same time minimizing political conflict that arises from competing claims over different forms of land use (see also Morrison and Reynolds Chapter 21).

An exclusively site-scale approach can also obscure the identification of threats that originate at a great distance from existing protected areas. Influencing public policy at a regional scale in anticipation of the effects of agriculture, mining, migration, and infrastructure policies can permit their adaptation into a conservation plan at low cost compared with the higher costs of repairing damage from fragmentation, flooding, and landslides caused by these policies (Noss and Daly Chapter 23).

There is, therefore, a need for a two-scaled approach. Local or proximate factors can be addressed at the site scale, and comprehensive landscape planning for both economic development and biodiversity conservation can be carried out at the regional scale. This is not to be confused with Integrated Conservation and Development Projects (ICDPs) which are essentially a site approach. Experience has shown that the attempt to confine solutions to two apparently contradictory objectives within small areas is usually frustrated.

Conservation planning at a regional scale can include implementation options that allow for flexibility in matching funding for the most important areas for conservation with those that have the lowest opportunity costs. The opportunities at specific sites may appear to be so restricted as to make the social and economic costs prohibitive. In essence, the intention of biodiversity conservation corridor planning is to engage both conservation needs and inevitable economic development by finding mutually beneficial interventions that do not necessarily occur within the buffer zones of protected areas. This may include new protected areas to protect watersheds, tourism value-added from landscape management, and the use of tradable development rights and easements to promote development compatible with the movement of species between protected areas.

DEFINING BIODIVERSITY CONSERVATION CORRIDORS

The word "corridor" is used in a number of different contexts. Urban corridors are densely populated strips of land connecting two or more major cities; communications corridors are major transportation routes; and industrial corridors concentrate industrial complexes and businesses. In conservation biology and landscape ecology, a corridor has been defined as a "strip of land or water that differs from the adjacent landscape on both sides" (Levin 2000). They are also called *biological* or *wildlife* corridors and allow for movement of individuals between habitat patches. The idea for the need of corridors between reserves arose as a corollary of the theory of island biogeography (MacArthur and Wilson 1967; Diamond 1975, 1976).

Dispersal has been one of the most difficult aspects of population dynamics to investigate, but the effectiveness of corridors has been demonstrated in nature (Fahrig and Merriam 1985, 1994; Dunning *et al.* 1995; Haas 1995; Rosenberg *et al.* 1997; this volume). Recent large-scale experiments have demonstrated the role of corridors in expanding the effective area of otherwise isolated habitats, in facilitating the exchange of individuals, and enhancing plant–animal interactions, such as pollination and seed dispersal (Tewksbury *et al.* 2002). Even when lacking connection, an "archipelago" of proximate remnants facilitates the movement of many species. A landscape matrix of less intensive land uses may contain "stepping stones" of different kinds between protected areas and function as a biological corridor. A large-scale biodiversity conservation corridor may contain several "archipelagos" or complexes of isolated or semi-isolated habitats, protected areas, and other reserves, each of which constitutes a "nucleus" of habitats connected at various degrees through biological corridors.

The "corridor" terminology has been used to describe other conservation planning concepts. Although varying somewhat in terminology and precise definition, the *landscape corridor* (Soulé and Terborgh 1999), the *ecological corridor* (Ayres *et al.* 1997), and the *conservation corridor* (Sanderson and Harris 1998) all refer to large expanses of land under integrated strategies. The use of the term "corridor" for these recent large-scale conservation blueprints has been met with surprising acceptance in public discussion and in policy arenas. The concept of the biodiversity conservation corridor developed here adds explicit biodiversity conservation targets to the overall process of corridor planning and implementation.

We formally define a *biodiversity conservation corridor* as a biologically and strategically defined subregional space, selected as a unit for large-scale conservation planning and implementation purposes. In this space, conservation action can be reconciled with the land-use demands of economic development freed from the need to find viable solutions within the confines of existing and often small protected areas and their buffer zones. Within biodiversity conservation corridors, conservationists seek to put irreplaceable biodiversity areas under strict protection, allocate economically important areas to development, and identify areas that support both goals through sustainable use and direct incentives for conservation. A biodiversity conservation corridor, therefore, comprises a network of parks and reserves (the most important elements of the conservation strategy), interspersed with areas sustaining varying degrees of human occupation where management is integrated to ensure the survival of the largest possible spectrum of species and specifically avoiding the extinction of threatened species of regional, national, and global value. Identifying all species in peril, as measured by such tools as the IUCN Red List of Threatened Species compiled by the Species Survival Commission (SSC) of the World Conservation Union (Hilton-Taylor 2002), constitutes one of the principal and most fundamental building blocks of a biodiversity conservation corridor.

Biodiversity conservation corridor planning specifically demands the identification of land-use mosaics that are mutually beneficial to economic development and biodiversity. This is possible where protected areas also generate economic value in environmental services for local economies, where "viewsheds" are compensated for their production of tourism value, and where compatible development adds consumer value to practices that promote the movement of species between protected areas. Implementation mechanisms include the institutionalization of tradable development rights and easements, as well as best practices for tourism, industry, and agroforestry that contribute to the survival of threatened and restricted range species in buffer zones and habitats that provide connectivity between protected areas across the landscape.

Biodiversity conservation corridors are both a management response to the critical problem of habitat loss and fragmentation and a proactive response to foreseeable development scenarios for largely unbroken wilderness. Human activities have transformed entire landscapes with agricultural, industrial, and urban activities. The biodiversity conservation corridor concept permits a comprehensive approach to try to halt fragmentation trends in pristine ecosystems, and restore connectivity to maintain biological diversity in highly impacted areas, such as biodiversity hotspots.

The primary purpose of a biodiversity conservation corridor is to prevent the loss of any components of biodiversity and to ensure the perpetuation of ecological and evolutionary processes. The essence of the strategy is its biological objectives, which do not derive from the potential economic benefits to local communities, but provide the framework for the identification of opportunities to achieve development goals. Such a framework provides the basis for the selection of the needed financial mechanisms necessary for addressing the local, national, and global services that these communities will be engaged in protecting. The scale of the corridor is also partially defined by the scale and dynamics of the threats to biodiversity conservation objectives. To conserve biodiversity, conservation planners must understand and attempt to maintain or reconstruct the optimum arrangement of land uses in a landscape where both dynamic human activities and ecological and evolutionary processes coexist (CABS and IESB 2000).

DESIGNING BIODIVERSITY CONSERVATION CORRIDORS

Humans have altered most natural habitats, sometimes significantly. In places such as biodiversity hotspots (Myers *et al.* 2000), natural areas now occur only within a matrix dominated by human activities (Harris and Scheck 1991). Thus, the conceptualization of biodiversity conservation corridors integrating natural and human-altered landscapes should not be imagined as entirely new creations, but rather as the "stitching together" of what once occurred naturally. Conservation actions in this context are meant to restore landscapes that have been negatively altered by human activities to the point where naturally occurring processes have been compromised. Restoring ecosystems for biodiversity and ecosystem processes has also been defined as "rewilding" of the landscape (Soulé and Terborgh 1999). The reverse of this scenario can be found in wilderness areas (Mittermeier *et al.* 2002). There, more ambitious and cost-effective corridors can be planned and proactive conservation made possible by advancing concrete action ahead of the advancement of agriculture and development frontiers.

Elements of a biodiversity conservation corridor

Biodiversity conservation corridors are landscapes of considerable size, perhaps extending for thousands of square kilometers, where policies and actions emphasize the maintenance of biological diversity and other more benign resource use management schemes. Biodiversity conservation corridors must include at least three main elements. First, a *protected area system* is needed that is representative of all species of conservation concern, particularly the most threatened ones; that contains representative and viable samples of different habitats; and that, if sufficiently connected, is large enough to allow for functioning ecosystems services that are vital for the protection of target species. Protected areas form the core areas, or anchors, within biodiversity conservation corridors.

Second, a *connectivity network* is required that is formed by biological corridors or stepping-stone habitat that allows for the full range of spatial and temporal dynamics to take place. *Compatible land uses* and human settlements comprise the third element. The compatibility of "low-impact" land uses with a regional biodiversity conservation corridor is based on their contribution to the maintenance of viable populations of vulnerable species in the region and their contribution to overall ecosystem health. What constitutes compatible land uses is not easily definable, and will vary from region to region and with the conservation objective in question. For instance, carnivores are able to use many more types of land use than restricted forest-dwelling species. Particular land uses that result in sink habitats for most species may nonetheless offer positive contributions for watershed conservation objectives that indirectly benefit the overall landscape strategy. As a general rule, the best guide to what constitutes a land use of high value within a corridor or nucleus of a corridor, deserving of conservation investment, is provided by the requirements of the species most vulnerable to human pressures.

Without establishing the contribution of "low-impact" land use with the survival of a region's species most vulnerable as a result of human actions, a system cannot be said to be "compatible" with biodiversity conservation. An agroforestry system friendly to bird diversity may not be a key element for a corridor if the species for which they are "friendly" are common or invasive species. In this sense, not all species need be treated as equal in a corridor plan (Noss 2002).

A biodiversity conservation corridor can be considered as a logical extension of the biosphere reserve concept (IUCN 1998). In theory, biosphere reserves include one or more core areas and a buffer zone of ecologically sound human activities surrounding them. The objective of the buffer zone is to protect the core area. In a biodiversity conservation corridor, the core areas are also protected areas, but they are nested in a much larger landscape of managed land uses than that traditionally designated as a buffer zone. Clusters of biosphere reserves can be considered as the implementation units within a biodiversity conservation corridor.

Protected area system

Protected areas are the anchors of a biodiversity conservation corridor. If they fail, the entire biodiversity conservation corridor will collapse. They should include viable populations of threatened and restricted range endemic species, besides a viable and representative sample of the regional ecosystems.

Although virtually all protected areas were created in response to desires originating at different sectors of society, including those designed to protect unique cultural artifacts, places of unusual natural beauty, or watershed values, their relative contribution to biodiversity conservation objectives varies widely (Pulliam 1988; Pulliam and Danielson 1991). When choosing priority investments in existing and new protected areas, care should be taken to target those that are most likely to contribute to the conservation of species and habitats of conservation concern.

It is probable that the bulk of the investment in the formation of a corridor will be dedicated to protected areas. Parks and reserves have proven time and again to represent the best tool for the achievement of biodiversity conservation objectives. Parks and protected areas in tropical countries have been effective in deterring most immediate threats to their integrity despite a lack of funding (Bruner *et al.* 2001). Indeed, naturally functioning landscapes such as those that occur in many protected areas cannot be compromised. Nonetheless, many protected areas exist only on maps. In the long run, these areas remain vulnerable to ecological and socio-economic dynamics, and will very likely collapse unless landscape-scale measures are adopted to include conservation measures in the larger matrix.

Connectivity network

The connectivity network is composed of biological corridors and stepping-stone habitats that are also elements of the landscape mosaic or matrix. The network may be present or require restoration to functional levels. The width of biological corridors may range from a hedgerow to several kilometers of forest, including naturally occurring riparian vegetation. For some species with better dispersal abilities, stepping-stone habitat may function in a fashion analogous to biological corridors.

Biological corridors are impacted by adjacent land uses much more than the nuclei of protected areas. Abiotic and biotic edge effects influence corridor quality by reducing the area of interior forest. Changes in temperature, humidity, wind speed, species composition, and ecological processes all affect the quality of the corridor at the edges (Murcia 1995; Gascon et al. 2000). The distance from the edge which is affected varies widely, from 15 to 600 m. However, the major changes are within the first 50 m (Murcia 1995; Knutson and Naef 1997). High-quality corridors should be wide enough to be buffered from external influences: at least 200 or 300 m, and probably much more.

The quality of corridor habitat will vary widely depending on the species. As a general rule, the composition and structure of the vegetation in a corridor should be similar to that of the undisturbed habitats. The size of a functioning biological corridor will depend on the area requirement of the different species. For those with very small area requirements, the corridor can be equated with viable habitat (source habitat).

Riparian habitats, such as gallery forests, are priority candidates for the connectivity network. They are areas of high species diversity, and have been identified as movement corridors for a number of species (Redford and Fonseca 1986). Gallery forests also provide important environmental services such erosion control, water purification, and flooding control. Furthermore, they are frequently associated with recreational and esthetic aspects (Knutson and Naef 1997).

While a major objective of biodiversity conservation corridors is to work towards restoring ecological connectivity, there may be regions that have been isolated for long periods of time, in some cases for thousands of years. In these areas the distribution of one or several species abuts a different group of species. Often, there are related species or subspecies on either side of the geographical barrier. These areas may be good places to set up the limit of biodiversity conservation corridors, or to define corridor implementation nuclei within a larger corridor. Corridor strategies should not, therefore, attempt to link all habitats within a corridor. Even when connectivity has been lost in the recent past, there will always be priorities in restoring links between key protected areas and remaining habitat.

Compatible land uses

Protected areas and biological corridors are the backbone of the bio-diversity conservation corridor, but the matrix has a great influence on its functioning and the probability of long-term persistence of the patches and the species they protect (Taylor *et al.* Chapter 2). Furthermore, in many cases the possibilities to extend the protected area system and to restore the biological corridors will be limited. The only alternative to maintain viable populations may be through the incorporation of biodiversity objectives across the landscape at large.

Compatible land uses are those human activities that, to various degrees, aid the maintenance of regional biodiversity (Gemmill 2002). Particular land uses may, for instance, mimic the composition of the original habitat and as such help maintain populations of some target species. Shaded coffee, shaded cocoa (*cabruca*), and other agroforestry systems may bear some degree of resemblance to the structure of natural forest and provide more benign alternatives to harsher modes of land use such as monocrops and cattle pastures. Conservation investments to support compatible land uses should be evaluated according to whether their location effectively expands the size, or adds missing pieces to, the protected area system and the connectivity network. However, much more research is needed to understand and evaluate the effectiveness of different land uses as to the extent to which they support biodiversity conservation objectives.

Virtually all land within a given corridor has an active or presumed owner or manager. To achieve conservation goals — which can be trans-lated as securing as many irreplaceable and viable fragments as possible within the corridor (ideally contiguous units) — these objectives must be seen in light of other land uses that are interspersed with the con-servation network. Therefore, biodiversity requirements of species or ecological processes within the corridor will determine what can be considered compatible land uses from a management perspective. Defining who owns and manages what units of land, and existing and potential land uses, is fundamental to corridor design once biological targets are defined. In general, priority should be given to securing land units with clear ownership, with significant potential compatibility with biological targets, and which represent large units with one owner that has authority over what happens within the landholding. Examples of such large-scale private landholdings are forestry and mineral concessions, secondary growth or plantation forests, ranches, and agroforestry plantations. Many indigenous territories are managed by an indigenous authority that grants use rights to individuals or communities; thus, they meet these criteria. Other publicly held lands should also be evaluated: easements, military reserves, and land held by municipal, state, or national agencies may contribute to corridors more easily than other types of landholdings.

The definition of the relative allocation of different land uses, from protected areas to compatible forms of productive systems, is a vital step in the process of corridor design and implementation. Participatory processes have been employed to reach agreement on priorities. Participatory negotiations are also important to identify sectors or communities that might be economically affected by the introduction of certain conservation practices and to cost out the need to provide for equitable compensation and other incentives for improving local livelihoods. Expanding the extent of the area under analysis through the adoption of corridor-scale strategies allows for better solutions to reconcile competing conservation and development needs. As an example, region-wide agricultural suitability maps are important layers to understand where the best areas for agricultural intensification reside, usually also a function of proximity to road networks and markets.

KEY CONSIDERATIONS IN CORRIDOR DESIGN

Four principal considerations should guide the selection and definition of biodiversity conservation corridors: species, protected areas, landscapes, and socio-economic dynamics. While a corridor is a region defined by multiple criteria whose relative weights vary across hotspots and wilderness areas, the selection of the area and determination of boundaries is an extremely important step that may influence the likelihood of success of any corridor strategy. Although boundaries can be adjusted and corrected as the strategy and implementation progresses, maps can have lasting effects on how conservation and development proposals are perceived by different sectors of society, and how they may build up support or opposition to the objectives they represent.

Identifying conservation priorities

A biodiversity conservation corridor should ideally encompass all areas of high biological value. These can be identified by an analysis of a series of biological criteria (i.e., endemism, threatened species, species richness, critical habitats, ecological and evolutionary processes), and threats (e.g., land-use changes). Most importantly, the first tier of priorities should include all irreplaceable habitats and the species they contain. The analyses should be done preferably at a large enough scale (hotspot-wide), to identify sub-regions requiring the integrated management of a pro-tected area system with its surrounding landscape. The process of deter-mining conservation priorities must include the identification of clusters of globally threatened species since the major objective of the biodiversity conservation corridor is to prevent extinction. Once these clusters have been identified the area requirements for viable populations of these species must also be estimated.

Expert-based participatory processes involving multiple stakeholders such as non-governmental organizations (NGOs), government agencies, academia, and the private sector must complement the biological assess-ment of priorities using systematic conservation planning tools (see Margules and Pressey 2000). Several biodiversity conservation corridors have already been identified through the participatory priority setting process and have gained national consensus.

Delineating corridor boundaries

Corridors are conservation planning spaces in which both biodiversity and socio-economic objectives are intertwined (see also Morrison and Reynolds Chapter 21). The delineation of the corridor should observe the objectives of maintaining natural processes and biotic assemblages, while also encompassing an arena of conservation action that is strategic for addressing socio-economic threats and opportunities. Therefore, boundary delineation inevitably involves complex compromises. Delineation of corridors is usually an iterative process, based on core principles, but in which adjustments are made as better information becomes available about biogeographic patterns, biological dynamics, and socio-economic factors. While strict criteria have yet to be developed to encapsulate these characteristics, numerous conservation corridors, landscapes, and "seascapes" have already been defined and provide an effective starting point. Boundaries represent compromises between having areas large enough and adequately positioned to encompass the principal drivers of conservation and development, and delineating an area that may end up being too big to be effectively planned and managed. Setting boundaries is a complex endeavor and boundary lines are never to be taken as permanent and immutable.

Landforms provide an approximation for some landscape processes on how boundaries can be identified. Geomorphic processes such as weathering, erosion, deposition, and water catchments are influenced by landforms. In turn, environmental gradients of temperature, humidity, soil depth, and productivity are also influenced by landforms. Geomorphologic features such as mountain chains, drainage basins, valleys, and floodplains may offer a way to begin corridor design.

Remote sensing tools that make use of satellite images and aerial photography, aided by the analytical power of geographical information systems (GIS), are highly recommended mechanisms one can employ in corridor design. Remotely acquired images provide invaluable information on the extent of vegetation cover, current land uses and even on the history of change in a particular region. Digital elevation models (DEM) supply information on landform changes and degree of habitat heterogeneity. A number of valuable statistics obtained from the DEM can be used as an approximation for biodiversity turnover (a measure of how species and communities are substituted along the landscape) and ecosystem services (e.g., erosion rates and catchment regimes). Videography and aerial photography are also important techniques to assist with vegetation and land-use analysis at a more detailed level, particularly in selecting priority areas within corridors.

For some ecosystems, remote sensing provides a first approximation of the conservation status of a region by identifying forest cover continuity. When viewed from space the continuity of natural ecosystems can provide some clues for corridor boundaries. Often, large and continuous forest remnants are included within the corridor. However, disjoint patches in transformed landscapes may need to be included and later restored. When available, information on the spatial distribution of ecosystem types is important to delineate boundaries. Major issues are the inclusion of large continuous ecosystems, of unique ecosystems, and of ecosystem gradients. If not available, some approximation of environmental diversity can be obtained from variables such as topographic diversity, position, and elevation gradients obtained from DEMs.

Ideally a corridor should be as large as possible, certainly large enough to include viable populations of the target species. Corridors on islands may be relatively small and may just encompass environmental gradients of particular importance, such as elevation zones. However, as the size of the proposed corridor increases, the concept begins to lose its power as a useful operational unit. Even small countries such as Costa Rica (50 000 km^2) have problems creating a national network of protected areas and sustainable landscapes. The size of a biodiversity conservation corridor thus depends on what species are present, on the species' life-cycle requirements, the diversity of ecosystems, and the human activities that take place in the corridor. Unlike biological corridors, which are often shaped as elongated strips of habitat linking parks and other areas, there are no particular shapes or spatial arrangement that should characterize the majority of biodiversity conservation corridors. Conservation and development dynamics dictate the final format of a landscape-scale corridor.

An emerging consideration in corridor boundary delineation is the dynamics of climate change. Species ranges and ecosystem distributions are shifting as changes in global climate occur, a pressure that is even more severe if we include changes that act synergistically with habitat degradation, modifying even further regional climatic regimes, and consequently affecting the boundaries of the species' optimal "climatic envelope." Predictive models can facilitate investigation of the effect of potential changes on biodiversity, the alternatives for connectivity requirements, and best options for locating protected areas (Hannah *et al.* 2002).

The obvious place to begin corridor design is examining the existing protected areas. Provided that biodiversity conservation outcomes are being met, corridors can be created by encompassing existing protected areas and by creating new protected areas. While we have emphasized the design of corridors according to natural criteria, corridors remain an abstract human construct composed of a combination of different considerations including socio-economic and geopolitical ones. Human actions are now driving forces of environmental change and effective corridor design can only achieved within the human context. For instance, we may want to publicly restrict the boundary of a particular corridor to a political boundary at the risk of bringing the strategy to a halt because of existing geopolitical concerns.

Addressing socio-economic factors

Landscapes consist of a mosaic of land uses including agriculture, timber extraction, cattle pastures, urban and rural settlements, protected areas, and communication networks. Some of these land uses may have a major impact on regional biodiversity or may be future threats. These root causes of biodiversity loss should be included within the corridor boundaries to share the conservation agenda with the main actors and stakeholders.

Corridor design must respond to differing socio-economic threats and opportunities in biodiversity hotspots and in wilderness areas. Threats to hotspots are typically manifested by increasing levels of fragmentation requiring corridor design to encompass the protection and reconnection of remaining habitat, while threats to wilderness areas are typically manifested in both the potential loss of globally significant carbon sequestration and climate regulation services, as well as extinction threats from progressive habitat reduction and frag-mentation along development frontiers. Corridor delineation antici-pates threats and opportunities, which for hotspots means planning that prioritizes irreplaceable habitat fragments and the potential for habitat restoration, while for wilderness areas the focus will reside along incipient development pathways and large pristine areas with the full complement of large vertebrates still largely unaffected by hunting pressures.

Therefore, economic, legal, and social assessments that confront the social dimensions of biodiversity conservation are an important tool for corridor design and implementation. Using available informa-tion on the profitability of current activities, the opportunity cost of conservation can be mapped. Analysis leading to a spatial projection of future land use premised on "business-as-usual" management is key to the identification of threats and opportunities. The expansion and contraction of land-occupying industries, road construction, tourism development, urban expansion, and especially the projected trends in deforestation can show the likely state of the landscape in 20 years should conservation effort stagnate. The business-as-usual scenario requires the collaboration of resource economists and remote-sensing specialists.

Macro-economic policy analysis helps to discern the role of govern-ment economic policies (e.g., tax structure, subsidies, trade barriers) in contributing to current threats and also to understand the implications of economic development plans in shaping threats. Together these analyses explain and address the macro-economic implications of large-scale conservation and help in the definition of biodiversity conservation corridor boundaries.

IMPLEMENTING CORRIDORS: CHALLENGES AND OPPORTUNITIES

Many challenges face planners working to design and implement effective biodiversity conservation corridors. These challenges range from harmonizing the diverse interests of multiple stakeholders, to coordinating complex institutional arrangements, to managing overlapping institutional mandates and balancing competing economic interests. This complex work takes place in areas that span multiple countries, cultures, and languages. Despite these difficulties, however, working at large scales is absolutely necessary to effectively conserve biodiversity across landscapes.

Managing and coordinating corridor initiatives

Although challenges unquestionably exist, the activity of designing biodiversity conservation corridors also offers great opportunities for stakeholders — from both the development and conservation communities — to achieve their various goals. In many cases, the establishment of multi-stakeholder management committees can coordinate among efforts for protected areas and low-impact uses. Ideally a desirable outcome of this process would be for governments to adopt institutions to coordinate the management of protected areas that are within their corridor regions, so that parks administration and economic policies for the region that impact land use are evaluated together.

Governance

Conservation outcomes depend on government actions: therefore, government offices at local, provincial, regional, and national levels must be effectively engaged in planning biodiversity conservation corridors. The role of government offices is critical in part because public protected areas currently account for over 90% of all protected natural habitats worldwide. Most of the additional terrestrial and marine habitats that are priorities for conservation remain under public ownership. Governments also play the critical roles in providing the majority of conservation funding, enforcing conservation and environmental protection laws, and creating and regulating markets in environmental goods and services.

Government actions can also have serious negative impacts on biodiversity. For example, major public infrastructure projects in wilderness areas can rapidly deforest millions of hectares, as can public subsidies for agriculture and logging. Government subsidies to the fishing industry, together with the failure to control access to fish stocks, have decimated the world's fisheries and damaged marine habitats and ecosystems. Market and policy failures are the two main drivers of overuse and underprovision of environmental assets including biodiversity. Because of government influence on these factors, good governance coupled with transparent and efficient governmental organization is key to effective conservation (World Bank 2003).

Achieving conservation at the regional scale is often constrained by the lack of coherence and transparency among stakeholders, including government agencies. Supportive relationships between NGOs and government agencies is another element essential to implementing corridor designs, though the loci of authority for various aspects of land management are often unclear, as are the official relationships between the offices and agencies involved.

Land-use policy often rests at different levels in different government agencies. For example, the national and/or provincial agencies in charge of forestry, agriculture, roads, and parks may all have responsibility for some aspect of land use in the same localities. Often, responsibility for the enforcement of parks and forestry codes is divided, and with agencies receiving little funding or public support, there is great opportunity for private gain by illegal loggers and others who exploit the gaps in interdiction and prosecution. As corridors are developed, planners should map out the jurisdictions and responsibilities of relevant government agencies and offices to ensure that all are involved at appropriate phases in the development of corridor objectives. An accurate picture of institutional authority and responsibility will greatly enhance the achievement of the goals of biodiversity conservation corridors. In addition to greater transparency of government responsibilities, civil society can also act as a powerful means of improving the accountability of the public sector. NGOs, for example, can contribute to conservation and the effectiveness of government implementation of land-use policies by providing supplementary information on the costs and benefits of proposed projects and implementation plans.

Property rights and community values

Economic incentives are critical to encouraging stakeholders to find long-term solutions to natural resource problems. Unlike tradable goods in the marketplace, public goods, like biodiversity, by their very nature do not create incentives for their own maintenance. As a result, public goods that have no restrictions on access tend to be overused.

Markets can allocate resources to public goods efficiently only if the external costs from their overuse are incorporated into costs that must be paid for them. Without enforcement of rules requiring the incorporation of external costs, individuals and firms will maximize revenue by using public resources without considering the exhausting impact on the reproductive capacity of the resource. This understanding of the incentives to resource use means that individuals easily find it rational to overuse resources. One method of countering the failure of the marketplace to restrict overuse is the development of incentives encouraging sustainable use of resources including negative incentives that impose taxes or penalties for overuse, and positive incentives permitting underusers to sell underuse "credits" to overusers.

The absence of property rights over common pool resources eliminates incentives that communities might otherwise have to self-regulate use. Even with such property rights in place, technological advances in extractive techniques and growing desires for greater income in the community can lead to purposeful destruction of ecosystem resources. Communities can self-regulate use of public resources (Ostrom 1999) when there exist social rules establishing clear boundaries, limits, and appropriate consequences for overuse. Corridor planning can incorporate social regulation where this is possible, and identify situations of rapid social change requiring the clarification and assignment of property rights to resources, giving economic value to biodiversity and ecosystem processes.

IMPLEMENTING CORRIDORS: TOOLS AND APPROACHES

Corridor planners are faced with many challenges in creating innovative approaches to management, as well as technical tools and policy options that create incentives for the conservation of biodiversity within a region. The tools below have already proven useful in the efforts to achieve biodiversity conservation within corridors.

Alliances

Successful alliances across institutions and regions are essential to building successful biodiversity conservation corridors (see also Beier *et al.* Chapter 22). Corridor strategies must define clear roles and niches for institutions so that the strengths and capacities of each institution can be most effectively utilized, including human and financial resources, technical expertise, and the mandates and jurisdiction of organizations within the corridor. At the same time that institutions need to have clear separate roles and responsibilities, it is also important that partners work together, in activities ranging from data-sharing to joint publication. Such collaboration builds trust and tolerance, all essential ingredients in the success of reaching long-term goals and sustaining long-term partnerships.

Work on large-scale corridors requires, in many cases, national-scale initiatives (awareness, policy action) that will require national-scale agreements within the conservation leadership. At other times bilateral cooperation is necessary to achieve conservation outcomes in specific regions. As a result, there is a need for alliances at a variety of scales to perform some very distinct functions.

Institutional capacity-building

In many cases the institutional and technical capacity to undertake effective conservation action at the biodiversity conservation corridor scale is not sufficient to address regional biodiversity challenges. Government departments and agencies at the local, district, and national scales often lack the financial, human, and technical capacity to engage in and implement conservation action effectively. Limited career opportunities, fixed short-term contracts, and low wages together also deter highly qualified people from careers within the local environmental NGO sector and contribute to the sector's weak capacity in many regions. Strengthening local institutions and investing in human resources in fields relevant to conservation is another important component that can ensure the success of conservation efforts over the long term. Skills in communication, conflict resolution, and negotiation become increasingly important as institutions work together to expand conservation planning from the site scale to the scale of biodiversity conservation corridors.

Multi-stakeholder management institutions and committees

Government officials, the private sector, and local NGOs often have diverse goals related to conservation. Their organizations often reinforce disciplinary and institutional segregation of expertise in park management, indigenous affairs, plantation forestry, commercial fisheries, water, and road engineering from expertise in conservation planning. To move beyond the superficial "naming" of corridors, large-scale corridor planning will depend on these actors' recognition of mutual benefits from interagency, interdisciplinary, and participatory approaches to problem-solving. At the regional scale, NGOs can be critical players in the creation of cross-sector committees that are constructed to facilitate corridor-scale activities. These committees help focus investments that support conservation, prioritizing resources towards private or public protected areas, enforcement of forestry and fishery codes, or easements and incentives for verifiable biodiversity-friendly uses. Investments may originate from domestic tax incentives and credits, multilateral sources, the Clean Development Mechanism, or private funding earmarked for biodiversity conservation, or through leveraging investments in tourism, agriculture, or mining sectors.

International contributions of funds to such domestic landscape management institutions can often meet both international and domestic goals, while keeping land ownership and management firmly in domestic hands. This method of managing funds can avoid conflicts over issues of sovereignty. The domestic institution can assess local goals and priorities, set up transparent rules for providing and distributing incentives, set up compliance and enforcement mechanisms, and receive domestic and international financing, both public and private. Domestic institutions can often be integrated with regional development authorities and use funding to address poverty alleviation needs that are only indirectly tied to land use, but are perceived as being part of a comprehensive vision of sustainable local development. Having such an institution in place as a precondition for international conservation finance would allay international fears that the promise of funding would perversely induce greater habitat destruction.

Zoning

The need to expand area coverage of conservation often conflicts with urgent economic pressures. To minimize this conflict, methods of zoning for agroecological uses have been developed to regulate land use over large areas. Zoning aims to direct development to areas of high agricultural potential while restricting land use in ecologically significant and sensitive areas.

The experience with zoning, however, has been disappointing, since zoning goals can only be achieved when zoning regulations are enforced. Zoning enforcement has typically relied on a command and control approach rather than on economic incentives. In practice, enforcement has been problematic where zoning imposes potentially large costs on private actors and where political support is lacking (Mahar and Ducrot 1998). The lack of success in using zoning tools suggests a need for a deeper economic analysis of the basis of zoning, as well as further study of instruments and institutions that can reconcile zoning objectives and landholder incentives.

Role of incentives and enforcement

Historically, restrictions such as laws, regulations, economic policies, executive orders, and other kinds of directives have been the principal approach to environmental policy. These tools have often used negative incentives, through prohibiting overuse, mandating emission limits, or loading zoning requirements with fines and other penalties for non-compliance. Biodiversity conservation corridors will certainly need restrictive tools for managing protected areas, forestry practices, land uses, and buffer zones. However, corridor implementation plans are unlikely to be effective or sustainable if they exclusively use negative incentives. Society may be able to achieve change in the costs of social and economic development more easily through economic instruments, which can be more flexible (Chomitz and Gray 1996; Simpson and Sedjo 1996).

For example, emerging market-based (economic) instruments for conservation financing may make it possible to reconcile landholder incentives with forest conservation. These could include, for example, the potential sale of emissions reductions under the Kyoto Protocol, and the potential of establishing tradable development rights, whereby landholders in environmentally sensitive areas could sell development rights to those in non-sensitive areas. Both these instruments could, in theory, mobilize substantial funds and induce the conservation of areas with high environmental value and low economic value.

Conservation planners and policy-makers must be careful to consider issues of equity, as well as the economic costs and benefits of flexible land-use regulation. For example, sometimes individuals or firms that have contributed most to resource degradation may be the ones most eligible for benefits to protect what is left. This may exacerbate inequality, since compensation for conservation effort is often scaled to the size of the area protected in societies with highly concentrated income based on land ownership. The efficiency of compensating some user rights may also be more apparent than real, because some community-use rights may never have had legal standing or enforcement, resulting in the underestimation of their real opportunity cost. In objective terms, flexible policy implementation has the potential to reduce the cost of compliance in achieving a desired societal objective. Greater economic efficiency from policies institutionalizing markets that reduce resource use where it costs the productive economy the least can free resources that permit society to address issues of equity. Policy reform must evaluate equity issues explicitly, however, to ensure against penalizing the poorest in the process.

A critical problem facing conservation, particularly in the tropics, is the lack of effective enforcement of regulations related to protected areas and natural resource management. When there are no mechanisms for accurately identifying and punishing those who destroy or mismanage aspects of the environment, transgressors have no incentives to stop their behavior. When people believe they are unlikely to be caught or prosecuted for illegal deforestation, land use, or hunting, they will continue to engage in those activities. In addition, the positive impacts of local communities who use their resources responsibly are negated if the activities of outsiders are not controlled. Effective enforcement of regulations that protect natural resources and protected areas is essential for economic instruments and other positive incentives to advance conservation.

The potential for enforcement reform to succeed should not be under-estimated. Often policy-makers know how to make policy implementation more effective, but are prevented from using their know-how. Astute politicians have sometimes risked "untying" this know-how to achieve results for which they can claim credit (Tendler 1998). Stakeholders with resources eligible for compensation and who depend on credible enforcement to gain access to conservation compensation markets also provide a powerful incentive to improved governance of natural resources.

Spatial modeling

Efforts to conserve biodiversity continue to be hampered because conservation goals are not integrated into regional development plans. Conservationists will be best able to work with development authorities if they have explicit plans for biodiversity conservation on a landscape scale (see Groves *et al.* 2002; Groves 2003). Such plans would need to map out how and where conservation and socio-economic goals can be addressed together. Government authorities and other planners can use specific data and criteria to understand the trade-offs between the placement of specific infrastructure projects and the potential impact of that placement on threatened species. Such mapping of information can allow planners to design actions and policies that avoid extinctions for the entire region, indicate economic development opportunities, and minimize the costs of long-term conservation.

REVEALING AND USING THE ECONOMIC VALUE OF CONSERVATION

Over time, conservation planners have learned that to achieve conservation goals, a wide variety of financial tools must be used to engage local, regional, and national communities in conservation actions. Financial tools that translate the value of conserving biodiversity into concrete economic terms are necessary tools for all who are committed to achieving conservation goals. Many organizations have invested significant effort to developing economic models and methods related to conservation; some of these efforts have yielded evidence of the concrete economic value that environmental services provide across landscapes, some have created mechanisms for compensating stakeholders for activities that support conservation goals, and some have designed strategies for attracting investments for achieving specific biodiversity conservation outcomes. As corridors are initially designed, an initial set of financial mechanisms will be brought into play to engage those who, for whatever reasons, require direct or indirect financial benefits from conservation activities. However, since corridors are dynamic entities that change over time, the financial tools best suited to sustain the corridors will change as well. Those who are charged with managing biodiversity conservation corridors will need to remain attentive to the financial dimensions of a corridor, ever conscious of opportunities to use new or adapt existing financial methods to keep stakeholders engaged and committed to conservation goals.

Both corridor developers and managers can use direct and indirect financial approaches. For example, when secure tenure and markets for land exist, direct approaches may pay for land to be protected, including purchase or lease, easements, and conservation concessions, in which conservation organizations bid against timber companies or developers for the right to use government-owned land. An effective approach in one area might be to link tax incentives for municipal jurisdictions to the size of land they have set aside as parks (May et al. 2002). An effective tactic at another scale might be to offer compensation for taking land out of production, as when US farmers were offered bids for letting land lie fallow that were scaled according to the contribution of that land to an environmental conservation objective (Cooper and Klein 1995; Simpson and Sedjo 1996; Wu and Babcock 1996).

Indirect economic activities can also be designed to result in protection of habitat. For example, subsidies to eco-friendly commercial ventures (e.g., ecotourism and bioprospecting) can be used to construct facilities, train staff, or aid marketing or distribution. Mechanisms to provide payments for other ecosystem services such as carbon sequestration, flood and erosion protection, or water purification can also serve both to maintain these services and shelter biodiversity. Financial incentives can also be designed to direct human resources away from activities that degrade habitats. By diverting pressure away from sensitive areas, this approach provides assistance for activities such as intensive agriculture or off-farm employment. These activities may not be eco-friendly, but their expansion in less biologically significant or degraded areas can reduce local incentives to exploit more pristine and irreplaceable habitat (Ferraro and Simpson 2001).

Compensating landowners for conservation can, in many cases, provide significantly better prospects for income than other opportunities available to them. For example, conservation concessions – a lease on land granted by a government to be used for a specific purpose – use financial incentives to compensate local resource owners and users for conservation actions (Bruner *et al.* 2001). Conservation concessions can provide a dependable cash flow equal to what would be paid for destructive and unsustainable use of the land, such as logging. Monies paid to lease land for conservation will not be vulnerable to the vagary of commodity markets, such as timber, or the caprice of changing local currency. Community-owned forest concessions diversify the local economic base, and allow for community participation and control. In addition, local stakeholders can manage concessions more effectively than remote stakeholders. Conservation concessions can be used to protect large areas of land and water in a wider context that provides flexibility in regional corridor planning.

As an example of conservation concessions, in 2000, Conservation International leased a 200 000-ha tract of pristine forest deep in the heart of Guyana, in northern South America. Conservation International used an initial 3-year opportunity to demonstrate that just as harvesting forests can make money, protecting forests can also be profitable. As part of the deal, Conservation International also committed funds for managing the tract as a natural reserve. Guyana and Conservation International are working to find sustainable and non-destructive ways to use forest resources, such as by developing ecotourism or by selling carbon sequestration credits. Although the lease is renegotiable, all involved appear committed to keeping the land safe from destructive commercial interests.

The business sector can also play a critical role in shaping the market for conservation-compatible land use. For example, Starbucks, the largest coffee retailer in the USA, is promoting conservation coffee by providing economic incentives to those farms that maintain traditional coffee production practices. These practices promote growth of coffee under the shade of trees, maintaining a complex vertical habitat structure beneficial to a diverse community of birds. The Tour Operators' Initiative, a global network of tourism operators, banded together to promote the business benefits of sound environmental practices in the tourism industry.

Biodiversity has economic value, by default, because it exists on the same planet with humans. Conservation researchers, planners, practitioners, and policy-makers must highlight every aspect of that value and use it consistently and inventively, but most importantly non-invasively and sustainably, to create and sustain biodiversity conservation corridors.

ACKNOWLEDGEMENTS

This document has been shaped by many contributors, and provides an example of collaboration at its best. All contributors are gratefully acknowledged including Eustace Alexander, Redempto Anda, Artemio Antolin, Mário Barroso, Philippa Benson, Charlotte Boyd, Thomas Brooks, Roberto Cavalcanti, Bernard De Souza, Claude Gascon, Frank Hawkins, Elizabeth Kennedy, Cecília Kierulff, Reinaldo Lourival, François Martel, James-Christopher Miller, Russell Mittermeier, John Musinsky, Efraín Niembro, Elizabeth O'Neill, James Peters, Jaime Salazar, Carly Vynne, Stacy Vynne, and Iwan Wijayanto.

REFERENCES

Ayres, J. M., G. A. B. da Fonseca, A. B. Rylands, *et al.* 1997. *Abordagens Inovadoras para Conservação da Biodiversidade do Brasil: Os Corredores Ecológicos das Florestas Neotropicais do Brasil – Versão 3.0*, Programa Piloto para a Proteção das Florestas Neotropicais, Projeto Parques e Reservas. Brasília: Ministério do Meio Ambiente, Recursos Hídricos e da Amazônia Legal (MMA), Instituto Brasileiro do Meio Ambiente e dos Recusos Naturais Renováveis (Ibama).

Bruner, A. G., R. E. Gullison, R. E. Rice, and G. A. B. da Fonseca. 2001. Effectiveness of parks in protecting tropical biodiversity. *Science* 291:125–128.

CABS and IESB. 2000. *Designing Sustainable Landscapes*. Washington, DC: Center for Applied Biodiversity Science (CABS), Conservation International, and Ilhéus, BA, Brazil: Instituto de Estudos Sócio-Ambientais do Sul da Bahia (IESB).

Chomitz, K. M., and D. A. Gray. 1996. Roads, land use and deforestation: a spatial model applied to Belize. *World Bank Economic Review* 10:487–512.

Cooper, J. C., and R. W. Klein. 1995. Incentive payments to encourage farmer adoption of water quality protection practices. *American Journal of Agricultural Economics* 78:54–66.

Diamond, J. M. 1975. The island dilemma: lessons of modern biogeography studies for the design of natural reserves. *Biological Conservation* 7:129–146.

Diamond, J. M. 1976. Island biogeography and conservation: strategy and limitations. *Science* 193:1027–1029.

Dunning, J. B., R. Borgella, K. Clements, and G. K. Meffe. 1995. Patch isolation, corridor effects, and colonization by a resident sparrow in a managed pine woodland. *Conservation Biology* 9:542–550.

Fahrig, L., and G. Merriam. 1985. Habitat patch connectivity and population survival. *Ecology* 66:1762–1768.

Fahrig, L., and G. Merriam. 1994. Conservation of fragmented populations. *Conservation Biology* 8:50–59.

Ferraro, P. J., and R. D. Simpson. 2001. Cost-effective conservation: a review of what works to preserve biodiversity. *Resources for the Future* 143:17–20.

Fonseca, G. A. B., da, J. M. Ayres, A. B. Rylands, *et al.* 1998. *A new vision for the conservation of biological diversity of Brazilian Rainforest: The ecological corridors concept – version 3.0*, Projeto Parques e Reservas PPR-PP/G7. Brasília: Ministério do Meio Ambiente.

Galindo-Leal, C., and F. Bunnell. 1995. Ecosystem management: implications and opportunities of a new paradigm. *Forestry Chronicle* 71:601–606.

Gascon, C., G. B. Williamson, and G. A. B. da Fonseca. 2000. Receding forest edges and vanishing reserves. *Science* 288:1356–1358.

Gemmill, B. 2002. *Managing Agricultural Resources for Biodiversity Conservation: A Guide to Best Practices*. Nairobi, Kenya: UNEP/UNDP Environment Liaison Centre International.

Groves, C. R. 2003. *Drafting a Conservation Blueprint: A Practitioner's Guide to Regional Planning for Biodiversity*. Washington, DC: Island Press.

Groves, C., D. B. Jensen, L. L. Valutis, *et al.* 2002. Planning for biodiversity conservation: putting conservation science into practice. *BioScience* 52: 499–512.

Haas, C. A. 1995. Dispersal and use of corridors by birds in wooded patches on an agricultural landscape. *Conservation Biology* 9:845–854.

Hannah, L., G. L. Midgley, T. Lovejoy, *et al.* 2002. Conservation of biodiversity in a changing climate. *Conservation Biology* 16:264–268.

Harris, L. D., and J. Scheck. 1991. From implications to applications: the dispersal corridor principle applied to the conservation of biological diversity. Pp. 189–220 in D. A. Saunders, and R. J. Hobbs (eds.) *Nature Conservation*, vol. 2, *The Role of Corridors*. Chipping Norton, NSW, Australia: Surrey Beatty and Sons.

Hilton-Taylor, C. 2002. *2002 IUCN Red List of Threatened Species*. Gland, Switzerland: IUCN. Available online at http://www.redlist.org

IUCN. 1998. *Biosphere Reserves: Myth or Reality?* Proceedings of a Workshop at the 1996 IUCN World Conservation Congress, Montreal, Canada. Gland, Switzerland: IUCN.

Knutson, K. L., and V. L. Naef. 1997. *Management Recommendations for Washington's Priority Habitats: Riparian*. Olympia, WA: Washington Department of Fish and Wildlife.

Levin S. A. (ed.) 2000. *Encyclopedia of Biodiversity*. New York: Academic Press.

MacArthur, R. H., and E. O. Wilson. 1967. *The Theory of Island Biogeography*. Princeton, NJ: Princeton University Press.

Mahar, D., and C. E. H. Ducrot. 1998. *Land-Use Zoning on Brazil's Tropical Frontier: Emerging Lessions from the Brazilian Amazon*, Economic Development Institute Case Study 19674. Washington, DC: World Bank.

Margules, C. R., and R. L. Pressey. 2000. Systematic conservation planning. *Nature* **405**: 243–253.

May, P. H., F. Veiga Neto, V. Denardin, and W. Loureiro. 2002. Using fiscal instruments to encourage conservation: municipal responses to the "ecological" value-added tax in Paraná and Minas Gerais, Brazil. Pp. 173–200 in S. Pagiola, J. Bishop, and N. Landell-Mills (eds.) *Selling Forest Environmental Services: Market-Based Mechanisms for Conservation and Development*. London: Earthscan Publications.

Mittermeier, R. A., C. G. Mittermeier, P. Robles, *et al.* 2002. *Wilderness: Earth's Last Wild Places*. Mexico City: CEMEX, S.A.

Murcia, C. 1995. Edge effects in fragmented forests: implications for conservation. *Trends in Ecology and Evolution* 10:58–62.

Myers, N., R. A. Mittermeier, C. G. Mittermeier, G. A. B. da Fonseca, and J. Kent. 2000. Biodiversity hotspots for conservation priorities. *Nature* **403**:853–858.

Newmark, W. D. 1985. Legal and biotic boundaries of western North American national parks: a problem of congruence. *Biological Conservation* **33**:197–208.

Newmark, W. D. 1995. Extinction of mammal populations in western North American National Parks. *Conservation Biology* 9:512–526.

Noss, R. 2002. Context matters: considerations for large-scale conservation. *Conservation Biology in Practice* 3:3.

Ostrom, E. 1999. Coping with tragedies of the commons. *Annual Review of Political Science* 2:493–535.

Pulliam, H. R. 1988. Sources, sinks, and population regulation. *American Naturalist* **132**:652–661.

Pulliam, H. R., and B. J. Danielson. 1991. Sources, sinks, and habitat selection: a landscape perspective on population dynamics. *American Naturalist* **137**:S50–S66.

Redford, K. H., and G. A. B. da Fonseca. 1986. The role of gallery forests in the zoogeography of the Cerrado's non-volant mammalian fauna. *Biotropica* 18: 126–135.

Rosenberg, D. K., B. R. Noon, and E. C. Meslow. 1997. Biological corridors: form, function, and efficacy. *BioScience* **47**:677–687.

Salwasser, H., C. Schönewald-Cox, and R. Baker. 1987. The role of interagency cooperation in managing for viable populations. Pp. 159–173 in M. Soulé (ed.) *Viable Populations for Conservation*. Cambridge, UK: Cambridge University Press.

Sanderson, J. G., and L. D. Harris (eds.) 1998. *Landscape Ecology*. Boca Raton, FL: Lewis Publishers.

Simpson, R. D., and R. A. Sedjo. 1996. Paying for the conservation of endangered ecosystems: a comparison of direct and indirect approaches. *Environment and Development Economics* 1:241–257.

Soulé, M. E., and J. Terborgh (eds.) 1999. *Continental Conservation*. Washington, DC: Island Press.

Tendler, J. 1998. *Good Government in the Tropics*. Baltimore, MD: Johns Hopkins University Press.

Tewksbury, L. T., R. A. Casagrande, B. Blossey, M. Schwarzlaender, and P. Haefliger. 2002. Potential for biological control of *Phragmites australis* in North America. *Biological Control* 23:191–212.

World Bank. 2003. *Sustainable Development in a Dynamic World: Transforming Institutions, Growth, and Quality of Life*. Washington, DC: World Bank.

Wu, J. J., and B. A. Babcock. 1996. Contract design for the purpose of environmental goods from agriculture. *American Journal of Agricultural Economics* 78:935–946.

Zavaleta, E. S., R. J. Hobbs, and H. A. Mooney. 2001. Viewing invasive species removal in a whole-ecosystem context. *Trends in Ecology and Evolution* 16:454–459.

25

The role of connectivity in Australian conservation

MICHAEL E. SOULÉ, BRENDAN G. MACKEY,
HARRY F. RECHER, JANN E. WILLIAMS,
JOHN C. Z. WOINARSKI, DON DRISCOLL,
WILLIAM C. DENNISON, AND MENNA E. JONES

INTRODUCTION

In Australia and globally, nature and society face a historically unprecedented wave of extinction and ecological degradation (Wilson 2002). Although large ecological reserves are an essential core component of any biodiversity conservation program, protected areas comprise only about 6–12% of the land globally (IUCN 2003) and nationally (Mackey et al. 2006) and are typically widely dispersed and isolated. This percentage of strictly protected land is too small – by a factor of five or ten, even if the reserves were optimally distributed (Soulé and Sanjayan 1998).

In response, critics of conventional conservation (e.g., Soulé and Terborgh 1999) often suggest that long-term prospects for biodiversity will be enhanced the more the entire landscape, irrespective of tenure, is managed as a conservation (rather than a production) matrix. Such a transformation, however, will demand a bolder and more systematic approach to nature protection. This will require increases in the area protected, enhanced biotic and abiotic connections between core protected habitat areas, and reconsideration of the economic and recreational activities on lands where native ecosystems still dominate.

Connectivity Conservation eds. Kevin R. Crooks and M. Sanjayan. Published by Cambridge University Press. © Cambridge University Press 2006.

In North America and elsewhere, it has been recognized that existing conservation initiatives fail to provide sufficient area and ecological connectivity to accommodate the key, large-scale, long-term ecological processes necessary to sustain natural systems (Soulé and Terborgh 1999; this volume). Neither do they allow for evolutionary adaptation to environmental change. The current situation for biodiversity in Australia is similar (Australian Government 2001). During the last two centuries 19 vertebrate species have become extinct and a further 10 have disappeared from the mainland (Australian Museum Online 2004). If current trends in land use and degradation continue unabated, the future will be as grim or worse (compare Recher 1999; Garnett and Crowley 2000). Not only is vertebrate biodiversity at risk from the intensification of land use (especially logging, grazing, and cropping) and invasive species, but most conservation responses to threatening processes do not consider the necessity of large-scale connectivity processes.

In response, in 2000 the Wilderness Society Australia launched the WildCountry Project (WildCountry) in partnership with other non-government organizations, government at state and local levels, industry and private landowners, and the Wildlands Project USA (Wildlands). Mackey *et al.* (2006) present the scientific and technological framework for this project, including the need for a more extensive system of protected (core) areas. The continent-wide, WildCountry conservation plan will be developed and implemented through cooperative regional projects and partnerships. The goal is to comprehensively address and halt the continuing degradation of Australia's biotic diversity by providing a positive, detailed, science-based vision for the integration of existing conservation programs into an expanded, interconnected system of core reserves, and the compatible management of off-reserve lands and waters. The purpose of this chapter is to encourage discussion of just one of the essential elements of such a project – the maintenance and restoration of large-scale ecological connectivity at landscape, regional and continental scales.

CONNECTIVITY FOR BIODIVERSITY PROTECTON

Change and heterogeneity at all geographic scales are as much a cause of natural diversity as they are products. In other words, dynamic ecological processes or flows are essential for both the evolution and the persistence of species and ecosystems. Ecosystems are open systems and will decay if cut off from continuous or episodic inputs of many kinds or if barriers prevent biotic and abiotic flows. Thus, ecosystem integrity and resilience require ongoing exchanges of energy, water and nutrients. Some inter-changes or flows require locations to be contiguous or adjacent. Alternatively, such flows may involve tele-connections – aerial exchanges and flows between distant locations. The terms landscape connectivity and landscape permeability are frequently employed in conservation biology to remind us of this ecological imperative for interchange.

Interchanges of plants and animals (or their propagules) that maintain species diversity occur at many temporal and spatial scales, from local to intercontinental and from daily to decadal or much longer. Organisms must be able to move in order to forage, migrate, and disperse to locate new territory or other habitat resources. In Australia, the ubiquity of rela-tively infertile soils and extreme temporal and spatial variability in rainfall and productivity periodically requires many species to move long distances (Nix 1976; Morton et al. 1995). At a local scale the persistence of popula-tions may be compromised in the absence of export and import of both individuals (for demographic rescue) and genetic material to maintain heterozygosity and minimize inbreeding and genetic drift (Soulé 1980). Finally, species diversity can depend on the presence of effective numbers of highly interactive species (see below) that often require landscape permeability at regional scales (Terborgh et al. 1999; Soulé et al. 2003). Long-term ecological resilience, therefore, requires all of these kinds and scales of movement (Dobson et al. 1999).

We assume, therefore, that the maintenance of movements and flows at all scales is a critical component of any conservation strategy. The term "connectivity" is generally used to convey this idea, though it lacks specifi-city. Some of the terms in use for the planned or designated conservation elements that allow for connectivity or the persistence of essential move-ments and flows are landscape linkages, wildlife or ecological corridors, and stepping stones. Because the term "corridor" is used in many other contexts, including for utility rights of way, recreational routes, networks of protected areas, and roads, and because the term is colloquially associated with narrow passageways in the built environment, it is falling out of favor in the conservation biology literature. The phrase "land-scape permeability" is often substituted for connectivity, in part because (1) it suggests the importance of dynamic processes, (2) it reminds us of the species-specific nature of obstacles to movements, (3) it requires conservationists to consider the landscape (including the "matrix" of unprotected country) as a whole, rather than focusing on narrow, defined corridors.

CONNECTIVITY FOR EXTENSIVE ECOLOGICAL PROCESSES IN AUSTRALIA

The major objective of this chapter is to briefly describe the ecologically extensive processes in Australia most relevant to the conservation of biodiversity. We identify seven such connectivity-related phenomena. Our premise is that conservation in Australia cannot succeed unless conservation planning addresses these phenomena at all relevant spatial and temporal scales. The following subsections briefly describe this "set of seven."

1. Critical species interactions

Species with relatively high per capita interaction strengths have been referred to as keystone species (Paine 1969; Ledec and Goodland 1988; Power and Mills 1995) or strongly interacting species (Soulé *et al.* 2003). Among the ecologically important activities of such species are the creation of structures such as cavities, burrows, and dams, and interactions such as predation, pollination, and competition. While climate and climatically driven primary productivity are ultimate determinants of productivity and the structural qualities of vegetation, species themselves often play a major role in regulating species diversity and how energy, water, and nutrients are distributed in an ecosystem (Soulé *et al.* 2003). The interactions of animal species, for example, can have profound effects on the number of trophic levels and on the distribution, abundance, and population dynamics of species in the same and in other levels (Hairston *et al.* 1960).

The disappearance of relatively interactive species, therefore, often causes profound simplification and restructuring of ecosystems, and can initiate ecological chain reactions, or trophic cascades that may lead to the disappearance of entire ecosystems, causing a rapid decrease in species diversity (Paine 1969; Arnold and Wassersug 1978; McNaughton *et al.* 1989; Hall *et al.* 1992; Gillespie and Hero 1999; Oksanen and Oksanen 2000; Terborgh *et al.* 2001; Soulé *et al.* 2003). This is why so much emphasis is placed on the wolf and other large carnivores by conservation planners in North America (Carroll Chapter 15; Noss and Daly Chapter 23). Australian ecosystems are unique in that the native marsupial carnivore fauna has been largely replaced by introduced placental predators (dingoes *Canis lupus dingo*, foxes *Vulpes vulpes* and cats *Felis catus*), yet the interactions among these exotic species can be critical for the survival of many of the persisting marsupials (Lundie-Jenkins *et al.* 1993;

Corbett 1995; Risbey *et al.* 2000; O'Neill 2002). Thus, it is essential that conservation networks be designed so that major ecological players persist in core areas. In other words, the landscapes that surround core areas must be permeable to dispersing and migrating individuals of relatively interactive species. In the absence of such permeability, the risk of local extirpation of such species in core areas is high.

Among highly interactive species are mycophagus mammals (Johnson 1996); honeyeaters (Paton *et al.* 2000); water birds (Roshier *et al.* 2001); frugivores, granivores, and other insectivores (NLWRA 2002); pollinators and animal dispersers of seeds and fungal spores. Conservation biologists and planners should identify as many of these species as possible, and determine the threats to their dispersal or migration routes. The potential impacts on such species of deleterious interactions, such as those identified between exotic grasses and altered fire regimes in northern Australia, also need to be considered.

2. Long-distance biological movement

Conservation planning must explicitly consider long-distance biological movement. Both vertebrates and invertebrates can have stages in their life cycles that are associated with large-scale movement (Isard and Gage 2001). Anywhere between 30% and 60% of Australian woodland and open-forest birds are non-residents and their persistence in a region may depend on large-scale movements that occur either seasonally (migratory) or from year to year (episodic or dispersive) (Griffioen and Clark 2002; Recher and Davis 2002). The propagules of all plants disperse, with the scale of movement depending on life-history attributes and the extent to which dispersal is aided by wind, animal vectors, or water flow.

Long-distance animal movement is strongly associated with temporal variability in primary productivity and associated food resources (Nix 1976). While parts of the continent experience relatively high levels of seasonally reliable primary productivity, the entire continent is subject to extreme year-to-year variability in precipitation (Hobbs *et al.* 1998). This year-to-year variability, coupled to the semi-arid and arid climatic regimes that dominate around 70% of the continent, has been a core factor in the evolution of Australia's wildlife and on the commonness of dispersive life-history characteristics.

It follows that, among other things, habitat loss, fragmentation, and modification reduce the likelihood of wildlife finding suitable resources, and thereby decrease the probability of reproductive success and survival. Large areas incorporating an interconnected network of patches are essential for many species in such dynamic systems. In addition, the removal of any particular patch – including those that serve as "stepping stones" for long-distance movements – may affect the whole system, with consequences far beyond the proportional loss of habitat in the system as a whole. Over time, the cumulative effects of patch removal can lead to widespread extirpation of species, in part because some patches are more critical than others owing to their precise locations and the resources they provide (Woinarski *et al.* 1992, 2000; Price *et al.* 1999).

The present and future obstacles to long-distance movements need to be evaluated when considering the long-term viability and ecological effectiveness of interactive species and the potential for speciation as discussed below. Such obstacles may be caused by processes outlined below, including deleterious fire regimes (Mackey *et al.* 2002), the construction of barriers like roads, dams, and fences, the failure of hydroecological processes, global climate change and drought, and the degradation of appropriately spaced habitat resources such as water, food, and resting sites. Finally, the protection of refugia that provide resources to dispersive species during times of stress is of paramount importance (Mackey *et al.* 2002). Conservation plans must ensure that the landscape is permeable to movements in and out of remote refugia, even if the intervals between the episodic events that provoke such movements may be on the order of centuries.

3. Disturbance at local and regional scales

Many categories of disturbance, both natural and anthropogenic, affect landscape permeability. Among these are fire, vegetation clearance, livestock grazing, foraging by feral carnivores and herbivores, weed invasion, and built structures such as roads and dams (Hobbs 2003). Disturbance is natural and inevitable, but anthropogenic disturbance often exceeds the historic range of variability and intensity of natural disturbance regimes. In systems fragmented by vegetation clearance and modification, such as many of the woodlands and grasslands in eastern and southern Australia, broad landscape processes have been disrupted for many decades leading to altered fire regimes (Gill and Williams 1996; Hobbs 2002). Moreover, species and life stages respond idiosyncratically to disturbance. Conservation plans, therefore, must create scenarios for every possible kind and degree of disturbance to ensure that the projected network of protected areas remains permeable at appropriate spatial and temporal scales to all native species. All categories of disturbance need to be considered independently and in combination.

Fire, because it can be an important tool in management, and because it interacts with many other categories of disturbance, is heuristically useful for grasping the complexity of landscape permeability. Fire affects the permeability of landscapes for individual species (Williams *et al.* 1994; Whelan *et al.* 2002), and different kinds and schedules of burning in space and time can affect migration, dispersal, and other kinds of animal movements. For example, the continued depletion of old-growth vegetation by frequent burning has been identified by Woinarski (1999) as a process threatening a number of fire-sensitive bird species in Australia due to the loss of habitat. These birds have relatively limited dispersal ability and low reproductive rates, exacerbating their vulnerability to frequent fire (Woinarski 1999). Whether fire enhances or inhibits landscape connectivity for native flora and fauna depends on the geographic scale of analysis, the ecological context, and the characteristics of species.

4. Global climate change

In coming decades, it is likely that human-forced global climate change will contribute massively to the extirpation of species and ecosystems (Howden *et al.* 2003; Thomas *et al.* 2004). The general basis of this statement is that the climatic envelope within which species currently persist will either (a) cease to be found anywhere or (b) shift geographically such that species are unable to disperse and relocate to a landscape that supports an essential resource.

On the other hand, climate change may enable many species to disperse into landscapes where current competitors and diseases cannot follow (Dobson *et al.* 2003; Johnson and Cochrane 2003). Many species will be able to expand their geographic ranges, with implications for spatially extensive evolutionary processes and for the spread of invasive species (McKenney *et al.* 2003), including infectious diseases (Williams *et al.* 2002). The ecological effects of climate change in the oceans is likely to be as great as those elicited in terrestrial systems (Seibel and Fabry 2003).

Maintaining connectivity in the face of major climate changes will prove a formidable challenge, and decision-makers are already being called upon to mitigate negative effects wherever possible (National Task Group on the Management of Climate Change Impacts on Biodiversity 2003). Conservation planners, too, must consider climate change scenarios in developing plans for the persistence of biodiversity. First, major, climatically driven biome changes cannot be accommodated by small or isolated protected areas. Large, contiguous areas are needed to accommodate essential movements and flows. Moreover, even small relictual assemblages, such as patches of rainforest in northern Australia (Russell-Smith *et al.* 1992), may be the sources of species that will constitute future plant communities when climate change leads to novel conditions (Nix 1982; Hopkins *et al.* 1993). Second, proposed natural resource management strategies should, where necessary, include the translocation of threatened species and the maintenance of habitat linkages to promote species migration and dispersal (Hannah and Salm 2003).

Nevertheless, the resilience of Australia's biota should not be underestimated. The ability of species to evolve and persist on a continent subject to extraordinary climatic variability over millions of years may have preadapted them to overcome some kinds of distributional obstacles.

5. Hydroecology

Hydroecology (Mackey *et al.* 2001) describes the role that vegetation plays in regulating surface and subsurface hydrological flows at local and regional scales, and the importance of water availability to ecosystems and animal habitat. The significance of hydroecology in Australia is amplified by high year-to-year variability in rainfall. Because water is so scarce in most of Australia, attention at all scales to catchment processes, particularly the influence of vegetation cover on infiltration and evaporation, is critical for maintaining perennial springs and waterholes, river base flows, and perennial and seasonal stream flows.

Interruptions of hydroecological processes, whether natural or artificial, can impede regional- and continental-scale phenomena (Pringle Chapter 10). For example, estuarine food abundance for migrating birds may depend on water catchment processes occurring hundreds of kilometers from the ocean (Tracey *et al.* 2004). Hydroecological processes are critical for biodiversity conservation at local, regional, and continental scales because they underpin landscape primary productivity and habitat values. Though such processes have been more often discussed in the context of land degradation and salinity problems (e.g., Littleboy *et al.* 2003), conservation planners and managers must attend to whole-of-catchment dynamics that influence water flow and quality, especially the extent and condition of the vegetation cover and other factors affecting groundwater recharge and discharge.

6. Coastal zone fluxes

Inland and coastal human settlements, agriculture, aquaculture, and industry have increased the inputs of nutrients and biologically active chemicals in coastal ecosystems (Pringle Chapter 10). The consequences – including pesticide-related extirpation of bird populations, heavy metal pollution of fishes, and offshore, anoxic "dead zones" – are well documented (Crowder and Norse 2005). Phytoplankton blooms in rivers, bays, and littoral habitats are increasing in severity and frequency, often to the detriment of benthic communities. Coral reefs are particularly sensitive to nutrient enrichment (Koop et al. 2001).

Natural flows and movements in coastal areas are often interrupted or curtailed by freshwater impoundments and diversions, dredging, chemical pollution, commercial boat traffic, the use of estuaries for aquaculture and recreational boating, and the construction of jetties and breakwalls. These activities and structures can interfere with many ecological and behavioral processes, including reproduction in terrestrial, freshwater, and marine species and the movements of inorganic and organic materials that support estuarine biota and that provide sand and sediments for beaches, estuaries, and bays. The fitness of animals such as migrating shorebirds is affected by coastal zone fluxes of pollutants and by the construction of barriers to water flow such as dams. Many of the activities mentioned above create barriers – sensory, physical, and chemical – to the movements of organisms and their propagules in the water column and in benthic and reef communities.

In the absence of explicit planning and protection, the continuing creation of anthropogenic barriers to natural flows and movements in coastal regions may prove catastrophic for human and natural communities. Given the inevitability of commercial development in the coastal zone, a coastal zone conservation planning framework that explicitly incorporates fluxes of energy and matter and animal movements is a conservation and economic priority for Australia.

7. Spatially dependent evolutionary processes

Biodiversity protection must attend to the conditions necessary for continuing evolution, particularly the potential for adaptation to changing environmental conditions and for speciation (Frankel and Soulé 1981). Ultimately, evolutionary processes require the movement of organisms over relatively long distances. Not only is gene flow (a major source of genetic variability) dependent on connectivity, but landscape permeability is a requisite for range expansion, often a key stage in evolutionary differentiation and speciation (Avise 2000).

Range expansion serves to spread new genetic variants across the landscape (Moritz 1991; Kearney et al. 2003) and to deliver genetically continuous populations into areas that may later become isolated and differentiated. The latter process is exemplified by closely related (sister) taxa in southeastern and southwestern Australia (e.g., Roberts and Maxson 1985), and by the occurrence of mesic-adapted plant and animal communities in isolated pockets within the arid zone (Bowman 1996). Habitat fragmentation in the future could preclude these processes. For species with limited mobility, including many amphibians, even local habitat destruction and the resulting deterioration in landscape permeability can militate against natural evolutionary processes (Driscoll 1998). Genetic differentiation and evolutionary diversification of populations depend on the maintenance of habitat integrity on a regional basis.

For these reasons conservation strategies need to accommodate shifts and expansion of geographic ranges so that evolutionary processes can operate over millennia at local to continental scales (Soulé 1980; Moritz *et al.* 2000). Widespread habitat loss and modification may preclude gene flow and range expansion (Woinarski and Ash 2002; Driscoll 2004) and probably will effectively eliminate speciation and adaptive evolutionary processes. Habitat reconstruction, strategic habitat management, and the restoration of landscape linkages are needed to reinstate natural evolutionary processes.

CROSS-CUTTING CONNECTIVITY ISSUES IN AUSTRALIAN CONSERVATION

The current distribution of protected areas is too sparse and poorly connected to adequately protect Australia's biodiversity in perpetuity. Unless these deficits are corrected soon, the hemorrhaging of biodiversity can only accelerate. The "set of seven" connectivity-related phenomena described in the previous section is just the first step in developing a useful connectivity analysis to inform planning. In part, this is because the interactions among these phenomena are at least as important as the phenomena themselves. When the concern is the long-term persistence and accessibility of refugia, for example, planners must model how an increase in fire size and intensity affects habitat and the long-distance movements of terrestrial animals. For example, refugia for Leadbeater's possum *Gymnobelideus leadbeateri* and other arboreal mammals were found to depend on vegetation structure, which in turn is influenced by the pattern of fire in space and time (Mackey *et al.* 2002).

The twin purposes of this section are to provide a foundation for the integration of the seven categories into a comprehensive biodiversity protection strategy and a preliminary tabular synthesis (Table 25.1)

Table 25.1. *Probable relationships^a between the seven connectivity phenomena and the cross-cutting issues*

Cross-cutting issues	Connectivity-related phenomena						
	(1) Species interactions	(2) Long-distance movement	(3) Disturbance	(4) Climate change	(5) Hydro-ecology	(6) Coastal zone fluxes	(7) Evolutionary processes
Scale and context	Yes	Yes	Yes	Yes	Yes	Yes	Yes
Conservation phase	P, M	P, m	P, M	P, m	P, m	P, M	P, m
Core area impacts	S, l	L	S, l	s, L	L	S, L	S, L
Matrix impacts	D	D	B	D	D	D, B	B
Spatial and temporal analyses	?	Yes	Yes	Yes	Yes	Yes, ?	?
Change	C, A	C, A	T, C, A	T, A	T, C, A	T, C, A	T, C, A

^aYes, clear utility or relevance; P, highly relevant to planning and design; M, highly relevant to management and stewardship; m, often relevant to management and stewardship; S, highly relevant to scale issues; s, sometimes relevant to scale issues; L, highly relevant to location issues; l, sometimes relevant to location issues; D, potential for "sink" problems; B, potential for "barrier" problems; ?, utility unknown or doubtful; T, sensitive to technological innovation; C, sensitive to climate; A, sensitive to intensification or other changes in agriculture.

of interactions among the seven categories so that conservation workers can better grasp the complex spatial, temporal, biological, and social dimensions of the conservation challenge. Table 25.1 illustrates the complexity and interactions of the seven phenomena in the context of seven cross-cutting themes that affect the planning, efficacy, and long-term persistence of conservation networks.

We use the term "cross-cutting themes" as a rubric for this synthesis, because from the perspectives of planning and management, it is essential to avoid oversimplified, single-factor analyses. The seven connectivity-related phenomena are referred to below by their numbers in parentheses in order to facilitate cross-referencing. We recognize that this brief synthesis is not comprehensive. In particular, the symbols (words, letters) used in the cells of Table 25.1 are meant only to alert conservationists to potential interactions and problems. We hope, however, that the table will be useful as a guide and checklist for analysis, planning, and management of conservation systems, including protected area networks.

Scale and context

The first row in Table 25.1 reminds us that scale and context issues are relevant to all of the seven connectivity issues. For example, the conservation value of any particular habitat patch or site depends on its size and its ecological and geographic contexts (5, 6). In addition, propinquity of sites generally increases their utility for maintaining ecological flows and dispersal of plants and animals (2), including relatively interactive species (1). Nevertheless, even relatively isolated patches, such as pockets of forest, wetlands, or estuaries, may serve a critical stepping-stone function at regional and continental scales for migrating or dispersing biota (2), for the persistence of metapopulations, and for tracking ecosystem characteristics during climate change (4). Thus, a small and isolated reserve in the highly cleared wheatbelt in southwest Western Australia will have a higher conservation priority than otherwise realized if it coincides with a dispersal route for birds between the arid and mesic zones (2). Moreover, the value of particular sites can increase over time due to the cumulative effects of habitat degradation (3) and patch extirpation. Finally, whether a type of disturbance, such as fire, is beneficial or harmful to connectivity depends on the habitat and species under consideration. It would be hard to exaggerate the significance of context in conservation, particularly when analyzed from the perspective of appropriately long temporal scales, including those relevant to evolutionary change and diversification (7).

Conservationists must think about connectivity at multiple geographic scales as well. For example, connectivity for less vagile organisms, such as amphibians, is usually a local issue, whereas connectivity for migratory birds can be regional, continental, and inter-continental (2). Both temporal and spatial scales must be considered when the concern is the potential for recolonization following local extirpation or for speciation. Planners should avoid the trap of the "generic" or all-purpose, "one size fits all" corridor, and must depend on specialists to alert them to species-specific connectivity issues (7).

Disturbance of any kind alters the ecological context, affecting ecological flows and movement. For example, fire regimes are altered when systems are fragmented by vegetation clearance and modification (3), such as in woodlands and grasslands in eastern and southern Australia (Gill and Williams 1996; Hobbs 2002). A landscape that is fragmented by roads may be prone to frequent, anthropogenic burning and thus to local extinction of fire-sensitive species. In addition, frequent fires for fuel reduction can adversely affect the persistence of plant species that require long intervals between fires (Mackey *et al.* 2002). It is even possible that revegetation could be hazardous to native species because it might increase the probability of fires spreading into undisturbed habitat (Wasson 2003). At the opposite extreme, local vegetation clearing may result in the absence of fire over long periods. This can occur where the fire-management goals for agriculture differ from those for native flora (Keith *et al.* 2002). Thus, context determines whether fire increases or decreases connectivity, or whether it benefits or reduces biodiversity locally or regionally.

Conservation phases

As indicated in the second row of Table 25.1, conservation is carried out in phases, including an assessment/analytical phase, a design and planning phase (both of which are indicated by "P" in Table 25.1), and a management/stewardship phase (indicated by "M" or "m" in Table 25.1). For example, the significance of episodic, long-distance dispersal (2) and future shifts in geographic range due to climate variability and change (4) may be more germane to the identification of potential core areas and stepping-stone habitat patches during the design phase than for day-to-day management. Similarly, the locations and qualities of barriers to movement (2), such as dams or highways, and the role of evolutionary phenomena (7) should be addressed during the analysis, planning, and design phases. Proposals for future dams or highways are likely to be both design and management concerns. On the other hand, the design and implementation of appropriate fire regimes (3), both in core areas and on matrix (or compatible use) lands, are likely to be a perennial issue for management.

The planning phase is critical because mistakes of omission and lack of vision may not be correctable in the future. In coastal areas, for example, planners must consider potential conflicts between the necessity of habitat permeability for many species and the likelihood of intensive development (6). The virtual certainty of future growth and disturbance (3) in heretofore undeveloped coastal areas such as in northern Australia could severely curtail fresh water flows (5) and animal migrations that are critical for the integrity of coastal ecosystems (2).

Core area enhancement

All seven of the connectivity phenomena affect the conservation value, long-term viability, species diversity, and resilience of core areas. The phenomena inform decisions about issues of size and scale ("S" or "s" in the third row of Table 25.1) of core areas and their location ("L" or "l"). Depending on the kinds and likelihoods of threats, each potential linkage zone connecting core areas should be evaluated for its contribution to the viability of highly interactive species (1), and the probable conservation utility under various scenarios of disturbance (3), development, climate change (4), and other perturbations. In particular, the interaction of connectivity and disturbance regimes (3) needs be deeply embedded into planning processes (Williams 2003) and integrated across all land tenures (Esplin *et al.* 2003). Such systematic analyses will also be of value for developing management plans for core areas and compatible-use lands (see below) and waters (Hale and Lamb 1997; Lindenmayer and Recher 1998; Lindenmayer and Franklin 2002).

Potential core areas should be carefully examined with regard to the impact of land-use changes that compromise connectivity. Wetlands, for example, are likely to be selected as core areas, but they are highly susceptible to water diversion or draining, agricultural development, land clearing (3), and other land-use changes. Economic development of estuaries and their catchments endangers subsistence economies and reduces the quality of life for people living in coastal areas (5, 6). Only those estuaries in remote regions of tropical Australia or western Tasmania are in a nearly pristine state (Tracey *et al.* 2004). In addition, many coastal fisheries have disappeared because of overfishing (which perturbs species interactions) and habitat disturbance (3) locally and in distant catchments (Crowder and Norse 2005).

Freshwater habitats and the core areas that contain them are particularly sensitive to changes in hydroecological processes that can often operate over long distances (5). Among the many examples that illustrate this is the increasing level of water extraction from the Great Artesian Basin. The Basin covers 1 117 000 km² − 22% of the Australian continent − and it has persisted for tens of thousands of years. The Basin stores around 8700 million megaliters and feeds over 1000 remote springs and soaks that support wetland ecosystems, including many endemic species and communities (GABCC 2003), and that are refugia (2) during droughts. Widespread extraction of water from the Great Artesian Basin threatens to decrease the number and size of natural springs, thereby reducing population sizes and increasing the risk of extinction of endemic species (Ponder *et al.* 1995; Tyre *et al.* 2001). In tropical Australia, maintenance of the integrated subsurface/surface hydrological processes is essential to the biology and ecology of the plants and animals of the region (Horn 1995; Horn *et al.* 1995).

The matrix

The long-term prospects for biodiversity will be enhanced the more the entire landscape, irrespective of tenure, is managed as a conservation (rather than a production) land. Advancing this objective, however, will require (a) the mitigation of threats posed by matrix lands and waters to biodiversity, (b) better linking and buffering of core areas, and (c) changed land use and management to promote landscape permeability for ecological flows. Row 4 of Table 25.1 suggests how connectivity to non-core, matrix areas can affect the integrity of species and ecosystems in cores; "D" indicates the potential for deleterious effects related to "sink-like" qualities in matrix areas or to harmful disturbances emanating from such areas; "B" indicates processes in matrix areas can create barriers to movements and flows.

Some "matrix" or non-core areas in the vicinity of high-value core sites may provide connectivity that helps to sustain populations of vulnerable species, even if such areas lack the qualities necessary for permanent residency. On the other hand, lands and waters under intensive economic uses such as irrigated agriculture and aquaculture may entrain "sink-like" conditions such as high mortality rates (*sensu* Pulliam and Danielson 1991), and create barriers that threaten natural ecological flows and movements. In any case, all non-core areas should be examined systematically for their current and potential connectivity conservation opportunities and threats to biodiversity. For instance, it should not be assumed that presumptive compatible-use areas, including pastoral or forestry lands, would benefit particular taxa. Plantations, though they may provide shelter, foraging, and nest sites, can have elevated mortality rates for some or all life-history stages such as nestling birds. Other sink-like qualities of matrix areas can include low insect productivity and toxicity caused by herbicide (for understory) and insecticide use (R. Hobbs, pers. comm.). Planners must also assume that the kinds of economic uses on matrix lands will change and possibly intensify over time.

Wide-ranging species (2) may be particularly vulnerable to the sink-like qualities of unreserved, matrix lands where survival rates of foraging or dispersing individuals are low. This is particularly problematic for highly interactive species such as predators (1). The tolerance of matrix land managers for dingoes, for example, may affect their densities in core areas and indirectly affect the persistence of small marsupials (O'Neill 2002). Dingoes control feral pigs (*Sus scrofa*), kangaroos (*Macropus* spp.), and emus (*Dromaius novaehollandiae*) (Pople *et al.* 2000; Newsome *et al.* 2001); they also may determine the local distribution, numbers, and predatory impacts of feral cats and foxes (Corbett 1995; Edwards *et al.* 2002; O'Neill 2002), which can cause local decline and extinction of the smaller marsupials, including native carnivores (Lundie-Jenkins *et al.* 1993; Risbey *et al.* 2000; Morris *et al.* 2003). Conservation planners need to know more about these potential benefits for native vegetation and marsupials. If such benefits occur, it would behoove planners to consider whether dog fences, poisoning, and access to water sources help or hinder the protection of native ecosystems.

Spatial and temporal analyses

The large-scale connectivity processes discussed here are essential for biodiversity assessments and planning. Data are often lacking, however, to conduct the necessary space/time studies at the required scales. For example, climate change (4) in the past has altered the composition of ecological "communities" and species that are associated now may no longer be sympatric in the future (Graham and Lundelius 1984). Thus interspecies interactions (1) that we take for granted today, such as pollination, seed dispersal, predator–prey relations, herbivory, and plant–microbial symbioses (Hughes 2003), will be less predictable and are beyond the capacity of current models to predict.

Nevertheless, certain phenomena may be amenable to analysis by geographical information systems (GIS) and remote-sensing tools. It is possible to analyze threatening processes such as land clearing and overgrazing (3), the likely locations of refugia (2) and aspects of hydro-ecology (5). Analysis of other phenomena, including the habitats of threatened species and the roles of predators as top–down regulators of ecosystems (1), will typically rely on new field surveys (perhaps complemented by remotely sensed information) or may require deliberate field experimentation at local or regional scales.

Each issue that arises during any of the conservation phases mentioned above should be scanned with a checklist of available methodologies in order to prevent gaps in analytical rigor. For example, advances in GIS, environmental modeling, and remote sensing enable the classification, mapping, and tracking of the temporal variability in the distribution and availability of primary production and hence food resources (Landsberg and Waring 1997; Austin *et al.* 2003; NASA 2003). These analytical capabilities add to existing technologies and aid in identifying core habitat, together with dispersal and migration linkages (2) at local, landscape, regional, and continental scales (Mackey *et al.* 1988, 1989, 2001; Lesslie 2001; Mackey and Lindenmayer 2001).

Anticipating change

The ecological, economic, and social systems in which conservation operates are dynamic and difficult to predict, although we can be quite certain of some changes in Australia. One of these is that human populations will continue to grow in coastal areas. Row 6 of Table 25.1 suggests certain obvious categories of change that are likely to exacerbate landscape permeability and flows during the next few decades; these categories are changes in technology ("T"), changes in climate ("C"), and agricultural intensification ("A").

Technology will continue to be a major driver of changes on the land. The increasing rate of technological innovation will exacerbate development pressures in heretofore-intact country. One of the most threatening technologies to Australia's biodiversity is the desalinization of seawater. Fresh water produced by desalinization, even if relatively expensive, may open up vast areas for marina and resort development (6), and even expensive fresh water may open the flood gates to intensive forms of farming and aquaculture. Another threatening technology that will emerge in the next decade or so is all-season, all-terrain vehicles capable, for instance, of carrying people and goods throughout seasonally flooded regions such as northern Australia. Such transport will accelerate economic development, tourism, and habitat fragmentation.

Dramatic changes in climate are also likely (4). The average global surface temperature is projected to increase by about 2.5 °C to 3 °C by this century's end (Kerr 2004) with the projected rate of warming very likely to be without precedent during the last 10 000 years (Taylor 1999). Sea levels are predicted to rise between 0.5 and 2.0 m. The future climate suggested by regionally scaled global change models for Australia is detailed in CSIRO (2001). Annual average temperatures are projected to be 1.0−6.0 °C warmer over most of Australia by 2070. By 2070, the range of predicted change in precipitation is −20 to +20%, with locally unpredictable consequences for the biota.

The future of non-protected areas is uncertain given current rates of land conversion, population growth, agricultural intensification, and species introductions, not to mention rapid, unpredictable technological innovations that facilitate access to intact country, human aspirations in the poorer nations, and the desires of investors to maximize profits using ecologically unsustainable practices. Therefore, some of the current discussion about the conservation value of unreserved or small areas (e.g., Daily *et al.* 2001; Rosenzweig 2003), if interpreted too broadly, can lead non-ecologists to a false sense of security about the utility and compatibility of matrix or off-reserve lands for biodiversity protection. Improved long-term conservation outcomes will not occur on non-protected areas by accident. Rather, careful long-term planning is an imperative. Planning for the long-term conservation of biodiversity must assume worst-case scenarios and take an unashamedly cautionary approach. The current limited knowledge about most of the ecological connectivity issues discussed here is further impetus for a precautionary stance (see Crooks and Sanjayan Chapter 1).

CONCLUSIONS

This chapter has identified a set of seven ecological processes and phenomena that require connectivity at continental, regional, and landscape scales. This "set of seven" is part of the preliminary scientific framework for the WildCountry Programme − a new approach to conservation assessment and planning in Australia initiated by The Wilderness Society. The overarching goal of WildCountry is to protect Australia's biodiversity by creating an expanded system of core reserves, sustained by ecologically permeable landscape linkage zones and compatible management of off-reserve lands and waters.

The next step is to examine these seven processes and their interactions in the context of particular regions to determine how they will be regulated and optimized — singly and in combination — to maintain native biodiversity in perpetuity at all spatial and temporal scales. This effort will require both systematic research and broad consultation through informal contacts, literature reviews, workshops, conferences, and partnerships across all sectors. It will also require a more realistic, rigorous approach to conservation in general.

Culturally, one of the major impediments to effective conservation, worldwide and in Australia, is the ignorance of connectivity's role in sustaining ecological dynamics and diversity. Though society has been relatively successful in protecting scenic landscapes and isolated intact country and wild rivers, most of these successes will be pyrrhic victories in a few decades if greater attention is not paid to connectivity. Knowledge about the phenomena related to natural flows and movements has been increasing, and there is a growing effort to attend to the regional and continental scales in conservation (Soulé and Terborgh 1999; this volume), but nowhere have these phenomena been systematically integrated into conservation assessment and planning on regional and continental scales.

Effective ecological connectivity for biodiversity conservation will be an ongoing research and development challenge if for no other reason than all ecosystems will be subject to climate change, exotic species introduction, and new kinds of landscape-altering technologies that must elicit a "futuristic attitude" in conservationists. Conservation planners must assume attempts will be made to exploit for private benefit virtually every landscape or natural resource on or near the continent using technologies that cannot even be imagined today. The key actions we can take now to enable biodiversity to survive are to (a) conserve in perpetuity large, contiguous areas to promote the integrity of natural processes across regionally scaled climatic gradients (ensuring that the landscape remains permeable to all beneficial ecological processes), (b) design such systems to ensure effective movements and fluxes under all imaginable scenarios, (c) protect regionally anomalous ecosystems or refugia as the possible sources of species for ecosystems under future climate, and (d) implement natural resources management practices that do no harm to native biodiversity and allow its continuing evolution.

We must assume, however, that conservation networks will always be a work in progress, needing to adapt to changing environmental and cultural circumstances. The creation of networks of protected areas designed with appropriate kinds and levels of connectivity is just the beginning of a millennial project to protect the unique flora and fauna of Australia.

ACKNOWLEDGEMENTS

We are grateful to the Wilderness Society for their encouragement and logistical support. We also acknowledge the editorial advice of Kevin Crooks and the assistance of Leon Barmuta. This chapter is reprinted here (with minor changes) from the December 2004 issue of *Pacific Conservation Biology* (**10**:266–279).

REFERENCES

Arnold, S. J., and R. J. Wassersug. 1978. Differential predation on metamorphic anurans by garter snakes (*Thamnophis*): social behaviour as a possible defence. *Ecology* **59**:1014–1022.

Austin, J. M., B. G. Mackey, and K. P. Van Niel. 2003. Estimating forest biomass using satellite radar: an exploratory study in a temperate Australian eucalyptus forest. *Forest Ecology and Management* **176**:575–583.

Australian Government. 2001. *Australia State of the Environment 2001 Report*. Canberra, ACT, Australia: Australian Government, Department of the Environment and Heritage.

Australian Museum Online. 2004. *Extinct Mammals*. Available online at http://www.amonline.net.au/mammals/collections/extinct/index.cfm/

Avise, J. C. 2000. *Phylogeography: The History and Formation of Species*. Cambridge, MA: Harvard University Press.

Bowman, D. 1996. Diversity patterns of woody species on a latitudinal transect from the monsoon tropics to desert in the Northern Territory, Australia. *Australian Journal of Botany* **44**:571–580.

Corbett, L. K. 1995. *The Dingo in Australia and Asia*. Sydney, NSW, Australia: University of New South Wales Press.

Crowder, L., and E. Norse. 2005. *Marine Conservation Biology: The Science of Maintaining the Sea's Biodiversity*. Covelo, CA: Island Press.

CSIRO. 2001. *Climate Change Projections for Australia*. Melbourne, Vic, Australia: CSIRO Atmospheric Research. Available online at http://www.dar.csiro.au/publications/projections2001.pdf/

Daily, G., P. R. Ehrlich, and G. A. Sánchez-Azofeifa. 2001. Countryside biogeography: use of human-dominated habitats by the avifauna of southern Costa Rica. *Ecological Applications* **11**:1–13.

Dobson, D., K. Ralls, M. Foster, *et al.* 1999. Reconnecting fragmented landscapes. Pp. 129–170 in M. E. Soulé and J. Terborgh (eds.) *Continental Conservation: Scientific Foundations for Regional Conservation Networks*. Washington, DC: Island Press.

Dobson, A., S. Kutz, M. Pascual, and R. Winfree. 2003. Pathogens and parasites in a changing climate. Pp. 33–38 in L. Hannah and T. E. Lovejoy (eds.) *Climate Change and Biodiversity: Synergistic Impacts.* Washington, DC: Conservation International.

Driscoll, D. A. 1998. Genetic structure, metapopulation processes and evolution influence the conservation strategies for two endangered frog species. *Biological Conservation* **83**:43–54.

Driscoll, D. A. 2004. Extinction and outbreaks accompany fragmentation of a reptile community. *Ecological Applications* **14**:220–240.

Edwards, G. P., N. De Preu, I. V. Crealy, and B. J. Shakeshaft. 2002. Habitat selection by feral cats and dingoes in a semi-arid woodland environment in central Australia. *Austral Ecology* **27**:26–31.

Esplin, B., A. M. Gill, and N. Enright. 2003. *Report of the Inquiry into the 2002–2003 Victorian Bushfires.* Melbourne, Vic, Australia: State Government of Victoria.

Frankel, O. H., and M. E. Soulé. 1981. *Conservation and Evolution.* Cambridge, UK: Cambridge University Press.

GABCC. 2003. *Great Artesian Basin Consultative Committee: Fact Sheet.* Available online at http://www.gab.org.au/index.html/

Garnett, S., and G. Crowley. 2000. *The Action Plan for Australian Birds 2000.* Canberra, ACT, Australia: Environment Australia.

Gill, A. M., and J. E. Williams. 1996. Fire regimes and biodiversity: the effects of fragmentation of southeastern eucalypt forests by urbanization, agriculture and pine plantations. *Forest Ecology and Management* **85**:261–278.

Gillespie, G. R., and J. M. Hero. 1999. Potential impacts of introduced fish and fish translocations on Australian amphibians. Pp. 131–144 in A. Campbell (ed.) *Declines and Disappearances of Australian Frogs.* Canberra, ACT, Australia: Environment Australia.

Graham, R. W., and E. L. Lundelius, Jr. 1984. Coevolutionary equilibrium and Pleistocene extinctions. Pp. 223–249 in P. S. Martin and R. G. Klein (eds.) *Quaternary Extinctions: A Prehistoric Revolution.* Tuscon, AZ: University of Arizona Press.

Griffioen, P. A., and M. F. Clarke. 2002. Large-scale bird-movement patterns evident in eastern Australian atlas data. *Emu* **102**:99–125.

Hairston, N. G., F. E. Smith, and L. B. Slobodkin. 1960. Community structure, population control, and competition. *American Naturalist* **94**:421–425.

Hale, P., and D. Lamb (eds.) 1997. *Conservation outside Nature Reserves.* Brisbane, Qld, Australia: Centre for Conservation Biology, University of Queensland.

Hall, C. A. S., J. A. Stanford, and F. R. Hauer. 1992. The distribution and abundance of organisms as a consequence of energy balances along multiple environmental gradients. *Oikos* **65**:377–390.

Hannah, L., and R. Salm. 2003. Protected areas and climate change. Pp. 91–100 in L. Hannah and T. E. Lovejoy (eds.) *Climate Change and Biodiversity: Synergistic Impacts,* Advances in Applied Conservation Science No. 4. Washington, DC: Conservation International.

Hobbs, J. E., J. A. Lindesay, and H. A. Bridgman. 1998. *Climates of the Southern Continents: Past, Present and Future.* Chichester, UK: John Wiley.

Hobbs, R. J. 2002. Fire regimes and their effects in Australian temperate woodlands. Pp. 305–326 in R. Bradstock, J. E. Williams and A. M. Gill (eds.) *Flammable Australia: Fire Regimes and the Biodiversity of a Continent.* Cambridge, UK: Cambridge University Press.

Hobbs, R. 2003. How fire regimes interact with other forms of ecosystem disturbance and modification. Pp. 421–436 in I. Abbott and N. Burrows (eds.) *Fire in Ecosystems of Southwest Western Australia: Impacts and Management.* Leiden, The Netherlands: Backhuys.

Hopkins, M. S., J. Ash, A. W. Graham, J. Head, and R. K. Hewett. 1993. Charcoal evidence of the spatial extent of the *Eucalyptus* woodland expansions and rainforest contractions in North Queensland during the late Pleistocene. *Journal of Biogeography* 2:357–372.

Horn, A. M. 1995. *Surface Water Resources of Cape York Peninsula.* Canberra, ACT, Australia: Cape York Peninsula Land Use Strategy (CYPLUS), Queensland and Commonwealth Governments.

Horn, A. M., E. A. Derrington, G. C. Herbert, R. W. Lait, and J. R. Hillier. 1995. *Groundwater Resources of Cape York Peninsula.* Canberra, ACT, Australia: Cape York Peninsula Land Use Strategy (CYPLUS), Queensland and Commonwealth Governments.

Howden, M., L. Hughes, M. Dunlop, *et al.* (eds.) 2003. *Climate Change Impacts on Biodiversity in Australia,* outcomes of a workshop sponsored by the Biological Diversity Advisory Committee, 1–2 October 2002. Canberra, ACT, Australia: Commonwealth of Australia.

Hughes, L. 2003. Ecological interactions and climate change. Pp. 45–49 in L. Hannah and T. E. Lovejoy (eds.) *Climate Change and Biodiversity: Synergistic Impacts,* Advances in Applied Conservation Science No. 4. Washington, DC: Conservation International.

Isard, S. A., and S. H. Gage. 2001. *Flow of Life in the Atmosphere: An Airscape Approach to Understanding Invasive Organisms.* East Lansing, MI: Michigan State University Press.

IUCN. 2003. *Proceedings of the 5th World Parks Congress: Benefits beyond Boundaries,* IUCN Bulletin No. 2. Gland, Switzerland: IUCN. Available online at http://www.iucn.org/bookstore/Bulletin/Vth-WPC.htm/

Johnson, C. N. 1996. Interactions between mammals and ectomycorrhizal fungi. *Trends in Ecology and Evolution* 11:503–507.

Johnson, E.A., and M.A. Cochrane. 2003. Disturbance regime interactions. Pp. 39–44 in L. Hannah and T. E. Lovejoy (eds.) *Climate Change and Biodiversity: Synergistic Impacts,* Advances in Applied Conservation Science No. 4. Washington, DC: Conservation International.

Kearney, M., A. Moussalli, J. Strasburg, D. Lindenmayer, and C. Moritz. 2003. Geographic parthenogenesis in the Australian arid zone. I. A climatic analysis of the *Heteronotia binoei* complex Gekkonidae. *Evolutionary Ecology Research* 5:953–976.

Keith, D., J. E. Williams, and J. Woinarski. 2002. Fire management and biodiversity conservation: key approaches and principles. Pp. 401–428 in R. Bradstock, J. E. Williams, and A. M. Gill (eds.) *Flammable Australia: Fire Regimes and the Biodiversity of a Continent* Cambridge, UK: Cambridge University Press.

Kerr, R. A. 2004. Three degrees of Consensus. *Science* 305:932–934.

Koop, K., D. Booth, A. Broadbent, *et al.* 2001. ENCORE: the effect of nutrient enrichment on coral reef – synthesis of results and conclusions. *Marine Pollution Bulletin* 42:91–120.

Landsberg, J. J., and R. H. Waring. 1997. A generalized model of forest productivity using simplified concepts of radiation-use efficiency, carbon balance and partitioning. *Forest Ecology and Management* 95:209–228.

Ledec, G., and R. Goodland. 1988. *Wildlands: Their Protection and Management in Economic Development*. Washington, DC: World Bank.

Lesslie, R. G. 2001. Landscape classification and strategic assessment for conservation: an analysis of native cover loss in far south-east Australia. *Biodiversity and Conservation* 10:427–442.

Lindenmayer, D. B., and J. F. Franklin. 2002. *Conserving Forest Biodiversity: A Comprehensive Multiscaled Approach*. Washington, DC: Island Press.

Lindenmayer, D. B., and H. F. Recher. 1998. Aspects of ecologically sustainable forestry in temperate eucalypt forests: beyond an expanded reserve system. *Pacific Conservation Biology* 4:4–10.

Littleboy, M., R. Vertessy, and P. Lawrence. 2003. An overview of modelling techniques and decision support systems and their application for managing salinity in Australia. *Proceedings of 9th National Produce Use and Rehabilitation of Saline Land (PURSL) Conference*, Queensland, Australia.

Lundie-Jenkins, G., L. K. Corbett, and C. M. Phillips. 1993. Ecology of the Rufous Hare-Wallaby, *Lagorchestes hirsutus* Gould (Marsupialia, Macropodidae), in the Tanami Desert, Northern Territory III. Interactions with introduced mammal species. *Wildlife Research* 20:495–511.

Mackey, B. G., and D. B. Lindenmayer. 2001. Towards a hierarchical framework for modeling the spatial distribution of animals. *Journal of Biogeography* 28:1147–1166.

Mackey, B. G., H. A. Nix, M. F. Hutchinson, J. P. McMahon, and P. M. Fleming. 1988. Assessing representativeness of places for conservation reservation and heritage listing. *Environmental Management* 12:501–514.

Mackey, B. G., H. A. Nix, J. Stein, E. Cork, and F. T. Bullen. 1989. Assessing the representativeness of the Wet Tropics of Queensland World Heritage Property. *Biological Conservation* 50:279–303.

Mackey, B. G., H. A. Nix, and P. Hitchcock. 2001. *The Natural Heritage Significance of Cape York Peninsula*, A Report to the Queensland Environmental Protection Agency. Available online at http://www.env.qld.gov. au/environment/environment/capeyork/

Mackey, B. G., D. B. Lindenmayer, M. Gill, M. McCarthy, and J. Lindesay. 2002. *Wildlife, Fire and Future Climate: A Forest Ecosystem Analysis*. Melbourne, Vic, Australia: CSIRO Publishing.

Mackey, B.G, M.E. Soulé, H. A. Nix, *et al.* 2006. Towards a scientific framework for the WildCountry Project. Pp. 121–134 in J. Wu and R. J. Hobbs (eds.) *Key Topics and Perspectives in Landscape Ecology*. Cambridge, UK: Cambridge University Press.

McKenney, D., A. Anthony, K. Hopkin, *et al.* 2003. Opportunities for improved risk assessments of exotic species in Canada using bioclimatic modelling. *Environmental Monitoring and Assessment* 88:451–461.

McNaughton, S. J., M. Oesterheld, D. A. Frank, and K. J. Williams. 1989. Ecosystem-level patterns of primary productivity and herbivory in terrestrial habitats. *Nature* 341:142–144.

Moritz, C. 1991. The origin and evolution of parthenogenesis in *Heteronotia binoei* (Gekkonidae): evidence for recent and localized origins of widespread clones. *Genetics* 129:211–219.

Moritz, C., J. L. Patton, C. J. Schneider, and T. B. Smith. 2000. Diversification of rainforest faunas: an integrated molecular approach. *Annual Reviews of Ecology and Systematics* 31:533–563.

Morris, K., B. Johnson, P. Orell, A. Wayne, and G. Gaikorst. 2003. Recovery of the threatened chuditch (*Dasyurus geoffroii* Gould, 1841): a case study. Pp. 345–351 in M. E. Jones, C. R. Dickman, and M. Archer (eds.) *Predators with Pouches: The Biology of Carnivorous Marsupials*. Melbourne, Vic, Australia: CSIRO Publishing.

Morton, S. R., J. Short, and R. D. Barker (with an Appendix by Griffin, G. F. and Pearce, G.) 1995. *Refugia for Biological Diversity in Arid and Semi-Arid Australia*, a report to the Biodiversity Unit of the Department of Environment, Sport and Territories. Canberra, ACT, Australia: CSIRO.

NASA. 2003. *The Moderate Resolution Imaging Spectroradiometer (MODIS)*. Available online at http://modis.gsfc.nasa.gov/

National Task Group on the Management of Climate Change Impacts on Biodiversity. 2003. *Developing a National Biodiversity and Climate Change Action Plan*. Canberra, ACT, Australia: Department of Environment and Heritage.

Newsome, A. E., P. C. Catling, B. D. Cooke, and R. Smyth. 2001. Two ecological universes separated by the dingo barrier fence in semi-arid Australia: interactions between landscapes, herbivory and carnivory, with and without dingoes. *Rangeland Journal* 23:71–98.

Nix, H. A. 1976. Environmental control of breeding, post-breeding dispersal and migration of birds in the Australian region. In *Proceedings of the 16th International Ornithological Congress*, 1974, pp. 272–305.

Nix, H. A. 1982. Environmental determinants of biogeography and evolution in Terra Australis. Pp. 47–66 in W. R. Barker and P. J. M. Greensland (eds.) *Evolution of the Flora and Fauna of Arid Australia*. Sydney, NSW, Australia: Peacock Publications.

NLWRA. 2002. *Australian Terrestrial Biodiversity Assessment*. Canberra, ACT, Australia: National Land and Water Resources Audit

Oksanen, L., and T. Oksanen. 2000. The logic and realism of the hypothesis of exploitation ecosystems. *American Naturalist* 155:703–723.

O'Neill, A. 2002. *Dingoes*. Annandale, NSW, Australia: Envirobook.

Paine, R. T. 1969. A note on trophic complexity and community stability. *American Naturalist* 103:65–75.

Paton, D. C., A. M. Prescott, R. J. Davies, and L. M. Heard. 2000. The distribution, status and threats to temperate woodlands in South Australia. Pp. 57–85 in R. J. Hobbs and C. J. Yates (eds.) *Temperate Eucalypt Woodlands in Australia: Biology, Conservation, Management and Restoration*. Chipping Norton, NSW, Australia: Surrey Beatty and Sons.

Ponder, W. F., P. Eggler, and D. J. Colgan. 1995. Genetic differentiation of aquatic snails (Gastropoda: Hydrobiidae) from artesian springs in arid Australia. *Biological Journal of the Linnean Society* **56**:553–596.

Pople, A. R., G. C. Grigg, S. C. Cairns, L. A. Beard, and P. Alexander. 2000. Trends in the numbers of red kangaroos and emus on either side of the South Australian dingo fence: evidence for predator regulation? *Wildlife Research* **27**:269–276.

Power, M. E., and L. S. Mills. 1995. The Keystone cops meet in Hilo. *Trends in Ecology and Evolution* **10**:182–184.

Price, O. F., J. C. Z. Woinarski, and D. Robinson. 1999. Very large area requirements for frugivorous birds in monsoon rainforests of the Northern Territory, Australia. *Biological Conservation* **91**:169–180.

Pulliam, H. R., and B. J. Danielson. 1991. Sources, sinks and habitat selection: a landscape perspective on population dynamics. *American Naturalist* **137**:S50–S66.

Recher, H. F. 1999. The state of Australia's avifauna: a personal opinion and prediction for the new millenium. *Australian Zoologist* **31**:11–27.

Recher, H. F., and W. E. Davis. 2002. Foraging profile of a Salmon Gum woodland avifauna in Western Australia. *Journal of the Royal Society of Western Australia* **85**:103–111.

Risbey, D. A., M. C. Calver, J. Short, J. S. Bradley, and I. W. Wright. 2000. The impact of cats and foxes on the small vertebrate fauna of Heirisson Prong, Western Australia. II. A field experiment. *Wildlife Research* **27**:223–235.

Roberts, J. D., and Maxson, L. R. 1985. The biogeography of southern Australian frogs: molecular data reject multiple invasion and Pleistocene divergence models. Pp. 83–89 in G. Grigg, R. Shine, and H. Ehmann (eds.) *Biology of Australasian Frogs and Reptiles*. Chipping Norton, NSW, Australia: Surrey Beatty and Sons.

Rosenzweig, M. L. 2003. *Win–Win Ecology*. New York: Oxford University Press.

Roshier, D. A., P. H. Whetton, R. J. Allan, and A. I. Robertson. 2001. Distribution and persistence of temporary wetland habitats in arid Australia in relation to climate. *Austral Ecology* **26**:371–384.

Russell-Smith, J., N. L. McKenzie, and J. C. Z. Woinarski. 1992. Conserving vulnerable habitat in northern and north-western Australia: the rainforest archipelago. Pp. 63–68 in I. Moffatt and A. Webb (eds.) *Conservation and Development Issues in Northern Australia*. Darwin, NT, Australia: North Australia Research Unit.

Seibel, B. A., and V. J. Fabry. 2003. Marine biotic responses to elevated carbon dioxide. Pp. 59–67 in L. Hannah and T. E. Lovejoy (eds.) *Climate Change and Biodiversity: Synergistic Impacts*. Advances in Applied Conservation Science No. 4. Washington, DC: Conservation International.

Soulé, M. E. 1980. Thresholds for survival: criteria for maintenance of fitness and evolutionary potential. Pp. 151–170 in M. E. Soulé and B. M. Wilcox (eds.) *Conservation Biology: An Evolutionary-Ecological Perspective*. Sunderland, MA: Sinauer Associates.

Soulé, M. E., and M. Sanjayan. 1998. Conservation targets: do they help? *Science* **279**:2060–2061.

Soulé, M. E., and J. Terborgh (eds.) 1999. *Continental Conservation: Scientific Foundations of Regional Reserve Networks*. Washington, DC: Island Press.

Soulé, M. E., J. Estes, J. Berger, and C. Martinez del Rio. 2003. Ecological effectiveness: conservation goals for interactive species. *Conservation Biology* 17:1238−1250.

Taylor, K. 1999. Rapid climate change. *American Scientist* 87:320−327. Available online at http://waiscores.dri.edu/Amsci/taylor.html/

Terborgh, J., J. A. Estes, P. C. Paquet, *et al.* 1999. Role of top carnivores in regulating terrestrial ecosystems. Pp. 39−64 in M. E. Soulé and J. Terborgh (eds.) *Continental Conservation: Scientific Foundations of Regional Reserve Networks*. Washington, DC: Island Press.

Terborgh, J, L. Lopez, P. Nuñez, *et al.* 2001. Ecological meltdown in predator-free forest fragments. *Science* 294:1923−1925.

Thomas, C. D., A. Cameron, R. E. Green, *et al.* 2004. Extinction risk from climate change. *Nature* 427:145−149.

Tracey, D., L. Turner, J. Tilden, and W. C. Dennison. 2004. *Where River Meets Sea: Exploring Australia's Estuaries*. Brisbane, Qld, Australia: Coastal CRC.

Tyre, A. J., B. Tenhumberg, D. Niejalke, and H. P. Possingham. 2001. Predicting risk to biodiversity as a function of aquifer pressure in GAB mound springs. Pp. 825−829 in *Proceedings of the International Congress on Modelling and Simulation*, 10−13 December 2001.

Wasson, R. J. 2003. Connectivity. Pp. 166−169 in G. Cary, D. Lindenmayer, and S. Dovers (eds.) *Australia Burning: Fire Ecology, Policy and Management Issues*. Melbourne, Vic, Australia: CSIRO Publishing.

Whelan, R. J., L. Rodgerson, C. R. Dickman, and E. F. Sutherland. 2002. Critical life cycles of plants and animals: developing a process-based understanding of population changes in fire-prone landscapes. Pp. 94−124 in R. Bradstock, J. E. Williams, and A. M. Gill (eds.) *Flammable Australia: Fire Regimes and the Biodiversity of a Continent*. Cambridge, UK: Cambridge University Press.

Williams, E. S., T. Yuill, M. Artois, J. Fischer, and S. A. Haigh. 2002. Emerging infectious diseases in wildlife. *Revue Scientifique et Technique de l'Office International des Epizooties* 21:139−157.

Williams, J. 2003. Making the invisible visible. Pp. 26−31 in G. Cary, D. Lindenmayer and S. Dovers (eds.) *Australia Burning: Fire Ecology, Policy and Management Issues*. Melbourne, Vic, Australia: CSIRO Publishing.

Williams, J. E., R. J. Whelan, and A. M. Gill. 1994. Fire and environmental heterogeneity in southern temperate forest ecosystems: implications for management. *Australian Journal of Botany* 42:125−137.

Wilson, E. O. 2002. *The Future of Life*. New York: Alfred E. Knopf.

Woinarksi, J. 1999. Fire and Australian birds: a review. Pp. 55−111 A. M. Gill, J. C. Z. Woinarksi, and A. York (eds.) in *Australia's Biodiversity: Responses to Fire: Plants, Birds and Invertebrates*. Canberra, ACT, Australia: Environment Australia.

Woinarski, J., G. Conners, and D. Franklin. 2000. Thinking honeyeater: nectar maps for the Northern Territory, Australia. *Pacific Conservation Biology* 6:61−80.

Woinarski, J., P. Whitehead, D. Bowman, and J. Russell-Smith. 1992. Conservation of mobile species in a variable environment: the problem of reserve design in the Northern Territory, Australia. *Global Ecology and Biogeography Letters* 2:1–10.

Woinarski, J. C. Z., and A. J. Ash. 2002. Responses of vertebrates to pastoralism, military land use and landscape position in an Australian tropical savanna. *Austral Ecology* 27:311–323.

The future of connectivity conservation

ANDREW F. BENNETT, KEVIN R. CROOKS, AND
M. SANJAYAN

The present threat to Earth's biodiversity from the human enterprise is unprecedented in historic time. Understanding the consequences of environmental change, and developing effective strategies to maintain plant and animal species and the ecological processes on which all of life hinges, present enormous challenges. The growing awareness of environmental change has been mirrored, albeit with a lag, by a shift in focus of scientific endeavors in the ecological sciences. The last three decades have seen strong growth in disciplines that emphasize the importance of using scientific knowledge and skills to address threats to the future of ecosystems throughout the world.

Conservation biology emerged in the 1980s as a "mission-oriented" crisis discipline (Soulé and Wilcox 1980; Soulé 1985). It was to be a "new rallying point for biologists wishing to pool their knowledge and techniques to solve problems" (Soulé and Wilcox 1980). Rapid growth in this field has been accompanied by new journals, such as *Animal Conservation*, *Biodiversity and Conservation*, *Conservation Biology*, *Ecological Applications*, *Ecology and Society*, and *Pacific Conservation Biology*, in which setting out the conservation implications of the published research is regarded as a necessary and important part of the contribution. Likewise, rapid growth in the discipline of landscape ecology has been based on the premise that conceptual advances and empirical studies of the ways in which spatial pattern affects ecological processes will deliver insights for improved land management (Forman 1995; Turner *et al.* 2001; Wu and Hobbs 2002). Indeed, Hobbs (1997) charged that the products of

Connectivity Conservation eds. Kevin R. Crooks and M. Sanjayan. Published by Cambridge University Press. © Cambridge University Press 2006.

landscape ecology (i.e., theory, methodology, models) should be judged by the impact that they have on the planning and management of real landscapes.

The role of connectivity in nature conservation, the theme of this book, is characteristic of the type of issues addressed by this new direction in conservation science. It illustrates the potential benefits that can be gained by applying scientific knowledge and understanding to conservation problems, but also the complexities and difficulties in successful inter-action between scientists and practitioners and between theory and practice (Saunders and Hobbs 1991; Saunders et al. 1995). It is a topic of direct relevance to land-use planning and land management, and the decisions made will have profound implications for the future status of biodiversity. Connectivity is also a topic of strong scientific interest and, like any active research field, has its share of controversy and debate (Crooks and Sanjayan Chapter 1).

Connectivity is important from a scientific perspective because move-ments of organisms and continuity of processes are fundamental to a conceptual understanding of how species persist in landscapes modified by human activities. Whether derived from island biogeography theory, metapopulation theory, or concepts and models of land mosaics, the degree to which organisms can move or their propagules can disperse between different parts of the landscape is central to a theoretical under-standing of survival in heterogeneous environments. Scientific under-standing of connectivity has come a long way in the last 30 years from the first simple recommendations for the benefits of corridors arising from the equilibrium theory of island biogeography (e.g., Diamond 1975). Attention has moved from "corridors" to the broader concept of "connectivity." Much intellectual effort has been devoted to defining connectivity, its relationship to the scale of investigation and organism (or process) being studied, and ways to measure different components of connectivity (Merriam 1991; Taylor et al. 1993; Tischendorf and Fahrig 2000; Crooks and Sanjayan Chapter 1; Taylor et al. Chapter 2; Moilanen and Hanski Chapter 3; Talley et al. Chapter 5; Fagan and Calabrese Chapter 12). The rapid increase in the number of publications relating to "corridors" and "connectivity" (Crooks and Sanjayan Chapter 1; Pringle Chapter 10; Haddad and Tewksbury Chapter 16), and the breadth of scientific research illustrated in this book, testify to an active and dynamic research field. Reviews have periodically summarized and evaluated current knowledge, critically appraised scientific approaches and assump-tions, and highlighted new questions (e.g., Simberloff and Cox 1987;

Hobbs 1992; Simberloff *et al.* 1992; Rosenberg *et al.* 1997; Beier and Noss 1998; Bennett 1999; this volume). Along the way, the process of discovery and learning has been greatly assisted by a range of empirical studies of how real organisms respond to landscape structure (e.g., Mansergh and Scotts 1989; Bennett 1990; Prevett 1991; Beier 1993; Thomas and Jones 1993; Haas 1995; Tracey Chapter 14; Carroll Chapter 15; Theobald Chapter 17; Clevenger and Wierzchowski Chapter 20).

Contributions to this volume illustrate that connectivity has also become a key issue in practical land management in countries throughout the world (also see Harris and Scheck 1991; Saunders and Hobbs 1991; Bennett 1999; Soulé and Terborgh 1999). On a day-to-day basis, land managers face many conservation issues arising from the destruction, subdivision, degradation, or isolation of habitats critical to maintaining plant and animal populations. Examples of the diverse circumstances, reviewed in this volume, in which an understanding of connectivity will inform practice include mitigating the effects of roads as barriers to wildlife (Clevenger and Wierzchowski Chapter 20); evaluating the potential impacts of urban and exurban development (Tracey Chapter 14; Theobald Chapter 17; Beier *et al.* Chapter 22); managing habitats for long-distance migrants (Marra *et al.* Chapter 7; Harrison and Bjorndal Chapter 9); control of invasive species (Crooks and Suarez Chapter 18) and infectious diseases (McCallum and Dobson Chapter 19); facilitating gene flow and minimizing inbreeding in fragmented populations (Frankham Chapter 4; Neville *et al.* Chapter 13); restoring degraded stream systems (Pringle Chapter 10; Neville *et al.* Chapter 13) and coastal zones (Talley *et al.* Chapter 5; Soulé *et al.* Chapter 25); facilitating ecosystem services such as crop pollination in agricultural landscapes (Ricketts *et al.* Chapter 11); and setting priorities for the location of new conservation reserves in marine (DiBacco *et al.* Chapter 8; Harrison and Bjorndal Chapter 9) and terrestrial (Morrison and Reynolds Chapter 21; Beier *et al.* Chapter 22; Noss and Daly Chapter 23; Sanderson *et al.* Chapter 24; Soulé *et al.* Chapter 25) systems. Further, and perhaps most troubling, is the added spectre of climate disruption and the rapidity with which plant and animal populations are responding (Noss and Daly Chapter 23; Soulé *et al.* Chapter 25). For example, using meta-analysis on long-term data sets, Parmesan (2005) has estimated that the average movement in response to climate change for a wide variety of taxa is over 6 km per decade. Without the ability to move as connectivity is lost, climate disruption will have a profound and negative consequence on the diversity of life on Earth. We will lose species that are unable to traverse human-dominated landscapes

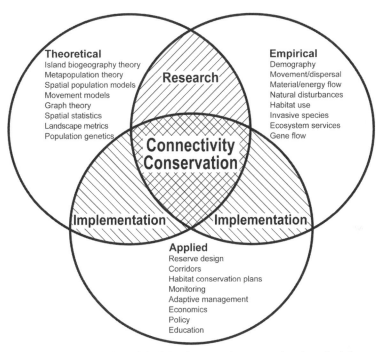

Fig. 26.1. Conceptual model of the theoretical, empirical and applied domains of connectivity conservation (taken from Crooks and Sanjayan Chapter 1). Synthesis and breakthroughs in connectivity conservation largely occur at the intersection of the three domains (hatched).

and unable to adapt to a bioclimatic zone that is changing underneath them.

The contributions in this book highlight a broad range of theoretical, empirical and applied approaches to understanding landscape connectivity (Fig. 26.1) (Crooks and Sanjayan Chapter 1), and also convey the enthusiasm with which scientists are responding to the need for solutions to threats posed by human land use. Recognizing the imperative for conservation science to be "mission-oriented," we contend that the greatest advances will be made where the diverse insights from theoretical, empirical and applied studies can be brought together in a complementary manner (Fig. 26.1). Further efforts are needed to define and extend this zone of overlap (Fig. 26.2). We outline below key areas in which integration of efforts will achieve useful advances, but first it is useful to briefly review the distinctive way in which connectivity contributes to conservation science and strategy.

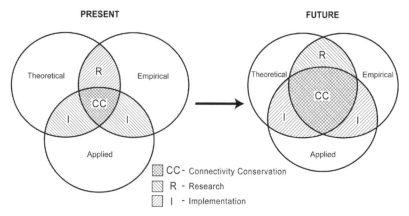

PRESENT   FUTURE

CC - Connectivity Conservation
R - Research
I - Implementation

Fig. 26.2. The key to the future of connectivity conservation (CC) is to enlarge the area of overlap, the synergy between domains, of our theoretical, empirical, and applied efforts in connectivity research and implementation. (Taken from Crooks and Sanjayan Chapter 1.)

THE DISTINCT ROLE OF CONNECTIVITY IN CONSERVATION STRATEGY

Environmental changes resulting from human land uses are diverse, but the destruction and fragmentation of natural habitats has particularly far-reaching effects. The subsequent replacement of natural vegetation with new land uses (e.g., agricultural, urban, industrial) further changes the land mosaic. These new land uses support different assemblages of plants and animals and influence the conservation context of remnant natural or semi-natural habitats.

Conservation managers undertake many kinds of activities to protect and improve the status of plant and animal species in land mosaics, but in general such actions fall into one of four main categories (Bennett 1999) (Fig. 26.3).

1. Increase the total extent of habitat that is protected or managed for the conservation of the native biota. This can be achieved, for example, by expanding the size of existing conservation reserves or by establishing additional new reserves, by greater protection of other tracts of natural habitat (such as on private land, in managed forests), or by restoration and re-establishment of vegetation to expand the size and overall extent of habitats.

2. Improve the quality of existing habitats to better provide the resources that species require, such as for foraging, shelter, and breeding.

This may involve greater control of land uses that degrade habitats, managing the level of extraction of natural resources, or managing disturbance regimes (such as fire or logging) that influence the extent of successional stages of vegetation.

3. Reduce detrimental impacts on habitats from surrounding land uses and species typical of adjacent land. This may require control of invasive plants and animals, the modification or removal of particular land uses, or the use of buffer zones in areas outside of reserves.

4. Promote increased connectivity of habitats to facilitate the movement of organisms and the continuity of ecological processes across the landscape. This can be achieved by: (a) managing specific types of connecting habitats (corridors, stepping stones, streams, and waterways), or (b) managing land uses within the overall landscape mosaic.

Two key points concerning the role of connectivity in nature conservation emerge from considering these general approaches. First, increasing connectivity is not the only response to habitat destruction and fragmentation, but is part of a broader "tool box" of options to respond to landscape change (Fig. 26.3). That is, as emphasized by authors in this volume

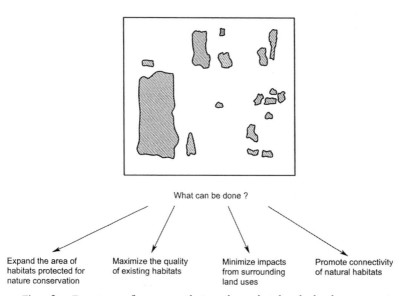

Fig. 26.3. Four types of measures that can be undertaken by land managers to counter the effects of habitat fragmentation. (After Bennett 1999.)

(e.g., Taylor *et al.* Chapter 2; Noss and Daly Chapter 23), establishing habitat corridors, stepping stones, or other links that increase movements of organisms is not a universal panacea, although should be universally considered in developing comprehensive strategies for conservation. Indeed, current evidence suggests that, at least in the short term, the total amount of habitat often may be a more important determinant of the status and persistence of species in modified landscapes than the spatial pattern or configuration of habitats (Trzcinski *et al.* 1999; Fahrig 2003). All four types of management response (Fig. 26.3) have a useful role and, in practice, many conservation programs implement all of these to varying extent.

Second, undertaking actions to enhance connectivity offers a fundamentally different and distinctive approach to land management and conservation. The first three types of measures (Fig. 26.3) are actions that increase the conservation value of *individual* blocks of habitat – by enlarging them, improving their quality, or reducing external threats. Enhancing connectivity creates the opportunity to achieve conservation goals by building interconnected *systems* of habitat. By increasing the flow of organisms or continuity of processes between parts of the landscape mosaic, there is potential to build habitat networks that integrate conservation efforts at multiple levels, including local, landscape, regional, continental, and even global scales – a particularly important strategy for large-scale regional or global changes such as climate disruption. Developing strategies for conservation networks is now at the forefront of conservation planning throughout the world (Jongman 1995; Saunders *et al.* 1995; Verboom *et al.* 2001; Groves 2003; this volume).

LOOKING TO THE FUTURE

What is the future direction and vision for connectivity conservation? There are many directions to pursue (Fig 26.1), and not surprisingly like other emerging fields, there are different views on which are currently the most important and pressing issues. Those contributing to this field most actively hail from widely disparate disciplines and when asked to nominate and rank their three highest priorities for the future, contributors to this book ($n = 28$ respondents) selected virtually all (21 of the 24) topics listed within the applied, empirical, and theoretical domains outlined in Fig. 26.1. Chapter authors, persons currently working on connectivity research and application, believe that practical efforts to actually implement connectivity conservation will be a critical direction for the future, with nearly half (47.5%) of all nominations within the applied domain, followed by the empirical (35% of nominations) and theoretical (17.5% of nominations) spheres (Fig. 26.4). This is hardly surprising given that connectivity conservation arose out of a practical need to maintain populations of species in highly modified human-dominated landscapes. Of particular note was the strong desire for further research into determining empirical measurements of movement and dispersal; this category received 22% of all nominations, and a third of the contributors picked it as the strongest single theme for future work, resulting in the high overall ranked score for the empirical domain (Fig. 26.4). Design and implementation of corridors was identified as the next highest priority, receiving 13% of all nominations. Other priorities included empirical studies of demography and population viability (6% of nominations), and applied efforts regarding reserve design (7%), habitat conservation plans (5%), monitoring (6%), and policy (6%). The strong showing for empirically determining measures of movement and dispersal may be a reaction from conservationists who are increasingly being asked to design and implement corridor plans without good data on which to base these designs.

Clearly, research studies investigating connectivity, and practical implementation of connecting landscapes, promise many new outcomes and challenges in coming years. Here, we highlight four key challenges that focus on ways to more effectively meet the practical task of maintaining connections for nature, and which also take advantage of the synergies between theoretical, empirical, and applied aspects of conservation science.

1. Beyond species: connectivity of communities, ecological processes, and ecological flows

As evident from the contributions to this volume, much of the research on the conservation benefits of connectivity has been based on empirical or modeling studies of single or a few species (also see Henein and Merriam 1990; Mabry and Barrett 1992; Beier 1993; Bennett et al. 1994; Haas 1995; Andreassen et al. 1996; Sutcliffe and Thomas 1996; Goodwin and Fahrig 2002). Such studies have yielded many insights, particularly concerning the influence of landscape structure on the capacity of individuals to move, the frequency of movements, the types of individuals that move, and the potential benefits of such movements. Application of such research to land management is based on the generality of the principles identified, or on the extent to which the species studied are believed to act as surrogates for

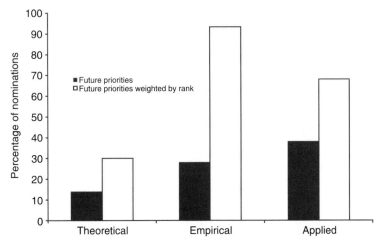

Fig. 26.4. Priorities for future direction in connectivity research and implementation. Contributors to this volume were asked to nominate and rank their three highest priorities based on the topics listed in Fig. 26.1. Numbers represent percentage of total nominations (closed bars) and the percentage of total nominations weighted by the rank (1, 2, or 3) provided by contributors (open bars) that fell within the theoretical, empirical, and applied domains of connectivity conservation.

the requirements of other species that use similar habitat types. However, we also need to develop a broader perspective on connectivity – to extend from our understanding of species to the implications of connectivity for assemblages of plants or animals, for the ecological interactions that underpin ecosystem function in communities, and for other ecological "flows" in land mosaics.

There are increasing efforts to establish "landscape linkages," broad swaths of habitat between large natural areas such as conservation reserves (Baranga 1991; Harris and Scheck 1991; O'Donnell 1991; Smith 1993) and even continental-scale links (Soulé 1995; Soulé and Terborgh 1999; Noss and Daly Chapter 23; Sanderson *et al.* Chapter 24; Soulé *et al.* Chapter 25). While charismatic species may be used as a "flagship" for such projects (e.g., large carnivores: Paquet *et al.* Chapter 6; Tracey Chapter 14; Carroll Chapter 15; Theobald Chapter 17; Clevenger and Wierzchowski Chapter 20), generally the goal is to ensure long-term connectivity for whole assemblages of species. As emphasized by Haddad and Tewksbury (Chapter 16), to date little theoretical attention has been given to the connectivity of assemblages or communities, and relatively little is known about the form, habitat requirements, and spatial dimensions of linkages to ensure connectivity and genetic interchange for assemblages. A working hypothesis is that if the landscape link provides residential habitat for species in the assemblage, then, in the long term, there is likely to be interchange among core areas through normal dispersal processes. Questions such as the following are pertinent:

- What is the relative importance of spatial dimensions, landscape context, and habitat quality of linkages, core areas, and the intervening matrix on the structure of assemblages?
- What factors influence the rate of movement and exchange of genes through landscape links for organisms of differing size and mobility?
- Which species or groups within faunal assemblages are most likely and least likely to be effectively linked in this way?

We are also faced with limited knowledge of how landscape structure and connectivity affect major processes that involve interactions between species — such as pollination, seed dispersal, parasitism, competition, or predation — that will be modified or disrupted if one of the participants declines in abundance or disappears due to isolation effects. The potential consequences for the structure and function of communities may be far-reaching, particularly with changes in highly interactive species (e.g., see Soulé *et al.* Chapter 25), and time lags in effects of such changes are likely. Loss of effective seed dispersal, for example, may not be fully apparent until the parental cohort of plants disappears. There are also likely to be further "flow-on" effects (i.e., trophic cascades) to other aspects of community structure. Ecosystem services on which humans depend, such as pollination of crops, regulation of insect pests, and health of trees in farmland, may also be affected (Ricketts *et al.* Chapter 11). Relevant issues include:

- What types of ecological interactions are most vulnerable to changes in landscape connectivity?
- Which species tend to be highly interactive?
- What are the spatial and structural requirements for the movement of key vectors in particular processes?
- Is there a threshold type of response in the way that changes in connectivity affect ecological interactions and associated ecosystem services?

A third sphere for expanding understanding of landscape connectivity beyond individuals and species involves the wider range of ecological "flows" that determine landscape function. Here, less emphasis is given to movement between the same type of (suitable) habitat, and greater attention to interactions between different types of elements in the landscape (see Talley *et al.* Chapter 5). The topic of ecological flows within land mosaics was recently identified as a key frontier for research in landscape ecology (Wu and Hobbs 2002). Examples from this volume relevant to the conservation of biota in modified landscapes include:

- The role of wind flow and water flow in the passive dispersal of plants and animals (or their propagules).
- The role of disturbance regimes, both natural and anthropogenic, and their movement across landscapes.
- The factors that influence movements of different species among "matrix" habitats in land mosaics.
- The migratory pathways of species and their spatial variability and intensity of use through time.
- Factors that influence movements across the land–water interface in oceans, streams and wetlands.
- The pathways and rates of invasion of pest plants and animals and disease agents from one landscape element to another.

2. Integration of biological and socio-political issues in implementing connectivity

A sound scientific understanding of connectivity, including the potential benefits and how they can best be achieved, is an essential basis for conservation action. The consequent implementation of actions to enhance connectivity depends not only on this knowledge, but also on the capacity and commitment of people and organizations to carrying out the proposed actions. Such action occurs within a social and political context – the human dimension of conservation. Organizational dynamics, adequate human and financial resources, ownership and access to land, and other such considerations often pose a similar, or greater, challenge to the effective implementation of corridor projects than the scientific knowledge (e.g., Morrison and Reynolds Chapter 21; Sanderson *et al.* Chapter 24). In other words, despite the best biological plans, achieving connectivity within a network of reserves is impossible without considering the socio-economic context. Table 26.1 summarizes a number of key issues in the biological and socio-political domains, respectively, that need to be considered during implementation (Bennett 1999).

Table 26.1. *Considerations for achieving effective connectivity in human-dominated landscapes*

Biological issues	Socio-political issues
Biological purpose of the linkage	Status and tenure of the land
Ecology and behavior of species	Management requirements and responsibility
Dimensions and design of linkages	Adequacy of resources (financial and human skills)
Location	Support and involvement of local communities
Edge effects	Integration with other land management programs
Monitoring the use and benefits of linkages	Community education and awareness
	Strategic approach to planning

Source: Adapted from Bennett (1999).

Socio-political issues may be less familiar to scientific researchers, but are well recognized by land managers. The critical importance of the human dimension in conservation was highlighted by the proceedings of a conference "Nature Conservation: The Role of Networks" (Saunders *et al.* 1995), in which contributors identified the following types of issues: interacting with indigenous peoples, community participation, land ownership, functions of social networks, role of trust in social networks, the types of language used in communication, how to influence decision-making, and extension and education programs (see various contributors in Saunders *et al.* 1995).

Examples of what this could mean for conservation biologists include:

- A stronger emphasis in spatial modeling on spatially explicit scenarios, especially including the flexibility to respond to constraints imposed by land tenure and other factors.
- Empirical research that seeks to understand the potential consequences of different land ownership and management practices on the responses of species, communities, and processes.
- Seeking insights from social scientists and undertaking collaborative research to integrate biological and social perspectives. It is instructive, for example, that the staff employed for the development of the Talamanca–Carribbean landscape link in Costa Rica included an anthropologist, two lawyers, a geographer, a forest engineer, and a biologist (see Bennett 1999, pp. 144–146).

3. Learning from experience

A large number of projects are being implemented to maintain or restore connectivity in landscapes modified by human activities. These include tunnels, underpasses, and overpasses to assist wildlife to cross local barriers; restoring local connections such as hedges, fencerows, and roadside vegetation; protecting and managing streams, rivers, and their riparian vegetation to promote natural connectivity throughout catchments (watersheds); setting aside major vegetated corridors between conservation reserves; and planning for national and continental-scale links. There is great scope to make better use of these activities to test ideas and to learn more about connectivity. Too much emphasis has gone into the design of corridors without the subsequent test of their efficacy. In addition, resources for conservation action are limited and so it is essential that we learn from experience to avoid repeating mistakes or wasting resources on inefficient practices. A number of steps can be taken to increase learning outcomes from projects that are planned or underway.

First, communication and cooperation between researchers and land managers can help to create management manipulations that are more amenable to scientific evaluation and testing. In Britain, for example, management of linear "rides" favored by birds and butterflies in woodlands have been improved as a result of forest managers and scientists cooperatively testing different practices (such as width, shape and orientation of rides) to develop optimum wildlife habitat (Ferris-Kaan 1995). Haddad and Tewksbury (Chapter 16) describe similar opportunities for researchers and forest managers to work together to design timber harvesting protocols that allow stronger tests of connectivity along tracks and cleared lines.

Evaluating the effectiveness of habitat links in facilitating the movements of biota and in enhancing the status of connected populations and communities depends on monitoring. Few projects are monitored rigorously and consequently only limited evaluation of what has been achieved is possible. We stand to learn much more from current experience if land managers and scientists can work cooperatively to:

- Develop monitoring protocols, especially those that are resource efficient and operate at low intensity for the longer-term.
- Choose the appropriate metrics to monitor connectivity at multiple scales and levels of organization, including genetic, individual, population, community, and landscape levels.

- Make a commitment to long-term monitoring, because outcomes may take years or decades to be achieved.
- Document and communicate monitoring outcomes in a way in which others can benefit from the experience gained.

Frequently, projects are carried out to achieve multiple objectives. That is, the projects are intended to simultaneously achieve gains for nature conservation through movement and interchange of species, and also production of resources or other benefits for people. "Greenway" projects, for example, are often designed to provide benefits for conservation while also serving human recreation and esthetics (e.g., Little 1990; Smith and Hellmund 1993). Similarly, revegetation and restoration projects in agricultural environments are frequently perceived as providing agricultural benefits (e.g., shade, shelter, barrier to stock), while also serving as wildlife "corridors" or habitat. Do these projects actually achieve multiple benefits? Are conservation goals compromised by other activities? It would be useful to learn from these experiences, the contexts in which both conservation and other goals *can* be achieved, and those circumstances in which resources might be better directed elsewhere.

Finally, implementation of projects to enhance connectivity is done in a variety of ways. Some projects involve a few individuals or a single agency, others may involve multiple agencies and many landowners and concerned individuals. There is much to be learned by reflecting on the social process – the experience of how the project was undertaken, and the strengths and limitations of its implementation. What types of circumstances or activities assist the process of implementation, and what factors hinder or prevent it? What lessons can be learned that will improve the process for similar activities? Such evaluations are seldom reported (but see Saunders *et al.* 1995) but offer potential for learning and improvement.

4. Integrating connectivity into strategic planning for conservation

The process of developing a strategic plan for nature conservation for a designated region can be a very effective way to bring together complementary insights from theoretical, empirical, and applied approaches (Groves 2003). The task of planning for a specific region requires setting objectives, developing a conceptual framework, and bringing together empirical knowledge of the biota and ecosystem. Limitations and gaps can be identified in a "real-world" context and used to direct further research. Conservation planning for ecological networks in the Netherlands offers a good example (e.g., Opdam *et al.* 1995; Verboom *et al.* 2001): metapopulation theory, empirical data on the demography and mobility of species, and applied experience in reserve and corridor design are brought together to develop nature conservation plans. Likewise, strategic planning in southern California is guided not just by connectivity science, supported by empirical data on animal movement and tools such as least-cost path and individual-based movement models (Tracey Chapter 14; Beier *et al.* Chapter 22), but also the development of broad coalitions (Beier *et al.* Chapter 22) and assessments of the social, political, and economic feasibility of connectivity implementation (Morrison and Reynolds Chapter 21).

Strategic planning is also important because protecting and restoring connectivity takes time, and critical sections of links may be threatened or lost in the meantime. This is particularly relevant when connections are established in a reactive sense, as a localized response to "development" proposals that involve the destruction and isolation of natural environments. In areas where continued pressure will be exerted on the natural environment, such as outer suburban areas of major cities, there is an urgent need for connectivity of habitats to be incorporated in strategic forward planning. The tyranny of case-by-case decision-making and the associated incremental loss of habitats preclude a regional perspective, and may foreclose options for the future. Scientific knowledge and experience can contribute to strategic planning in a variety of ways.

- Applying the best available knowledge to planning processes *now*, recognizing that present understanding of connectivity is incomplete and further research will bring new insights. Waiting until further research is complete may be too late, and future options may be destroyed. This requires scientists to bring theory and empirical knowledge to the planning table and to engage with land-use planners.
- Gaining greater understanding of the conservation values of present opportunities for habitat connections. Many landscape features that add to present structural connectivity were retained by default or for other purposes. These include hedges, fencerows, riparian vegetation, roadside vegetation, and other vegetated connections, broad and narrow. A better knowledge of their conservation values and ecological function will help in deciding whether they are important to incorporate in strategic plans, or whether their incremental loss will have little consequence.

- Applying skills in spatial modeling to identify and evaluate alternative scenarios for future land use. Careful a priori consideration of alternative options can help identify critical components and pathways in habitat networks and those most vulnerable to loss, thus identifying priorities for protection and planning before crises develop.
- Undertaking research into restoration ecology as a basis for effective restoration or re-creation of habitat connections. This includes identifying the key components of habitats that facilitate connectivity, and working with practitioners to ensure that these can be incorporated in restoration programs.

The rapidity of change and the severity of threats to biodiversity are such that in many situations today practitioners must make decisions now, whether or not specific scientific advice is available to guide the actions they take. There is an urgent need for sound guidance on how a scientific understanding of connectivity can be translated into effective measures to ensure the long-term protection of biodiversity. Issues such as the most suitable location and the optimum dimensions of habitat linkages, the types of species that will benefit, and simple ways to monitor outcomes are the types of questions at the forefront of the minds of the growing cadre of conservation professionals. The scientific community must rise to the challenge and respond to this need. Ultimately, the future of conservation lies in the reconciliation of the human presence with the maintenance of Earth's biological diversity. Connectivity conservation offers a great deal towards understanding this reconciliation and providing solutions to make it a reality.

REFERENCES

Andreassen, H. P., S. Halle, and R. A. Ims. 1996. Optimal width of movement corridors for root voles: not too narrow and not too wide. *Journal of Applied Ecology* **33**:63–70.

Baranga, J. 1991. Kibale Forest Game Corridor: man or wildlife? Pp. 371–375 in D. A. Saunders and R. J. Hobbs (eds.) *Nature Conservation*, vol. 2, *The Role of Corridors*. Chipping Norton, NSW, Australia: Surrey Beatty and Sons.

Beier, P. 1993. Determining minimum habitat areas and habitat corridors for cougars. *Conservation Biology* **7**:94–108.

Beier, P., and R. F. Noss. 1998. Do habitat corridors provide connectivity? *Conservation Biology* **12**:1241–1252.

Bennett, A. F. 1990. Habitat corridors and the conservation of small mammals in a fragmented forest environment. *Landscape Ecology* **4**:109–122.

Bennett, A. F. 1999. *Linkages in the Landscape: The Role of Corridors and Connectivity in Wildlife Conservation*. Gland, Switzerland: IUCN.

Bennett, A. F., K. Henein, and G. Merriam. 1994. Corridor use and the elements of corridor quality: chipmunks and fencerows in a farmland mosaic. *Biological Conservation* **68**:155–165.

Diamond, J. M. 1975. The island dilemma: lessons of modern biogeographic studies for the design of natural reserves. *Biological Conservation* 7:129–146.

Fahrig, L. 2003. Effects of habitat fragmentation on biodiversity. *Annual Reviews of Ecology, Evolution and Systematics* **34**:487–515.

Ferris-Kaan, R. 1995. Management of linear habitats for wildlife in British forests. Pp. 67–77 in D. A. Saunders, J. L. Craig, and E. M. Mattiske (eds.) *Nature Conservation*, vol. 4, *The Role of Networks*. Chipping Norton, NSW, Australia: Surrey Beatty and Sons.

Forman, R. T. T. 1995. *Land Mosaics: The Ecology of Landscapes and Regions.* Cambridge, UK: Cambridge University Press.

Goodwin, B. J., and L. Fahrig. 2002. How does landscape structure influence landscape connectivity? *Oikos* **99**:552–570.

Groves, C. R. 2003. *Drafting a Conservation Blueprint: A Practitioner's Guide to Planning for Biodiversity.* Washington, DC: Island Press,.

Haas, C. A. 1995. Dispersal and use of corridors by birds in wooded patches on an agricultural landscape. *Conservation Biology* **9**:845–854.

Harris L. D., and J. Scheck. 1991. From implications to applications: the dispersal corridor principle applied to the conservation of biological diversity. Pp. 189–220 in D. A. Saunders and R. J. Hobbs (eds.) *Nature Conservation*, vol. 2, *The Role of Corridors*. Chipping Norton, NSW, Australia: Surrey Beatty and Sons.

Henein, K., and G. Merriam. 1990. The elements of connectivity where corridor quality is variable. *Landscape Ecology* **4**:157–170.

Hobbs, R. J. 1992. The role of corridors in conservation: solution or bandwagon? *Trends in Ecology and Evolution* **7**:389–391.

Hobbs, R. 1997. Future landscapes and the future of landscape ecology. *Landscape and Urban Planning* **37**:1–9.

Jongman, R. H. G. 1995. Nature conservation planning in Europe: developing ecological networks. *Landscape and Urban Planning* **32**:169–183.

Little, C. E. 1990. *Greenways for America*. Baltimore, MD: The Johns Hopkins University Press.

Mabry, K. E., and G. W. Barrett. 2002. Effects of corridors on home range sizes and interpatch movements of three small mammal species. *Landscape Ecology* **17**:629–636.

Mansergh, I. M., and D. J. Scotts. 1989. Habitat continuity and social organisation of the mountain pygmy-possum restored by tunnel. *Journal of Wildlife Management* **53**:701–707.

Merriam, G. 1991. Corridors and connectivity: animal populations in heterogeneous environments. Pp. 133–142 in D. A. Saunders and R. J. Hobbs (eds.) *Nature Conservation*, vol. 2, *The Role of Corridors*. Chipping Norton, NSW, Australia: Surrey Beatty and Sons.

O'Donnell, C. F. J. 1991. Application of the wildlife corridors concept to temperate rainforest sites, North Westland, New Zealand. Pp. 85–98 in D. A. Saunders and R. J. Hobbs (eds.) *Nature Conservation*, vol. 2, *The Role of Corridors*. Chipping Norton, NSW, Australia: Surrey Beatty and Sons.

Opdam, P., R. Foppen, R. Reijnen, and A. Schotman. 1995. The landscape ecological approach in bird conservation: integrating the metapopulation concept into spatial planning. *Ibis* **137**:S139–S146.

Parmesan, C. 2005. Biotic response: range and abundance changes. Pp. 41–61 in T. E. Lovejoy and L. Hannah (eds.) *Climate Change and Biodiversity*. New Haven, CT: Yale University Press.

Prevett, P. T. 1991. Movement paths of koalas in the urban-rural fringes of Ballarat, Victoria: implications for management. Pp. 259–272 in D. A. Saunders and R. J. Hobbs (eds.) *Nature Conservation,*, vol. 2, *The Role of Corridors*. Chipping Norton, NSW, Australia: Surrey Beatty and Sons.

Rosenberg, D. K., B. R. Noon, and E. C. Meslow. 1997. Biological corridors: form, function and efficacy. *BioScience* **47**:677–687.

Saunders, D. A., and R. Hobbs (eds.) 1991. *Nature Conservation,*, vol. 2, *The Role of Corridors*. Chipping Norton, NSW, Australia: Surrey Beatty and Sons.

Saunders, D. A., J. L. Craig, and E. M. Mattiske (eds.) 1995. *Nature Conservation*, vol. 4, *The Role of Networks*. Chipping Norton, NSW, Australia: Surrey Beatty and Sons.

Simberloff, D. S., and J. Cox. 1987. Consequences and costs of conservation corridors. *Conservation Biology* **1**:63–71.

Simberloff, D., J. A. Farr, J. Cox, and D. W. Mehlman. 1992. Movement corridors: conservation bargains or poor investments? *Conservation Biology* **6**:493–504.

Smith D. S. 1993. Greenway case studies. Pp. 161–208 in D. S. Smith and P. C. Hellmund (eds.) *Ecology of Greenways*. Minneapolis, MN: University of Minnesota Press.

Smith, D. S., and P. C. Hellmund (eds.) 1993. *Ecology of Greenways*. Minneapolis, MN: University of Minnesota Press.

Soulé, M. E. 1985. What is conservation biology? *BioScience* **35**:727–734.

Soulé M. E. 1995. An unflinching vision: networks of people defending networks of lands. Pp. 1–8 in D. A. Saunders, J. L. Craig, and E. M. Mattiske (eds.) *Nature Conservation*, vol. 4, *The Role of Networks*. Chipping Norton, NSW, Australia: Surrey Beatty and Sons.

Soulé, M. E., and J. Terborgh. 1999. *Continental Conservation: Scientific Foundations of Regional Reserve Networks*. Washington, DC: Island Press.

Soulé, M. E., and B. A. Wilcox (eds.) 1980. *Conservation Biology: An Evolutionary – Ecological Perspective*. Sunderland, MA: Sinauer Associates.

Sutcliffe, O. L., and C. D. Thomas. 1996. Open corridors appear to facilitate dispersal by ringlet butterflies (*Aphantopus hyperantus*) between woodland clearings. *Conservation Biology* **10**:1359–1365.

Taylor, P. D., L. Fahrig, K. Henein, and G. Merriam. 1993. Connectivity is a vital element of landscape structure. *Oikos* **68**:571–573.

Thomas, C. D., and T. M. Jones. 1993. Partial recovery of a skipper butterfly (*Hesperia comma*) from population refuges: lessons for conservation in a fragmented landscape. *Journal of Animal Ecology* **62**:472–481.

Tischendorf, L., and L. Fahrig. 2000. On the usage and measurement of landscape connectivity. *Oikos* **90**:7–19.

Trzcinski, M. K., L. Fahrig, and G. Merriam. 1999. Independent effects of forest cover and fragmentation on the distribution of forest breeding birds. *Ecological Applications* **9**:586–593.

Turner, M. G., R. H. Gardner, and R. V. O'Neill. 2001. *Landscape Ecology in Theory and Practice: Pattern and Process.* New York: Springer-Verlag.

Verboom, J., R. Foppen, P. Chardon, P. Opdam, and P. Luttikhuizen. 2001. Introducing the key patch approach for habitat networks with persistent populations: an example for marshland birds. *Biological Conservation* **100**:89–101.

Wu, J., and R. Hobbs. 2002. Key issues and research priorities in landscape ecology: an idiosyncratic synthesis. *Landscape Ecology* **17**:355–365.

Index